OPTICAL FIBER TELECOMMUNICATIONS IIIA

OPTICAL FIBER TELECOMMUNICATIONS IIIA

Edited by

IVAN P. KAMINOW
Lucent Technologies, Bell Laboratories
Holmdel, New Jersey

THOMAS L. KOCH
Lucent Technologies, Bell Laboratories
Holmdel, New Jersey

San Diego London Boston
New York Sydney Tokyo Toronto

This book is printed on acid-free paper. 

ACADEMIC PRESS
525 B Street, Suite 1900, San Diego, CA 92101-4495, USA
1300 Boylston Street, Chestnut Hill, MA 02167, USA
http://www.apnet.com

Academic Press Limited
24–28 Oval Road, London NW1 7DX, UK
http://www.hbuk.co.uk/ap/

Library of Congress Cataloging-in-Publication Data

Optical fiber telecommunications III / [edited by] Ivan P. Kaminow, Thomas L. Koch.
p. cm.
Includes bibliographical references and index.
ISBN 0-12-395170-4 (v. A) — ISBN 0-12-395171-2 (v. B)
1. Optical communications. 2. Fiber optics. I. Kaminow, Ivan P. II. Koch, Thomas L.
TK5103.59.H3516 1997 96-43812
621.382'75—dc20 CIP

Printed in the United States of America
97 98 99 00 01 MP 9 8 7 6 5 4 3 2 1

For our dear grandchildren:
Sarah, Joseph, Rafael, Nicolas, Gabriel, Sophia, and Maura — IPK

For Peggy, Brian, and Marianne — TLK

Contents

Contributors

Neal S. Bergano (Ch. 10), AT&T Laboratories, 101 Crawfords Corner Road, Holmdel, New Jersey 07733

Joseph E. Berthold (Ch. 2), Bellcore, 331 Newman Springs Road, Room 3Z331, Red Bank, New Jersey 07701

Vincent W. S. Chan (Ch. 3), Massachusetts Institute of Technology, Lincoln Laboratory, 244 Wood Street, Room D-309, Lexington, Massachusetts 02173

A. R. Chraplyvy (Ch. 8), Lucent Technologies, Bell Laboratories, Crawford Hill Laboratory, 791 Holmdel-Keyport Road, Holmdel, New Jersey 07733

Thomas E. Darcie (Ch. 14), AT&T Laboratories–Research, Communication Infrastructure Research Laboratory, 791 Holmdel-Keyport Road, Room R219, Holmdel, New Jersey 07733

David J. DiGiovanni (Ch. 4), Lucent Technologies, Bell Laboratories, 600 Mountain Avenue, Room MH 6C-312, Murray Hill, New Jersey 07974

Chungpeng Fan (Ch. 9), Lucent Technologies, Bell Laboratories, 101 Crawfords Corner Road, Room HO 2G-601A, Holmdel, New Jersey 07733

F. Forghieri (Ch. 8), AT&T Laboratories–Research, Crawford Hill Laboratory, 791 Holmdel-Keyport Road, Holmdel, New Jersey 07733

Nicholas J. Frigo (Ch. 13), AT&T Laboratories–Research, Crawford Hill Laboratory, 791 Holmdel-Keyport Road, Room HOH L-175, Holmdel, New Jersey 07733

A. H. Gnauck (Ch. 7), AT&T Laboratories–Research, Crawford Hill Laboratory, 791 Holmdel-Keyport Road, Room HOH-L129, Box 400, Holmdel, New Jersey 07733

J. P. Gordon (Ch. 12), Lucent Technologies, Bell Laboratories, 101 Crawfords Corner Road, Holmdel, New Jersey 07733

Donald P. Jablonowski (Ch. 4), Lucent Technologies, Bell Laboratories, 2000 Northeast Expressway, Room GAE870 G-030, Norcross, Georgia 30071

Kenneth W. Jackson (Ch. 5), Lucent Technologies, Bell Laboratories, 2000 Northeast Expressway, Room 1C-51, Norcross, Georgia 30071

R. M. Jopson (Ch. 7), Lucent Technologies, Bell Laboratories, 791 Holmdel-Keyport Road, Room HOH R-107, Holmdel, New Jersey 07733

Ivan P. Kaminow (Ch. 1, 15), Lucent Technologies, Bell Laboratories, 791 Holmdel-Keyport Road, Holmdel, New Jersey 07733

J. P. Kunz (Ch. 9), Lucent Technologies, Bell Laboratories, 101 Crawfords Corner Road, Room 4B-611, Holmdel, New Jersey 07733

P. V. Mamyshev (Ch. 12), Lucent Technologies, Bell Laboratories, 101 Crawfords Corner Road, Holmdel, New Jersey 07733

T. Don Mathis (Ch. 5), Lucent Technologies, Bell Laboratories, 2000 Northeast Expressway, Room 1D-51, Norcross, Georgia 30071

L. F. Mollenauer (Ch. 12), Lucent Technologies, Bell Laboratories, 101 Crawfords Corner Road, Room 4C-306, Holmdel, New Jersey 07733

Jonathan Nagel (Ch. 6), AT&T Laboratories–Research, Crawford Hill Laboratory, Holmdel-Keyport Road, Room L-137, Holmdel, New Jersey 07733

Kinichiro Ogawa (Ch. 11), Lucent Technologies, Bell Laboratories, 9999 Hamilton Boulevard, Breinigsville, Pennsylvania 18031

Yong Kwan Park (Ch. 11), Lucent Technologies, Bell Laboratories, 9999 Hamilton Boulevard, Breinigsville, Pennsylvania 18031

P. D. Patel (Ch. 5), Lucent Technologies, Bell Laboratories, 2000 Northeast Expressway, Room 1D-40, Norcross, Georgia 30071

Mary R. Phillips (Ch. 14), ATx Telecom Systems, 1251 Frontenac Road, Naperville, Illinois 60563

Craig D. Poole (Ch. 6), EigenLight Corporation, 8 Wemrock Drive, Ocean, New Jersey 07712

Eiichi Sano (Ch. 11), NTT LSI Laboratories, Atsugi, Japan

Manuel R. Santana (Ch. 5), Lucent Technologies, Bell Laboratories, 2000 Northeast Expressway, Room 1D-32, Norcross, Georgia 30071

Phillip M. Thomas (Ch. 5), Lucent Technologies, Bell Laboratories, 2000 Northeast Expressway, Room 1C-55, Norcross, Georgia 30071

R. W. Tkach (Ch. 8), AT&T Laboratories–Research, Crawford Hill Laboratory, 791 Holmdel-Keyport Road, Holmdel, New Jersey 07733

Liang D. Tzeng (Ch. 11), Lucent Technologies, Bell Laboratories, 9999 Hamilton Boulevard, Breinigsville, Pennsylvania 18031

Man F. Yan (Ch. 4), Lucent Technologies, Bell Laboratories, 600 Mountain Avenue, Room MH 6C-308, Murray Hill, New Jersey 07974

Chapter 1 | Overview

Ivan P. Kaminow
AT&T Bell Laboratories (retired), Holmdel, New Jersey

History

Optical Fiber Telecommunications, edited by Stewart E. Miller and Alan G. Chynoweth, was published in 1979, at the dawn of the revolution in lightwave telecommunications. This book was a stand-alone volume that collected all available information for designing a lightwave system. Miller was Director of the Lightwave Systems Research Laboratory and, together with Rudi Kompfner, the Associate Executive Director, provided much of the leadership at the Crawford Hill Laboratory of Bell Laboratories; Chynoweth was an Executive Director in the Murray Hill Laboratory, leading the optical component development. Many research and development (R&D) groups were active at other laboratories in the United States, Europe, and Japan. The book, however, was written exclusively by Bell Laboratories authors, although it incorporated the global results.

Looking back at that volume, I find it interesting that the topics are quite basic but in some ways dated. The largest group of chapters covers the theory, materials, measurement techniques, and properties of fibers and cables — for the most part, multimode fibers. A single chapter covers optical sources, mainly multimode AlGaAs lasers operating in the 800- to 900-nm band. The remaining chapters cover direct and external modulation techniques, photodetectors and receiver design, and system design and applications. Still, the basic elements for the present-day systems are there: low-loss vapor-phase silica fiber and double-heterostructure lasers.

Although a few system trials took place beginning in 1979, it required several years before a commercially attractive lightwave telecommunications system was installed in the United States. This was the AT&T Northeast Corridor System operating between New York and Washington, DC,

OPTICAL FIBER TELECOMMUNICATIONS,
VOLUME IIIA

ISBN: 0-12-395170-4

that began service in January 1983, operating at a wavelength of 820 nm and a bit rate of 45 Mb/s in multimode fiber. Lightwave systems were upgraded in 1984 to 1310 nm and about 500 Mb/s in single-mode fiber in the United States, as well as in Europe and Japan.

Tremendous progress was made during the next few years, and the choice of lightwave over copper for all long-haul systems was ensured. The drive was to improve performance, such as bit rate and repeater spacing, and to find other applications. A completely new book, *Optical Fiber Telecommunications II* (*OFT II*), edited by Stewart E. Miller and me, was published in 1988 to summarize the lightwave design information known at the time. To broaden the coverage, we included some non-Bell Laboratories authors, including several authors from Bellcore, which had been divested from Bell Laboratories in 1984 as a result of the court-imposed "Modified Final Judgment." Corning, Nippon Electric Corporation, and several universities were represented among the contributors. Although research results are described in *OFT II*, the emphasis is much stronger on commercial applications than in the previous volume.

The early chapters of *OFT II* cover fibers, cables, and connectors, dealing with both single- and multimode fiber. Topics include vapor-phase methods for fabricating low-loss fiber operating at 1310 and 1550 nm, understanding chromatic dispersion and various nonlinear effects, and designing polarization-maintaining fiber. Another large group of chapters deals with a wide geographic scope of systems for loop, intercity, interoffice, and undersea applications. A research-oriented chapter deals with coherent systems and another with possible local area network applications, including a comparison of time-division multiplexing (TDM) and wavelength-division multiplexing (WDM) to effectively utilize the fiber bandwidth. Several chapters cover practical subsystem components, such as receivers and transmitters, and their reliability. Other chapters cover the photonic devices, such as lasers, photodiodes, modulators, and integrated electronic and integrated optic circuits, that compose the subsystems. In particular, epitaxial growth methods for InGaAsP materials suitable for 1310- and 1550-nm applications, and the design of high-speed single-mode lasers are discussed.

The New Volume

By 1995, it was clear that the time for a new volume to address the recent research advances and the maturing of lightwave systems had arrived. The contrast with the research and business climates of 1979 was dramatic. System experiments of extreme sophistication were being performed by building on the commercial and research components funded for a proven multibillion-dollar global industry. For example, 10,000 km of high-performance fiber was assembled in several laboratories around the world for NRZ (non-return-to-zero), soliton, and WDM system experiments. The competition in both the service and hardware ends of the telecommunications business was stimulated by worldwide regulatory relief. The success in the long-haul market and the availability of relatively inexpensive components led to a wider quest for other lightwave applications in cable television and local access network markets. The development of the diode-pumped erbium-doped fiber amplifier (EDFA) played a crucial role in enhancing the feasibility and performance of long-distance and WDM applications.

In planning the new volume, Tom Koch and I looked for authors to update the topics of the previous volumes, such as fibers, cables, and laser sources. But a much larger list of topics contained fields not previously included, such as SONET (synchronous optical network) standards, EDFAs, fiber nonlinearities, solitons, and passive optical networks (PONs). Throughout the volume, erbium amplifiers, WDM, and associated components are common themes.

Again, most of the authors come from Bell Laboratories and Bellcore, where much of the research and development was concentrated and where we knew many potential authors. Still, we attempted to find a few authors from elsewhere for balance. Soon after laying out the table of contents and lining up the authors, however, a bombshell and a few hand grenades struck. AT&T decided to split into three independent companies, Bellcore was put up for sale, and several authors changed jobs, including Tom Koch and I. The resulting turmoil and uncertainty made the job of getting the chapters completed tougher than for the earlier volumes, which enjoyed a climate of relative tranquillity.

In the end, we assembled a complete set of chapters for *Optical Fiber Telecommunications III*, and can offer another timely and definitive survey of the field. Because of the large number of pages, the publisher recommended separating the volume into two sections, A and B. This format should prove more manageable and convenient for the reader. The chapters are numbered from Chapter 1 in each section, with this Overview repeated as Chapter 1 in both sections A and B to accommodate users who choose to buy just one book.

Survey of Volumes IIIA and IIIB

The chapters of Volumes IIIA and IIIB are briefly surveyed as follows in an attempt to put the elements of the book in context.

VOLUME IIIA

SONET and ATM (Chapter 2)

The market forces of deregulation and globalization have driven the need for telecommunications standards. Domestically, the breakup of AT&T meant that service providers and equipment suppliers no longer accepted *de facto* standards set by "Ma Bell." They wanted to buy and sell equipment competitively and to be sure that components from many providers would interoperate successfully. The globalization of markets extended these needs worldwide. And the remarkable capability of silicon integrated circuits to perform extremely complex operations at low cost with high volume has made it possible to provide standard interfaces economically.

The digital transmission standard developed by Bellcore and employed in all new domestic circuit-switched networks is SONET, and a similar international standard is SDH (synchronous digital hierarchy). In the same period, a telecommunications standard was devised to satisfy the needs of the data market for statistical multiplexing and switching of bursty computer traffic. It is called *ATM (asynchronous transfer mode)* and is being embraced by the computer industry as well as by digital local access providers. The basics of SONET, SDH, and ATM are given in Chapter 2, by Joseph E. Berthold.

Information Coding and Error Correction in Optical Fiber Communications Systems (Chapter 3)

The ultimate capacity of a communication channel is governed by the rules of information theory. The choice of modulation format and coding scheme determines how closely the actual performance approaches the theoretical limit. The added cost and complexity of coding is often the deciding factor in balancing the enhanced performance provided by this technology. So far, coding has not been required in high-performance lightwave systems. However, as the demands on lightwave systems increase and the performance of high-speed electronics improves, we can expect to see more uses of sophisticated coding schemes. In particular, forward error-correcting codes (FECs) may soon find applications in long-distance, repeaterless undersea systems. A review of coding techniques, as they apply to lightwave systems, is given by Vincent W. S. Chan in Chapter 3.

Advances in Fiber Design and Processing (Chapter 4)

The design and processing of fibers for special applications are presented in Chapter 4, by David J. DiGiovanni, Donald P. Jablonowski, and Man F. Yan. Erbium-doped silica fibers for amplifiers at 1550 nm, which are described in detail in Chapter 2, Volume IIIB, are covered first. Rare-earth-doped fluoride fibers for 1300-nm amplifiers are described later, as are fibers for cladding-pumped high-power fiber amplifiers.

Dispersion management is essential for the long-haul, high-speed systems described in later chapters. The design and fabrication of these fibers for new WDM installations at 1550-nm and for 1550-nm upgrades of 1310-nm systems are also reviewed.

Advances in Cable Design (Chapter 5)

Chapter 5, by Kenneth W. Jackson, T. Don Mathis, P. D. Patel, Manuel R. Santana, and Phillip M. Thomas, expands on related chapters in the two previous volumes, *OFT* and *OFT II.* The emphasis is on practical applications of production cables in a range of situations involving long-distance and local telephony, cable television, broadband computer networks, premises cables, and jumpers. Field splicing of ribbon cable, and the division of applications that lead to a bimodal distribution of low and high fiber count cables are detailed.

Polarization Effects in Lightwave Systems (Chapter 6)

Modern optical fibers possess an extremely circular symmetry, yet they retain a tiny optical birefringence leading to polarization mode dispersion (PMD) that can have severe effects on the performance of very long digital systems as well as high-performance analog video systems. Systems that contain polarization-sensitive components also suffer from polarization-dependent loss (PDL) effects. In Chapter 6, Craig D. Poole and Jonathan Nagel review the origins, measurement, and system implications of remnant birefringence in fibers.

Dispersion Compensation for Optical Fiber Systems (Chapter 7)

Lightwave systems are not monochromatic: chirp in lasers leads to a finite range of wavelengths for the transmitter in single-wavelength systems, whereas WDM systems intrinsically cover a wide spectrum. At the same time, the propagation velocity in fiber is a function of wavelength that can be controlled to some extent by fiber design, as noted in Chapter 4. To avoid pulse broadening, it is necessary to compensate for this fiber chromatic dispersion. Various approaches for dealing with this problem are presented in Chapter 7, by A. H. Gnauck and R. M. Jopson. Additional system approaches to dispersion management by fiber planning are given in Chapter 8.

Fiber Nonlinearities and Their Impact on Transmission Systems (Chapter 8)

Just a few years ago, the study of nonlinear effects in fiber was regarded as "blue sky" research because the effects are quite small. The advance of technology has changed the picture dramatically as unrepeatered undersea spans reach 10,000 km, bit rates approach 10 Gb/s, and the number of WDM channels exceeds 10. In these cases, an appreciation of subtle nonlinear effects is crucial to system design. The various nonlinearities represent perturbations in the real and imaginary parts of the refractive index of silica as a function of optical field. In Chapter 8, Fabrizio Forghieri, Robert W. Tkach, and Andrew R. Chraplyvy review the relevant nonlinearities, then develop design rules for accommodating the limitations of nonlinearities on practical systems at the extremes of performance.

Terrestrial Amplified Lightwave System Design (Chapter 9)

Chungpeng (Ben) Fan and J. P. Kunz have many years of experience in planning lightwave networks and designing transmission equipment, respectively. In Chapter 9, they review the practical problems encountered in designing commercial terrestrial systems taking advantage of the technologies described elsewhere in the book. In particular, they consider such engineering requirements as reliability and restoration in systems with EDFAs, with dense WDM and wavelength routing, and in SONET–SDH rings.

Undersea Amplified Lightwave Systems Design (Chapter 10)

Because of their extreme requirements, transoceanic systems have been the most adventurous in applying new technology. EDFAs have had an especially beneficial economic effect in replacing the more expensive and less reliable submarine electronic regenerators. Wideband cable systems have reduced the cost and improved the quality of overseas connections to be on a par with domestic communications. In Chapter 10, Neal S. Bergano reviews the design criteria for installed and planned systems around the world.

Advances in High Bit-Rate Transmission Systems (Chapter 11)

As the transmission equipment designer seeks greater system capacity, it is necessary to exploit both the WDM and TDM dimensions. The TDM limit is defined in part by the availability of electronic devices and circuits. In Chapter 11, Kinichiro Ogawa, Liang D. Tzeng, Yong K. Park, and Eiichi Sano explore three high-speed topics: the design of high-speed receivers, performance of 10-Gb/s field experiments, and research on devices and integrated circuits at 10 Gb/s and beyond.

Solitons in High Bit-Rate, Long-Distance Transmission (Chapter 12)

Chromatic dispersion broadens pulses and therefore limits bit rate; the Kerr nonlinear effect can compress pulses and compensate for the dispersion. When these two effects are balanced, the normal mode of propagation is a soliton pulse that is invariant with distance. Thus, solitons have seemed to be the natural transmission format, rather than the conventional NRZ format, for the long spans encountered in undersea systems. Still, a number of hurdles have manifested as researchers explored this approach more deeply. Perhaps the most relentless and resourceful workers in meeting and overcoming these challenges have been Linn Mollenauer and his associates. L. F. Mollenauer, J. P. Gordon, and P. V. Mamyshev provide a definitive review of the current R&D status for soliton transmission systems in Chapter 12. Typical of a hurdle recognized, confronted, and leaped is the Gordon–Haus pulse jitter; the sliding filter solution is described at length.

A Survey of Fiber Optics in Local Access Architectures (Chapter 13)

The Telecommunications Act of 1996 has opened the local access market to competition and turmoil. New applications based on switched broadband digital networks, as well as conventional telephone and broadcast analog video networks, are adding to the mix of options. Furthermore, business factors, such as the projected customer *take rate*, far outweigh technology issues.

In Chapter 13, Nicholas J. Frigo discusses the economics, new architectures, and novel components that enter the access debate. The architectural proposals include fiber to the home (FTTH), TDM PON, WDM PON, hybrid fiber coax (HFC), and switched digital video (SDV) networks. The critical optical components, described in Volume IIIB, include WDM lasers and receivers, waveguide grating routers, and low-cost modulators.

Lightwave Analog Video Transmission (Chapter 14)

Cable television brings the analog broadcast video spectrum to conventional television receivers in the home. During the last few years, it was found that the noise and linearity of lightwave components are sufficiently good to transport this rf signal over wide areas by intensity modulation of a laser carrier at 1310, 1060, or 1550 nm. The fiber optic approach has had a dramatic effect on the penetration and performance of cable systems, lowering cost, improving reliability, and extending the number of channels. New multilevel coding schemes make rf cable modems an attractive method for distributing interactive digital signals by means of HFC and related architectures. Thus, cable distribution looks like an economic technology for bringing high-speed data and compressed video applications, such as the Internet, to homes and offices. Now, in the bright new world of deregulation and wide-open competition, cable may also carry telephone service more readily than telephone pairs can carry video. In Chapter 14, Mary R. Phillips and Thomas E. Darcie examine the hardware requirements and network architectures for practical approaches to modern lightwave cable systems.

Advanced Multiaccess Lightwave Networks (Chapter 15)

The final chapter in Volume IIIA looks at novel architectures for routing in high bit-rate, multiple-access networks. For the most part, the emphasis is on wavelength routing, which relies on the novel wavelength-sensitive elements described in Volume IIIB. Such networks offer the prospect of "optical transparency," a concept that enhances flexibility in network design. Commercial undersea and terrestrial networks are already incorporating preliminary aspects of wavelength routing by the provision of WDM add–drop multiplexing. Further, the proposed WDM PON networks in Chapter 13 also employ wavelength routing.

Chapter 15, however, considers a wider range of architectures and applications of this technology. After reviewing optical transparency, it treats WDM rings for local networks, metropolitan distribution, and continental undersea telecommunications (AfricaONE). Then it reviews several multiaccess test beds designed by consortia organized with partial support from DARPA (Defense Advanced Research Projects Agency).

VOLUME IIIB

Erbium-Doped Fiber Amplifiers for Optical Communications (Chapter 2)

A large part of the economic advantage for lightwave systems stems from the development of the diode-pumped EDFA, which replaced the more expensive and limited electronic regenerators. By remarkable coincidence, the EDFA provides near noise-free gain in the minimum-loss window of silica fiber at 1550 nm. It provides format-independent gain over a wide WDM band for a number of novel applications beyond its original use in single-frequency, long-haul terrestrial and undersea systems.

Important considerations in the basics, design, and performance of EDFAs are given in Chapter 2, by John L. Zyskind, Jonathan A. Nagel, and Howard D. Kidorf. Designs are optimized for digital terrestrial and undersea systems, as well as for applications to analog cable television and wavelength-routed WDM networks, which are covered in Chapters 13, 14, and 15 in Volume IIIA. Performance monitoring and the higher order effects that come into play for the extreme distances encountered in undersea systems are also discussed.

Transmitter and Receiver Design for Amplified Lightwave Systems (Chapter 3)

Chapter 3, by Daniel A. Fishman and B. Scott Jackson, defines the engineering requirements for transmitters and receivers in amplified systems, mainly operating at 2.5 Gb/s and satisfying the SONET–SDH standards. Topics that are essential for commercial networks, such as performance monitoring, are included.

Laser Sources for Amplified and WDM Lightwave Systems (Chapter 4)

As lightwave systems have become more sophisticated, the demands on the laser sources have become more stringent than those described in Chapter 13 of *OFT II.* The greater fiber spans and the introduction of EDFA and WDM technologies require both improved performance and totally new functionality. In Chapter 4, Thomas L. Koch reviews lasers and subsystems designed for low-chirp applications, employing direct modulation, external modulation, and integrated laser–modulators. He also covers a variety of laser structures designed to satisfy the special needs of WDM systems for precise fixed wavelengths, tunable wavelengths, and multiple wavelengths. These structures include fixed DFB (distributed feedback) lasers, tunable DBR (distributed Bragg reflector) lasers, multifrequency waveguide grating router lasers (MFL), and array lasers.

Advances in Semiconductor Laser Growth and Fabrication Technology (Chapter 5)

Some of the greatest advances in laser performance in recent years can be traced to advances in materials growth. In Chapter 5, Charles H. Joyner covers such advances as strained quantum wells, selective area growth, selective etching, and beam expanded lasers.

Vertical-Cavity Surface-Emitting Lasers (Chapter 6)

The edge-emitting lasers employed in today's lightwave systems are described in Chapter 4. In Chapter 6, L. A. Coldren and B. J. Thibeault update progress on a different structure. Vertical-cavity surface-emitting lasers (VCSELs) are largely research devices today but may find a role in telecommunications systems by the time of the next volume of this series. Because of their unique structure, VCSELs lend themselves to array and WDM applications.

Optical Fiber Components and Devices (Chapter 7)

Although fiber serves mainly as a transmission line, it is also an extremely convenient form for passive and active components that couple into fiber transmission lines. A key example is the EDFA, which is described in Chapter 4, Volume IIIA, and Chapter 2, Volume IIIB. In Chapter 7, Alice E. White and Stephen G. Grubb describe the fabrication and applications of UV-induced fiber gratings, which have important uses as WDM multiplexers and add–drop filters, narrow band filters, dispersion compensators, EDFA gain equalizers, and selective laser mirrors.

Special fibers also serve as the vehicles for high-power lasers and amplifiers in the 1550- and 1310-nm bands. High-power sources are needed for long repeaterless systems and passively split cable television distribution networks. Among the lasers and amplifiers discussed are 1550-nm Er/Yb cladding-pumped, 1300-nm Raman, and Pr and Tm up-conversion devices.

Silicon Optical Bench Waveguide Technology (Chapter 8)

A useful technology for making passive planar waveguide devices has been developed in several laboratories around the world; at AT&T Bell Laboratories, the technology is called *silicon optical bench* (*SiOB*). Waveguide patterns are formed photolithographically in a silica layer deposited on a silicon substrate. In Chapter 8, Yuan P. Li and Charles H. Henry describe the SiOB fabrication process and design rules suitable for realizing a variety of components. The planar components include bends, splitters, directional couplers, star couplers, Bragg filters, multiplexers, and add–drop filters. Different design options are available for the more complex devices, i.e., a chain of Fourier filters or an arrayed waveguide approach. The latter technique has been pioneered to Corrado Dragone of Bell Laboratories (Dragone, Edwards, and Kistler 1991) to design commercial WDM components known as *waveguide grating routers* (*WGRs*) serving as multiplexers and add–drop filters.

Lithium Niobate Integrated Optics: Selected Contemporary Devices and System Applications (Chapter 9)

More than 20 years have passed since the invention of titanium-diffused waveguides in lithium niobate (Schmidt and Kaminow 1974) and the associated integrated optic waveguide electrooptic modulators (Kaminow, Stulz, and Turner 1975). During that period, external modulators have competed with direct laser modulation, and electrooptic modulators have competed with electroabsorption modulators. Each has found its niche: the external modulator is needed in high-speed, long-distance digital, and high-linearity analog systems, where chirp is a limitation; internal modulation is used for economy, when performance permits. (See Chapter 4 in Volume IIIB.)

In Chapter 9, Fred Heismann, Steven K. Korotky, and John J. Veselka review advances in lithium niobate integrated optic devices. The design and performance, including reliability and stability, of phase and amplitude modulators and switches, polarization controllers and modulators, and electrooptic and acoustooptic tunable wavelength filters are covered.

Photonic Switching (Chapter 10)

Whereas Chapter 9 deals with the modulation or switching of a single input, Chapter 10 deals with switching arrays. These arrays have not yet found commercial application, but they are being engineered for forward-looking system demonstrations such as the DARPA MONET project (Multiwavelength Optical Network), as mentioned in Chapter 15, Volume IIIA. In Chapter 10, Edmond J. Murphy reviews advances in lithium niobate, semiconductor, and acoustooptic switch elements and arrays. Murphy also covers designs for various device demonstrations.

References

Dragone, C., C. A. Edwards, and R. C. Kistler, 1991. Integrated optics N × N multiplexer on silicon. *IEEE Photon. Techn. Lett.* 3:896–899.

Kaminow, I. P., L. W. Stulz, and E. H. Turner. 1975. Efficient strip-waveguide modulator. *Appl. Phys. Lett.* 27:555–557.

Schmidt, R. V., and I. P. Kaminow. 1974. Metal-diffused optical waveguides in $LiNbO_3$. *Appl. Phys. Lett.* 25:458–460.

Chapter 2 | SONET and ATM

Joseph E. Berthold
Bellcore, Red Bank, New Jersey

I. Background

The mid-1980s to the mid-1990s has seen unprecedented change in communications technology. This chapter deals with two great technological advances. The first is the synchronous digital hierarchy (SDH), an International Telecommunications Union (ITU) standard, and the closely related synchronous optical network (SONET), an American National Standards Institute (ANSI) standard. Although there are differences between the two standards, for the purposes of this chapter the distinctions are unimportant. We use the terms *SDH* and *SONET* interchangeably. The second is the asynchronous transfer mode (ATM), which also is standardized in the ITU. These are the major standards that specify how to make use of optical communications technology, in that they specify how information is carried and manipulated through optical networks.

Optical communications technological advances were necessary, but not sufficient, for the creation of the SDH and ATM. They arose as a result of the fundamental changes that shook the communications industry beginning in the early 1980s. They were also stimulated by the needs of businesses that were beginning to understand and apply information technology, and to do so in the context of companies that spanned national boundaries.

This chapter gives an overview of SONET and ATM, mainly to show what else is needed beyond high-capacity optical links to create broadband networks. We first examine the forces in the marketplace that helped bring these concepts forth and the technological advances that made them possible. We then see some of the specific needs that the SDH and ATM sought to meet. We present a high-level view of some key concepts in networking to better understand where SONET and ATM fit in. We also see how these new standards have redefined what is needed to make major advances in networking technology. This should be instructive as researchers to seek to create the next breakthrough in optical communications networking.

OPTICAL FIBER TELECOMMUNICATIONS,
VOLUME IIIA

ISBN: 0-12-395170-4

A. EMERGING COMPETITION IN TELECOMMUNICATIONS

The 1980s and 1990s have seen continued, fundamental change in the communications industries throughout the world. In the United States, the 1982 agreement to divest AT&T of its local-exchange business fractured the Bell System into eight independent companies. The seven local-exchange companies formed had nonoverlapping service areas. They were not allowed to transport signals across their entire service areas, but were restricted to stay within local access and transport areas (LATAs), of which several hundred were formed. The local-exchange companies were required to offer equal access to their LATA networks to interexchange telephone companies. AT&T was the largest interexchange company in the United States, but it had a growing number of competitors. What had been a nationwide network largely designed, built, and operated by a vertically integrated company, AT&T, was to be transformed into a nationwide network composed of many competing but interoperable networks. In this new environment, neither the network operators nor the network equipment suppliers were willing to have the new AT&T unilaterally make the technology decisions for the U.S. communications industry. The original SONET proposal was a direct outcome of the AT&T divestiture. There was a need to have a standardized, high-capacity interface, because so many interfaces were now required.

Although AT&T divestiture is often used as the prime example of the changing telecommunications business environment, many similar upheavals are in progress throughout the world. Nippon Telephone and Telegraph (NTT) is the process of privatization, and competition has begun in the Japanese market. In Europe, the nationally owned telecommunications organizations, the PTTs (posts, telephone, and telegraphs), are on a timetable for privatization and competition. England was among the first to introduce competition, and British Telecom now has competition for local telephone service by companies that also offer cable television services.

B. TECHNOLOGICAL FORCES THAT INFLUENCED SONET AND ATM

Besides market demand, major advances were made in three key technologies: optical transmission technology, high-speed integrated circuit technology, and microprocessor technology. All these technologies had a major impact on the final outcome of the SDH and ATM standards.

The history of optical communications has been one of ever-increasing performance and increasing performance–cost ratio. The measure of transmission performance, the distance–bit-rate product, has seen exponential growth during several decades, and this growth is expected to continue well into the future. The challenge to networking systems architects, researchers, and developers is to provide the networking technology needed to exploit the capability of optical communications technology — that is, to create the correct signal structures and networking principles that will make the most of optical transmission performance now and in the future. Later in this chapter, we see that the SONET and ATM standards are distinguished from most previous data networking and telecommunications standards in that they are scalable in bit rate. They were designed with that relentless exponential in mind!

Another impact of the expected continuing growth of optical transmission performance is that bandwidth was no longer viewed as a scarce commodity, one to be carefully guarded. Efficiency was in some cases sacrificed to keep a simple, scalable multiplexing structure. Bandwidth was also dedicated to support operations functions.

Besides optical transmission technology, high-speed very large scale integration (VLSI) semiconductor technology has had a fundamental impact on the basic structure of SONET and ATM. VLSI makes highly complex signal structures possible and cost-effective, because whatever can be accomplished "on a chip" can be mass-produced at low cost. If high levels of integration are achievable, it is possible to make trade-offs in the execution of signal processing functions by choosing between high-speed serial and parallel logic, or by choosing the appropriate combination of the two. Functions that were associated previously with serial implementations, such as high-speed signal scrambling (used to make clock recovery in a digital receiver easier and more reliable), can also be done by using parallel algorithms. With the capability of doing all the complex logic required in parallel, the only required high-speed electronics function in an SDH multiplexer is the final bit-interleaving process that creates the high-speed transmission signal. The VLSI technology used for signal processing may even have adequate speed for all multiplexing functions, including serialization, for some signal rates. This comes about as a by-product of semiconductor technology evolution to increase circuit density. As circuit dimensions decrease, gate delays also decrease. For the highest line rates, the relatively few logic gates required for the sterilization function may be included on a separate chip, perhaps the same chip responsible for optical signal modulation.

Many assumptions went into the development of SONET and ATM that were not manifest directly in the digital signal formats but are nonetheless groundbreaking in the context of previous systems. One of the most important other technological impacts was that of the microprocessor. Microprocessor and semiconductor memory technologies are on the same relentless exponential performance-improvement curves as optical transmission technology is. Early transmission systems were relatively static. They were monitored and controlled, whenever control was possible, by centralized management systems. Systems in the future could be foreseen whereby a local processor could be as powerful as one of the processors in the older central monitoring systems. Although this technology will have a profound impact on the way in which management processes are accomplished in the future, and in the architecture of network management systems, it did also have an impact on the signal format for SONET, through the addition of an embedded data communications channel in the signal to allow remote management and the download of software into network elements. Distributing intelligence and control also raised new security and network reliability issues.

II. Network Solutions

A. *APPLICATION NEEDS*

Before we discuss the details of the SDH and ATM, it is instructive to consider some examples of applications that should be supported by wide-area networks. The ones chosen here are in use in private networks, predominantly in local area networks (LANs). The choices of the following four applications, listed in Table 2.1, were made to illustrate their differences and to show that a network optimized for specific services is unlikely to be flexible enough to meet the needs of as-yet undiscovered applications.

The first example is *distributed computing,* a broad term that is meant to include the execution of tasks by computers where there is a need to share processing power or data across a network of computers. The traffic generated by distributed computing applications is characterized by the transfer of variable-sized packets of information. In some cases they will be extremely short, in other cases very long. The traffic patterns are unpredictable, except for a very specific task. The system designer would most like to have low latency — that is, low delay from when data are inserted into the network and when they are delivered. Systems can deal with any level of delay, but the designer prefers it to be as low as possible. If the

Table 2.1 **Characteristics of Applications and the Requirements They Place on Networks**

Application	*What Is Transported*	*Desired Delay*	*Data Loss Impact*	*Symmetrical Traffic?*
Distributed computing	Variable-length packets	"As low as possible"	Retransmit	Yes
High-resolution image delivery	Large data blocks	Moderate (~1 s)	Retransmit	No
Audio and video database access	Bursts of continuous data	Moderate, bounded (~1 s)	Accept impairment, conceal	No
Interactive audiovisual communications	Continuous data	Low (~50 ms)	Accept impairment, conceal	Yes

network introduces errors in the data, it will be discovered and the data will need to be retransmitted. Because traffic is so unpredictable, it is necessary to assume that network demands will be bidirectional and symmetrical.

High-resolution image delivery is characterized by the transfer of large data blocks. A 2000- by 2000-pixel image with 24-bit pixel resolution requires 100 Mb of data. If it were transmitted over a 100-Mb/s network, it would consume the full bandwidth for 1 s. Assuming that the image retrieval is done by a person, delays of several seconds would be tolerable. Any transmission errors would require retransmission of the image, or part if it was segmented for delivery. If the source of the images was a data archive, the outbound bandwidth would be much larger than the inbound bandwidth, which would consist mainly of short request packets.

Audio and video database access brings the first real-time requirement to the network. The traffic is a large block of data, but it is data that must be delivered to a user continuously once it has begun. The continuous data rate required may vary over many orders of magnitude, depending on the audio and video fidelity. The audio and video must be synchronized. The delay to start playback can be on the order of seconds, which allows a portion of the data stream to be stored locally to compensate for later delays in network transmission. If errors are introduced, there is no time for retransmission and the signal would appear impaired unless the decoder employs mechanisms to disguise the loss, such as replaying the previous video frame. The traffic pattern would be asymmetrical.

The final example involves human interaction, and psychology comes into play. In an audiovisual conference, absolute delay is important. Researchers found that a delay in the communications path is noticeable when it becomes significantly longer than 50 ms. Satellite telephone links were found to be objectionable to many users. The delays between questions or statements and reactions to them can be interpreted as less than frank and honest responses, which can lead to a lack of trust. Again, loss of data must be tolerated, because it would be impossible to retransmit errored data in time for it to be useful. Techniques for error mitigation in codecs may lessen the impact of errors.

There was no single networking approach that was able to support these example services over the wide area cost-effectively. We had a mixture of bursty data, some with low delay requirements and others without, and continuous media requirements where guaranteed bandwidth and low delay were essential. Not all applications were symmetrical, so a symmetrical network might waste half the network resources. There was a need for a flexible approach that could also handle the mix of application demands efficiently.

B. *MAJOR NETWORKING PARADIGMS: CIRCUITS, PACKETS, AND CELLS*

1. Circuit Networks

Circuit technology is used in telecommunications networks worldwide. Circuits are connections between communication endpoints and provide users with access to a network resource for the duration of the connection. Some connections are long-lived and are called *permanent connections.* They are in contrast to shorter lived connections, or *switched connections.* Permanent connections are provided by dedicating a transmission circuit, and they are managed through transport facilities rather than switching facilities. A very common use of permanent connections is for circuits used by businesses to interconnect their local computing facilities across wide areas. The Internet makes use of permanent circuits leased from telecommunications carriers to interconnect their switches. Permanent circuits are leased for relatively long periods, months to years. Switched connections are established on demand, on the basis of end-user signaling. They are provided through service switches, and the networkwide connection is established on the basis of available resources, according to a routing policy.

In early days, a telephony circuit was just a pair of wires, dedicated or connected through a switch. With the advent of digital technology and time-division multiplexing (TDM), many circuits could share a transmission medium, but even so, they were a dedicated resource, a time slot. If a circuit was left connected, even though no data flowed through it, the transmission resource could not be used by another connection that had data to send.

SONET and the SDH are circuit networks, and they are administered as *permanent* circuit networks. It would be possible to create SDH circuit switches where connections could be set up in response to user signaling as our telephone circuits are today, and proposals to do so were debated at some length in the late 1980s. However, the advent of ATM as a networking solution that could support the same type of application needs, but was also more flexible in bit rate and could additionally support packet traffic, relegated SONET and the SDH to the role of a transport rather than a service switching infrastructure.

2. Packet Networks

Packet networks grew out of the computing world and used computing technology for their switching. Packet networks rely on circuit networks for transmission, and packets provide a means of sharing the circuit bandwidth. Most simply put, a packet is a block of information with some overhead information attached. The overhead information is called a *header,* and its most important purpose is to distinguish one packet from another so that they can be delivered to the correct destination. Another important function included in the header is a mechanism for error checking. The header needs to have error protection so that errors do not lead to misrouted packets. The information content needs error checking so that the computers can be sure that the information is transmitted error free. In the event that errors are introduced in transmission, the packets can be retransmitted.

There are two important types of packet networks: connectionless packet networks and connection-oriented packet networks. Connectionless packet networks operate in a way analogous to the post office, and their packets are sometimes referred to as *datagrams.* Connectionless networks offer best effort packet-delivery service. Connectionless packets are launched into the network with a large and powerful header that contains all the information necessary to deliver the packet, whatever its destination. As a by-product of their connectionless nature, it is not possible to guarantee performance or reserve network resources for any particular communication, so connectionless networks do not support real-time services reliably. The switches in the network are stateless. When the network is subjected to overload, packets are buffered for later delivery. If the buffer capacity is exceeded, the packets are discarded. In connectionless networks there is no guarantee that all packets will traverse the same network path and be subject to the same delays. If a communication requires the transmission of a series of packets, it is up to the end stations to ensure that the sequence is preserved.

The Internet is an example of a connectionless packet network, because it is based on the Internet protocol (IP), which is a connectionless protocol. IP packets have a complete source and destination address in their header capable of delivering information to anyone on the Internet around the world.

Connection-oriented packet networks have similarities with both circuit networks and packet networks. They are like circuit networks in that a connection, called a *virtual circuit,* must be established before data can be transported. Network nodes are prepared to support virtual circuits, and they devote resources to them. The resources include buffer memory and link bandwidth. Unlike circuits, when packets are not submitted to the network for some of the virtual circuits, the resources can be devoted to carry other virtual circuits. This is called *statistical multiplexing.* Assumptions are made about the traffic likely to be contributed by all the sources, and the network may overload if all sources begin sending packets at their peak rate. Occasional pileups of traffic are to be expected and are handled by maintaining local memory to buffer packets until it is possible to transmit them. The network is not stateless, as a connectionless packet network is, but stores the state of all connections that are present. When all the resources are committed, the network may deny requests to establish new virtual circuits. The headers of connection-type packet networks can be shorter than those of connectionless networks because the addressing portion of the header needs only to distinguish each virtual circuit from each other established on common links and switch ports.

3. Cell Networks

Cell networks, sometimes called *fast packet networks,* were designed to support both continuous high-bandwidth data streams and bursty packet traffic. Cell networks are a special type of connection-oriented packet network with a protocol so simple that it can be easily placed in hardware to allow very fast operation. The ATM is a cell-based networking protocol. Its cells are short and of a fixed length, an important simplification. There is a connection setup process before traffic can be sent, and as a part of that process there can be a negotiation of resources, which in the ATM case is link and switch port bandwidth. Depending on how the connection setup process is managed by a network operator, resources can be allocated to guarantee that all established circuits can continually transmit at their peak allowable bit rates. In this case, it is a circuit network that is freed from the rigid bit rates of the TDM hierarchies. It can also be administered as a network that allows statistical multiplexing, where different connections vie for a share of the bandwidth available. There are a number of service classes supported that would allow both high-priority circuits and loss- or delay-tolerant packets to coexist. ATM is not tied to any physical transmission medium, although it was designed with the assumption of low error rate fiber optics transport. We describe ATM in more detail later.

C. NEEDS LEADING TO SONET AND THE SDH

There are a number of advantages that would result from a new system of multiplexing digital signals. These needs that the SDH sought to meet are as follows:

- *Internationally standard broadband signals.* The digital transmission standards that preceded the SDH were lacking in several ways. They arose from a need to time-division multiplex voice signals and were therefore based on a unit of bandwidth of 64 kb/s. No other rate had any special significance from a service perspective. TDM systems were deployed first in the United States, and created a 24-channel multiplex. European systems came somewhat later, and they were able to achieve a larger multiplex because of advances in electronics technology in the intervening period. Different multiplex hierarchies grew, but none were universal. In the United States, standards were in use for signal levels up to 45 Mb/s, but for levels above that, different vendors diverged in their approaches. The result was that there was no internationally standardized broadband signal, and this would become a major cost impediment to the availability of economic broadband services.

- *Scalable system capacity.* Optoelectronics technology was progressing very rapidly, with system capacity doubling every few years, and this trend will continue indefinitely. We needed a set of standards that could scale with technology so that we would not need to reconvene standards groups for each generational change.
- *Lower cost access to signal components.* Digital systems are clocked, but we need ways to distribute the clocks and synchronize systems. We had an asynchronous multiplexing hierarchy above a certain level, and bit stuffing was used to make up for imperfect timing information. Bit stuffing, along with parsimonious multiplexing formats, resulted in the need to fully demultiplex the signal to identify the payload information. There was a desire to create a system where each byte could be simply identified and manipulated. Elimination of back-to-back multiplexing, and the creation of an add–drop multiplexer (ADM), would offer major cost savings to network operators.
- *Survivable network architectures.* As system capacity continued to grow with improvements in optoelectronics technology, the adverse impact of equipment or link failures grew proportionally. Service survivability had become a high priority for many network customers, so there was strong motivation to create architectures that were robust to failures, and cost-effective. One of the major benefits of the SDH is the introduction of self-healing ring network architectures.
- *Standardized automated operations.* Network operators need to automate their operations to the greatest extent possible, to improve both their costs and their service quality. Automated operations enable rapid service activation and rapid service restoration following faults. Further economies result if operators can share the same operations staff to monitor and control their networks, independent of vendor equipment. Ultimately there is a desire to automate the interworking of network control and management across network boundaries, so that customers can have service activated quickly and repaired quickly even when the end-to-end service is supplied by several operators.

D. NEEDS LEADING TO ATM

The ATM originally received the support of the telecommunications community as a method to be used in the broadband integrated services digital network (BISDN). The other alternatives that were debated at the time were TDM circuit approaches employing various mixes of SDH channels and subchannels. One of the channels could be used to support packets for signaling or packet data. This is analogous to the approach taken for the ISDN, where the basic rate service called for two 64-kb/s channels plus the use of a 16-kb/s packet channel for signaling or packet data transfer. ATM was embraced by the private data networking community as a way to upgrade LAN performance, support real-time services, and eventually integrate with wide-area public networks. ATM received such widespread support because it met the following needs:

- *Efficient and scalable support for data traffic.* The history of data networking is strikingly different from the history of telecommunications in that data networks emerged after computers were entrenched in the marketplace. Networking of computers was as proprietary and heterogeneous as computers themselves were. It was difficult for any two manufacturers' computers to interwork. In contrast, the telecommunications industry was created to support universal communications. The Internet arose to internetwork heterogeneous computers. Still, the Internet did not specify physical or link-level protocols. Interworking was achieved by inserting relatively complicated network equipment between LAN segments. As the data needs of individual computers grew, it could be foreseen that the shared media LANs would become congested. An approach that was simple, that supported more parallelism (switching), and that could work up to arbitrarily high data rates was very attractive.
- *Bandwidth guarantees for real-time applications.* There was no satisfactory technique in the data networking industry to provide support for real-time applications, such as video communications or video database access. Circuit approaches would yield guaranteed bandwidth and low delay, but which rates should be standardized? Circuit approaches were thought to be too difficult to manage and too rigid. Coding technology was moving rapidly, and today's rates and standards might be obsolete quickly. Besides, several video coding techniques yielded inherently bursty data streams, and some of

their proponents believed that the industry would be able to achieve statistical multiplexing gains while still preserving service quality.

- *Integration of data and real-time capabilities.* The future applications expected to appeal most to the mass market were expected to be rich in high-resolution full-color images, high-fidelity sound, and high-quality video, along with textual data. It would be desirable for a network to work comfortably with data signals that varied in time from bursts of continuous data streams to bursty data traffic, to low bit-rate but low-delay voice communications.
- *Flexible and more powerful management of networks.* With the networking needs of computers growing rapidly, it was becoming very difficult to plan and manage data networks. The data networking industry was already evolving the topology of shared media (bus) LANs, such as ethernets, to physical star topologies. New networking devices called *hubs* entered the networking marketplace, soon to evolve into switched hubs. This evolution was driven mainly by operations and maintenance requirements of LANs. The environment was ripe for a technique that could both provide bandwidth where it was needed in a network and prevent the actions of one user or set of users from impairing the performance of the entire network.

III. Synchronous Transmission Networking: SONET and the SDH

A. *SDH NETWORKING ARCHITECTURES AND NETWORK ELEMENTS*

There are four major categories of network architectures for SONET networks. The simplest is the point-to-point transmission link. Next comes the chain architecture, which is made possible by the introduction of the add/drop multiplexer (ADM). ADMs can be further enhanced and used to create ring architectures with automatic protection switching, or self-healing, capabilities. The last architecture is a mesh network of switches. The SONET transport facilities' switches are called *digital cross-connect switches* (*DCSs*). We discuss each of these in turn next.

1. Point-to-Point Links

The simplest transmission networking architectural element is the point-to-point transmission link. Even such an element is quite complex when looked at in detail. A point-to-point link comprises an end-to-end path, a number of multiplex sections, and a number of regenerator sections. The example in Figure 2.1 illustrates a link between communication endpoints that require a dedicated channel. The term *path* is used to describe the overall circuit between the endpoints. The signal includes all the appropriate overhead information as well as the information payload. The equipment that creates and terminates the SONET path-level signal is referred to as *path-terminating equipment.* A path-level signal may be combined with many other path-level signals as it transits an intervening portion of the network. The term *multiplex section* is used to describe the portion of the network where the signals are combined and transported at the higher line rate. Because a multiplex section may traverse a long distance, there may be a need for digital regeneration of the line-rate signal. The term *regenerator section* refers to the portion of the network between adjacent regenerators. In general, an end-to-end path may pass through many multiplex sections, and many regenerator sections in each multiplex section. Later we describe some of the signal overhead functions that correspond to paths, multiplex sections, and regenerator sections.

Because high-speed fiber optic transmission systems can support so much traffic, there are heightened concerns about network reliability and service survivability. Failures can occur not only in electronic and optoelectronic transmission equipment, but also in fiber optic cables. Protection against electronic equipment failure is achieved through redundancy, with the most robust systems being ones with fully duplicated electronics. Protection against cable failures can be obtained by redundancy as well, with the most robust systems being ones that use physically diverse routes for the working and protection cables. Figure 2.2 illustrates a fully duplicated, diversely routed point-to-point transmission link.

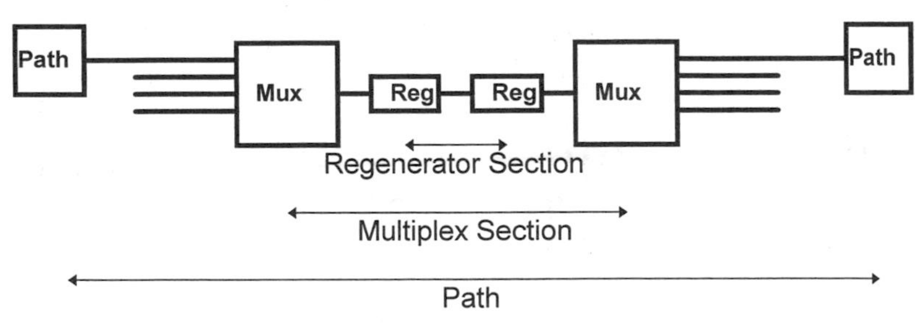

Fig. 2.1 Composition of point-to-point SONET transmission links.

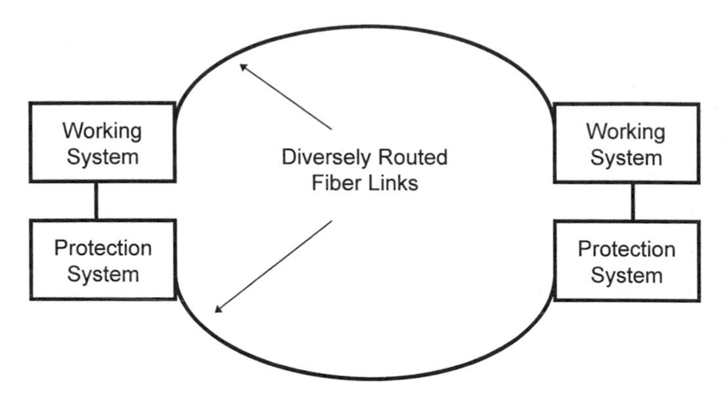

Fig. 2.2 Protected point-to-point transmission link.

2. Chains

Another network element very similar to the terminal multiplexer is the ADM. The intent of the ADM is to allow access to a portion of the bandwidth of the optical fiber communications link. In the example in Fig. 2.3, we see a chain of multiplexers, with terminal multiplexers at the ends of the chain and ADMs at intermediate points along the chain. The chain shares the bandwidth to provide the communication needs of the traffic demand sets A, B, and C. As shown in Fig. 2.3, demand pair A exists between the terminal multiplexer on the left and an ADM along the chain. Demand pair B exists between the two terminal multiplexers at the ends of the chain, and demand pair C exists between two of the ADMs along the chain. As traffic patterns change, the terminal and add–drop multiplexers can be reconfigured to carry the new demands. Without ADMs there would need to be point-to-point systems with terminal multiplexers

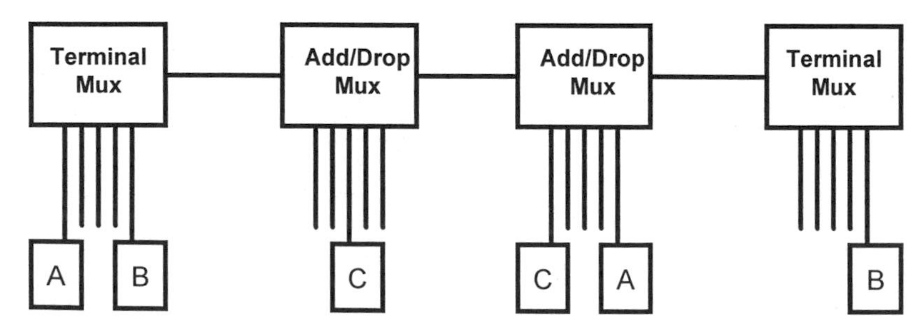

Fig. 2.3 Multiplexer chains with terminal multiplexers and add–drop multiplexers.

between each pair of network nodes that had traffic demands. Those point-to-point systems would potentially be lower speed, but even so represent a more costly and less flexible solution.

3. Self-Healing Rings

Self-healing rings provide the same flexibility as SONET chains in their ability to handle a mix of traffic demand patterns among all the equipment interconnected. They also provide the service survivability of diverse, fiber-routed point-to-point systems. In fact, a ring with just two nodes looks very similar to a diversely routed, protected point-to-point system. Rings protect against both cable cuts and node failures. When a cable is cut, the electronic equipment is configured to change the routing of traffic to bypass the cut. If there are multiple cuts on a ring, an unlikely scenario, the ring is broken into two segments, each of which can continue to communicate between themselves. In the case of a node failure, the traffic is again reconfigured to avoid the failed node, and the remaining nodes continue to function with their common traffic. The traffic to the failed node is interrupted. Figure 2.4 illustrates a self-healing SONET ring. SONET rings themselves come in three main types, each of which offers advantages in different network situations. They differ in the direction of traffic flow, whether they are unidirectional or bidirectional, and whether they use two fibers or four. A detailed description of SONET rings is beyond the scope of this chapter.

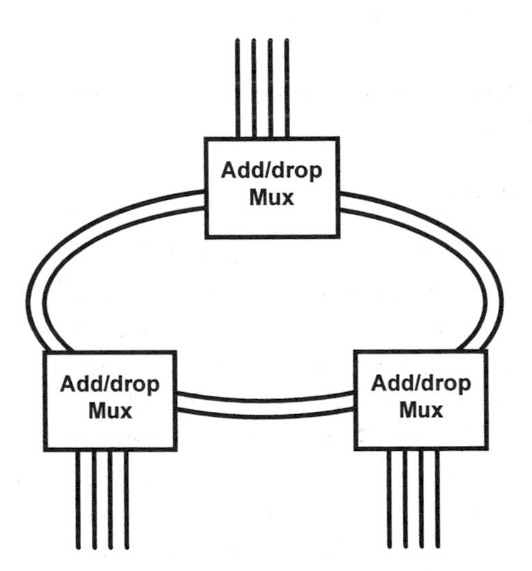

Fig. 2.4 Self-healing SONET ring.

4. Switch Networks and SDH Cross-Connects

A mesh of switches interconnected in an arbitrary pattern in the final architecture that we discuss for SONET networks. As mentioned previously, SONET switches are referred to as *digital cross-connect switches (DCSs)*. They differ from telephony service switches in that they do not support signaling, but are administered by a management and control operator or an automatic operations system. The configurations established in DCSs are generally long-lived.

Switches can grow to arbitrary numbers of port counts. They can support interfaces at a number of SONET signal rates. Internally, they are designed to do switching that spans a number of levels of the SONET hierarchy. In Fig. 2.5 we show a schematic of a DCS that is composed of line interface modules and a switch fabric. We show interfaces that are bidirectional, with modules for 155 Mb/s, 622 Mb/s, and 2.5 Gb/s. Internally, the switch fabric will rearrange the components of these signals. For example, a 155-Mb/s signal on one of the interfaces can be interconnected to one of the 155-Mb/s channels of a 2.5-Gb/s signal.

DCSs are most often used in the backbone network, where large amounts of traffic are aggregated. National scale core networks can be made up of a mesh of cross-connects, with rings used for more local traffic management and distribution to end customers. DCSs can provide interconnection among a large number of ring networks. They are the most flexible and highest capacity of the architectural elements.

B. BASIC MULTIPLEXING STRUCTURE AND SIGNAL HIERARCHY

1. SDH Signal Rates

The base signal of the SDH has a rate of 155.52 Mb/s and is called the *synchronous transport module, level 1 (STM-1) signal.* Higher rate signals are exact multiples of this rate and increase by factors of four. The SDH and SONET signals are compatible with each other. Before there was international agreement on the SDH, the SONET hierarchy was being standardized in the United States. The base-level SONET signal, called *STS-1* for synchronous transport signal, level 1, was created to efficiently carry the *digital signal, level 3 (DS-3) signal,* which has a rate of 44.736 Mb/s. The SONET hierarchy was subsequently modified to reach international agreement on the SDH. Because of this history, the term *SONET* is still

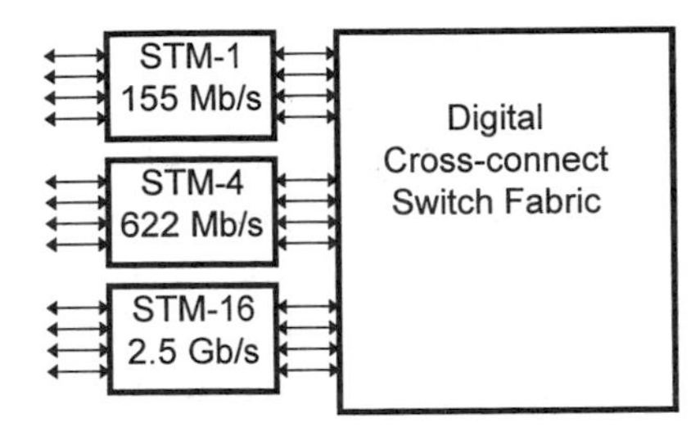

Fig. 2.5 Conceptual diagram of a SONET digital cross-connect switch.

used in North America. The SONET hierarchy starts at exactly one-third the rate of the SDH base signal. In order to make transport systems compatible with the existing signal hierarchies, a number of mappings are defined. At this point, the highest rate with available equipment is STM-64. The signal rates in most common use are shown in Table 2.2.

Higher rate signals in the hierarchy either can be a multiplex of lower rate signals or can themselves constitute a new, high-capacity single channel. The term *concatenated signal* is used to indicate a signal that is not composed of multiple component signals. Concatenated signals are referred to with a *C* after the hierarchy number, such as STM-4C, or STS-12C, to denote a signal with a single payload that must be delivered intact. This feature allows for the transport of future high-capacity signals as the need for them arises.

2. The SDH Frame Structure

The TDM signal is periodic, with a period of 125 μs. This corresponds to the period of a voice sample from telephony. One octet serves to carry an 8-bit digital sample for one voice channel, and these samples must be taken every 125 μs to faithfully reproduce the frequencies in the 3-kHz analog

Table 2.2 **Selected Signal Rates for the SDH and SONET**

Signal Rate	*SDH Signal*	*SONET Signal*
155.52 Mb/s	STM-1	STS-3
622.08 Mb/s	STM-4	STS-12
2.488 Gb/s	STM-16	STS-48
9.953 Gb/s	STM-64	STS-192

bandwidth voice signal. The SDH was created to serve the needs of many services besides voice, but the bandwidth dedicated to other services will be multiples of an octet every 125 μs, or 64 kb/s.

The SDH signal on a high-speed transmission line is a sequence of octets that form a pattern that repeats every 125 μs. The term *frame structure* refers to the rules for assigning data and other functionality to octet positions. The majority of octets carry subscriber information, but a portion of them, about 3–4%, are dedicated to system overhead, the name given to capacity set aside for system administration, control, and management.

The high-speed or line-rate signal for a multiplexer is created by assembling the overhead information and placing it in the appropriate locations with the information-carrying octets. Octets, rather than bits, are interleaved. This provides a signal where octets are always kept together rather than having their bits interleaved with other signals. A particular octet in the frame can be located, once the frame boundaries are determined, by counting and simple logic functions. When the SDH was being defined, a conscious decision was made to create a regular structure that would easily scale to a higher bit rate by interleaving signals from lower rates. Even though this regular structure might render some of the overhead redundant and unusable in higher rate systems, the complexity required to continually repack the transmission system to get a little more efficiency was deemed too high a price to pay.

Figure 2.6 illustrates the SDH frame structure of the base STM-1 signal. It is represented as a table with 9 rows and 270 columns of octets, yielding

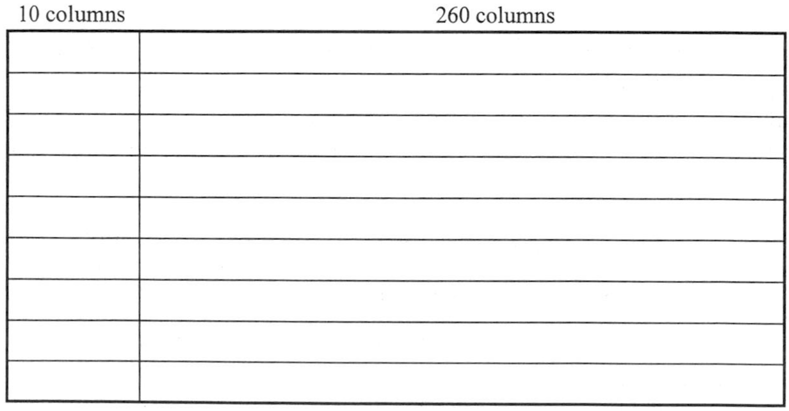

Fig. 2.6 Frame structure of the synchronous digital hierarchy STM-1 signal and the SONET STS-3 C signal.

a total of 2430 octets that repeats every 125 μs. The representation of the frame structure in a table is convenient to explain its logical structure. The sequence of octets on the transmission line is obtained by traversing the table row by row, moving from left to right. The line rate is 155.52 Mb/s. The overheads are contained in the first 10 columns and have three separate components: one for the path level, one for the multiplex section level, and one for the regenerator section level. The first 9 columns contain the regenerator section and the multiplex section overheads. The regenerator section is contained in the first 3 rows, whereas the multiplex section overhead is contained in the remaining 6 rows. The 10th-column contains the path-level overhead, and in some situations it is not placed in the 10th column but is located in a different column of the remaining 260.

3. Overview of Selected SDH Overheads

Following is a summary of some of the important SDH overheads:

- *Framing.* The framing bytes are composed of a specific bit pattern that is repeated in the same place in each frame. If these patterns continue to be present at the appropriate time, there is an extremely high probability that the signal is the framing signal and that the alignment of all other bytes is known. To make it difficult for a data pattern to mimic the framing pattern and disrupt the frame alignment process, the data as well as a number of the overheads are scrambled.
- *Payload pointer.* The payload pointer is one of the most significant innovations in the SONET multiplexing structure. In previous systems, the low-speed input signals to a multiplexer could arrive at an arbitrary position relative to their own frame boundaries. The input data would have to be buffered to allow all them to be aligned with the frame of the high-speed multiplex signal. These buffers were also necessary to allow for slight differences in clocks in the transmission lines that fed the multiplexer. The payload pointer eliminated the need for these buffers, sometimes referred to as *elastic stores.* The pointer is a specific set of overhead bytes whose value can be used to determine the offset of the path overhead and payload signal from the frame boundary.
- *Performance monitoring.* Octets in the path, multiplex section, and regenerator section portions of the overhead are used for perfor-

mance monitoring. This mechanism allows for continuous performance monitoring and the possibility of proactive fault management. Fault localization is made possible by the two section overheads, and the path-level performance monitoring provides a mechanism whereby customers can be assured of the quality of the service that they are receiving. Performance monitoring uses a technique called *bit-interleaved parity checking,* whereby a parity calculation is done on a portion of the signal before transmission and the result is recorded in a designated octet. At the receiver, the calculation is repeated and any discrepancy is taken as evidence of an error condition.

- *Protection switching.* Two octets are allocated to control the automatic protection switching function. These are in the multiplex section of the overhead.
- *Data communications channels.* Three octets are set aside in the regenerator section level of the overheads and create a 192-kb/s channel for communications among regenerators. In the multiplex section level, there are nine octets set aside, which create a data communications channel with a capacity of 576 kb/s. The purpose of the data communications channel is to allow remote control and management. It also can be used to support remote software downloads.
- *Engineering order wire.* This is an octet set aside for voice communications by technicians over the transmission system.

This has been a very brief overview of the SDH and SONET concepts. For more detailed information, the reader should consult the references by Sexton and Reid[1] and Wu[2] at the end of this chapter.

IV. ATM Networking

A. *ATM NETWORK ARCHITECTURES AND NETWORK ELEMENTS*

Whereas SONET is used as an underlying transport network infrastructure, ATM is used both for service switching and as a transport infrastructure. ATM is carried on SONET transport facilities in many cases.

The major network architecture for ATM consists of a mesh of switches. ATM virtual circuit switches support services to end users, whereas ATM virtual path swtiches are analogous to SDH–SONET DCSs. Virtual circuits and virtual paths can be considered two levels of a transport hierarchy. A virtual path can contain a large number of virtual circuits. Virtual path and virtual circuit identification information is contained in the ATM cell headers, as we see later. Some network equipment may recognize and act on only virtual path information, whereas others may recognize both virtual circuit and virtual path information. Virtual path switching can be thought of as a bulk switching capability. Figure 2.7 illustrates the relationship of virtual path and virtual circuit switches. Virtual circuits are requested between the three virtual circuit switches shown in the figure. Two virtual paths are already in place, ready to satisfy requests for switch pairs 1–2 and 1–3. In this example we have one fiber, and possibly even only one SONET channel, connected to switches 1, 2, and 3. Fiber is connected from the three virtual circuit switches to the virtual path switch in the center. The virtual path switch serves to establish connectivity among the three virtual circuit switches, and the number and capacity of the virtual paths can be varied to suit anticipated traffic needs.

There are a number of ways in which virtual paths are used in private and public networks. A large business may use ATM services to connect two distant locations. By purchasing a large-bandwidth virtual path from a service provider, customers may fill it with a dynamic mix of connections, as long as they do not exceed the capacity of the path. In this situation, the service provider's switch will be instructed to operate only on the virtual path identifier field and to ignore the virtual circuit identifier field. It will

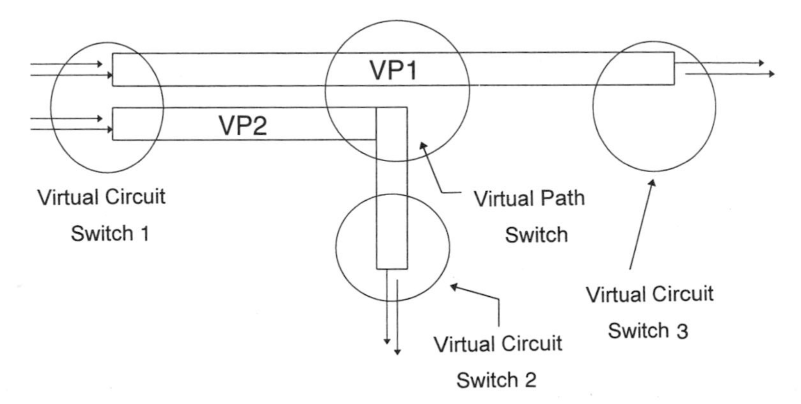

Fig. 2.7 ATM virtual path and virtual circuit switching.

not intercept and act on any of the signaling messages that cross the network between the customer's endpoint switching equipment.

Network service providers may also use virtual paths for their own internal purposes. They may construct a network with ATM virtual circuit switches interfacing to customers, and then interconnect the switches with a set of ATM virtual paths. The virtual path switch is configured to support all the interswitch connections needed in a cluster of ATM service switches by supporting dedicated virtual paths between switch pairs. The network that results is logically a mesh network, but physically a star or hub network, as shown in Fig. 2.8. This example is similar to the one in Fig. 2.7. If the network were configured to support a physical mesh, it would require $n(n - 1)/2$ bidirectional connections and $n(n - 1)$ switch ports. Using a virtual path requires n bidirectional links and $2n$ switch ports. The mesh situation grows as n^2 as the number of switches increases, while the alternative grows linearly in n. The bandwidth dedicated to each virtual path can be set to match the expected traffic demand.

The concept of an ATM virtual path switch or cross-connect is analogous to that of the SDH cross-connect. Both are considered elements of a transport layer, a flexible, reconfigurable layer that supports the transmission requirements of a wide variety of network services. The ATM cross-connect has SDH transmission interfaces, but they are higher capacity than ATM service switch customer interfaces on average. As ATM traffic demands grow, it will be common that the aggregate traffic demands between pairs of service switches will be many hundreds of megabits per second to several gigabits per second, so the network-side switch interfaces are likely to become STM-4 (622 Mb/s), STM-16 (2.5 Gb/s), or even STM-64

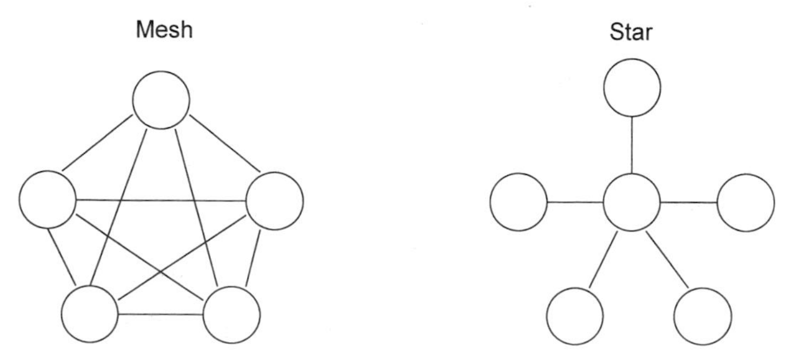

Fig. 2.8 Virtual circuit mesh network provided by a virtual path switch. The logical network (*left*) is compared with the physical network (*right*).

(10 Gb/s). Within a single transport interface there may be a number of virtual paths, each carrying traffic to a different service switch. The ATM cross-connect can selectively route the virtual paths to their destinations.

Another architectural element expected to enter the marketplace soon is the ATM virtual path ADM. This element is analogous to the SDH ADM, which is used in chain or ring topologies. An ATM ADM can also be used in chain or ring topologies to share the bandwidth of a high-capacity transport facility among a number of network nodes. It will have SDH transport-level interfaces, but add–drop functionality is achieved by making the add–drop decision based on the information in the virtual path identifier field of the ATM cell header. Figure 2.9 shows a situation where four ATM ADMs reside on a common ring, and each is connected to an ATM service switch.

The result achieved with the example in Fig. 2.9 is similar to the cases shown earlier in Fig. 2.8; that is, ATM service switches are interconnected in a logical mesh network. In Fig. 2.8 the physical network was a star, whereas in Fig. 2.9 the physical network is a ring. In the earlier case, the interconnection is achieved with an ATM virtual path cross-connect, whereas in this case it is achieved through an interconnecting high-capacity transport ring — for example, an STM-64 (10-Gb/s) virtual path ring carrying all its information in ATM cells. The choice for any specific situation depends on the amount of traffic that the application must support. In the case of the ring, the aggregated traffic cannot exceed the total ring capacity.

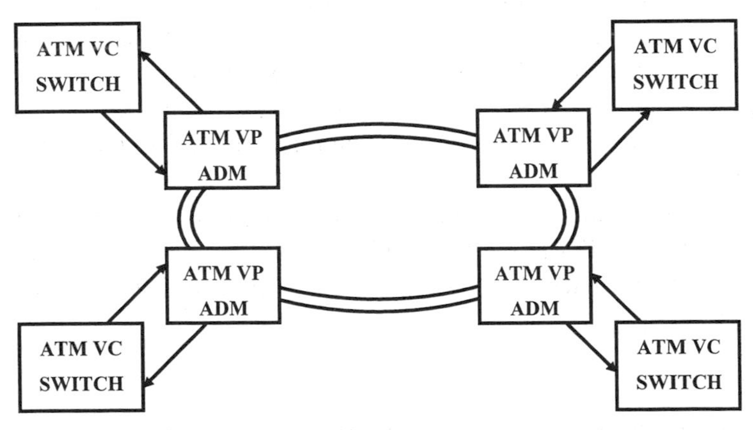

Fig. 2.9 Interconnection of a network of ATM virtual circuit (VC) switches by means of ATM add–drop multiplexers (ADMs) on an ATM virtual path (VP) ring.

Cross-connects can grow in capacity by adding more switch ports and increasing the dimension of the internal switch fabric.

B. ATM CELL STRUCTURE

ATM is a packet-oriented mode of information transport. The basic unit of information transport is called a *cell* to distinguish it from other types of packets. ATM cells are fixed-length packets, 53 octets long. The first 5 octets are the header, which contains overhead information for use by networks in delivering the cell to its destination. The remaining 48 octets are the information payload. The ATM cell format is illustrated in Fig. 2.10.

ATM is a connection-oriented packet network protocol. In connection-oriented packet networks, the network nodes along the communication path have been preloaded with information related to the connections that are established through them. Therefore, the network elements need enough information to distinguish only among the packets that will be routed through them, not all possible packets in the entire network. This is how it is possible to have such a small header.

ATM cell header formats are defined for two types of network interfaces. The first is called the *user–network interface (UNI),* and the second is called the *network–network interface (NNI).* One instance of a UNI exists between an individual workstation and an ATM switch. The NNI exists between swtiches. An important function of the UNI is signaling to set up switched connections, switched virtual circuits. Another function that may be required at the UNI is flow control, and at the switch side of the UNI it may be necessary to monitor the cell flow offered by the user and to take actions in the event that the cell rate exceeds the agreement. One function that must be supported at the NNI is signaling between switches to fulfill user requests for switched connections, including making decisions on routing of circuits.

There are a number of fields that make up the header for ATM cells, and these fields differ at the UNI and the NNI. The two address fields are of primary importance. They are the virtual path identifier (VPI) field and

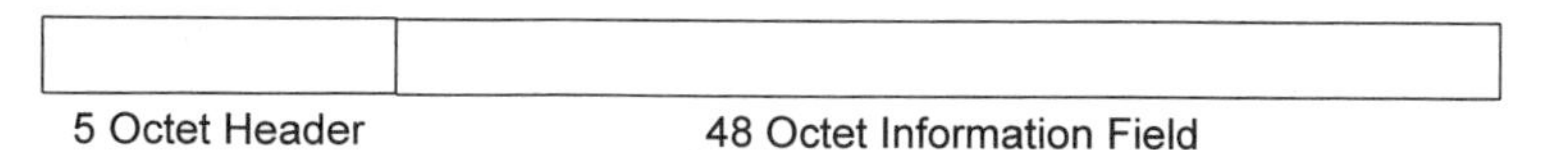

Fig. 2.10 ATM cell structure.

the virtual circuit identifier (VCI) field. Both these fields convey the identity of the cell to network equipment.

We now describe the structures for the ATM cell header for the UNI and the NNI. Figure 2.11 illustrates the header structure by placing the 5 octets as rows of a table. Each column of the table represents a bit position in the octet. At the UNI, the first 4 bits are the generic flow-control (GFC) field. The next 8 bits are dedicated to the VPI, providing for 256 possible virtual paths at an interface. The next 16 bits allow the differentiation of more than 65,000 virtual circuits. The payload type identifier (PTI) is used to distinguish between user cells and management cells. The last bit of the fourth octet is the cell-loss priority bit, to indicate cells that can be discarded under certain network congestion conditions. The final octet is the header error check (HEC) field, which can identify errored cell headers and fix any single-bit errors. The only difference between the UNI and the NNI is that the 4-bit GFC field is omitted at the NNI, and those bit positions are added to the VPI, allowing for 4096 possibilities.

Virtual path networking is simpler to implement than virtual circuit networking is. Virtual paths are provisioned by management processes, not by end-user signaling, so no signaling processing and signaling system management needs to be included. At the virtual circuit layer there may be a variety of service classes defined, with different queuing mechanisms and traffic-control mechanisms for these services. Virtual path networking elements have a bulk-delivery responsibility and no responsibilities that are specific to individual services or service classes. Both types of switches need to support physical level transmission interfaces and to do whatever processing is required by those interfaces. In Table 2.3, we show the

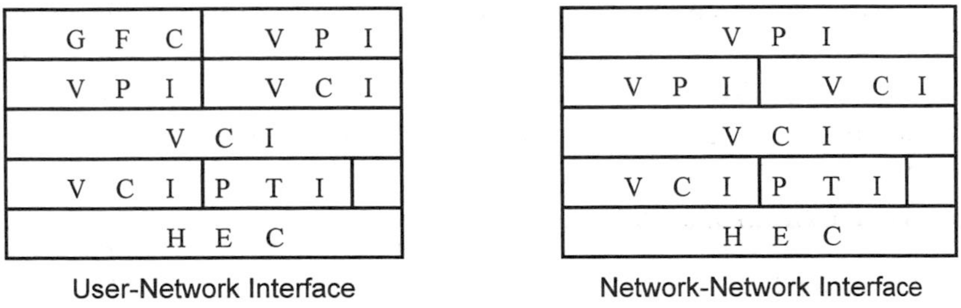

<table>
<tr><td>GFC</td><td colspan="2">VPI</td></tr>
<tr><td>VPI</td><td colspan="2">VCI</td></tr>
<tr><td colspan="3">VCI</td></tr>
<tr><td>VCI</td><td>PTI</td><td></td></tr>
<tr><td colspan="3">HEC</td></tr>
</table>

User-Network Interface

<table>
<tr><td colspan="3">VPI</td></tr>
<tr><td>VPI</td><td colspan="2">VCI</td></tr>
<tr><td colspan="3">VCI</td></tr>
<tr><td>VCI</td><td>PTI</td><td></td></tr>
<tr><td colspan="3">HEC</td></tr>
</table>

Network-Network Interface

Fig. 2.11 ATM cell header composition at the UNI and the NNI. GFC, generic flow control; VPI, virtual path indicator; VCI, virtual circuit indicator; PTI, payload type identifier; HEC, header error check.

Table 2.3 **SDH and ATM Functionality at the Interfaces of Virtual Circuit and Virtual Path Switches**

Functionality	*Virtual Circuit*	*Virtual Path*
SDH Functions		
Frame detection	X	X
Cell delineation and extraction	X	X
Error monitoring for links and paths	X	X
Management communications	X	X
ATM Functions		
Header error check	X	X
Address translation	X	X
Performance monitoring	X	X
Flow control, rate policing	X	
Service class-dependent queuing	X	
Signaling	X	

functionality required at the interfaces of ATM virtual path and virtual circuit switches. Table 2.3 compares some of the functions that will be included in virtual circuit switches and virtual path switches. Although there is much in common, the virtual path switch has a number of important simplifications that allow it to reach higher speeds more easily than virtual circuit switches.

The preceding discussion is not an exhaustive list of all the processing that must be done on the ATM cells at a switching point, but it does serve to convey the complexity required. The processing does not easily lend itself to parallel processing because it is essential that cell order be maintained within a virtual circuit. As line rates increase, the amount of time allowed for ATM cell processing decreases proportionally. Table 2.4 lists the ATM cell period, the maximum time allowed for ATM cell processing, for transmission line and switch port rates from 155 Mb/s to 10 Gb/s.

C. *ATM CELLS IN THE SDH*

ATM cell specifications make no reference to cell rate or physical transport medium. This is one of the major advantages of ATM over previous data communications protocols, such as ethernet or token ring, that need to be redefined or revised as higher speed versions are introduced. It is the

Table 2.4 **Available ATM Cell Processing Time as a Function of Switch Port Transmission Rate**

Signal	*Line Rate*	*ATM Cell Period*
STM-1	155.52 Mb/s	2.73 μs
STM-4	622.08 Mb/s	682 ns
STM-16	2.488 Gb/s	170 ns
STM-64	9.953 Gb/s	43 ns

primary intent for wide-area networks in North America and much of the world to carry ATM cells on SONET or SDH systems. Just as there are mappings between DS-1 and DS-3 digital signals into the SONET signals, there is also a mapping of ATM cells. This is shown schematically in Fig. 2.12. Cells are placed in the SDH information fields. The pointer in the SDH path overhead is used to identify the positions of cells in the information fields, and its value is equal to the number of octets between it and the beginning of the next ATM cell header in the same row.

Two approaches can be taken to increase the aggregate cell rate on an SDH transmission system. In the first case, cells can be placed in multiple STM-1 signals. As more capacity is needed, additional STM-1 signals can be dedicated and filled with cells. Following this method, an STM-4 signal can be composed of four STM-1 signals, each dedicated to the transport of ATM cells. An alternative is to make use of the concatenated signal. In this case, an STM-4C signal can dedicate its entire payload to support ATM cell transport. Given sufficient traffic demand, one single large container is preferable to multiple smaller containers because there is a larger resource to share among the constituent cell streams. It is to be expected that an STM-4 signal can carry more cells than four STM-1 signals can if there is

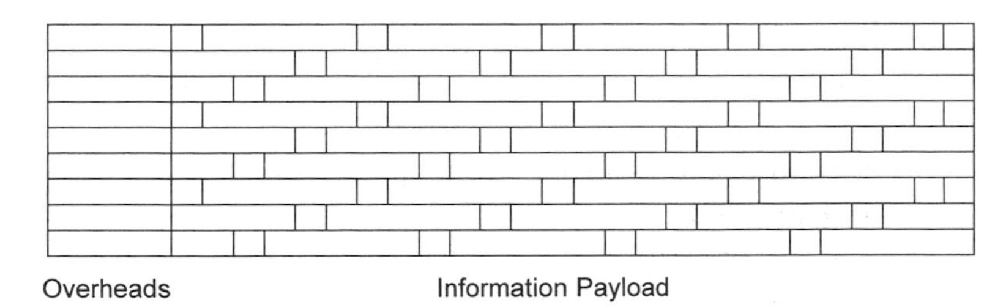

Fig. 2.12 ATM cells placed in an SDH–SONET frame.

statistical multiplexing, or for the same number of cells in both cases, the STM-4 signal can provide lower cell loss. For a more detailed description of ATM, the reader may want to consult de Prycker.[3]

V. SONET and ATM

We just completed an overview of the SDH and ATM. These standards have a complex relationship. ATM was introduced as a service switching standard, a universal bearer service that could meet the needs of current and future applications. It was designed to be carried over an SDH transport-level network. The same SDH transport-level network would carry aggregated voice traffic and leased circuits along with ATM, and provide an evolutionary path for the gradual introduction of broadband services. But ATM can also be considered competitive with the SDH in certain applications.

Let us consider the situation when ATM services, and not voice or low-speed leased lines, are the dominant traffic in networks. The SDH will continue to be the dominant line-rate transmission format, but the balance of multiplexing and switching will likely evolve from the TDM circuit-based SDH toward ATM. An advantage of ATM virtual paths that motivates this possible evolution is the flexibility that they provide, in comparison with the rigidity of the SDH–SONET TDM hierarchy. Virtual paths have no fixed size, and they can be very small or very large, depending on the bandwidth needs. In the SDH, there are fixed-size transport modules, such as the STM-1, STM-4, and STM-16 signals.

All this ATM flexibility comes at a price, and that price is the need to do comparatively more complex per-cell header processing in continually decreasing times as the line rates increase. With ATM one must also deal with the statistical nature of traffic. The rigidity of TDM in the SDH is a major strength when one strives to create high-speed networks that are right at the limits of optoelectronics technology. As the SONET world is introducing 10-Gb/s line-rate systems, the ATM community is planning for the introduction of 622-Mb/s signals. Thus it's impossible to see one standard or the other winning this competition, but a complementary relationship is likely going forward.

The SDH has been accepted worldwide as the standard for future fiber optic transmission systems. Even as multiwavelength transmission technology moves into commercial use, it is integrated within the SDH standards rather than viewed as a replacement technology that obsoletes the SDH.

A great deal of momentum has built up worldwide behind the ATM standards. Although ATM was advanced by the telecommunications industry, it was embraced by the data networking community to meet their needs for dramatic growth in bandwidth and the support of real-time services. It is seeing initial deployment in private networks, and it will later migrate to become a major force in wide-area public networks. Thus, it is likely that ATM will be one of the platforms used for future broadband data and multimedia networks.

With both the SDH and ATM we have a set of standards that meet many of the future transport networking needs. Both standards are flexible and scalable to arbitrarily high bit rates. Both have seen widespread acceptance worldwide. Both will be the major networking protocols that lower level optoelectronics technology will need to support into the indefinite future.

References

1. Sexton, M., and A. Reid. 1992. *Transmission networking, SONET and the SDH.* Norwood, MA: Artech House.
2. Wu, T-H. 1992. *Fiber network service survivability.* Norwood, MA: Artech House.
3. de Prycker, M. 1991. *Asynchronous transfer mode, solution for broadband ISDN.* Chichester, England: Ellis Horwood.

Chapter 3 | Coding and Error Correction in Optical Fiber Communications Systems

Vincent W. S. Chan

Massachusetts Institute of Technology, Lincoln Laboratory, Lexington, Massachusetts

3.1 Introduction

During the development of lightwave communications with fiber since the mid-1970s, the major emphasis for research has been on the technology of lightwave devices and subsystems. Fiber is one of the most predictable and stable communications channels ever developed and characterized. In lightwave systems, the modulation and demodulation subsystems are typically designed and manufactured with very low bit error probabilities. In fact, the extremely competent device designers and fabricators may have spoiled the system engineers, often creating pristine devices and subsystems with outstanding properties and tolerances. Examples are low-chirp laser diodes to prevent pulse spreading due to dispersion in the fiber, laser diodes with ultralow mode partition noise for good pulse detection performance, and highly sensitive positive-intrinsic-negative field-effect transistor (PIN-FET) and avalanche photodiode (APD) receivers for good receiver sensitivities. Although the lightwave systems available to date have detection sensitivities at least 10 dB away from theoretical limits, there has never been a compelling reason, economic or otherwise, to improve receiver performance. For a few years in the 1980s, coherent detection was explored to gain a few decibels of receiver performance, especially for undersea cables. With the advent of erbium-doped fiber amplifiers (EDFAs), coherent-system-like performance became possible with a preamplified direct detection receiver. Indeed, near-quantum-limit detection performance at 10 Gbps was demonstrated in 1996 [3.1]. Thus, it is not surprising that

OPTICAL FIBER TELECOMMUNICATIONS, VOLUME IIIA

ISBN: 0-12-395170-4

lightwave communications systems have not yet exploited what other communications systems have for years — forward error-correction coding and decoding to improve channel reliability.

As the field of telecommunications moves into the 21st century, lightwave systems will be more than point-to-point links with a highly controlled and predictable environment. For instance, the use of EDFAs will reduce the need for frequent regenerators, wavelength-division multiplexing (WDM) will increase single fiber capacities many fold, the use of optical routing and switching will introduce a new way of building optical networks, and the interconnection of many types of optical and nonoptical networks into a global network will result in a mix of networks of different designs and different generations interacting by means of an open network architecture. In these scenarios, applications of forward error-correction coding and decoding may substantively affect device fabrications, system designs, and their economics. Thus, it is worthy to consider how this technique can be profitably used in lightwave systems of the future.

The technique of forward error-correction coding has its theoretical foundations on the celebrated papers by Claude E. Shannon in the late 1940s [3.2, 3.3] and has seen many years of highly sophisticated mathematical development in the field of Information Theory. In the following brief discussion of the Noisy Channel Coding Theorem, the spirit of the main enabling result of those works is captured. For a more rigorous treatment of the subject, the reader should refer to a text on Information and Coding Theory [3.4–3.6].

3.1.1 NOISY CHANNEL CODING THEOREM

For a channel with known degradations (e.g., noise power and statistics) and restrictions on resources (e.g., signal power, bandwidth), a quantity called the *capacity* of the channel (usually expressed in bits per second) can be established such that

(1) For information transmission rates less than the capacity, there exists a forward error-correcting code that can be used to transmit data over the channel with arbitrary accuracies.

(2) For information transmission rates exceeding the capacity, no matter what the code may be, there will always be an unavoidable and nonzero error probability.

This result may seem counterintuitive at first glance, but realizing how it is practically implemented can greatly help understanding. Figure 3.1

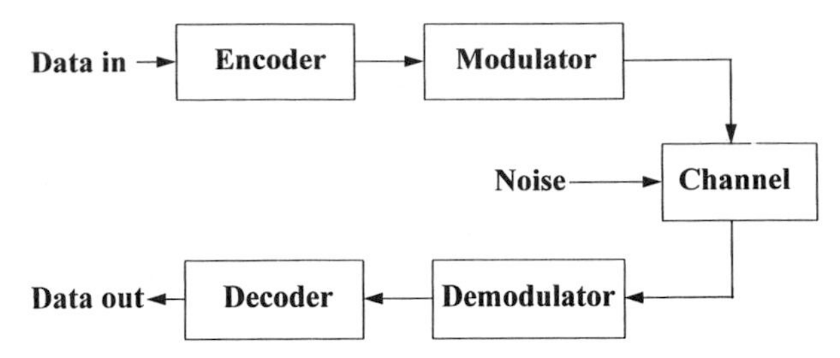

Fig. 3.1 Basic model of a digital communication system with forward error correction.

illustrates the essential building blocks of a coded digital communication system. The encoder introduces redundancy to the input data before transmissions. The decoder uses this redundancy to reconstitute the input sequence even in the presence of transmission errors, provided the frequency of errors is less than a threshold defined as the *power* of the code. The codes used in current systems are typically well designed to minimize the overhead (redundancy) required and usually have efficient decoding algorithms so that they can be implemented in modest hardware or software modules. For example, a code with a 25% increase in redundancy can be effective enough to substantively shift operating points of lightwave systems (e.g., from 10^{-12} to 10^{-6} channel bit error rates) and affect device specifications. This chapter provides some insight into the process of using forward error correction in noisy lightwave channels, and, in particular, indicates how lightwave system designs can be more efficient with the addition of coding. Some of these impacts can be significant. For example, WDM systems can pack more wavelengths into the EDFA band for more capacity, longer transmission distances can be attained with more cascaded EDFAs before signal regeneration, and optical devices such as routers, frequency translators, and amplifiers can function well with less stringent tolerance on cross talk and the amount of additive noise.

3.2 Common Modulation Formats for Fiber Systems — Direct Detection and Coherent Systems

Although there are many types of waveforms being used in current communication systems, the discussions in this chapter are restricted to the waveforms that are most commonly used in lightwave systems. For the purpose of understanding the fundamental limits of detection of these signals, the signals can be divided into three classes: *on–off signaling*, *orthogonal signaling*, and *antipodal signaling*. Most modulation formats used in lightwave systems are binary (i.e., each symbol will carry 1 bit of information). Higher order modulations (those that carry multiple bits per symbol) tend to be more power efficient but do so at the expense of more bandwidth required per bit of information transmitted. Because of the cost of high-speed optoelectronics, these schemes are seldom used in high-speed lightwave systems.

On–off signaling is the simple intensity modulation of the optical field. The presence of a pusle indicates the symbol *one* and the absence *zero*. This type of signaling is used in most lightwave systems because of its bandwidth efficiency and ease of implementation, such as direct laser current modulation or electrooptic waveguide modulation by a Mach–Zehnder interferometer or with an electroabsorption modulator. Soliton systems almost exclusively use this format because of its simplicity and because it is easier to implement at very high speeds.

Orthogonal signaling formats use modulation waveforms that are *orthogonal*, such as *frequency shift keying* (*FSK*), where different orthogonal frequencies denote different symbols, and *pulse position modulation* (*PPM*), where orthogonal time slots denote different symbols. Binary PPM is sometimes used in lightwave systems. These signaling formats are more commonly known as *Manchester Coding* (not to be confused with error-correction coding) in the lightwave community. *Orthogonal signaling* is common and well understood in classical radio frequency communications, but it is a misnomer with regard to lightwave communications. The reason is that although the classical waveforms of the optical fields corresponding to different symbols are orthogonal, at high receiver sensitivities where quantum effects are significant the quantum state representations of these signals are not orthogonal. In fact, unlike classical RF communications, in the absence of additive noise the detection error probability of orthogonal optical signals (in the classical sense) is not zero but a well-defined amount dictated by quantum mechanics.

Antipodal signals are almost never used in lightwave systems, although one was used in 1996 to achieve high detection sensitivity [3.1]. Because these are the quantum optimum signal sets for binary signaling, they are included here for reference. The simplest form of antipodal signaling is *phase shift keying* (*PSK*), where different phase shifts of the optical carrier denote different symbols. In *binary phase shift keying* (*BPSK*), the signals one and zero are separated by 180° in optical phase of the carrier.

Two classes of receiver structures are used to detect optical signals: *direct (incoherent) detection* and *coherent (heterodyne or homodyne) detection.* Optically preamplified direct detection is equivalent to heterodyne detection in terms of its fundamental quantum model and detection performance.

In direct detection, the received optical field is energy detected by means of a photodetector that usually provides gain. Examples are APDs or PMTs (photomultiplier tubes). Modulation schemes for direct detection systems are limited to intensity modulations such as on–off signaling and Manchester Coding.

In coherent detection, an optical local oscillator field is added to the received optical field, and the sum is detected by a photodetector. The resulting signal is further processed at base band (homodyne detection) or at an intermediate frequency (heterodyne detection). Phase and/or frequency tracking of the signal field by the local oscillator laser is required. The mixing of the weak signal field and the strong local oscillator field at the front-end of a coherent receiver provides linear amplification and converts the optical signal into an electrical output with gain (usually tens of decibels), raising the signal level well above the noise level of subsequent electronics. This is why a detector with gain is not required. Coherent detection can be used on any of the three classes of modulations discussed. However, because antipodal signaling is a class of signals that rely on phase modulation, it can be detected only by coherent detection, not by direct detection.

3.3 Fundamental Detection Performances and Deviations of Currently Available Systems

Often, sensitive lightwave receivers can be at or a few decibels from the quantum detection limit. Thus, understanding where the quantum limit lies is important to the system designer. For, if the performance of the designer's hardware is close, the design will be of diminishing return to try to improve system performance, whereas if the design is far from the quantum limit, there is a chance of significant payoffs if he or she spends more effort to improve the design.

The output of a laser well above threshold can be represented by a special quantum state called the *coherent state* [3.7]. A coherent state $|\boldsymbol{\alpha}\rangle$ is labeled by a complex number $\boldsymbol{\alpha}$, where the modulus corresponds to the amplitude of the optical field, and the phase of $\boldsymbol{\alpha}$ corresponds to the phase of the field. When direct detection is performed on such a state, the photon count is Poisson distributed with intensity (also known as the *rate parameter*) $|\boldsymbol{\alpha}|^2$, and when coherent detection is used, the output has the complex amplitude $\boldsymbol{\alpha}$, with added white Gaussian noise, which is actually a manifestation of quantum detection noise.

Suppose that a signal with two possible states, $\boldsymbol{\alpha}$ and $\boldsymbol{\beta}$, representing the symbol *zero* and the symbol *one*, respectively, is received by an optical receiver. The quantum optimum receiver can be found by optimizing over the set of all possible receivers, consistent with the laws of quantum mechanics. This set of receivers is usually known as *quantum observables*. The mathematical structure of the quantum optimum receiver and its performance were found in 1970 by Helstrom, Liu, and Gordon [3.8]. The probability of error can be found exactly:

$$Pr[\varepsilon] = \tfrac{1}{2}\left(1 - \sqrt{1 - |\langle \boldsymbol{\alpha} | \boldsymbol{\beta} \rangle|^2}\right), \tag{3.1}$$

where the inner product of the two coherent states $|\boldsymbol{\alpha}\rangle$ and $|\boldsymbol{\beta}\rangle$ is given by $\langle \boldsymbol{\alpha} | \boldsymbol{\beta} \rangle$. The magnitude of the inner product between the two coherent states is

$$|\langle \boldsymbol{\alpha} | \boldsymbol{\beta} \rangle|^2 = \exp\left(-|\boldsymbol{\alpha} - \boldsymbol{\beta}|^2\right). \tag{3.2}$$

The average number of photons in the field represented by the state $|\boldsymbol{\alpha}\rangle$ is simply given by $\mathbf{N}_\alpha = \langle \boldsymbol{\alpha} | \boldsymbol{\alpha} \rangle$. For small errors, the inner product is also small and the probability of error can be simplified to

$$Pr[\varepsilon] \approx \tfrac{1}{4}|\langle \boldsymbol{\alpha} | \boldsymbol{\beta} \rangle|^2 = \tfrac{1}{4}\exp\left(-|\boldsymbol{\alpha} - \boldsymbol{\beta}|^2\right) = \tfrac{1}{4}e^{-\boldsymbol{\theta}}, \tag{3.3}$$

where $\boldsymbol{\theta} = |\boldsymbol{\alpha} - \boldsymbol{\beta}|^2$ is the exponential parameter that describes the behavior of the system at low error rates. It is trivial to note that for on–off signaling, $\boldsymbol{\beta} = 0$, $\boldsymbol{\theta} = |\boldsymbol{\alpha}|^2 = \mathbf{N}_\alpha = 2\mathbf{N}_s$, where $\mathbf{N}_s$ is the average number of photons received per bit. It is also easy to see that under an energy constraint per bit, the signal set that minimizes the error probability is antipodal signaling, which maximizes the error exponent $\boldsymbol{\theta}$ at $4\mathbf{N}_s$.

Often the quantum optimum receiver is unrealizable with known techniques, or its implementation, even if it is known, is extremely complicated. Thus, simple receiver realizations are used as near-optimum compromises. For example, direct, or incoherent, detection is most commonly used in lightwave systems (note that the quantum model for preamplified direct detection is the same as that for heterodyne detection and should be viewed as equivalent to that of heterodyne detection rather than that of standard direct detection). The ideal form of direct detection is a photon-counting receiver (i.e., a receiver with enough electrical gain per photoelectron emitted by the detector surface such that individual photo events are detected and counted by subsequent electronics). APD and PIN-FET receivers, however, are generally at least 10 dB less sensitive. As stated before, modulations for such detection schemes are limited to intensity modulations.

In lightwave systems with high-performance single-mode lasers, the detected photo events can be well characterized by Poisson statistics. Given that one chooses binary signaling and direct detection, on–off signaling is the optimum signal set. The optimum receiver is a counting receiver with a threshold. When the count exceeds the threshold, a one is considered sent, with a zero otherwise. The probability of detection error can be well approximated by the bound

$$Pr[\varepsilon] \leq e^{-2\mathbf{N}_s}, \tag{3.4}$$

where $\mathbf{N}_s$ is the average number of photons detected per symbol. It can be shown that the exponent $\boldsymbol{\theta} = 2\mathbf{N}_s$ gives the tightest exponential bound (sometimes known as the *Chernoff Bound*) for the error probability and for all practical purposes (at low error rates) can be used as an excellent approximation of the actual error probability. If Manchester Coding or binary PPM is used, the exponent $\boldsymbol{\theta} = \mathbf{N}_s$, which means that PPM is 3 dB less efficient than on–off signaling with ideal direct detection.

When the local oscillator field intensity of a coherent receiver is high enough (which can be achieved in lightwave systems), the detector output is quantum noise limited. The output noise process can be modeled as an additive white Gaussian noise with power spectral density $\mathbf{N}_0/2 = \frac{1}{2}$ for heterodyning and $\frac{1}{4}$ for homodyning (normalized to the energy of a single photon and assuming unity quantum efficiency detectors). The communication performance of the detection of signals in additive white Gaussian noise is well known. Given that one chooses binary signaling, binary antipodal signaling (e.g., PSK) is the optimum signal set. The channel error probability for homodyne detection is given by

$$Pr[\varepsilon] = Q(\sqrt{2\mathbf{N}_s}) \leq \tfrac{1}{2}e^{-\boldsymbol{\theta}}, \qquad \boldsymbol{\theta} = 2\mathbf{N}_s, \tag{3.5}$$

where Q is the Gaussian error function and $\mathbf{N}_s$ is again the average received number of photons per symbol. On–off signaling and binary PPM are 3 dB less efficient with the exponent $\boldsymbol{\theta} = \mathbf{N}_s$. Heterodyne detection is again 3 dB less efficient than homodyne detection because the upper and lower sidebands present twice the amount of noise to the receiver, compared with that of the base-band signal of the homodyne receiver.

The performance of the various common signal sets and detection schemes is summarized in Table 3.1. The most common signaling format and detection scheme currently used in lightwave communications is on–off signaling and direct detection. The direct detection photon-counting receiver achieves quantum optimum performance with a sensitivity of 10^{-12} at 28 detected photons per bit. Current state-of-the-art commercial APD or PIN-FET receivers are 10–15 dB away. The degradation of performance at modest data rates (100 Mb/s to 1 Gb/s) is mostly due to the noisy avalanche process of an APD or the front-end noise of electronics after detection. This type of noise increases as the bandwidth of the receiver increases. Thus, higher rate receivers are noisier, with greater degradations in performance. At very high rates (10–100 Gb/s), in addition to electronics noise, there are the effects of bandwidth limitations of electronics and timing system jitters giving rise to intersymbol interference and perhaps also cross talk. With a suitable low-noise optically preamplified direct detection receiver such as an EDFA, several of the lost decibels can be recovered. As alluded to before, this receiver curiously has the same quantum mechanical model and detection limit as those of a heterodyne receiver. This limit has been approached within a factor of 2 [3.1] in an EDFA receiver at 10 Gb/s. In general, it would be difficult and/or costly to improve transmitter and receiver performance enough to approach quantum limited performance. With the advent of high-density and high-speed applications-specific

Table 3.1 **Receiver Performance Comparison: Probability of Detection Error, $Pr[\varepsilon]$, for Binary Signaling**[a,b]

Signal Set	*Direct Detection*	*Heterodyne Detection*	*Homodyne Detection*	*Quantum Optimum*
On–off signaling	$2\mathbf{N}_s$	$\mathbf{N}_s/2$	$\mathbf{N}_s$	$2\mathbf{N}_s$
Orthogonal signaling (PPM, FSK)[c]	$\mathbf{N}_s$	$\mathbf{N}_s/2$	$\mathbf{N}_s$	$2\mathbf{N}_s$
Antipodal signaling (PSK)[d]	Not applicable	$\mathbf{N}_s$	$2\mathbf{N}_s$	$4\mathbf{N}_s$

[a] Exponent $\boldsymbol{\theta}$ of tightest exponential bound, $Pr[\varepsilon] \leq e^{-\theta}$.
[b] $\mathbf{N}_s \equiv$ average number of detected photons per bit.
[c] PPM, pulse position modulation; FSK, frequency shift keying.
[d] PSK, phase shift keying.

integrated circuits (ASICs), the use of error-correcting codes is an attractive alternative to the development of highly accurate and pristine devices. The next section presents some of these opportunities.

3.4 Potential Role of Forward Error-Correcting Codes in Fiber Systems and Its Beneficial Ripple Effects on System and Hardware Designs

Forward error correction can be implemented simply in a lightwave system by encoding the information symbols into code words by means of an encoder (Fig. 3.1). At the receiver, two major types of decoding, *hard* and *soft decisions decoding*, can be used to recover the information bits. With hard decisions decoding, the receiver first makes tentative decisions on the channel symbols and then passes these decisions to the decoder, where errors are corrected. With soft decisions decoding, the receiver, in principle, would pass on to the decoder the analog signal at the output of the demodulator. The decoder would then make use of the information in the analog signal (rather than the hard decisions) to re-create the transmitted information bits. Soft decision decoding thus is always better performing than hard decisions decoding, usually by a couple of decibels.

With hard decision decoding, most lightwave channels can be modeled as *binary symmetrical channels* (*BSCs*) (Fig. 3.2). A BSC will make an error with the same probability p for inputs zero and one. The parameter p can be measured experimentally or derived using a model of the receiver via a similar process that leads to the expressions in Table 3.1. The *capacity* of the BSC is well known [3.4–3.6]:

$$C_{hard} = 1 + p \log_2 p + (1 - p) \log_2 (1 - p). \tag{3.6}$$

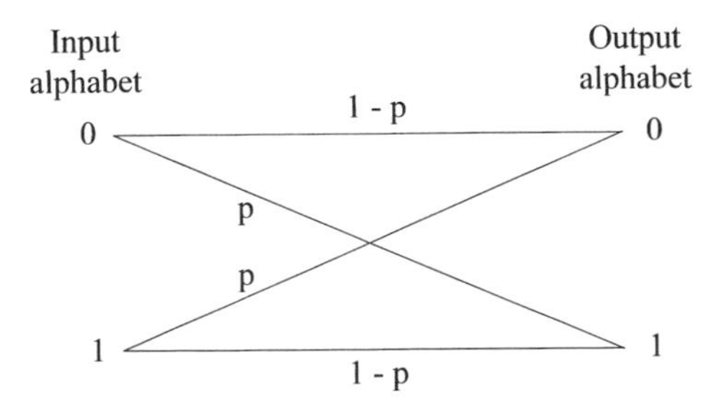

Fig. 3.2 Binary symmetrical channel.

The capacity C_{hard} is for each use of the channel (per channel bit transmitted for the BSC). The error probabilities given in Table 3.1 can be used to find C_{hard} for the various modulation and detection schemes.

Often it is difficult for implementable systems to work near capacity. A second quantity, R_{comp}, the *computation cutoff rate* of the channel, is used as a convenient measure of the performance of the overall communication channel. R_{comp} is usually less than the capacity of a channel and represents a soft upper limit of information rates for which moderate-size decoders are readily implementable. The R_{comp} for hard decisions decoding is

$$R_{comp,hard} = 1 - \log_2 [1 + 2\sqrt{p(1-p)}]. \tag{3.7}$$

$R_{comp,hard}$ is a realistically achievable performance to expect of a coded system with present-day electroncis technology. Later in this chapter, examples of practical coders and decoders are given.

To achieve the ultimate capacity of most communication systems, it can be shown that soft decisions decoding must be used. Soft decisions decoding is currently done for only modest-rate communication systems (e.g., 10 Mb/s) and is unlikely to be used soon in typically high-rate lightwave systems. For an appreciation of the potential gains, the capacity C_{soft} and the computation cutoff rate $R_{comp,soft}$ for binary PPM signaling (Manchester Coding) are given next:

$$\begin{aligned} C_{soft} &= 1 - \tfrac{1}{2}e^{-N_s} \\ R_{comp,soft} &= 1 - \log_2 (1 + e^{-N_s}). \end{aligned} \tag{3.8}$$

Figure 3.3 depicts a plot of the capacity and the cutoff rate in bits per use of the binary PPM channel for hard and soft decisions decoding. In most interesting regions of operations, $R_{comp,hard}$ is within 3–6 dB of C_{soft}. The added complexity of a soft decisions decoder makes it difficult to implement soft decisions decoding at high rates (>100 Mb/s). Thus, only hard decisions decoding is used in the examples given subsequently.

The ultimate performance limit of lightwave systems lies in nonbinary systems, where the signaling alphabet can be much larger than 2. The capacities and cutoff rates of direct and coherent detection systems are included in Table 3.2 for reference. Derivations can be found in Ref. 3.8, or from Eqs. (3.6) to (3.8). Note that the capacity for the direct detection channel with no additive noise and only quantum detection noise, given in Table 3.2, is infinite. This may sound counterintuitive at first, but this performance occurs in an unrealistic scenario, when the PPM signaling

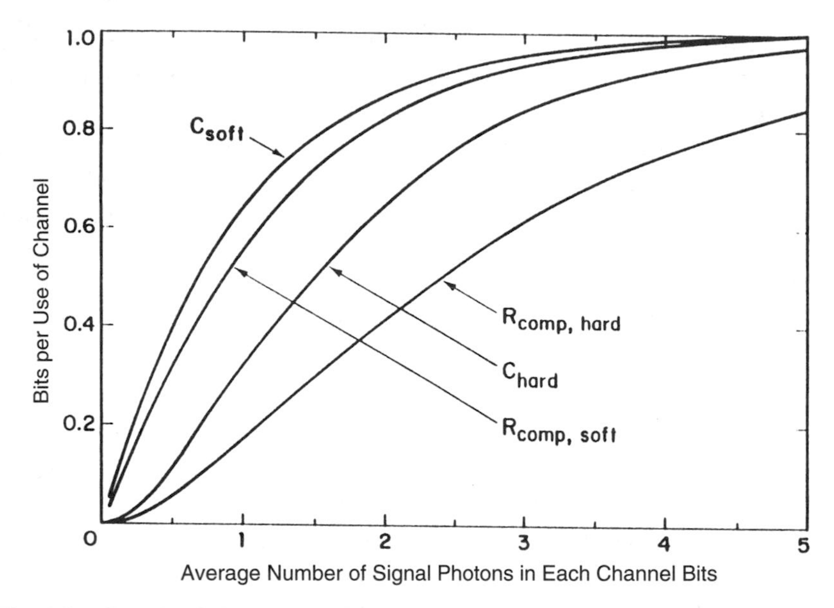

Fig. 3.3 C_{soft}, C_{hard}, $R_{comp,soft}$, and $R_{comp,hard}$ versus the average number of photons per channel bit for quantum limited performance.

symbol size and the energy in the pulse both approach infinity. As is evident, current lightwave systems are very far away (>20 dB) from these limits. Even for binary systems, current lightwave systems are about 10 dB away from the theoretical limits. To recover a few decibels of performance will require better optical devices and electronics, which can be expensive. Another way of recovering a few decibels (e.g., 5) is the use of forward error correction. Not only can error-correcting codes provide a few decibels of power efficiency, but they can also shift the operating point of a link from virtually error free for the uncoded channel to frequent errors for a coded channel. For example, a code with a modest coding gain of 3 dB

Table 3.2 **Receiver Performance Comparison: Computation Cutoff Rate R_0 and Capacity, C, of Coded Systems**[a]

Detection Scheme	*Direct Detection*	*Homodyne Detection*
Computation cutoff rate R_0	1 nat/photon	1 nat/photon
Capacity, C	∞	2 nat/photon

[a] 1 nat = $\log_2 e$ bits.

(i.e., it can transmit at the performance of the uncoded channel with a factor of 2 better in power efficiency) can operate at a raw link bit error rate of 10^{-6} but yields a delivered information bit error rate after decoding of 10^{-12}. The next section introduces some practical codes and a little more insight into the technique.

3.5 Some Practical Codes

The simplest code to understand is the *repeat code* — repeating the information bits several times and letting the receiver correct for errors by using majority logic (i.e., when there are disagreements on the binary bit sent among replicas, the symbol with more replicas is selected). This primitive error-correction code, although easy to implement, is inefficient, requiring a significant increase of bandwidth, and seldom provides any coding gain. A more efficient code called the *parity check code* is illustrated in Fig. 3.4. Its operation is easy to understand also. The data sequence is first broken into blocks of bits. Each block of information bits is arranged in a two-dimensional array with one parity check (exclusive-or, XOR) for each row and one for each column. The parity bit in the lower right-hand corner is a check on the parity bits. If a single error occurs during transmission, it can be detected and corrected by cross-checking the corresponding row and column parity bits. Generally, multiple errors cannot be corrected unless they occur at certain places. Although this code is effective for the occasional isolated error, it is especially weak against short bursts of errors. Because such bursts of errors can occur in lightwave systems (such as those caused by power dropouts due to laser partition noise), the use of this code

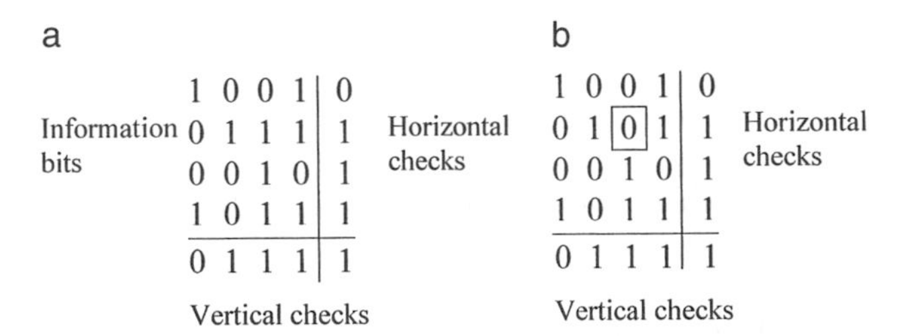

Fig. 3.4 Horizontal and vertical parity check code. (a) Coded message. (b) A single channel error (*boxed bit*) can be located and corrected.

is undesirable. However, the basic principle of the two-dimensional parity check code can be readily translated into more general parity check codes.

A more powerful set of codes operating on the same principle as the parity code is *linear block codes*. For binary input data, a linear code is constructed from a set of modulo-2 linear combinations of the data symbols. Again, the binary information sequence is segmented into message blocks of length k, denoted by $\mathbf{u}$. There are 2^k possible distinct messages. The encoder uses either a codebook or an encoding algorithm to map each input message $\mathbf{u}$ into another binary block $\mathbf{v}$ of length $n > k$ (see Fig. 3.5). Thus, there are $n - k$ binary bits added as redundancy. the set of 2^k n-tuple code words is called a *block code* (usually labeled as an $[n, k]$ block code to indicate the amount of redundancy). For a block code to be useful, the code length n should be large. The decoder looks at each block of length n and finds in the codebook the closest code word to the received block. The inverse mapping to an information block of length k recovers the information sequence. The ratio k/n is sometimes known as the *code rate* of the code and is an indication of the degree of redundancy in the code. Usually, the smaller the code rate, the more error-correction capability of the code, albeit at the expense of more channel bits transmitted per input bit.

Unless a block code has a special mathematical structure, it is difficult to encode and decode a long code. Almost all block codes currently used are linear block codes (just as parity check codes). Some commonly used block codes are the Reed–Solomon Code and the BCH (Bose–Chaudhuri–

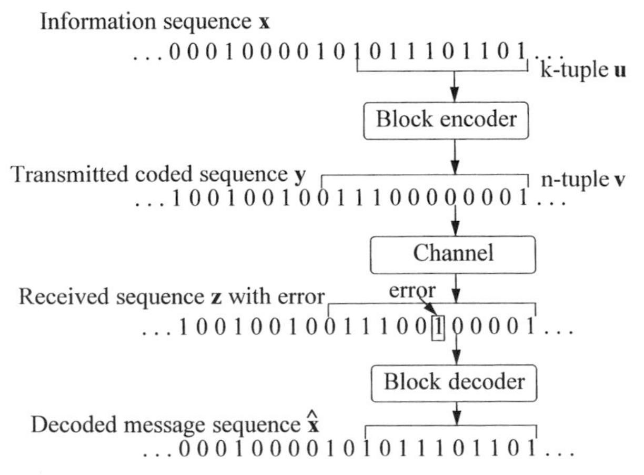

Fig. 3.5 Block coding and decoding with an (n, k) code.

Hocquenghem) Code. Coding gains of 3–5 dB can be attained. Of significant note is that the best linear block codes can perform about as well as the best block codes (without the linearity constraint) with the same lengths and code rates. The encoding and decoding processes of these codes would require extensive mathematical development before they could be presented here. Thus, instead, the reader should refer to Refs. 3.4 through 3.6.

The most common experiment on the use of error-correction code in lightwave systems to date is the application to the extension of long-haul repeaterless transmission with EDFAs. When multiple EDFAs are cascaded to extend repeaterless long-haul transmission, noise accumulation will ultimately limit the number of EDFAs that can be used in series before data regeneration. Reed–Solomon Codes have been used for such experiments [3.9–3.12]. When no forward error correction is used, the noise accumulated from EDFAs imposes a floor on the bit error rate performance. When error correction is used, the raw link can operate at much higher error rates, which renders the error rate floor inconsequential. As we see later in this chapter, not only can more EDFAs be used in cascade, but EDFA spacings can also be increased. This fact has significant implications on the design and costs of long-haul, especially transoceanic, systems. Because the decoding algorithm of Reed–Solomon Codes requires significant processing, these experiments have been realized with stream rates of less than 1 Gb/s and multiplexed up to higher rates through interleaving. Recent advances in GaAs and InP high-speed electronics promise decoders at base rates of a few to tens of gigabits per second in the future.

Convolutional Codes [3.4–3.6] are another set of commonly used error-correction codes. They differ from the codes discussed before in that they are not block codes. The continuous convolutional structure of the encoder has a property that facilitates hard and soft decision decoding and improves coding performance; thus it is the most popular coding scheme currently in use. Figure 3.6 shows a binary convolutional encoder with a K-stage register. The information sequence enters the encoder 1 bit at a time, and every bit in the register shifts right one position, with the last one on the extreme right discarded. The taps feeding into the modulo-2 (XOR) adders are selected for good coding performance. There can be v of such adders, so v channel bits are generated for each information bit, which makes the code rate $1/v$.

There are a number of ways to decode convolutional codes with various levels of decoding complexities. *Threshold* and *sequential decoding* schemes are easy to implement, providing modest but not best coding gains

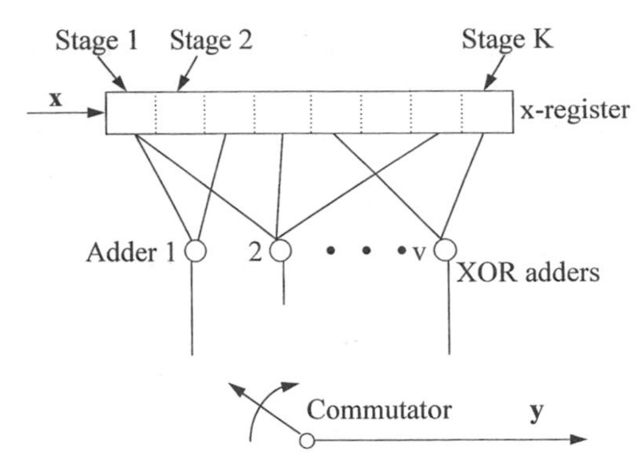

Fig. 3.6 Binary convolutional encoder with a *K*-stage shift register. XOR, exclusive-or.

(~3 dB). They are ideal for very high-speed lightwave systems [3.13]. A much more powerful scheme is called *Viterbi decoding*, especially the soft decision version. Viterbi decoding is the optimum decoding scheme, and coder–decoder chips are available commercially at speeds up to tens of Mb/s. All the coding schemes discussed thus far use hard decisions (i.e., tentative decisions on the channel bits are made before further processing by the decoder). Viterbi decoding has a natural way to accept analog (usually quantized to 3 bits for ease of operation) outputs from the matched filters and processes these signals to recover the information sequence. Additional coding gains of up to 2 dB can usually be realized. A common convolutional code using Viterbi decoding is a rate $\frac{1}{2}$, constraint length 7 code. Unfortunately, this decoding scheme is complicated, requiring many computations, and will not be useful for lightwave systems until ultrafast electronics technology improves drastically.

A recent example of a high-rate code that has been implemented in a lightwave system is a convolutional code at 2 Gb/s [3.13, 3.14]. It is a rate $\frac{4}{5}$ code with a constraint length (K) of 332. The rate $\frac{4}{5}$ code requires only 25% bandwidth expansion (i.e., requires 25% more transmitted bits). This is an important consideration for high-rate lightwave systems, not because fiber bandwidth is precious, but because high-speed electronics are expensive and generally have poorer performance. Therefore, requiring a lot of bandwidth expansion is the wrong direction to take for high-rate systems. The encoder and decoder are implemented as a single, custom, silicon bipolar integrated circuit gate array. Threshold decoding was chosen as the decoding algorithm because of its simple decoder design, which permits high-speed operations. Figure 3.7 is a graph of the decoded bit error rate

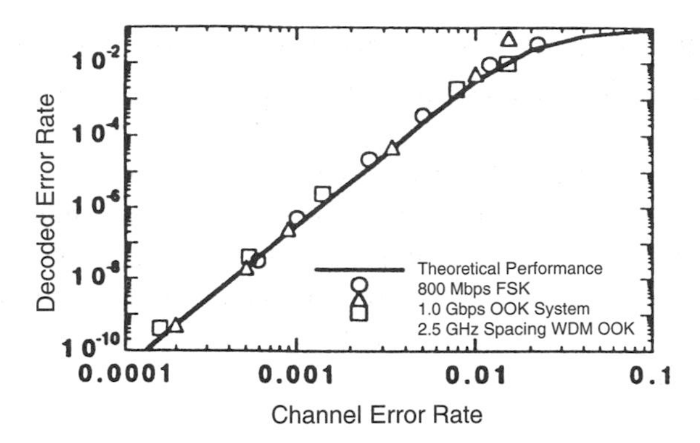

Fig. 3.7 Decoded bit error rate versus channel error rate for on–off keying (OOK) and frequency shift keying (FSK) experiments. WDM, wavelength-division multiplexed.

plotted against the channel error rate. Note that one phenomenon is common to all coded systems: the presence of a threshold raw channel bit error rate beyond which the code starts to take effect. In this particular example, the transition point is at a 4×10^{-2} bit error rate, below which there is a coding gain of approximately 3 dB (i.e., the exponent of the decoded bit error rate is approximately twice as large as the exponent of the raw channel). The raw channel uses NRZ (non-return-to-zero) signaling, achieving a system sensitivity of 87 received photons per bit at 10^{-12}. The coded system achieves the same performance at 37 received photons per information bit, realizing a gain of 3.7 dB (see Fig. 3.8).

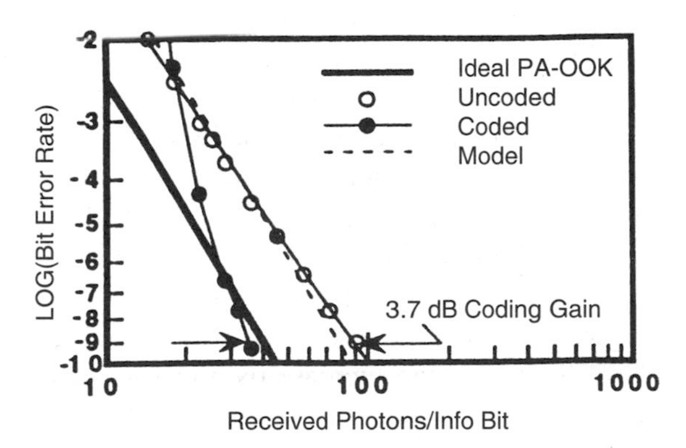

Fig. 3.8 Measured bit error rate performance of the codec at a 1.0-Gb/s data rate on a single OOK channel. Coded sensitivity is 37 photons per bit. PA-OOK, preamplified on–off keying.

3.6 Applications of Coding to Future Lightwave Systems Such as WDM Systems and Optical Networks

Gaining a few decibels of receiver sensitivity is not very important for present-day lightwave systems, especially at 1.5-μm wavelengths, where an EDFA preamplifier can readily improve receiver sensitivity. However, the application of forward error correction does have some significant implications in the design of future lightwave systems. Some of these implications are discussed next.

3.6.1 DECREASED SENSITIVITY TO DEVICE AND SUBSYSTEM FLAWS

A coded system can and will operate the raw lightwave channel at much higher error rates. Minute device and subsystem flaws do not usually become apparent until the error rate is extremely low. Hence, a coded system barely sees the difference between pristine and not-so-perfect devices. This system feature allows the use of less expensive and more mass-producible optoelectronics, which has significant implications on system costs. In addition, a system's lifetime can be extended as devices slowly degrade. Thus, a lightwave system using a laser with appreciable mode partition noise may never be able to achieve 10^{-12} raw channel error performance but will with the use of forward error correction. A detailed theoretical analysis and simulation of coded lightwave systems using a direct detection APD receiver can be found in Ref. 3.15. Performance gains up to 9 dB are predicted. Similarly, when significant laser phase noise is present, a coherent system will exhibit an error rate floor but can have perfectly acceptable performance when an error-correcting code is used [3.16]. To first order, in the limit of high signal-to-detection-noise ratio, the laser phase noise behaves like additive white Gaussian noise in a phase- or frequency-modulated system (see Ref. 3.16). The exponent θ, of the bit error rate floor $e^{-\theta}$ due to laser phase noise, increases linearly with the laser linewidth. Thus, the specification on laser phase noise in a coherent system can be significantly relaxed, by approximately the coding gain of the code used (e.g., if the coding gain is 3 dB, θ and hence the laser linewidth can increase by a factor of 2). Finally, when installed fiber plants exhibit reflections from fiber discontinuities, which occur more frequently than expected because of variations in the quality of work in the field, the raw channel may see error rate floors that a coded system may ride through and provide acceptable performance.

3.6.2 *INCREASED LINK DISTANCES FOR OPTICALLY AMPLIFIED FIBER SYSTEMS*

As alluded to before, forward error correction can be used to extend the link distances of optically amplified fiber systems. To quantify the potential gain, assume that the gain of each of the in-line optical amplifiers exactly equals the loss over the fiber between amplifiers [3.17]. Thus, it is easy to see that the amplifier noise accumulated over N amplifiers is $N.N_{sp}$, where N_{sp} is the spontaneous emission noise added by each amplifier. A good estimate of the maximum number of amplifiers, N_{max}, that can be used in series before regeneration can be derived from the condition that the accumulated amplifier noise, $N.N_{sp}$, cannot exceed the receiver detection noise, N_{det}. Obviously, under such conditions, when the receiver sensitivity is increased with the use of forward error correction, the maximum number of amplifiers increases proportionally. Because coding gains of 3–6 dB are achievable at lightwave system rates, the number of in-line amplifiers and thus the link distance before regeneration can be increased by a factor of 2–4. Also, one can see that the input optical power, P_{in}, into the fiber should be as large as possible without paying a severe penalty resulting from fiber nonlinearities [3.18]. In the next subsection, we see that the use of forward error correction can also alleviate degradations due to fiber nonlinearities. Hence, further gains in the ability to cascade links and in link distance can be realized. Because the explicit gains are complicated functions of the operating points of the system and fiber nonlinearity parameters, the exercise is left to the reader.

3.6.3 *INCREASED PACKING DENSITY OF WAVELENGTHS IN WDM ALL-OPTICAL NETWORKS*

WDM all-optical networks are becoming a reality as a result of rapid research and development worldwide. The use of optical amplifiers and wavelength multiplexing promises higher throughputs and more economical networks. The erbium amplifier bandwidth is of the order of 20 nm. There is much incentive to pack as many wavelengths into a fiber as possible. In addition to a larger capacity, an increased number of wavelengths provides increased connectivity for an all-optical network. When wavelengths are packed close together, there will be degradations due to spectrum spillover from one signal to an adjacent channel. This problem is particularly limiting when passive or active multiplexers–demultiplexers, routers, and switches

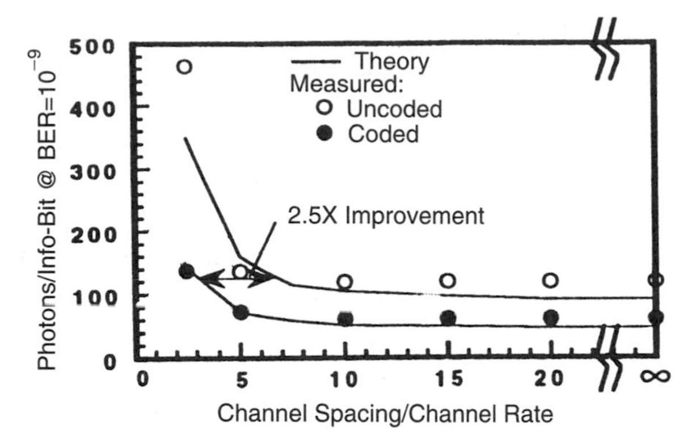

Fig. 3.9 Receiver sensitivity versus channel spacing. *Solid lines* are calculated from a model and indicate an expected improvement in channel spacing of approximately 2.5*X*. The measured improvement was 2*X*. BER, bit error rate.

are used. If, in addition, dispersion-shifted fiber is used for long-haul transmissions, fiber nonlinearities will also result in severe dgradations. The dominant nonlinear fiber effect is four photon mixing [3.18], the magnitude of which increases as the wavelengths move closer. Thus, both these effects limit wavelength packing density. The use of forward error correction shifts the operating point of the raw channel and allows more cross talk and nonlinear products. To first order, because the bit error rate curve is exponential in the signal-to-noise ratio, the gain in terms of allowable increase of these interference effects equals the coding gain. In Ref. 3.19, an increase of channel packing density of a factor of 2.5 has been demonstrated, using the convolutional code described in Section 3.5, which has a coding gain of 3.7 dB (see Fig. 3.9).

3.6.4 RANDOM ACCESS OF A SHARED FIBER SYSTEM VIA CODE-DIVISION MULTIPLEXING

When a transport medium is used as a shared broadcast medium, such as a star network, the benefit is that every user can hear the same information. The drawback is that every user signal acts as interference to other users' signals. There are several standard techniques employed to work around the interference problem. Time-division and frequency-division techniques are commonly used. Code-division multiplexing, first used in defense communications and more recently in cellular communications, is a potential candidate for lightwave networks. In this scheme [3.20], each user encodes his or her messages using a unique signature code and broadcasts the resulting signal into the medium. The receiver uses a decoder to sort out the intended user signal, treating all other user signals as noise. This random access scheme is particularly attractive when time synchronization is difficult, as in the case of a sizable all-optical network. Generally, there is a significant bandwidth expansion of the message rate (as much as the number of users sharing the medium), to accommodate many users in the network; thus, this method is less attractive for high-rate lightwave systems except when it is being used in the low-rate signaling channel for network management and diagnostics. The ability to work without time synchronization is an attractive feature for network management because of its ease of operation, particularly during network cold starts.

3.7 Summary

Although forward error correction has been used for decades in many communication systems, it did not proliferate into lightwave systems because of the advanced state of development of fiber and optoelectronics and the lack of high-rate coder–decoder technology. However, high-rate electronics technology has now finally caught up with the speed of lightwave systems, and single-chip coders–decoders are available. This development, coupled with novel new architectures, such as long-haul repeaterless transmissions with the use of EDFAs and WDM all-optical networks, will create new opportunities for the insertion of this powerful technique to enhance even more the performance of future lightwave systems and reduce costs significantly.

References

[3.1] Livas, J. 1996. High sensitivity optically preamplified 10 Gb/s receivers. In *OFC'96, San Jose, CA, February.* Postdeadline paper 4. Washington, DC: Optical Society of America.

[3.2] Shannon, C. E. 1948. A mathematical theory of communications. *Bell Syst. Tech. J.* 27:379–423, 623–656. Reprinted in 1949 in book form, with a postscript by W. Weaver, by the University of Illinois Press, Urbana.

[3.3] Shannon, C. E. 1948. Communications in the presence of noise. *Proc. IRE* 37:10–21.

[3.4] Gallager, R. G. 1968. *Information theory and reliable communications.* New York: Wiley.

[3.5] Lin, S., and D. Costello. 1983. *Error control coding: Fundamentals and applications.* Englewood Cliffs, NJ: Prentice-Hall.

[3.6] Viterbi, A., and J. Omura. 1979. *Principles of digital communication and coding.* New York: McGraw-Hill.

[3.7] Glauber, R. J. 1963. Coherent and incoherent states of the radiation field. *Phys. Rev.* 131:2766–2788.

[3.8] Helstrom, C. W., J. Liu, and J. Gordon. 1970. Quantum mechanical communication theory. *Proc. IEEE* 58:1578–1598.

[3.9] Yamamoto, S., H. Takahira, and M. Tanaka. 1994. 5Gbit/s optical transmission terminal equipment using forward error correcting code and optical amplifier. *Electron. Lett.* 30(3)254–255.

[3.10] Pamart, J. L., E. Lefranc, S. Morin, G. Balland, Y. C. Chen, T. M. Kissell, and J. W. Miller. 1994. Forward error correction in a 5 Gbit/s 6400 km EDFA based system. *Electron. Lett.* 30(4).

[3.11] Pamart, J. L., E. Lefranc, S. Borin, G. Ballard, Y. C. Chen, T. M. Kissell, and J. W. Miller. 1994. Forward error correction in a 5 Gbit/s 6400 km EDFA based system. In *OFC'94, San Jose, CA, February.* Postdeadline paper PD18-1. Washington, DC: Optical Society of America.

[3.12] Yamamoto, S., H. Takahira, E. Shibano, M. Tanaka, and Y. C. Chen. 1994. BER performance improvement by forward error correcting code in 5 Gbit/s 9000 km EDFA transmission system. *Electron. Lett.* 30(8).

[3.13] Castagnozzi, D. M., J. C. Livas, E. A. Bucher, L. L. Jeromin, and J. W. Miller. 1994. Performance of a 1 Gbit/s optically preamplified communication system with error correcting coding. *Electron. Lett.* 30(1):65–66.

[3.14] Castagnozzi, D. M., J. C. Livas, E. A. Bucher, L. L. Jeromin, and J. W. Miller. 1994. High data rate error correcting coding. *SPIE* 2123.

[3.15] Jeromin, L. L., and V. W. S. Chan. 1983. Performance of a coded optical communication system using an APD direct detection receiver. In *Proceedings of the IEEE International Conference on Communications, June, Boston, Massachusetts.* New York: IEEE.

[3.16] Jeromin, L. L., and V. W. S. Chan. 1986. M-ary FSK performance for coherent optical communication systems using semiconductor lasers. *IEEE Trans. Commun.* COM-34(4).

[3.17] Olsson, N. A. 1989. Lightwave systems with optical amplifiers. *J. Lightwave Tech.* 7(7).

[3.18] Tkach, R. W., A. R. Chraplyvy, F. Forghieri, A. H. Gnauk, and R. M. Derosier. 1995. Four-photon mixing and high-speed WDM systems. *J. Lightwave Tech.* 13(5).

[3.19] Livas, J. C., D. M. Castagnossi, L. L. Jeromin, E. A. Swanson, and E. A. Bucher. 1994. Forward error correction in a 1 Gbit/s/channel wavelength-division multiplexed system. In *IEEE/LEOS summer topical meeting on optical networks and their enabling technologies, Lake Tahoe, Nevada, 11 July.*

[3.20] Chan, V. W. S. 1979. Multiple-user random access optical communication system. In *ICC'79.*

Chapter 4 | Advances in Fiber Design and Processing

David J. DiGiovanni

Lucent Technologies, Bell Laboratories, Murray Hill, New Jersey

Donald P. Jablonowski

Lucent Technologies, Bell Laboratories, Norcross, Georgia

Man F. Yan

Lucent Technologies, Bell Laboratories, Murray Hill, New Jersey

I. Introduction

Fiber attenuation and dispersion are the major limiting factors in an optical network. These fiber attributes define the spans between regenerators and the signal transmission rates. In this chapter, we review the recent advances in fiber designs and processing technologies to overcome these limiting factors. In particular, we describe fiber processing for optical amplification, fiber laser devices, and dispersion management. For optical amplification at 1.55 μm, the erbium-doped fiber amplifiers (EDFAs) have been employed successfully in many undersea and terrestrial systems. We describe the appropriate index designs and processing technologies for active fibers that are key to their system implementation. Other emerging technologies are being developed for amplifier applications at 1.3 μm.

Dispersion management is crucial for high transmission rate operation. The conventional dispersion-shifted fibers are deployed successfully and adequately for optical networks operated at one single wavelength around 1.55 μm. The transmission capacity of an optical network can be substantially increased by wavelength-division multiplexing (WDM) several signal wavelengths in a transmission fiber. However, nonlinear effects in optical fibers impose new limitations on the multiple wavelength systems when the optical power exceeds a certain threshold. This limitation is alleviated by proper fiber designs to introduce a small and controlled amount of dispersion in an optical fiber. These fiber designs are described in this chapter.

OPTICAL FIBER TELECOMMUNICATIONS,
VOLUME IIIA

ISBN: 0-12-395170-4

There is also an emerging incentive to convert the installed 1.3 μm systems for 1.55 μm transmission such that EDFA can be used cost-effectively for signal repeater. In this case, dispersion-compensating fibers will be required to compensate for the dispersion at 1.55 μm in the non-dispersion-shifted fibers. Specialty connector fibers are being developed for optical attenuation as well as for low-loss coupling between fibers and light sources; and they will find important applications in optical networks.

II. Erbium Doped Fiber Amplifier

A. *APPLICATION*

The development and the deployment of optical fiber amplifiers have provided new and exciting alternatives for optical fiber systems. The fiber attenuation is no longer a limiting characteristic, and expensive regenerator stations can be eliminated. Optical amplification allows deployment of systems that can be upgraded in capacity simply by changing the terminal equipment and has enabled development of new architectures, such as more extensive WDM systems.

One of the first applications to make use of optical amplifiers was trans-oceanic systems. Previously, without optical amplifiers, expensive regenerators were required every 120 km. By using erbium-doped optical amplifiers every 45–85 km, designers have eliminated regenerators entirely from trans-oceanic systems (Trischitta *et al.* 1996; Barnett *et al.* 1996). Terrestrial systems similarly are employing optical amplifiers to upgrade existing installations (Li 1993). For every three existing regenerators, two are being removed and the third is being replaced by an optical amplifier. New terrestrial installations are being configured with 1000 km between regenerators, and continuing system experiments are pushing span distances even farther (Bergano *et al.* 1995).

To motivate the special design and fabrication issues for these active devices, a brief discussion of the principles and many advantages of optical amplification is presented here. A complete discussion of the operation and use of optical amplifiers and related components is presented in Chapters 9 and 10 of Volume IIIA and Chapters 2–5 and 7 of Volume IIIB.

B. *PRINCIPLES OF OPERATION*

An optical amplifier uses light at one wavelength as the energy source to amplify light at a second wavelength (Urquhart 1988). For telecommunications systems, 1.55 μm signals are amplified by passing them through a short section of optical fiber through which the energy source, the pump light, also propagates. The core of this special section contains the rare-earth element erbium, which acts as a storage medium for the transfer of energy from pump to signal. In an amplifier, emission is stimulated by a traveling wave, the signal. The amplifier section of optical fiber has a physical appearance similar to that of conventional fiber, but the core contains erbium and has a refractive index profile and composition tailored for optimum amplifier performance. As shown in Fig. 4.1, light at a wavelength of 980 nm is absorbed by the erbium ions; this absorption excites them to a higher energy level, $^4I_{11/2}$, which rapidly decays nonradiatively to a long-lived metastable state, $^4I_{13/2}$. The metastable state may also be populated by pumping the upper edge of the band at 1.48 μm. From this storage state, deexcitation will occur radiatively, emitting a photon around 1.5 μm. This process may occur spontaneously, but because the lifetime of this metastable state in silica is typically 10 ms, it is much more probable that emission will be stimulated by a traveling signal around 1.55 μm, providing signal amplification. The combination of long lifetime and absence of lower lying energy states renders the erbium system extremely efficient for amplification of telecommunications signals. Efficiencies very close to theoretical maximum are readily achieved (Pederson *et al.* 1991).

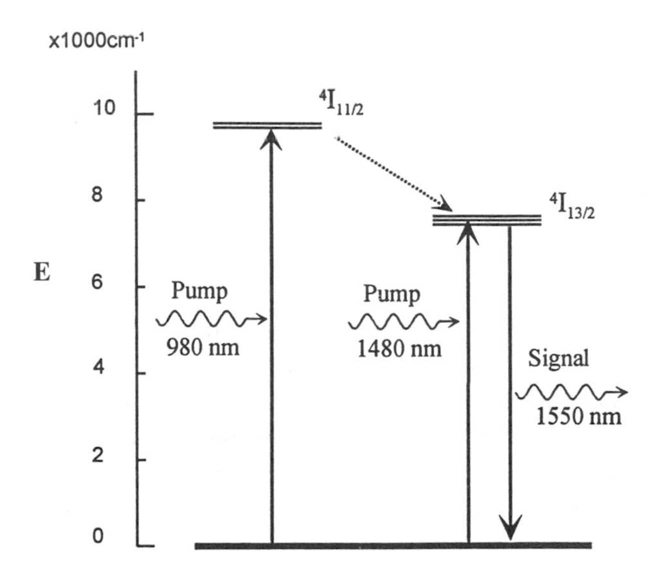

Fig. 4.1 Energy diagram of the three-level system in Er^{3+} shows the absorption of the 0.98 and 1.48 μm pump wavelengths and emission at 1.55 μm.

C. *FIBER DESIGN ISSUES*

The energy level diagram in Fig. 4.1 illustrates the important aspect that erbium forms a three-level system. That is, the ground state is also the terminal emission level; this causes the signal to be constantly reabsorbed. Without pump light to excite the ions, the fiber would be highly absorbing at thousands of decibels per kilometer. As pump power increases, however, the population of ions in the ground state, and hence the absorption, decreases. At the threshold power for a particular wavelength, the fiber is transparent, with just as many photons absorbed as emitted. This competition between absorption and emission implies that high optical pump power density will more fully invert the erbium population and increase the rate of emission (Armitage 1988). This is readily achieved either by reducing the mode field diameter or by confining the erbium to the central region of the core. The former design entails high-delta, small-core-diameter fiber. Deltas of 2%, compared with 0.35% for standard fiber, are typical and are realized simply by increasing the germanium concentration. In addition, for a maximum pump efficiency, the optimum cutoff wavelengths are 0.80 and 0.90 μm for pumping at 0.98 and 1.48 μm, respectively; these optimum cutoff wavelengths are independent of delta (Pederson *et al.* 1991).

A second criterion of erbium-doped fiber design centers around the low solubility of erbium ions in silica. Erbium ions enter the glass network as network modifiers that, because of their high field strength, require a solvation shell of up to eight nonbridging oxygens for charge balance. Because a pure silica network does not have nonbridging oxygens to accommodate the large coordination number of erbium, the ions tend to cluster at concentrations of less than 1% (Snitzer 1966). Even before outright crystallization occurs, these clusters exhibit several long-range optical interactions that cause loss of energy (Miniscalco 1991). The most serious effect is a cooperative up-conversion process called *pair-induced quenching* in which pairs of ions in the $^4I_{13/2}$ state combine their energy to promote one ion to a higher level while the other goes to the ground state (Delevaque *et al.* 1993). Decay back to the $^4I_{13/2}$ state is nonradiative and results in the loss of one photon. Aluminum dopant is the silver bullet used to homogenize the glass to inhibit scattering losses due to crystallization and inefficiency due to concentration quenching (Arai *et al.* 1986). Aluminum addition of only a fraction of a mole percent is effective in suppressing these effects for erbium concentrations less than about 500 mol ppm (Craig-Ryan *et al.* 1990). Aluminum has the added benefit of providing a more desirable emission wavelength spectrum. It increases the absorption at the pumping wavelength of 1.48 μm, and, as shown in Fig. 4.2, flattens the emission spectrum around 1550 nm, thereby decreasing the sensitivity to signal wavelength. Because the gain spectrum is flatter, multiple-wavelength or -channel operation becomes more feasible.

One significant drawback of the high-delta, small-core-diameter design is increased scattering attenuation (Sudo and Itoh 1990). In fibers doped only with germanium, an anomalous scattering mechanism is present that has a wavelength dependence flatter than the λ^{-4} of Rayleigh scattering but that increases with index above a delta of about 1.2% (Davey *et al.* 1989). Although the mechanism is as yet unidentified, it is very sensitive to fiber drawing conditions. Low-temperature draw can drastically reduce, but not eliminate, the anomalous loss (Davey *et al.* 1989). Although a typical required length of erbium-doped fiber is only about 20 m, losses can approach 0.01 dB/m, causing a noticeable penalty in performance. An increase in erbium concentration reduces the required fiber length, but at the expense of increased concentration quenching. This trade-off requires

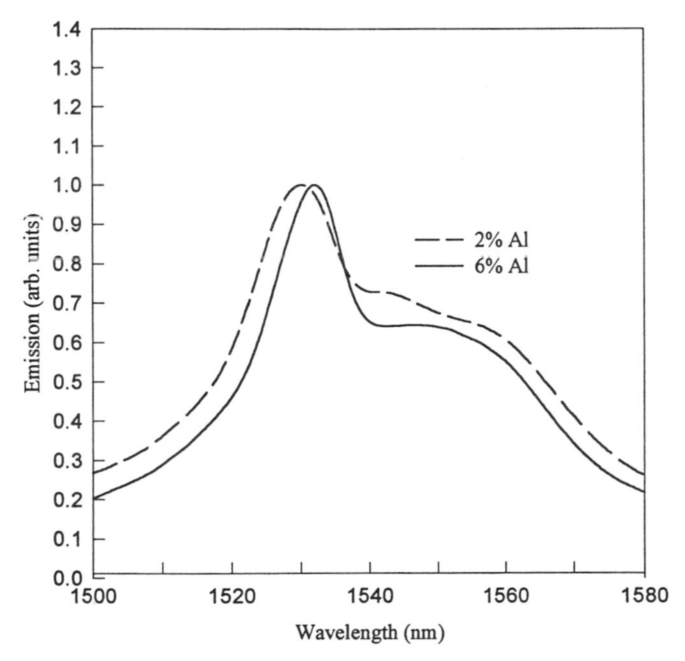

Fig. 4.2 Er^{3+} emission spectrum in silica fiber co-doped with two concentrations of aluminum.

tailoring the fiber design for each amplifier application and basically pits power efficiency against increased noise, the greatest component of which is amplified spontaneous emission (ASE). For example, because the primary objective of a power amplifier is to deliver high power, the amplifier efficiency is maximized at the expense of higher noise. Because the high efficiency is achieved by using a long device, this design requires a low delta to minimize scattering losses.

A second concern raised by the high-delta, small-core-diameter fiber is increased loss when this fiber is spliced to a standard fiber. Because of the small mode field diameter of erbium-doped fibers, typically less than 3 μm, losses of more than 3 dB may be incurred upon butt coupling to dispersion-shifted fibers. However, the mode field diameter can be altered during fusion splicing by extending the duration of the arc to tens of seconds. The extended time exposure to high-temperature arc allows significant diffusion of the highly doped core, which rapidly reduces the effective index and increases the core diameter. The mode field diameter of the standard fiber will also increase, but at a slower rate. For each pair of fibers, there is an optimum splicing schedule to minimize splice loss. Splice losses less than 0.1 dB are readily achieved.

Because the emission spectrum shown in Fig. 4.2 indicates gain over a very wide wavelength range, signals at different wavelengths can be amplified simultaneously. This provides a tremendous potential for increasing the transmission capacity of systems. For this application, additional restrictions are imposed on the shape of the emission spectrum. Although the emission spectrum shown in Fig. 4.2 is broad, it is not flat, so each channel will be amplified a different amount. In addition, because the emission spectrum changes with the power level in each channel, the shape of the spectrum will not be constant. In some cases, unequal gain can be compensated for by preemphasis of the launched power levels, but for a concatenated chain of amplifiers, or one in which the power level changes, the gain discrepancy becomes too large. An amplifier with a flatter spectrum is therefore desired. As mentioned, aluminum helps to a degree, but only limited modification may be expected in a silica host in which the few non-attenuation-inducing dopants are limited to concentrations of less than about 20%. This limitation is resolved by using other hosts, such as phosphates, borates, fluorides, and fluorophosphates. Among these hosts, fluorides are the most attractive because of fabrication and handling issues. In one test of a chain of three amplifiers, the discrepancy in output power among four channels was reduced from 15 dB for silica-based fibers to 4 dB for fluoride fibers (Clesca *et al.* 1994).

Fluoride fibers of the ZrF_4-BaF_2-LaF_3-AlF_3-NaF (ZBLAN) family have been investigated for many years (Poulain, Chanthanasin, and Lucas 1977) because they offer the potential of very low attenuation (<0.001 dB/km) due to reduced Rayleigh scattering (Lines 1988). The gain spectrum is inherently broader and flatter. Rare-earth elements may simply substitute for some of the LaF_3; therefore, solubility is not an issue as it is for silicates. Pair-induced quenching, discussed previously, still limits dopant levels to less than about 500 ppm, however. Fluoride fibers are discussed in more detail later in this chapter, in the context of amplifiers for signals at 1.3 μm.

D. FABRICATION ISSUES

This discussion on fabrication of rare-earth-doped fibers is restricted to issues related to silica-based fibers. Processing of conventional optical fiber uses the high vapor pressure chlorides of silicon, germanium, and phosphorus that enable extremely high purity and simple means of precursor delivery. However, because aluminum and erbium do not possess such high vapor pressure compounds, a host of doping methods have been developed for use with outside vapor deposition (OVD), vapor axial deposition (VAD), and modified chemical vapor deposition (MCVD) (Ainslie 1991; DiGiovanni 1990). The three most widely used techniques—solution doping, vapor delivery, and chelate delivery—are discussed here. Several other methods exist, such as sol-gel (DiGiovanni and MacChesney 1991), aerosol delivery (Morse *et al.* 1989), rod-in-tube (Yamashita 1989; Snitzer and Tumminelli 1989), seed fiber (Simpson *et al.* 1990), and various vapor delivery techniques (Ainslie 1991; Poole *et al.* 1987; Gozen 1988; Li 1985).

1. Solution Doping

The simplest and most widely used method for doping optical preforms with alternate dopants is solution doping, or molecular stuffing (Poole *et al.* 1987; Stone and Burrus 1973; Gozen 1988). In this process, when the glass that will form the preform core is still in porous form, it is soaked in a liquid that contains the desired dopant. The porous body is dried, dehydrated, and densified to lock the solute in the glass, and the process continues conventionally. The soot processes, VAD and OVD, lend themselves readily to solution doping because the glass boule is already in the form of a porous body. In MCVD, the process must be run at a lower temperature to deposit but not consolidate the core material. For erbium-doped fiber fabrication, germanium and phosphorus, if used, are incorporated from the gas phase, and aluminum and erbium are dissolved in the liquid. Virtually any solvent and any soluble precursor free of impurities may be used. A wide range of elements have been successfully incorporated.

Although solution doping is easy to implement, control of dopant concentration, which depends on solute concentration, soot density, and dehydration conditions, is difficult. For high-delta preforms, the high germanium concentration lowers the viscosity of the core glass and makes the process very sensitive to temperature variations. To achieve uniform doping, the solution must be drained and dried evenly, avoiding a buildup of material at the bottom or on one side of the body due to gravitational settling. Dehydration, which is accomplished by treating the body in an atmosphere containing about 10% Cl_2 at 800°C for approximately an hour (Townsend, Poole, and Payne 1987), may alter the composition by vaporizing the dopants as chlorides.

2. Vapor Phase

To maintain all-vapor delivery and avoid introduction of moisture into the system, erbium chloride, which melts at 780°C, may be heated to around 1000°C to generate sufficient vapor (MacChesney and Simpson 1984). Because the vapor pressure increases with temperature exponentially, however, such a process is very difficult to control. It is well known, though, that aluminum chloride readily complexes with many metal chlorides, forming a high vapor pressure complex. Thus, when aluminum chloride is passed over erbium chloride, the complex $Al_xErCl_{3(x+1)}$ is generated (Dewing 1970; Øye and Gruen 1969), which increases the delivery rate by seven orders of magnitude at 600°C and reduces the variation of vapor pressure with temperature. When this technique is used, many core layers may be deposited and dehydration is not required. However, it becomes complicated if multiple dopants are desired, such as for Er/Yb-doped fibers, which are discussed later in this chapter.

3. Chelates

Vapor phase delivery is also possible with certain metal-organic compounds of the rare-earth metals heated to about 200°C. Radiation-hardened fiber using high concentrations of cerium was fabricated with a $Ce(fod)_4$ complex using OVD (Thompson, Bocko, and Gannon 1984). This compound is a β-diketonate complex that is fluorinated to increase its volatility. To aid the processing of high Nd- and Yb-doped MCVD preforms, unfluorinated β-diketonate compounds have been used (Tumminelli, McCollum, and

Snitzer 1990). A carrier gas of helium or argon is used to transport the vaporized material to the OVD or MCVD preform being processed. This process is readily controlled and is suitable for multiple rare-earth dopants. With any metal-organic delivery, hydrogen is present in the precursor material, so the deposited glass must be dehydrated in the same manner as for solution doping.

In all three methods, because the dopant content of the core is much greater than that in standard telecommunications fiber, the process is more difficult to control and more susceptible to disturbances that might cause core ovality and variations in diameter and concentration. Amplifier performance is very sensitive to fiber design, so it is critical that such variations be suppressed. Few manufacturers throughout the world can achieve adequate uniformity to qualify their fiber for international use. Current issues in fabrication are to further broaden the gain spectrum to achieve WDM systems, to improve fiber uniformity, and to eke out the last fractions of a percentage in performance. Optimized amplifier designs and architectures utilizing erbium-doped fiber continue to evolve.

III. High-Power Fiber Amplifier

As amplifier designs and applications have evolved, the need for more power has grown. Higher signal power translates into more channels, longer distances, and higher splitting ratios. Because the output of an EDFA scales with pump power, higher signal power requires higher pump power (Zenteno 1994). To couple this power into a single-mode core, the pump laser must have a small spot size. Laser power is then limited by facet damage caused by high facet intensity in the diode. Single-mode InGaAs lasers at 980 nm and InGaAsP lasers at 1480 nm are currently limited to somewhat higher than 100 mW to meet reliability requirements. This figure is not expected to change significantly in the future. Higher pump power may be attained by multiplexing several pumps (Takenaka *et al.* 1992), but this is expensive and cumbersome. One method is to use a master oscillator power amplifier (MOPA) in which the oscillator is coupled to a flared large-area amplifier to reduce output facet intensity (Welch 1994). Several watts of power may then be imaged onto a single-mode fiber core using bulk optics. An alternative approach, described next, is to pump Yb^{3+} with an available high-power laser and transfer the energy to Er^{3+}, taking advantage of the excellent 1.5-μm emission of Er^{3+}. Because Yb^{3+} absorbs from 900 to 1100 nm, this greatly extends the pump range.

A. *Er/Yb*

Very efficient, high-power lasers may be constructed using Nd^{3+}, emitting at 1064 nm, and Yb^{3+}, emitting from 1060 to 1100 nm. These may be solid-state lasers made from crystals of Nd:YAG or Nd:YLF (Grubb *et al.* 1992), or they may be cladding-pumped lasers (Po *et al.* 1989), described later in this chapter. But because they do not emit at a wavelength absorbed by erbium, they cannot be used to pump erbium directly. However, these wavelengths are absorbed by Yb^{3+}, which can transfer the energy to erbium if the core contains both dopants. The 1.06-μm absorption by Yb^{3+} populates the $^2F_{5/2}$ level, as shown in Fig. 4.3. Because this level lies near the $^4I_{11/2}$ of erbium, this energy may be transferred to the erbium. Nonradiative relaxation of erbium populates the $^4I_{13/2}$ state from which signal amplification may occur. This sensitization process depends on several factors. The

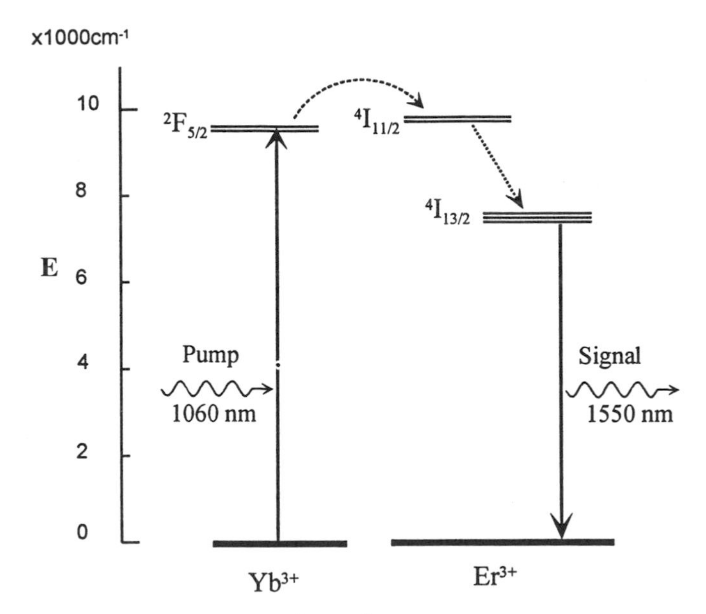

Fig. 4.3 Energy diagram of the Er^{3+}–Yb^{3+} system shows that the 1060 nm absorption by Yb^{3+} populates the $^2F_{5/2}$ level. Yb^{3+} in this level transfers energy to Er^{3+} and excites it to the $^4I_{11/2}$ level that locates at about 80 cm^{-1} higher than the Yb^{3+} excited state.

lifetime of the Yb^{3+} $^2F_{5/2}$ level must be long enough to make transfer more probable than radiative emission. Also, the lifetime of the Er^{3+} $^4I_{11/2}$ state must be short enough to inhibit back transfer. Most important, there must be enough phonon energy in the glass network to provide the transfer from the Yb^{3+} $^2F_{5/2}$ level into the higher Er^{3+} $^4I_{11/2}$ state. Such energy transfer was first demonstrated in 1965 by Snitzer and Woodcock in an alkali silicate host and later in a phosphate glass, which has higher phonon energy than silica. Addition of phosphorus to silica, though, can simulate the phosphate environment, allowing remarkable efficiency (Townsend *et al.* 1991). Output powers as high as 4 W have been obtained using a cladding-pumped fiber laser source, described next.

B. DOUBLE CLAD

To pump an Er/Yb fiber, light must be launched into the single-mode core, which again places a constraint on facet intensity if a single-mode diode is used. A clever solution to this limitation is to break the constraint on single-mode diode pumping. If the fiber cladding is surrounded with a second cladding that has a lower index than the inner cladding, the entire inner cladding region becomes a waveguide (Po *et al.* 1989). Pump light (807 nm) introduced into this region propagates along the fiber, occasionally crossing the core, where it is absorbed by a rare-earth dopant that can be made to lase. Such a design is called a *double-clad laser,* or *cladding-pumped laser* (Zenteno 1993), and was first demonstrated with neodymium lasing at a wavelength of 1060 nm. To collect the greatest amount of pump light, the second cladding should have as low an index as possible. This is best achieved by using a low-index polymer cladding directly on a silica fiber, for which the numerical aperture (NA) can approach 0.6. With a modest facet intensity of only 5 mW/μm^2, more than 700 W could theoretically be coupled into an 80 by 300 μm fiber with an NA of 0.6 (Zenteno 1994). Applications of these devices are described more fully in Chapter 7 of Volume IIIB.

An obvious application of this technology for amplifiers would be to create a laser at the pump wavelengths of erbium — 980 or 1480 nm. However, few rare-earth materials exhibit efficient emission at these wavelengths. One possible candidate is Yb^{3+}, which absorbs and emits strongly at 980 nm. Unfortunately, because the pump power is spread across the large-area inner cladding, its intensity is greatly reduced and the pumping rate is low. As a result, 980 nm emission is quickly reabsorbed. As this example illustrates, efficient cladding-pumped devices are limited to four-level systems — e.g., Nd^{3+} — which require a lower threshold pump power.

In a double-clad laser, the multimode pump power must cross the core to be effective. In circular fiber, this is inefficient because much of the light follows helical paths and is not absorbed. To prevent this, the fiber is made noncircular so that all rays eventually cross the core. Absorption of the pump scales with core area; therefore, it is important to keep the core diameter as large as possible. This places limits on the concentration of dopants such as aluminum and germanium. To reduce device length, the rare-earth dopant concentration is increased as high as possible before excess scattering due to clustering or crystallization occurs. It has been found that increasing the aluminum or germanium content is beneficial (Kirchhof *et al.* 1996), but at the expense of a higher delta and a longer device length.

In the desired design, then, the core contains a high concentration of Nd^{3+} or Yb^{3+}, has enough aluminum or germanium to inhibit scattering, and is as large and has as low a delta as allowed by single-mode operation; in addition, the silica cladding is noncircular and is coated by a polymer with as low an index as possible. Fabrication of the doped preform is similar to that for erbium-doped fiber, described previously. Processing of a noncircular fiber poses obvious problems in shaping the initially round preform, control of fiber dimensions during draw, and cleaving and splicing. Also, although there are several sources of high-NA plastic-clad silica fiber, there are few commercial sources of low-index polymer to be used for coatings.

IV. Amplification at 1.3 μm

Compared with the rapid development and impressive performance of erbium-doped amplifiers operating at 1.5 μm, little progress has been made in commercializing amplifiers in the second telecommunications window at 1.3 μm. Possible candidates are Pr^{3+} and Nd^{3+} fibers (analogous to Er^{3+}-doped fibers) and Raman amplifiers.

A. *NEODYMIUM DOPED FIBERS*

Initial investigation of amplifiers for the 1.3 μm window focused on Nd^{3+}, which has an emission peak around 1.36 μm in silica. The position of this peak in glass hosts is dominated by the overlap of emission with an excited state absorption (ESA) band at a slightly lower wavelength (Miniscalco 1988). Because the balance between gain and absorption is very sensitive to composition, much work was devoted to investigations of different dopants and different host materials. Phosphorus was found to be most effective in decreasing the emission wavelength in a silicate, but the peak drops to only 1.33 μm. Much greater variation in spectra can be achieved by abandoning silicate glasses and using phosphates and ZBLAN fluorides (Zemon *et al.* 1992). Predictions of performance (Dakss and Miniscalco 1990) indicate that even in the absence of ESA, competing transitions at 880 and 1064 nm reduce the branching ratio for the 1.3 μm emission to only about 10%. Thus, the small signal performance in neodymium-doped fiber amplifiers at 1.3 μm will be almost 100 times lower than that achieved in EDFAs at 1.55 μm. Better performance can be expected for power amplifiers, with photon conversion efficiencics as high as 75% in the absence of amplified emission from the competing transition at 1064 nm. Without suppression of this ASE, efficiency can drop to less than 10% (Pederson *et al.* 1992). Because of limitations imposed by ESA and ASE, attention has shifted mostly to Pr^{3+} dopcd amplifiers.

B. PRASEODYMIUM-DOPED FIBERS

Amplifiers based on Pr^{3+} show promise because of emission centered around 1.3 μm and the absence of competing transitions. As shown in Fig. 4.4, however, there are several lower energy levels to which the metastable $^1G_4$ state may decay through multiphonon relaxation. The probability of nonradiative decay varies exponentially with decreasing energy gap; therefore, it is essential that the host material have phonon energy less than about 600 cm^{-1}. Because of their low phonon energies, fluoride and chalcogenide hosts are currently the best choice.

In addition to nonradiative decay, Pr^{3+} exhibits the concentration-quenching mechanism of cross relaxation in which energy from one ion excited to $^1G_4$ is transferred to a nearby ion, with both ions terminating in intermediate states (Ohishi *et al.* 1991). Thus, the Pr^{3+} concentrations are limited to levels similar to those of Er^{3+}, less than about 500 ppm, with device lengths of tens of meters. Because of both ground state and excited state absorption around 1.4 μm, as shown in Fig. 4.4, this system is partially three-level and benefits from high optical intensity. Deltas of more than 3% are typical. Although the quantum efficiency of radiative emission from $^1G_4$ is only 3.4% (Ohishi *et al.* 1991), the predicted lifetime approaches

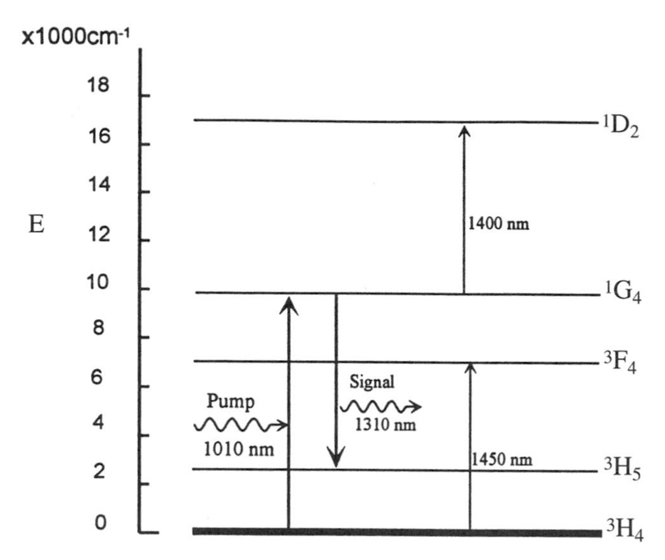

Fig. 4.4 Energy diagram of Pr^{3+} shows that below the metastable $^1G_4$ state, there are other energy levels to which multiphonon relaxation may occur, and this nonradiative decay competes with the radiative emission from the metastable state. Both the ground state and excited state absorptions at 1.4 μm are also shown in this diagram.

6 ms and performance could rival that of Er^{3+}. However, measured lifetimes are considerably lower, typically about 0.2 ms. Because of the short lifetime, the photon conversion efficiency will be limited to less than 50% (Pederson *et al.* 1992). Although performance does not match that of Er^{3+}-doped amplifiers, many systems tests have been performed and units are available commercially.

Several engineering problems have inhibited penetration of Pr^{3+}-doped amplifiers in commercial networks. Coupling to standard silica fiber must be mechanical and is problematic as a result of the high index of the Pr^{3+}-doped core. More important, though, low-loss fluoride fibers are very difficult to fabricate. The glasses are hygroscopic, which raises concerns about long-term reliability, and easily crystallize during fabrication, especially if contaminated with oxygen. In addition, manufacturing relies on batch processing, and forming preforms without the introduction of contaminants is difficult. To achieve the high clad–core ratio required for high-delta fibers, multiple rod-in-tube processes are used. Despite this handling, losses as low as 10 dB/km are typical. Given the mediocre performance and the difficulties with fabrication, it is questionable at this time whether Pr^{3+} doped amplifiers will see widespread use.

C. *RAMAN AMPLIFIER*

Recent advances in high-power diode lasers and Bragg gratings enable the use of stimulated Raman scattering to frequency-shift a pump wavelength to amplify optical signals at a longer wavelength (Grubb *et al.* 1994). In a 1.3 μm intracavity cascaded Raman amplifier, Bragg gratings were made, at both ends of a fiber loop, to selectively confine the pump and Stokes lines within the fiber loop. This configuration permits an efficient transfer over several Stokes shifts from the 1.064 μm pump wavelength to 1.3 μm for signal amplification. Most fibers employed in Raman amplifiers are standard germanium-doped fibers having a high delta and a small core diameter for power confinement. Because a relatively long (0.5- to 2-km) fiber loop is used, a reasonably low attenuation loss is required between the pump wavelength ($\sim$1 μm) and the signal wavelength (1.3 μm).

It has been observed that the frequency shift (1320 cm^{-1}) of the first-order Raman scattering peak in P-doped silica is about three times larger than that for Ge-doped and undoped silica (Suzuki, Noguchi, and Uesugi 1986). Recently, it has been demonstrated that strong Bragg gratings can be written in P-doped fibers sensitized by deuterium loading (Kitagawa *et al.* 1994; Malo *et al.* 1994; Strasser *et al.* 1995). Thus, P-doped fibers can permit the use of a fewer number of Stokes transitions between the pump wavelength and 1.31 μm for signal amplification. Because the Raman gain coefficient at 1320 cm^{-1} in P-doped fiber increases with the P-dopant concentration, it is desirable to have a high delta value. However, because of processing difficulties in incorporating a high concentration of phosphorus, the delta is usually less than 1.5–2% in fibers doped exclusively with phosphorus. A significant increase in attenuation loss is also observed in fibers containing a high P concentration. Thus, further improvement in fiber processing will be required to fully utilize the advantages of P-doped fibers for Raman amplifier applications.

V. Dispersion Management

A. *TRANSMISSION FIBER APPLICATION AT 1.55 μm*

Dispersion becomes a major factor in defining the span limits of optical fiber systems because attenuation loss is no longer an important issue after optical fiber amplifiers are deployed extensively. Bit rates become limited to approximately 10 Gb/s. However, systems designers are looking to other means that can extend the information carrying capacities of optical fiber systems. Because the erbium-doped optical fiber amplifiers have fairly wide gain profiles, they will boost the optical power over a fairly large wavelength range. Hence, it becomes possible to wavelength-division multiplex a group of signals on a system employing optical fiber amplifiers. The system's information carrying capacity is increased by a factor equal to the number of wavelengths that are multiplexed. A number of hero experiments have been reported that speak of unrepeatered systems hundreds of kilometers in length, operating at a combined 100+ Gb/s (Oda *et al.* 1995; Gnauck *et al.* 1995).

New limitations now govern the ultimate capacity of these new optical fiber systems. One key group of limitations is nonlinear optical effects. Detailed descriptions can be found in Chapter 8 in Volume IIIA. Briefly, these effects are associated with the interaction of high optical power with the optical fiber. Two of the effects, Brillouin and Raman scattering, are manifest when certain power density thresholds are exceeded. Other effects are produced when high powers induce changes to the index of refraction and thereby introduce phase differences. Important effects in single-wavelength systems are self-phase modulation, whereas in multiple-wavelength systems cross-phase modulation and four-wave mixing play an important role.

When the potential for using standard single-mode designs with optical amplifiers and dense WDM is examined, the limitations imposed by nonlinear effects become apparent. For the standard single-mode designs with zero dispersion at or near 1310 nm, carrier induced phase modulation will dominate because of the high dispersion of this fiber at 1540–1560 nm, the operating window of the optical amplifier. Dispersion compensation is required for spans greater than 50 km. This dispersion problem can be solved by considering standard dispersion-shifted fiber designs that have zero dispersion at 1550 nm. However, four-wave mixing causes interference between the discrete wavelengths and severely limits the span lengths. In both the standard and dispersion-shifted designs, it is desirable to keep the power density as low as possible in order to minimize the nonlinear effects. This can be accomplished by making the effective area of the fiber as large as possible while still retaining the desired waveguide and bending properties of the fiber.

A new fiber design has been introduced that overcomes the disadvantages of the standard dispersion-unshifted and -shifted fibers. Specifically, the design and application of fiber with a small, controlled amount of dispersion is found to suppress the four-wave mixing without introducing large amounts of dispersion that would limit span distances. The dispersion must be controlled to be either +2 or −2 ps/nm-km at 1550 nm. This new design provides the flexibility for easy control of the dispersion, together with a larger effective area to reduce power densities. The remainder of this section describes this new fiber and the considerations in choosing the specific design parameters.

1. Fiber Design Issues

The design intents for the new fiber are a larger effective area and easy control of dispersion in the window of 1540–1560 nm. A good set of actual targets would be an effective area greater than 50 μm^2 and a dispersion zero that could be easily adjusted between 1500 and 1600 nm. The effective area target can be restated in terms of a fiber mode field radius greater than 4.0 μm.

a. The Starting Point

We begin by examining a basic fiber design used by many as a dispersion-shifted fiber with a zero dispersion at 1550 nm. The generalized index profile is shown in Fig. 4.5 as a trapezoidal core surrounded by a depressed index cladding out to a b/a ratio of 8.

The design curves in Fig. 4.6 characterize the key parameters of mode field radius (for effective area) and zero-dispersion wavelength (for dispersion control) as a function of basic profile parameters of core delta (total) and core diameter. The s/a ratio that describes the shape of the trapezoid is constant at 0.20. The particular ranges of core and delta that are shown

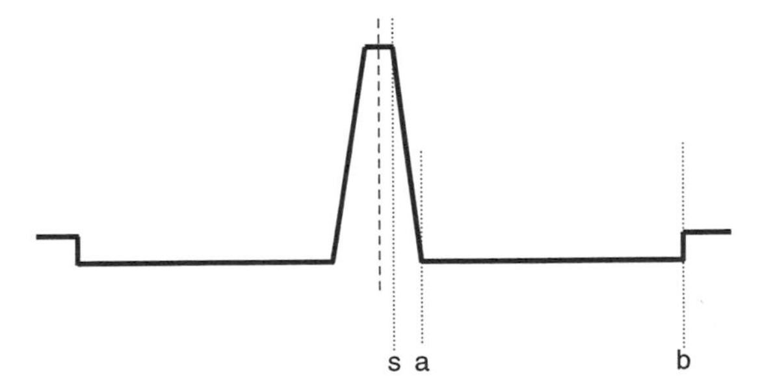

Fig. 4.5 In this generalized index profile of the conventional dispersion-shifted fiber, the trapezoidal core, defined by dimensions s and a, is surrounded by a depressed cladding region.

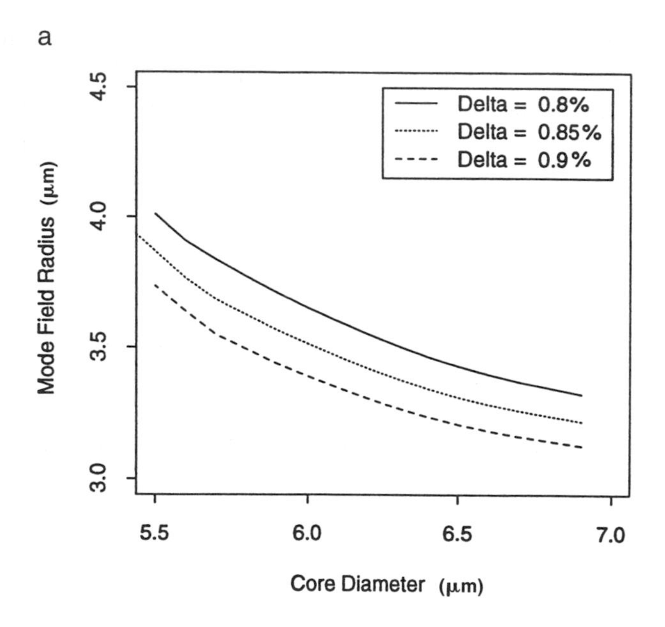

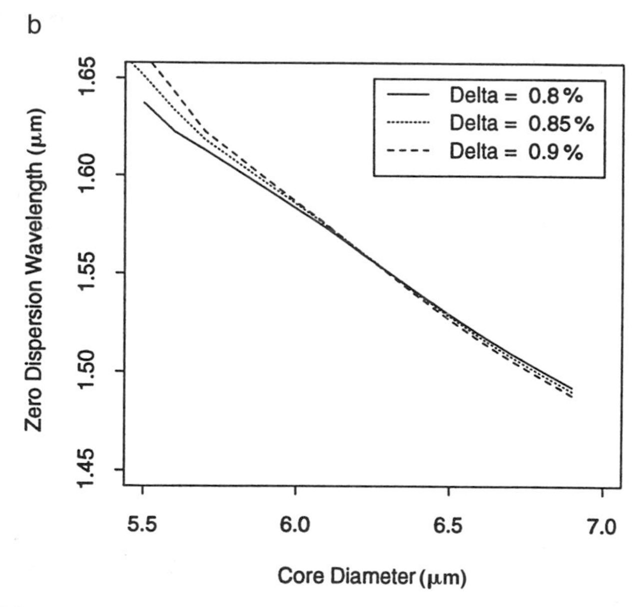

Fig. 4.6 (a) Mode field radius and (b) the zero-dispersion wavelength are plotted versus the fiber core diameter for the index profile shown in Fig. 4.5.

provide V numbers and effective indices that are high enough to allow for good bend performance. The use of a depressed index cladding allows for reduced amounts of dopant in the core, thereby improving intrinsic fiber loss. In general, the choice of fiber parameters was made so as to optimize the overall performance and provide zero dispersion at the 1550 nm wavelength. There was no consideration for its application in WDM in optical amplifier systems.

It can be noted from Fig. 4.6b that for a given core diameter there is very little variation of zero-dispersion wavelength with delta. From a zero-dispersion control standpoint, changes in core diameter could be used to effect a change in dispersion, moving from a low of 5.5 μm (for 1650 nm zero dispersion) to a high of 6.5 μm (for 1520 nm zero dispersion). However, to attain the goal of a mode field radius greater than 4.0 μm with an acceptable bend performance, Fig. 4.6a indicates that the delta must be adjusted downward. Before the design target of 4.0 μm can be realized, a point is reached where the lower V number and the reduced effective index result in a bend performance problem. It becomes apparent that additional flexibility is required to allow for the combined requirements of dispersion control, a larger mode field radius, and acceptable bend performance.

b. *The New Design*

The generalized index profile of a design that overcomes the drawbacks that were just discussed is described in Fig. 4.7. This design is characterized by a core with a trapezoidal shape, but now the core is surrounded by a raised index pedestal region of radius c. Outside the pedestal region, the deposited cladding index is slightly lower than that of fused silica for an unspecified distance. Beyond that, the cladding index matches that of fused silica. The shape of the trapezoidal core is again described by the ratio s/a. The ratios b/a, c/a and s/a play a secondary role of importance in this discussion. They are kept at constant values of 5.0, 3.5, and 0.35, respectively, for Fig. 4.8.

Design diagrams similar to those in Fig. 4.6 are depicted in Fig. 4.8. Because core parameters dominate the target parameters of waveguide dispersion and mode field, these diagrams basically represent extensions of Fig. 4.6 for lower core diameter and lower delta ranges. In comparison, it can be noted from Fig. 4.8b that zero dispersion is now a more sensitive function of delta than is core diameter; it is in fact affected very little by changes in core diameter. Also, Fig. 4.8a indicates that mode field radii

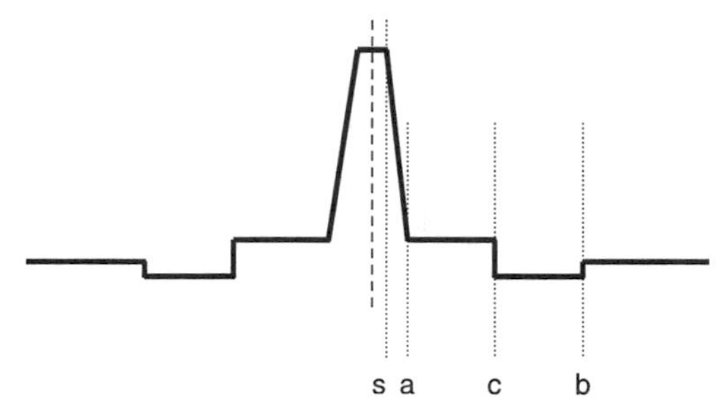

Fig. 4.7 The generalized index profile of the new fiber design yields a larger mode field radius and improved dispersion control. A raised index pedestal region is placed between the trapezoidal core and the depressed cladding.

greater than 4.0 μm can be obtained. From these curves, a delta between 0.65 and 0.85% and a core diameter of 5.0–6.0 μm will realize the target of zero dispersion between 1500 and 1600 nm with an effective area greater than 50 μm^2.

With the old design, these same parameters led us to unacceptable bend performance. However, the changes to the cladding in the new design prevent its deterioration. The first change is the reduction of the depressed index to a value near that of the undoped fused silica outer clad. This raises the effective index and prevents macrobending losses, but by itself this change is not sufficient to make up for lost ground. However, with the further addition of the pedestal with a delta of 0.06–0.08% the effective index is again increased, this time to a point where satisfactory bend performance is realized. The pedestal has a slight effect on the waveguide dispersion properties of the fiber, but it is secondary to the main components of delta and core.

As a result, a fiber design is realized that meets the needs of optical amplifier systems operating with WDM. Nonlinear optical effects are minimized by reducing power density with a much larger effective area, while easy dispersion control is available for eliminating four-wave mixing. The targets of an effective area greater than 50 μm^2 and zero dispersion between 1500 and 1600 nm have been realized in the manufacture of production quantities of fibers (Lucent TrueWave® fiber) that utilize these design parameters. The attenuation medians are 0.2 dB/km, and the bend performance has been verified with cabling tests and with actual cable production that indicates no added cabling loss.

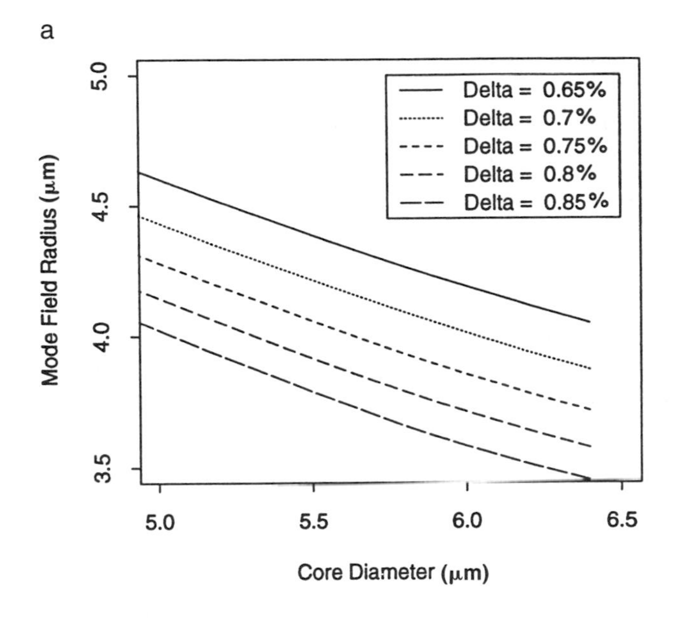

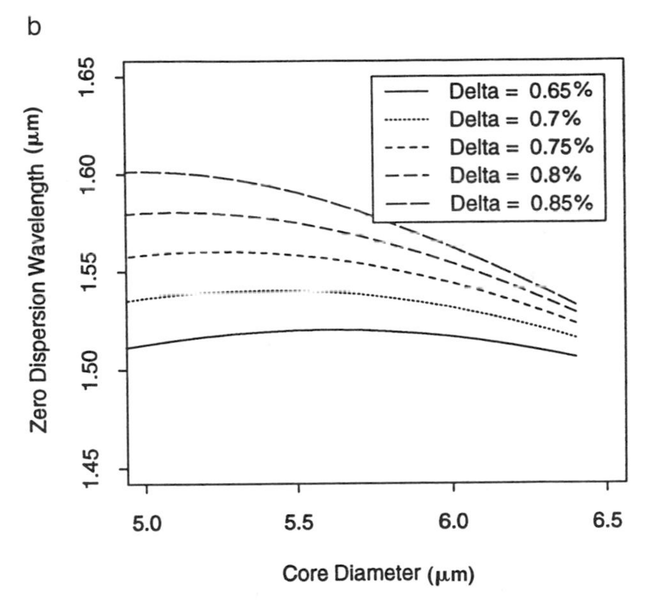

Fig. 4.8 (a) Mode field radius and (b) the zero-dispersion wavelength are plotted versus the fiber core diameter for the index profile shown in Fig. 4.7.

B. DISPERSION COMPENSATION FOR INSTALLED 1.3 μm SYSTEMS

Dispersion compensation will be required when the installed systems designed for 1.3 μm operation will be upgraded to take advantage of the recent advances in optical amplifiers operating at 1.55 μm. Although the installed fibers have a zero dispersion at 1.3 μm, the dispersion increases with the wavelength at a positive slope of 0.057 ps/nm^2-km to about 17 ps/nm-km at 1550 nm. For high-speed transmission, a relatively short length of dispersion-compensating fiber, which is designed to have a very negative dispersion value at 1550 nm, can be deployed in the network to compensate for the transmission fiber dispersion at this wavelength. Furthermore, the dispersion-compensating fiber should have a negative dispersion slope versus wavelength such that the compensated network has very little total dispersion over a wide spectral range (1530–1560 nm).

Theoretical analyses (Kawakami and Nishida 1974; Monerie 1982) have shown that the fundamental (LP_{01}) mode can have a finite cutoff wavelength in fiber designs containing a sufficiently deep down-doped cladding region around the core. More significant, these waveguide designs can introduce a very large negative dispersion value at wavelengths slightly below the LP_{01} cutoff. Using numerical methods (Lenahan 1983), Vengsarkar and Reed (1993) recently developed multilayer waveguide designs tailored to have an LP_{01} cutoff wavelength very close to the operating wavelength of 1.55 μm to result in both large negative dispersion values as well as negative slopes versus wavelength. Experimental measurements (Vengsarkar, Miller, and Reed 1993) of these designs showed a dispersion of −210 ps/nm-km at 1550 nm and a negative dispersion slope of about 0.43 ps/nm^2-km. These properties permit a compensation ratio of about 13 and a total link dispersion of less than ±0.5 ps/nm-km. The index profile and dispersion data are shown in Figs. 4.9 and 4.10, respectively.

Optimal dispersion-compensating designs require fiber cores having a small diameter (2–3 μm) and a high refractive index (Δ = 1–3%); such a combination often introduces a substantial increase in fiber attenuation (Davey *et al.* 1989). Details of fiber designs can also have a significant impact on their sensitivities to bending-induced attenuation. The required deeply down-doped cladding regions (Δ = −0.4 to −0.6%) can adversely increase the preform processing time. Furthermore, extreme sensitivities of the dispersion performance on fiber dimensions and details in the index

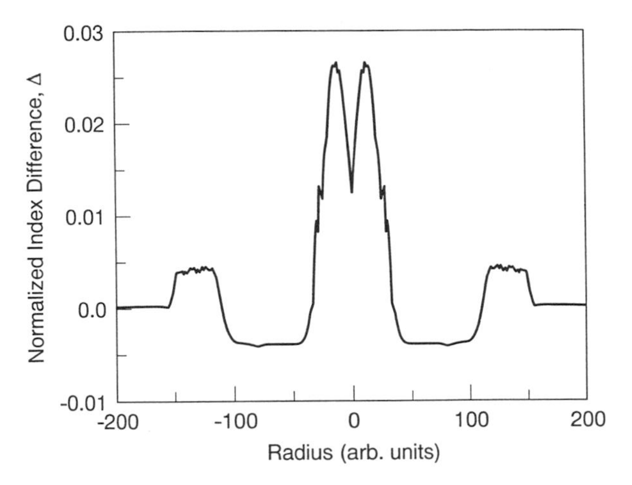

Fig. 4.9 Refractive index profile of the preform used for drawing the dispersion-compensating fiber.

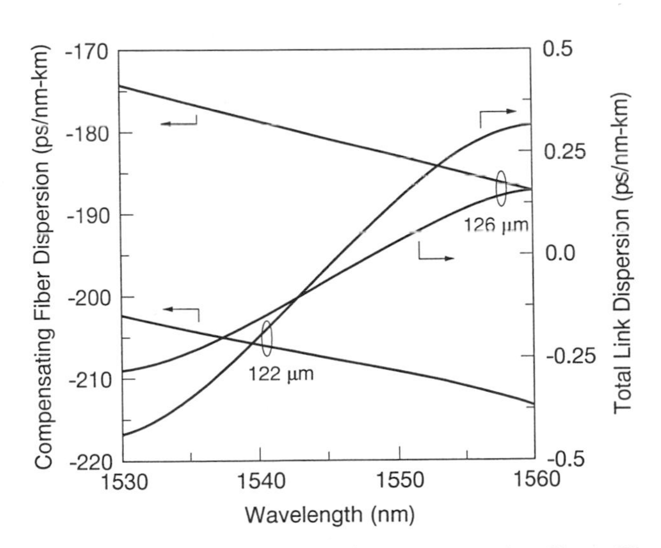

Fig. 4.10 Dispersion plots of two dispersion-compensating fibers. The total link dispersions are also shown when these fibers are used in the conventional 1.3 μm systems.

profile (Vengsarkar and Reed 1993; Vengsarkar, Miller, and Reed 1993) impose significant challenges on fiber processing.

VI. Specialty Connector Fibers

A. *ATTENUATORS*

A recent fiber design that has proven to be a useful component in optical devices is the fiber attenuator. This device is essentially a short length of fiber used to reduce the optical power level. Initial applications were as line build-out components to adjust power levels into or out of repeaters, but they have also found application as fixed attenuators in patch-cord assemblies for test sets. Previous designs used an offset splice between two fiber cores; the amount of offset controls the amount of attenuation. A more reliable method, though, is to use an absorbing length of fiber.

Transition metal impurities have long been the bane of optical fiber manufacture because even part-per-billion levels cause measurable optical attenuation (Schultz 1974). By intentionally doping the fiber with selected metals, however, one can induce losses from 1 to 1000 dB/m and create a useful device. At the high level, a 10 dB reduction for a patch-cord assembly requires only 1 cm of fiber mated to the connector. To achieve constant attenuation over a broad wavelength range, two or more transition metals may be added in the proper ratio.

B. *PIGTAILS*

New fiber designs and devices that depart radically from conventional transmission fiber result in high connector loss when they are incorporated into optical systems. In many cases, the splice loss between dissimilar fibers can be reduced significantly by optimizing the heating schedule during fusion splicing. However, for devices that cannot be fusion spliced, such as the diode pump laser for an erbium-doped amplifier, this is not an option. In this case, the spot size of the laser exceeds that of the high-delta, erbium-doped-fiber core, and direct coupling efficiency would be low. Pigtail fibers are specially designed to optimize coupling or performance at the device end while fusion splicing can accommodate any mode field mismatch with the transmission system. Such pigtail fibers are also used to provide single-mode propagation at nonsignal wavelengths, such as the 980 nm pump band for erbium-doped amplifiers. Proper choice of pigtail fiber is often critical for performance of devices such as pump lasers, optical isolators, and wavelength-division multiplexers.

VII. Conclusions

Recent advances in fiber processing technology enable the realization of many useful fiber designs and result in a significant improvement in optical networks to overcome limitations in attenuation and dispersion. The most important achievement is EDFA technology, which provides a very cost-effective means of optical amplification in both transoceanic and terrestrial communications systems. It also enables the development of more extensive WDM systems to further increase the transmission capacity to keep pace with the growth of telecommunications services. Fiber processing has also led to technological breakthroughs in high-power fiber amplifiers that can be used as new pump sources. Although a significant progress has been reported in optical amplification at 1.3 μm, difficulties in fiber processing are currently limiting the fiber amplifier application at this wavelength.

For high transmission rate operation, fiber dispersion management is a crucial factor. New developments in the designs and processing of 1.55 μm transmission fibers permit their use in WDM systems without the deleterious nonlinear effects associated with the interaction of the high optical power with the fiber. This new fiber design is epitomized by Lucent TrueWave fiber currently in production. A new dispersion-compensating fiber device was developed such that the EDFAs can be employed in the installed 1.3 μm systems. Specialty connector fibers have been developed for fiber interconnection and they will find useful applications in optical networks.

References

Ainslie, B. J. 1991. A review of the fabrication and properties of erbium doped fibers for optical amplifiers. *J. Lightwave Tech.* 9(2):220–227.

Arai, K., H. Namikawa, K. Kumata, T. Honda, Y. Ishii, and T. Handa. 1986. Aluminum or phosphorus codoping effects on fluorescence and structural properties of neodymium-doped silica glass. *J. Appl. Phys.* 59(10):3430–3436.

Armitage, J. R. 1988. Three level fiber laser amplifier. *Appl. Opt.* 27:4831.

Barnett, W. C., H. Takahira, J. Baroni, and Y. Ogi. 1996. The TPC-5 cable network. *IEEE Commun. Mag.* 34(2):36–40.

Bergano, N. S., C. Davidson, B. M. Nyman, S. G. Evangelides, J. M. Darcie, J. D. Evankow, P. C. Corbett, M. A. Mills, G. A. Ferguson, J. A. Nagel, J. L. Zyskind, J. A. Sulhoff, A. J. Lucero, and A. A. Klein. 1995. 40 Gb/s WDM transmission of eight 5 Gb/s data channels over transoceanic distances using the conventional NRZ modulation format. In *Conference on Optical Fiber Communication.* Paper PD19. Technical Digest. Washington, DC: Optical Society of America.

Clesca, B., D. Ronarc'h, D. Bayart, Y. Sorel, L. Hamon, M. Guibert, J. L. Beylat, J. F. Kerdiles, and M. Semenkoff. 1994. Gain flatness comparison between erbium doped fluoride and silica fiber amplifiers with wavelength-multiplexed signals. *IEEE Photon. Tech. Lett.* 6(4):509–512.

Craig-Ryan, S. P., J. F. Massicott, M. Wilson, B. J. Ainslie, and R. Wyatt. 1990. Optical study of low concentration Er^{3+} fibers for efficient power amplifiers. In *ECOC'90 Proceedings*, vol. 1, 571–574.

Dakss, M. L., and W. J. Miniscalco. 1990. Fundamental limits on Nd^{3+}-doped fiber amplifier performance at 1.3 μm. *IEEE Photon. Tech. Lett.* 2(9):650–652.

Davey, S. T., D. L. Williams, D. M. Spirit, and B. J. Ainslie. 1989. The fabrication of low loss high NA silica fibres for Raman amplification. *SPIE, Fiber Laser Sources and Amplifiers* 1171:181–191.

Delevaque, E., T. Georges, M. Monerie, P. Lamouler, and J-F. Bayon. 1993. Modeling of pair-induced quenching in erbium-doped silicate fibers. *IEEE Photon. Tech. Lett.* 5(1):73–75.

Dewing, E. W. 1970. Gaseous complexes formed between trichlorides $AlCl_3$ and $FeCl_3$ and dichlorides. *Metallurg. Trans.* 1:2169–2174.

DiGiovanni, D. J. 1990. Fabrication of rare earth doped optical fiber. Paper 01. *SPIE, Fiber Laser Sources and Amplifiers II* 1373.

DiGiovanni, D. J., and J. B. MacChesney. 1991. A new optical fiber fabrication technique using sol-gel dipcoating. In *Conference on Optical Fiber Communication.* Paper WA2. Technical Digest. Washington, DC: Optical Society of America.

Gnauck, A., R. M. Derosier, F. Forghieri, R. Tkach, A. M. Vengsarkar, D. W. Peckham, J. L. Zyskind, J. W. Sulhoff, and A. R. Chraplyvy. 1995. Transmission of 8 20-Gb/s channels over 232 km of conventional fiber. Paper PD23. Technical Digest. Washington, DC: Optical Society of America.

Gozen, T. 1988. Development of high Nd^{3+} content VAD singlemode fiber by molecular stuffing technique. In *Conference on Optical Fiber Communication.* Paper WQ1. Technical Digest. Washington, DC: Optical Society of America.

Grubb, S. G., T. Erdogan, V. Mizrahi, T. Strasser, W. Y. Cheung, W. A. Reed, P. J. Lemaire, A. E. Miller, S. G. Kosinski, G. Nykolak, P. C. Becker, and D. W. Peckham. 1994. 1.3 μm Raman amplification in Germand-silicate fiber. In *Optical amplifiers and their applications.* Paper PD3. Technical Digest Series, vol. 14. Washington, DC: Optical Society of America.

Grubb, S. G., P. A. Leilabady, K. L. Sweeney, W. H. Humer, R. S. Cannon, S. W. Vendetta, W. L. Barnes, and J. E. Townsend. 1992. High output power Er^{3+}/Yb^{3+} codoped optical amplifiers pumped by diode-pumped Nd^{3+} lasers. In *Optical amplifiers and their applications.* Paper FD1-1. Technical Digest Series, vol. 17. Washington, DC: Optical Society of America.

Kawakami, S., and S. Nishida. 1974. Characteristics of a doubly clad optical fiber with a low-index inner cladding. *IEEE J. Quantum Electron.* QE-10(12):879–887.

Kirchhof, J., S. Unger, V. Reichel, and A. Schwuchow. 1996. Background loss and devitrification in Nd-doped fiber laser glass. In *Conference on Optical Fiber Communication.* Paper TuL4. Technical Digest. Washington, DC: Optical Society of America.

Kitagawa, T., K. O. Hill, D. C. Johnson, B. Malo, J. Albert, S. Theriault, F. Bilodeau, K. Hattori, and Y. Hibino. 1994. Photosensitivity in P_2O_5-SiO_2 waveguide and its application to Bragg reflectors in single-frequency Er^{3+}-doped planar waveguide laser. In *Conference on Optical Fiber Communication.* Paper PD-17. Technical Digest. Washington, DC: Optical Society of America.

Lenahan, T. 1983. Calculation of modes in an optical fiber using the finite element method and EISPACK. *Bell Syst. Tech. J.* 62(9):2663–2694.

Li, T. Y. 1993. The impact of optical amplifiers on long-distance lightwave telecommunications. *Proc. IEEE* 81(11):1568–1579.

Li, T. Y., ed. 1985. *Optical fiber communications,* Vol. 1, *Fiber fabrication.* New York: Academic Press.

Lines, M. E. 1988. Theoretical limits of low optic loss in multicomponent halide glass materials. *J. Non-Cryst. Solids* 103:265.

MacChesney, J. B., and J. R. Simpson. 1984. Optical waveguides with novel compositions. In *Conference on Optical Fiber Communication.* Paper WH5, p. 100. Technical Digest. Washington, DC: Optical Society of America.

Malo, B., J. Albert, F. Bilodeau, T. Kitagawa, D. C. Johnson, K. O. Hill, K. Hattori, Y. Hibino, and S. Gujrathi. 1994. Photosensitivity in phosphorus-doped silica glass and optical waveguides. *Appl. Phys. Lett.* 65(4):394–396.

Miniscalco, W. J. 1988. 1.3 μm Fluoride fiber laser. *Electron. Lett.* 24(1):28–29.

Miniscalco, W. J. 1991. Erbium doped glasses for fiber amplifiers at 1500 nm. *J. Lightwave Tech.* 9(2):234–250.

Monerie, M. 1982. Propagation in doubly clad single-mode fibers. *IEEE J. Quantum Electron.* QE-18(4):535–542.

Morse, T. F., L. Reinhart, A. Killian, W. Risen, and J. W. Cipolla. 1989. Aerosol doping techniques for MCVD and VAD. *SPIE, Fiber Laser Sources and Amplifiers* 1171:72–79.

Oda, K., M. Fukutoku, M. Fukui, T. Kitoh, and H. Toba. 1995. 16-Channel $\times$ 10-Gbit/s optical FDM transmission over a 1000 km conventional single-mode fiber employing dispersion compensating fiber and gain equalizations. In *Conference on Optical Fiber Communication.* Paper PD22. Technical Digest. Washington, DC: Optical Society of America.

Ohishi, Y., T. Kanamori, T. Nishi, and S. Takahashi. 1991. A high gain, high output saturation power Pr^{3+} doped fluoride fiber amplifier operating at 1.3 μm. *IEEE Photon. Tech. Lett.* 3(8):715–717.

Øye, H. A., and D. M. Gruen. 1969. Neodymium chloride–aluminum chloride vapor complexes. *J. Am. Chem. Soc.* 91:2229.

Pederson, B., A. Bjaklev, J. Hedegaard, K. Dybdal, and C. Larsen. 1991. The design of erbium-doped fiber amplifiers. *J. Lightwave Tech.* 9(9):1105–1112.

Pederson, B., W. J. Miniscalco, S. Zemon, and R. S. Quimby. 1992. Neodymium and praseodymium doped fiber power amplifiers. In *Optical amplifiers and their applications.* Paper WB4. Technical Digest Series, vol. 17. Wahsington, DC: Optical Society of America.

Po, H., E. Snitzer, R. Tumminelli, L. Zenteno, F. Hakimi, N. M. Cho, and T. Haw. 1989. Double clad high brightness Nd^{3+} fiber laser pumped by GaAlAs phased array. In *Conference on Optical Fiber Communication.* Paper PD7-1. Technical Digest. Washington, DC: Optical Society of America.

Poole, S., D. N. Payne, R. J. Mears, M. E. Fermann, and R. I. Laming. 1987. Fabrication and characterization of low-loss optical fibers containing rare earth ions. *J. Lightwave Tech.* LT-4(7):870–876.

Poulain, M., M. Chanthanasin, and J. Lucas. 1977. New fluoride glasses. *Mater. Res. Bull.* 12:131.

Schultz, P. C. 1974. Optical absorption of the transition elements in vitreous silica. *J. Am. Ceram. Soc.* 57(7):309–313.

Simpson, J. R., L. F. Mollenauer, K. S. Kranz, P. J. Lemaire, N. A. Olsson, H. T. Shang, and P. C. Becker. 1990. A distributed erbium fiber amplifier. In *Conference on Optical Fiber Communication.* Paper PD19-1. Technical Digest. Washington, DC: Optical Society of America.

Snitzer, E. 1966. Glass lasers. *Appl. Opt.* 5(10):1487–1499.

Snitzer, E., and R. Tumminelli. 1989. SiO_2 clad fibers with selectively volatilized soft glass cores. *Opt. Lett.* 14(14):757–759.

Snitzer, E., and R. Woodcock. 1965. Yb^{3+}–Er^{3+} glass laser. *Appl. Phys. Lett.* 6(3):45–46.

Stone, J., and C. A. Burrus. 1973. Nd^{3+} doped SiO_2 lasers in end pumped fiber geometry. *Appl. Phys. Lett.* 23:388–389.

Strasser, T. A., A. E. White, M. F. Yan, P. J. Lemaire, and T. Erdogan. 1995. In *Conference on Optical Fiber Communication.* Paper WN-2. Technical Digest. Washington, DC: Optical Society of America.

Sudo, S., and H. Itoh. 1990. Efficient non-linear optical fibres and their applications. *Opt. Quantum Electron.* 22:187–212.

Suzuki, K., K. Noguchi, and N. Uesugi. 1986. Selective stimulated Raman scattering in highly P_2O_5-doped silica single mode fibers. *Opt. Lett.* 11(10):656–658.

Takenaka, H., H. Okuno, M. Fujita, Y. Odagiri, Y. Sunohara, and I. Mito. 1991. Compact size and high output power Er-doped fiber amplifier modules pumped with 1.48 μm MQW LDs. In *Optical amplifiers and their applications.* Paper FD2. Technical Digest Series, vol. 13. Washington, DC: Optical Society of America.

Thompson, D. A., P. L. Bocko, and J. R. Gannon. 1984. New source compounds for fabrication of doped optical waveguide fibers. *SPIE, Fiber Optics in Adverse Environments II* 506:170–173.

Townsend, J. E., K. P. Jadrezewski, W. L. Barnes, and S. G. Grubb. 1991. Yb^{3+} sensitised Er^{3+} doped silica optical fibre with ultrahigh transfer efficiency. *Electron. Lett.* 27(21):1958–1959.

Townsend, J. E., S. B. Poole, and D. N. Payne. 1987. Solution-doping technique for fabrication of rare-earth doped optical fibers. *Electron. Lett.* 23(7):329–331.

Trischitta, P., M. Colas, M. Green, G. Wuzniak, and J. Arena. 1996. The TAT-12/13 cable network. *IEEE Commun. Mag.* 34(2):24–28.

Tumminelli, R. P., B. C. McCollum, and E. Snitzer. 1990. Fabrication of high concentration rare-earth doped optical fibers using chelates. *J. Lightwave Tech.* 8(11):1680–1683.

Urquhart, P. 1988. Review of rare earth doped fiber lasers and amplifiers. *Proc. IEEE* 135(6):385–407.

Vengsarkar, A. M., A. E. Miller, and W. A. Reed. 1993. Highly efficient single-mode fiber for broadband dispersion compensation. In *Conference on Optical Fiber Communication.* Paper PD-13. Technical Digest. Washington, DC: Optical Society of America.

Vengsarkar, A. M., and W. A. Reed. 1993. Dispersion-compensating single-mode fiber efficient designs for first- and second-order compensation. *Opt. Lett.* 18(11):924–926.

Welch, D. F. 1994. High power laser diode. In *Conference on Optical Fiber Communication.* Paper WG1. Technical Digest. Washington, DC: Optical Society of America.

Yamashita, T. T. 1989. Nd^{3+} and Er^{3+} doped phosphate glasses for fiber lasers. *SPIE, Fiber Laser Sources and Amplifiers* 1171:291–297.

Zemon, S., W. J. Miniscalco, G. Lambert, B. A. Thompson, M. A. NewHouse, P. A. Tick, L. J. Button, and D. W. Hall. 1992. Nd^{3+} doped fluoroberyllate glasses for fiber amplifiers at 1300nm. In *Optical amplifiers and their applications.* Paper WB3. Technical Digest Series, vol. 17. Washington, DC: Optical Society of America.

Zenteno, L. A. 1993. High power double-clad fiber lasers. *J. Lightwave Tech.* 11:1435–1446.

Zenteno, L. A. 1994. Design of a device for pumping a double-clad fiber laser with a laser diode bar. *Appl. Opt.* 33(31):7282–7287.

Chapter 5 | Advances in Cable Design

Kenneth W. Jackson
T. Don Mathis
P. D. Patel
Manuel R. Santana
Phillip M. Thomas

Lucent Technologies, Bell Laboratories, Norcross, Georgia

5.1 Introduction

Since its introduction in the early 1970s, optical fiber has become the global medium of choice for broadband telecommunications networks. Cabled fiber composes the very backbone of these networks. The previous two volumes of this book, published under the auspices of Bell Telephone Laboratories, AT&T, and Bell Communications Research, include chapters that serve as excellent stand-alone references on the fundamentals of fiber optic cable design — the functional objectives, features, theory, and descriptions of the basic commercial cable designs (Miller and Chynoweth 1979; Miller and Kaminow 1988). Numerous other references are also available (Basch 1987; Mahlke and Gossing 1987; Murata 1988). In this chapter, we continue to expand the subject matter of previous volumes. We describe the new fiber optic cable designs, splicing technologies, and installation strategies that have emerged to meet the special economic and technical requirements for deployment of broadband fiber networks in the subscriber loop. We cover the evolution and advances in cable technology as it relates to outside plant, premises, and specialty cables.

5.2 Performance, Reliability, and Standards

Because even a temporary loss of high-bandwidth services can cause both network providers and users to quickly lose large amounts of revenue, added emphasis on cable design has emerged with the expanded deploy-

OPTICAL FIBER TELECOMMUNICATIONS,
VOLUME IIIA

ISBN: 0-12-395170-4

ment of high-capacity fiber systems in metropolitan area networks and in the subscriber loop. Specifically, attention has become more focused on reliability, survivability, restorability, and standardization issues. Also, there has been a trend toward increased deployment of higher fiber count cables. International, national, and local standards have emerged that now push the technological and economic limits with respect to geometric tolerances, fiber strength, polarization mode dispersion (PMD), and thermal- and mechanical-induced added loss (Bellcore 1994; IEC794-3 1994). Indeed, in many cases the stringency of the requirements contained in earlier standards has doubled. Contemporary standards address the conventional stranded loose tube, slotted-core, and loose fiber bundle cable designs as well as new specifications for the UV-bonded ribbon structure for compact cables.

The new network architectures require cable designs that are multifunctional with respect to routing and joining and that contain diverse fiber counts. Higher fiber count cables with higher packing densities, lower transmission and splice loss, and low PMD continue to be the leading performance indicators. Cables that are less than about 20-mm outside diameter (o.d.) and can contain more than 400 fibers (250-μm coated fiber) with transmission losses of 0.35 and 0.25 dB/km at 1310 and 1550 nm, respectively, represent achievable performance limits. At the other end of the design spectrum, very low fiber count cables for branching to subscribers and premises applications will present new challenges for economic manufacture.

5.3 Outside Plant Cables

5.3.1 HIGH FIBER COUNT CABLES

Background

Introduction of fiber into the subscriber loop has increased the installation of short cable lengths and has created a somewhat bimodal use distribution of high ($\geq$100) and low ($\leq$24) fiber counts per cable. Short cables have a large number of splices, and when they are used in a highly branched topology, the speed at which the cables can be placed, spliced, accessed, reconfigured, and restored becomes even more important. Installation in the loop is very labor intensive, with construction costs dominating the total cost of providing broadband access. Network providers have therefore welcomed new technologies that can reduce the deployment time and thus their initial cost. Compact, UV-bonded ribbon cable structures comprising fibers with high core-to-clad concentricity and mass fusion splicing technology have proven to be an enabling technology platform with significant advantages for rapid, low-loss, low-cost joining or restoration of high fiber count cables (Lindsay *et al.*, 1995). Cables with UV-bonded ribbons and mass fusion splicing machines suitable for field use became commercially available in the late 1980s. In 1989 and 1990, AT&T introduced 204- and 216-fiber central tube ribbon cables that were 40% more compact than AT&T's original adhesive sandwich ribbon (ASR) cable that was first introduced in the Chicago Lightwave Project in 1977 (Jackson *et al.* 1989). Concurrent with the UV-bonded ribbon development, a number of Japanese companies were perfecting portable mass fusion splicing machines that could be used by the craft to splice up to 12 fibers at a time with splice losses of a few hundredths of a decibel (Kawase *et al.* 1989). Since the mid-1980s, more than 2-million fiber kilometers of ribbon cable have been spliced, with sales growth on the order of 50% per year.

Ribbon Design

This section describes the basic designs and functional characteristics of fiber ribbons. Whereas the principal objective of fiber optic cable design is to protect and isolate the fibers from adverse environments, an additional objective of ribbon design is to organize the fibers for optimum space efficiency and ease of mass splicing and handling, while maintaining their inherent optical performance and mechanical reliability. Characteristics of an ideal ribbon include the following:

- It maximizes fiber packing density.
- It has low cabling–environmental added loss.
- It precisely organizes fibers for mass splicing or connectorization.
- It provides rapid, unique identification for single fibers.
- Its matrix and coating cleanly strip from glass.
- It is divisible into single fibers or robust subunits.
- Its fibers and subunits are accessible from either the end or midspan.
- It is mechanically robust.
- It maintains its properties with thermal aging in hydrocarbon filling compounds.

Bonded ribbons use fiber coating material and process technology to bond the color-coded fibers into a contiguous array, as contrasted to earlier ribbons in which the fibers were laminated between two pieces of adhesive-backed tape. Two basic ribbon structures have emerged for use with the different cable designs (Jackson *et al.*, 1993). As illustrated in Fig. 5.1, these structures are designated as *edge bonded* or *encapsulated* depending upon the amount of bonding matrix used. In the edge-bonded structure, the bonding matrix is applied predominantly in the interstices between the fibers. In contrast, in the encapsulated structure the matrix extends well beyond the outer boundary of the fibers, somewhat like a rectangular buffer tube. The edge-bonded ribbon, which is the preferred design and the Bellcore standard in the United States, is about 40% smaller than the ASR or encapsulated ribbons. The encapsulated ribbon is popular in Japan, where slotted-core-type cable designs are prevalent (Matsuoka *et al.* 1994). However, edge-bonded ribbons have also appeared recently in some very high fiber count (>500) slotted-core and stranded loose-tube-type cable designs (Kurosawa *et al.* 1992).

An extended development effort has been undertaken by several companies in Japan to make a more compact ribbon cable by downsizing the fiber coating. Prototype, thin-coated fiber ribbon cables capable of containing as many as 4000 fibers in a single 39-mm sheath have been made (Tomita *et al.* 1993). Although considerable progress has been made in designing thinner (~180-μm-o.d.) fiber coatings that exhibit microbending loss performance comparable to that of standard 250-μm-coated fiber, fiber processing and handling robustness issues must be resolved before thin-coated fibers are used commercially for ribbon fabrication (Kobayashi *et al.* 1995). An

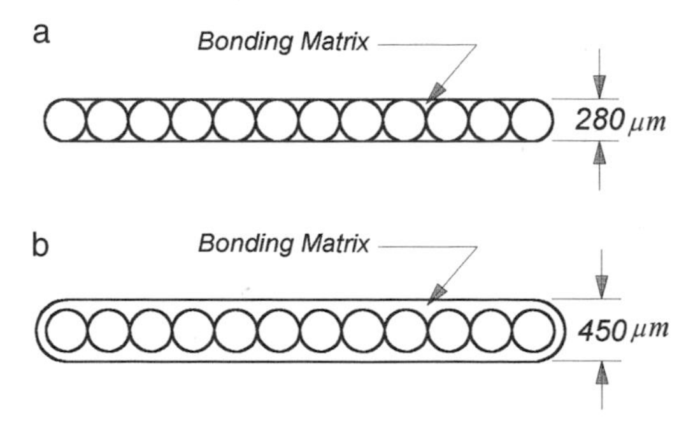

Fig. 5.1 Basic UV-bonded ribbon structures. (a) Edge-bonded ribbon and (b) encapsulated ribbon.

alternate approach for increasing the fiber packing density of central tube-type cables uses standard 250-μm-coated fiber and increases the fiber count per ribbon (Jackson *et al.* 1993). This approach has proven appropriate for central tube cables where the ribbon stack remains close to the neutral axis when the cable is bent. For stranded or slotted structures where the ribbon stack is located away from the neutral axis and is geometrically constrained, thicker ribbons have been used, although recent research indicates a trend toward thin ribbons (Kobayashi *et al.* 1995; Obi *et al.* 1995). Modular ribbons containing 12, 16, or 24 fibers have been demonstrated, and 16-fiber modular ribbon cables have been commercially deployed. Access to single fibers or modules can be gained from either the ribbon end or from midspan, with the modules retaining their structural integrity for routing, stripping, and mass fusion splicing. Figure 5.2 illustrates a typical modular ribbon cable, and Fig. 5.3 shows a quantile–quantile plot of average splice loss for 4-, 8-, and 12-fiber ribbon modules that have been separated from a 16-fiber ribbon and mass fusion spliced. The distribution of the data demonstrates that the splice loss of 4- and 8-fiber modules is comparable to that of a standard 12-fiber ribbon.

Ribbon Splicing

Because mass fusion splicing two factory-made ribbons in the field takes only slightly longer than splicing two individual fibers, large gains in splicing productivity are achievable. The productivity gains of mass fusion splicing

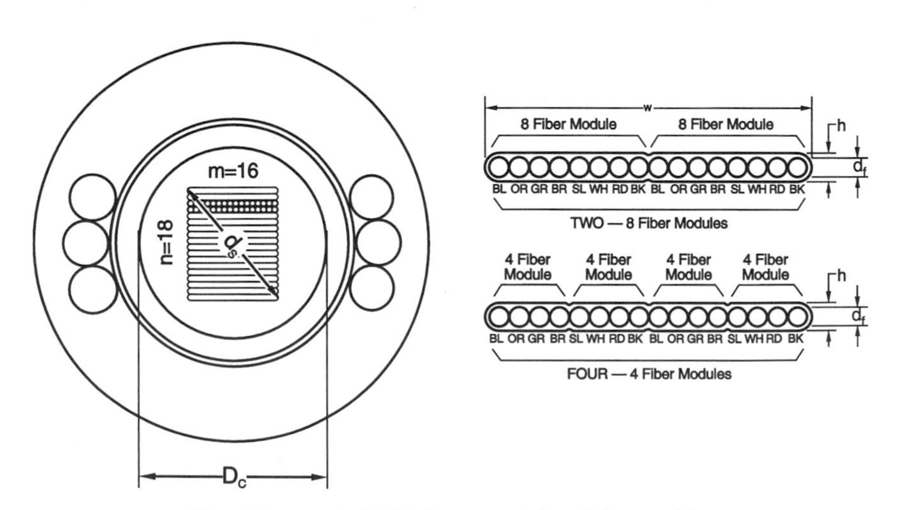

Fig. 5.2 Typical 288-fiber modular ribbon cable.

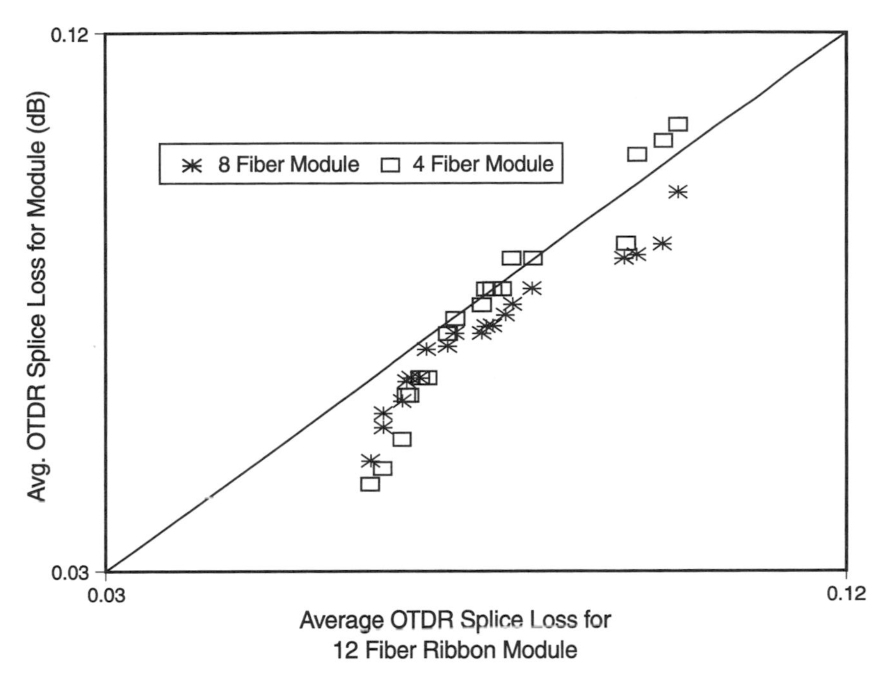

Fig. 5.3 Quantile–quantile plot for ribbon module splice loss. OTDR, optical time domain reflectometry.

can even justify ribbonizing individual fiber ends in the field for large fiber count cables comprising single fibers. Figure 5.4 compares the splicing time for mass fusion of factory- and field-made ribbons and for single-fiber fusion. The high productivity of mass splicing is critically linked to the fiber geometry, the ribbon, and the mass fusion splicing machines. High-yield, low-loss mass fusion splices can be consistently obtained in the field only from ribbons that are made from fibers having stringent geometric tolerances. Such fibers, with a cladding diameter of $125 \pm 1.0\ \mu m$ and a maximum core-to-clad concentricity error of $0.8\ \mu m$, are commercially available and are necessary for rapid low-loss splicing. Equally important are robust mass fusion splicing machines with highly optimized arc heating profiles and fiber alignment algorithms.

Cable Designs and Directions

Figure 5.5 illustrates some of the basic compact, high fiber count ribbon cable designs that have been proposed. The basic unit generally comprises a central tube and slotted cores or U channels. Scale-up from single-core

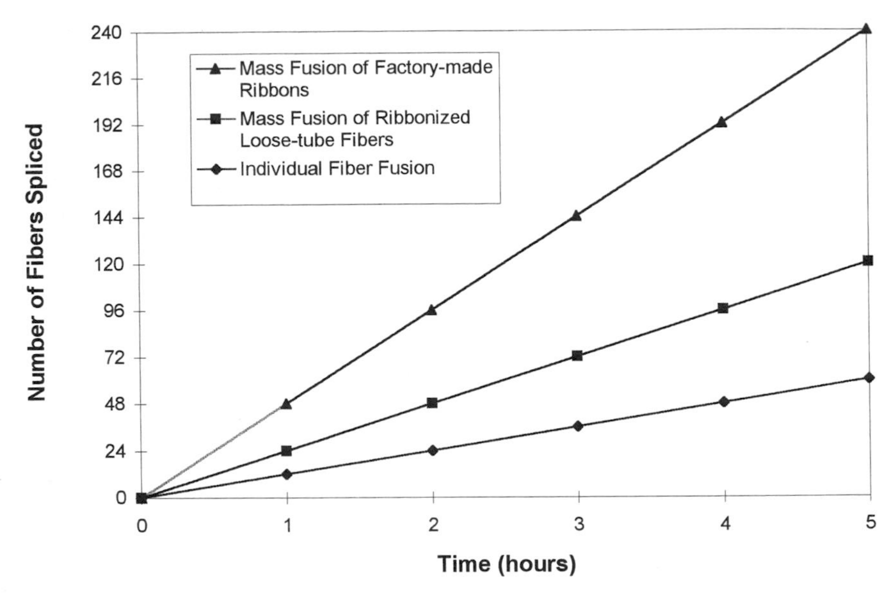

Fig. 5.4 Productivity gains with mass fusion splicing.

unit designs use multiple central tubes, slotted cores, or U channels that are stranded around a central member. Typically, the slotted-core-type cables use encapsulated ribbons, whereas the central tube designs use edge-bonded ribbons. Recent research has been directed to fiber-coating optimization to allow slotted-core cables to use the thinner edge-bonded-type ribbon (Kabayashi *et al.* 1995). High fiber count loose-tube-type cables are also available, as shown in Fig. 5.5e. Table 5.1 shows representative loss data for a 288-fiber central tube-type ribbon cable consisting of eighteen 16-fiber ribbons.

As fiber penetrates farther and farther into the loop, fiber counts per cable are expected to increase, as well as the frequency of midspan access. Emphasis will continue to be placed on ribbon cable designs with higher packing density and higher splicing and handling productivity.

5.3.2 LOW FIBER COUNT CABLES

Background

Historically metallic-based (twisted copper pair, coax, etc.) networks continue to be penetrated by optical fiber. Also, the unique transmission properties of optical fibers are contributing to (or even allowing) the introduction

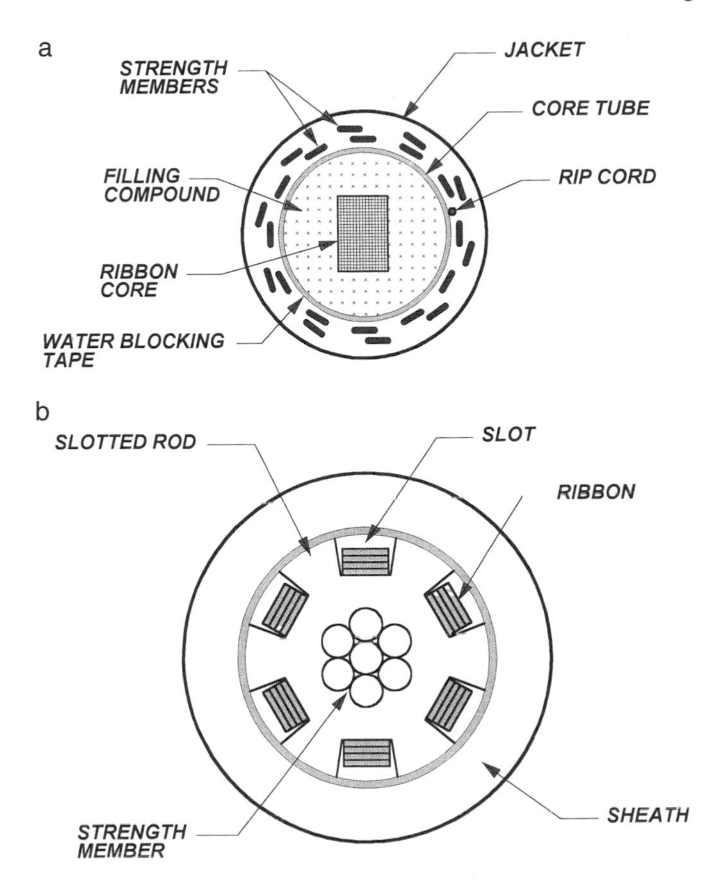

Fig. 5.5 High fiber count cable designs. (a) Central tube ribbon cable, (b) slotted-rod ribbon cable, (c) multiple loose-tube ribbon cable, (d) multiple-rod slotted-core ribbon cable, and (e) 288-fiber Mini Bundle® cable. (e, Courtesy of Siecor Corporation.)

of new fiber-based applications. These driving forces, among others, have led to a demand for fiber cables with a somewhat bimodal distribution of fiber counts — cables containing up to 24 fibers and cables containing hundreds of fibers.

Earlier cable designs, used primarily in long-haul networks, focused on providing flexibility in fiber counts within a given cable design or size. These designs were not optimized for size, installed cost, or flexibility in fiber types. Because low fiber count cables compose a large percentage of the total cable demand, designs that are optimized for low fiber counts (i.e., $\leq$24 fibers) are needed. Many cable manufacturers, realizing this

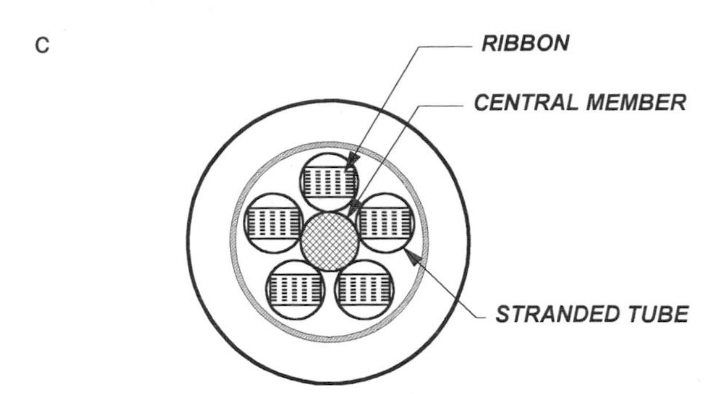

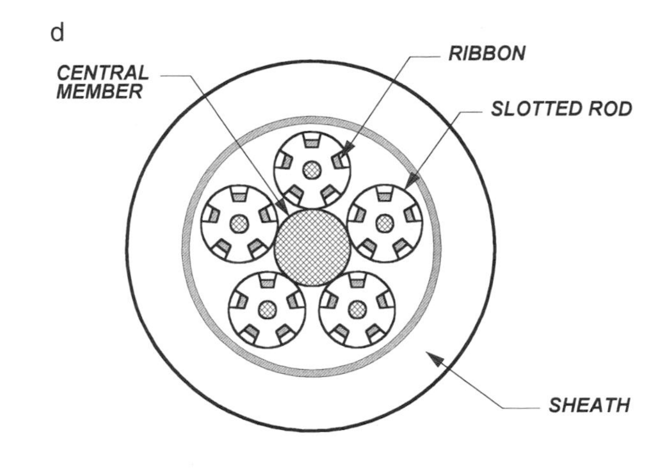

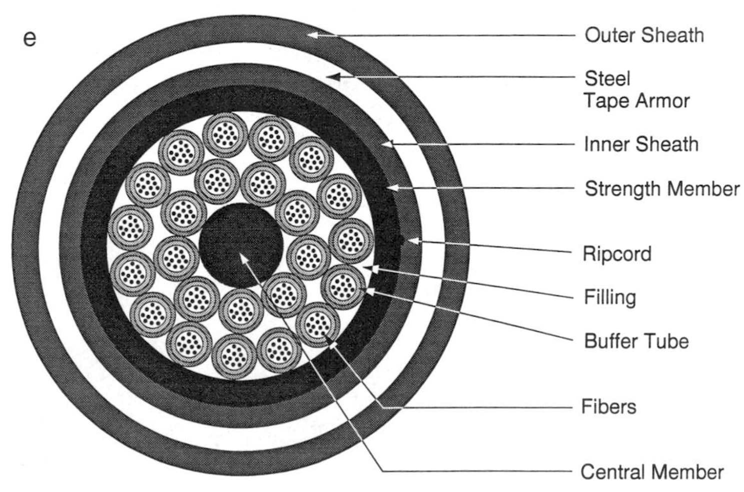

Fig. 5.5 (*Continued*)

Table 5.1 **Representative Loss Data for 288-Fiber Central Tube-Type Ribbon Cable**

		Cable Loss (dB/km)			
		Depressed Clad		**Matched Clad**	
Temperature		**1310 nm**	**1550 nm**	**1310 nm**	**1550 nm**
25°C	Mean	0.35	0.20	0.36	0.20
	Median	0.35	0.20	0.36	0.20
−40°C	Mean	0.36	0.26	0.37	0.26
	Median	0.35	0.22	0.36	0.23

demand, recently began to develop and offer specially designed low fiber count cables. One such design was recently developed that offers a variety of fiber configurations (Kinard *et al.* 1994); coated-fiber bundles, buffered fibers, and ribboned fibers can all be provided in the same small-diameter, central core tube. This family of products addresses the application needs of telephony, cable television, private, and long distance and broadband networks.

Characteristics

Cable designs that have been optimized for low fiber counts are characterized by their smaller size, increased flexibility in fiber types and configurations, and overall cost-effectiveness. The core tube (or unit tube) diameter is optimized for the given (low) fiber count to provide the maximum packing density while maintaining the desired mechanical and environmental performance. The overall cable size is correspondingly reduced to reflect the smaller core tube size.

Flexibility in fiber configuration is provided through the use of a cable core that accepts discrete coated or buffered fibers, bundles of coated fibers, or fiber ribbons. Fiber types that may be offered include single-mode fibers with depressed or matched cladding, or dispersion-shifted fibers. Excess fiber length (EFL) and packing density are varied in order to accommodate the different fiber configurations.

Typically, for both coated- and buffered-fiber configurations, the fibers are placed in a filled tube without intentional stranding. A controlled amount of EFL is introduced during manufacturing to ensure that fiber strain is kept at a safe level during cable installation. In the fiber ribbon configuration, ribbons are placed either with or without stranding depending on the size and number of ribbons. The small fiber count ribbons, however, are typically placed without stranding. In all configurations, the core tube is filled with a compound to prevent water migration along the core tube while minimizing restrictions to fiber movement.

Cost-effectiveness is achieved in the optimized low count cable design by virtue of the reduced cable size and increased flexibility in core configurations. The primary sacrifice in the optimized low count design is the limited range of fiber counts obtainable within the given core tube size.

Design Considerations

Careful consideration must be given to the design of the cable core in order to obtain acceptable mechanical and environmental behavior of the fibers in the completed cable. It is important that the cable design provide low fiber attenuation while maintaining high reliability. A major factor affecting both attenuation and reliability is fiber strain. High fiber strains can result in increased attenuation and reduced mechanical reliability.

Fiber strains can be induced by mechanical loads and thermal effects during manufacture, during installation, and during the service life of the cable. Some of the key parameters affecting these strains are as follows:

(1) Fiber packing density
(2) EFL
(3) Fiber unit stranding or twisting
(4) Cable bending
(5) Tensile load rating
(6) Temperature extremes

Fiber Packing Density

The two most common definitions used in the cable industry for packing density are (1) the ratio of total fiber cross-sectional area to cross-sectional area within the core tube and (2) the number of fibers per cable cross-sectional area. High packing density is required for compact designs; however, it can lead to higher fiber strains. Cable designs with up to 20% — or 4 f/mm^2 of cable cross section — packing density have been proposed and are discussed in the literature.

The critical dimension ratio, d_c/D, is a convenient design parameter to use when a cable is being optimized for highest packing density. For fiber bundles or buffered fibers, this ratio is the minimum diameter circumscribing the fibers, d_c, divided by the inside diameter (i.d.) of the core tube, D.

For fiber ribbons, this ratio can be assumed to be the diagonal of the ribbon stack divided by the core tube i.d.

Excess Fiber Length

It is generally desired that the length of fiber within the cable exceed that of the cable, and this EFL is expressed as a percentage of cable length. A higher EFL can be used to reduce the tensile strain in fibers or to allow a reduction in the tensile stiffness of the cable. However, the packing density and the EFL are interrelated and must be considered together. The higher values of EFL may induce undesirable bending losses in the fibers, especially at low temperatures.

EFLs of up to 1% are common in the cable industry. Analysis of a fiber bundle assuming a helical and a sinusoidal bundle deformation (Patel and Panuska 1990) shows that the sinusoidal model is most conservative because it predicts a smaller minimum bend radius than the helical model does. For the sinusoidal model, EFL can be expressed in the following form:

$$\mathrm{EFL} = 100[(1 + k^2)^{1/2}(1 - p) - 1], \qquad (5.1)$$

where

$$p = \tfrac{1}{4}\, q^2 + \tfrac{3}{64}\, q^4 + \tfrac{1}{256}\, q^6 + \ldots$$

$$q^2 = k^2/(1 + k^2)$$

$$k^2 = [1 - (d_c/D)]/(D_s/D)$$

and D_s is twice the minimum sinusoid radius.

Equation (5.1) relates EFL to the ratios involving fiber bundle or ribbon dimensions and fiber bend diameter. Figure 5.6 shows three curves for EFL as a function of the ratio of fiber bend diameter to core tube diameter, D_s/D, with the critical dimension ratio, d_c/D, as a parameter. For a given tube diameter and EFL, the curves predict smaller fiber bend diameters for larger values of d_c/D. These design curves can be used to minimize the core tube diameter for a given minimum fiber bend radius.

Fiber or Unit Stranding or Twisting

Within the core tube of a central core cable design, the fibers are placed either straight, stranded, or twisted during manufacture. The fibers within a loose tube cable are typically placed, without stranding, within the unit tubes, and the tubes are stranded about a central member. The stranding process can impart additional bending strain in the fiber.

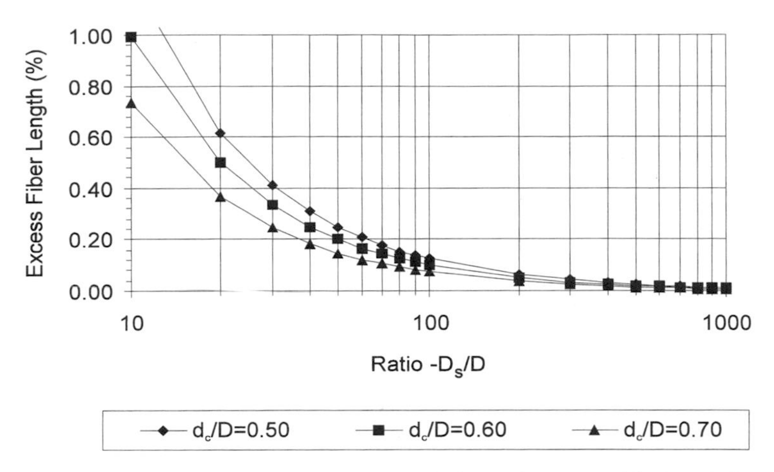

Fig. 5.6 Excess fiber length as a function of dimensionless cable bend diameter.

For helical stranding with pitch diameter D_p and pitch P_h, the radius of curvature of the fiber helix R_h is given by

$$R_h = \frac{D_p}{2}\left[\left\{\frac{P_h}{\pi D_p}\right\}^2 + 1\right]. \tag{5.2}$$

The maximum strain, ε_b, induced in a fiber with diameter d_f by the stranding process is then given by

$$\varepsilon_b = \frac{d_f}{2R_h} = \frac{d_f/D}{D_b/D}, \tag{5.3}$$

where D_b (i.e., $2R_h$) is the bend diameter of the fiber caused by the stranding process.

Equation (5.3) relates the bending strain to critical dimension and bend diameter ratios. Figure 5.7 shows three curves for the fiber bending strain as a function of the ratio of fiber bend diameter to core tube diameter with d_f/D as a parameter. These design curves, in conjunction with data correlating fiber reliability or added fiber loss with the critical dimension and bend diameter ratios, are useful in the design of an optimum cable core.

The bending strains due to helical stranding can be high and are left in the fiber for the service life of the cable. Because these strains affect both attenuation and reliability, they should be minimized in the optimized cable design.

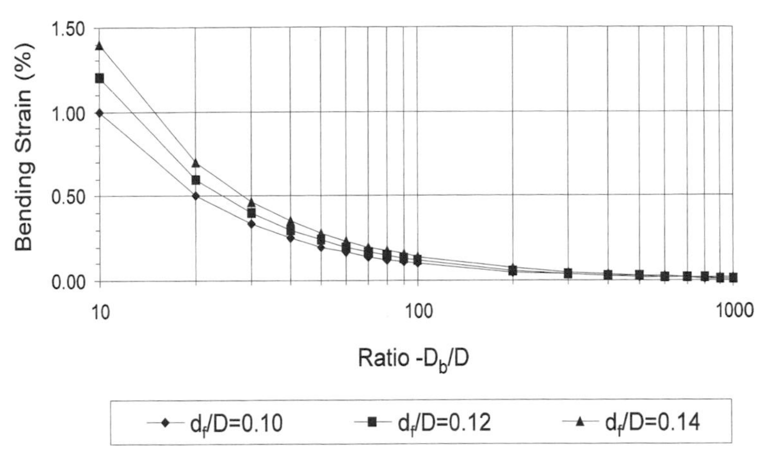

Fig. 5.7 Fiber bending strain as a function of dimensionless fiber bend diameter.

Cable Bending

For ease of installation, the cable should be flexible enough to bend to a specified radius without significantly increasing attenuation or reducing fiber reliability. Cable bends can induce two types of strain in the fiber: (1) bending of the fiber about its own axis, as discussed previously, and (2) bending of the fiber about the cable's neutral axis. The first strain is generally controlled by limiting the minimum bend radius of the cable. However, the second strain is dependent on the cable design. In configurations where the fibers are located well away from the neutral axis of the cable, a common design technique is to average this (cable-bending-induced) fiber strain by stranding or twisting the fiber units. Hence, strain averaging is particularly important in high fiber count (large critical dimension) designs. In low fiber count designs with small critical dimensions, strain averaging is of lesser importance and can be used to advantage.

Tensile Load Rating

The fibers must be able to survive the total strain caused by applying the rated tensile load to the cable. For example, Bellcore's objective is for cabled fibers not to be strained beyond 60% of the fiber proof strain. This design safety factor is based on static fatigue analysis, and it represents 0.3

and 0.6% maximum allowable fiber strains at 0.35- and 0.69-GPa proof stress levels. Most fiber manufacturers today offer 0.69-GPa proof-tested fibers.

Temperature Extremes

The cable, when exposed to temperature extremes, typically expands or contracts more than the fibers do, as a result of the difference in coefficients of thermal expansion between the cable and the fibers. Thus, at low temperatures an additional EFL is introduced in the cable, and at high temperatures the EFL is reduced. So, the EFL must be determined at both the minimum and maximum temperatures in order to ensure that (1) the fiber minimum bend radius is acceptable at the low temperature and (2) the fiber tensile strain is within design limits at the high temperature.

Cable Designs

Most optical fiber applications require the cable to withstand a wide range of installation and environmental conditions. For areas where rodent protection is desired, a robust armored cable design is needed, whereas a dielectric cable is preferred for lightning-prone areas or for areas where currents may be induced from nearby high-tension electric lines. Moreover, in hot and humid environments where termite infestation is a problem, a nylon overjacket is preferred. In special situations where both optical fiber and metallic media are required, a composite cable may be needed.

Linear Strength Member Cable Designs

Figure 5.8 shows a perspective view of the internal construction of a typical central core tube, linear strength member cable design. The metallic armored cable design (Patel *et al.* 1988) has been in use in the outside plant network since 1987. Variations of this design, optimized for low (Lachman and Rodder 1993) as well as high fiber counts, have been introduced by numerous cable manufacturers.

Cable designs utilizing dielectric linear strength members, similar in construction to the armored design depicted in Fig. 5.8, have also gained wide acceptance in the industry because of their compact size and ease of fiber access. Both metallic and dielectric cable designs are available in various fiber counts and tensile ratings.

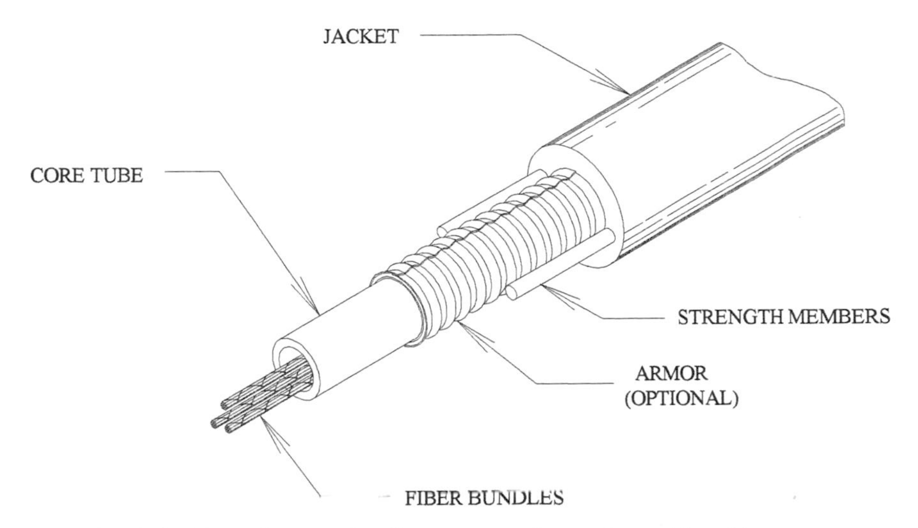

Fig. 5.8 Central core tube, linear strength member cable with armor.

Stranded Strength Member Cable Designs

Dielectric cable designs that use stranded strength members, placed in one or more layers over the core tube, have been introduced for small fiber counts (Ferguson and McCallum 1994). These designs typically use glass yarns, glass rovings, epoxy–glass rods, and aramid yarns in different combinations in order to obtain the desired mechanical properties in the finished cable. An all-dielectric design constructed with two layers of helically applied strength members (Arroyo *et al.* 1987) has been deployed in the outside plant since 1985.

Figure 5.9 shows a perspective view of the internal construction of a dielectric cable design that utilizes a single layer of stranded dielectric strength members applied over a central core tube. Designs incorporating a corrugated metallic armor, typically as an oversheath, over the dielectric core cable are also offered.

Optical Properties of Production Cables

The room-temperature fiber loss for one vendor's armored, linear strength member cable is presented in Fig. 5.10. The data represent more than six fiber megameters of cabled single-mode fiber. The mean cabled fiber loss

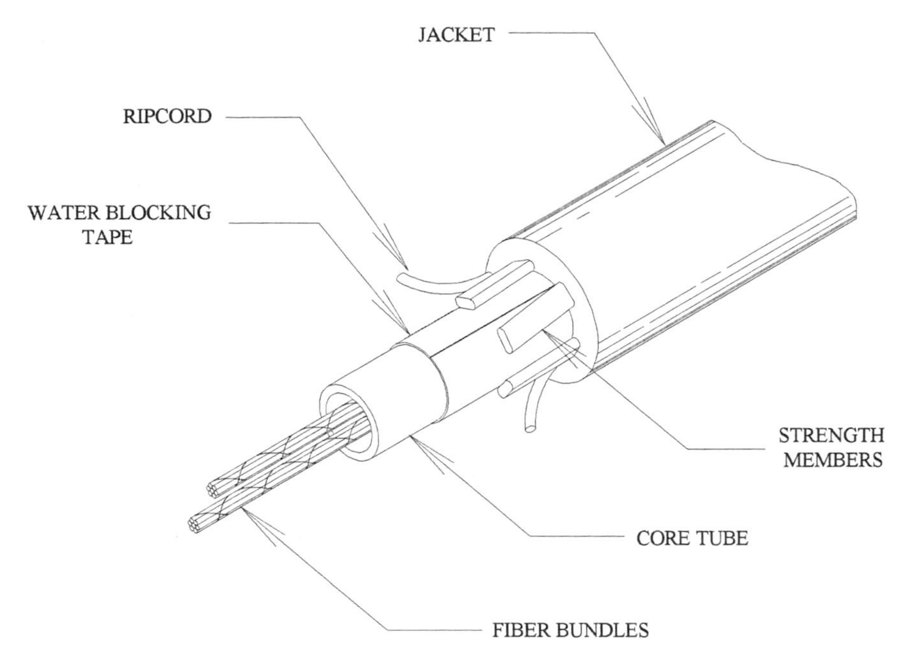

Fig. 5.9 Central core tube cable with stranded dielectric strength members.

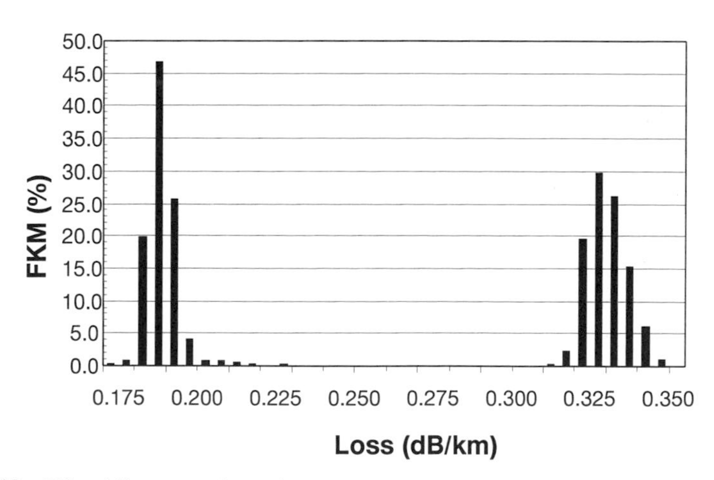

Fig. 5.10 Fiber kilometers (FKM) as a function of loss performance (1310/1550 nm) of linear strength member cables.

is 0.33 and 0.19 dB/km, with a sigma of 0.006 dB/km, at 1310 and 1550 nm, respectively.

5.3.3 PREMISES CABLES

Background

Premises cables are designed for use inside buildings. There are many applications that require designs using a variety of materials. Premises cables can be divided into categories of *general purpose*, *riser*, and *plenum*. These categories primarily define the materials and cable construction that must be used to pass the flame spread and smoke requirements in the National Electric Code Reaction to Fire Tests (UL 1581, UL 1666, and UL 910). These categories can be further subdivided into *high fiber count*, *jumper,* and *breakout cables.* These subcategories define the number and packaging of fibers, and the cable application.

Indoor–Outdoor Cables

An indoor–outdoor cable design must satisfy the mechanical requirements of an optical fiber cable that can be buried outside, and the indoor UL 1666 fire tests. Indoor–outdoor cables find application where it is desired that one cable length interconnect two buildings (using the cable as a riser inside the building and as a buried cable outside the building and then again as a riser inside an adjacent building). Typical performance requirements are listed next.

Low–high temperature bend	−20 and 50°C
Impact test	1.00 ft lbf
Compressive strength	150 lbf/in.
Tensile strength	400 lbf
Cable twist	±180°/2 m
Cyclic flexing	1000 cycles
Water penetration	1-m length/1-m head/1 h
Operating temperature	−20 to 65°C
Cable aging	−40 to 85°C
Flammability test	UL 1666

This is a difficult design because there are conflicting requirements. For example, a grease-filled cable might pass the water-penetration tests but

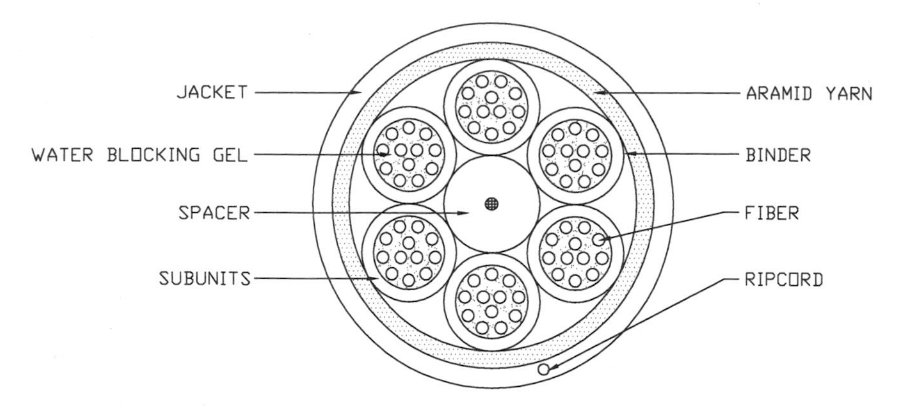

Fig. 5.11 Indoor–outdoor cable design.

Table 5.2 **Comparison of Cable Cross-Sectional Areas**

Diameter (mm)	*Cross-Sectional Area (mm^2)*
1.6	2.0
2.0	3.1
2.4	4.5

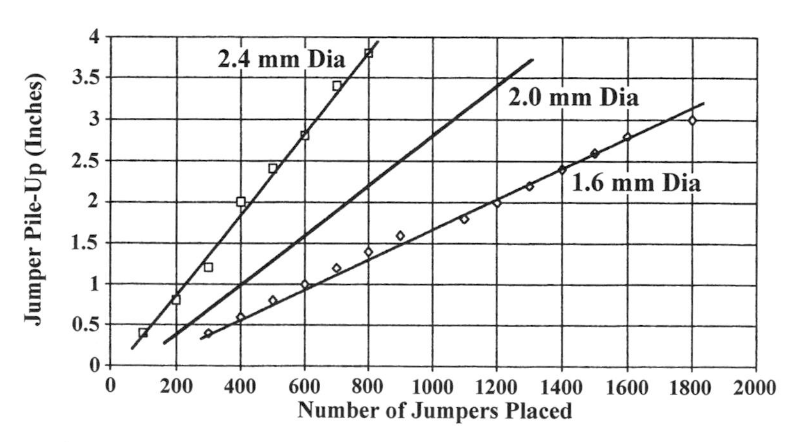

Fig. 5.12 Jumper pileup as a function of the number of jumpers.

Table 5.3 **Cable Density at 2-In. Pileup Depth**

	Cable Density (jumpers/in.^{2})		
Cordage (mm diam.)	**Theoretical No. Based on Dimensions**	**Empirical No. Based on Pileup Data**	**Avg. Packing Factor**[a]
2.4	112	35–45	0.36
2.0	205	65–85	0.40
1.6	252	100–125	0.45

[a] Avg. Packing Factor = empirical/theoretical.

fail the flammability test. An indoor premises cable should be flexible and easy to handle; yet, many cable materials that have high impact strength also have a high modulus. Such materials are inherently stiff and can be

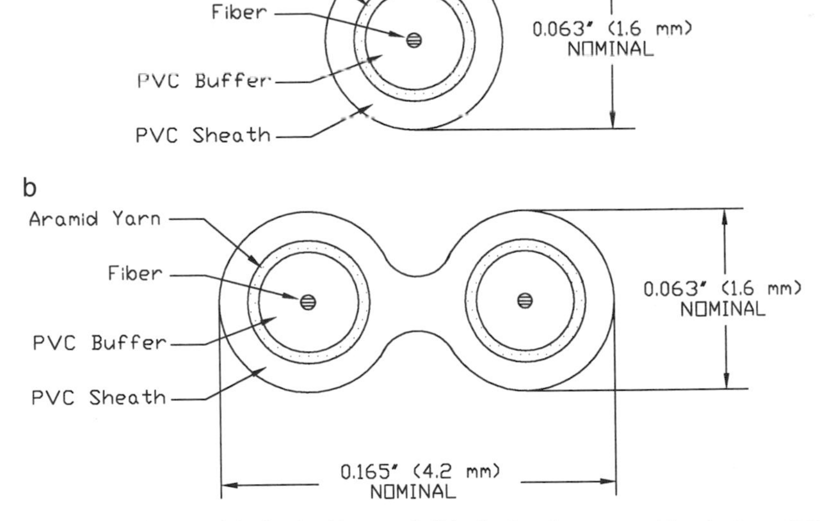

Fig. 5.13 Miniature (a) single-fiber and (b) duplex jumper cable designs. PVC, polyvinyl chloride.

difficult to use inside a building, where many sharp cable bends occur. A typical indoor–outdoor cable design is shown in Fig. 5.11.

Miniature Jumper Cables

As local exchange carriers add fiber to their systems, the central office becomes more congested at the fiber distribution frame (FDF). Small jumper cables are needed to reduce the congestion not only at the frame but also inside cable troughs. A simple way to compare the various cordage products available is by cross-sectional area, as shown in Table 5.2.

A 1.6-mm cordage takes up 55% less space than the 2.4-mm cordage and 35% less space than the 2.0-mm cordage. Figure 5.12 shows the jumper pileup as a function of the number of jumpers placed in a 4-in. wide cable trough.

It has been suggested that the pileup depth should not exceed 2 in. to reduce excessive stress to the jumpers during handling. Table 5.3 shows the calculated and empirical number of jumpers per square inch for various-diameter jumpers.

Figure 5.13 shows 1.6-mm-diameter cable in both single-fiber and duplex designs.

References

Arroyo, C. J., A. C. Jenkins, P. D. Patel, and A. J. Panuska. 1987. A high performance nonmetallic sheath for lightguide cables. In *Proceedings of the 36th IWCS*, 344–349.

Basch, E. E. B. 1987. *Optical-fiber transmission*. Indianapolis, IN: Howard W. Sams.

Bellcore. 1994. *Generic requirements for optical fiber and fiber optic cable, GR-20-CORE.* Piscataway, NJ: Bell Communications Research.

Ferguson, S. T., and W. J. McCallum. 1994. Development of dielectric fiber optic drop cable. In *Proceedings of the 43rd IWCS*, 219–226.

IEC.794-3. 1994. *International standard: Telecommunication cables—Sectional specification.* International Electrotechnical Commission.

Jackson, K. W., P. D. Patel, M. L. Pearsall, and J. R. Petisce. 1989. An enhanced ribbon structure for high fiber count cables in the loop. In *Proceedings of the 38th IWCS*, 569–574.

Jackson, K. W., N. W. Sollenberger, S. P. Gentry, R. J. Brown, J. R. Petisce, C. R. Taylor, and S. H. Webb. 1993. Optimizing ribbon structures for performance and reliability. In *Proceedings of the NFOEC*, 205–222.

Kawase, M., T. Fuchigama, M. Matsumoto, S. Nagasawa, S. Tomita, and S. Takashima. 1989. Subscriber single mode optical fiber ribbon cable technologies suitable for midspan access. *J. Lightwave Tech.* 7(11):1675–1681.

Kinard, M. D., T. D. Mathis, A. J. Panuska, and P. D. Patel. 1994. Low fiber count cable for emerging broadband networks. In *Proceedings of the 43rd IWCS*, 219–226.

Kobayashi, K., N. Okada, K. Mitsuhashi, K. Ishida, M. Miyamoto, and S. Araki. 1995. Coating design of thin-coated ribbons using 250 μm coated fibers. In *Proceedings of the 44th IWCS*, 607–614.

Kurosawa, Y., H. Sawano, M. Miyamoto, and N. Sato. 1992. Development of ribbon loose tube cable for optical subscriber loop. In *Proceedings of the 41st IWCS*, 151–157.

Lachman, D., and T. Rodder. 1993. Steel tube armoured fiber optic cable design. In *Proceedings of the 42nd IWCS*, 516–520.

Lindsay, R. G., S. Robbins, S. C. Mettler, C. F. Cottingham, and K. W. Jackson. 1995. Ribbon cable and mass fusion splicing technologies accrue extended benefits. *Lightwave* 12(June):40–46.

Mahlke, G., and Gossing. 1987. *Fiber optic cables: Fundamentals, cable engineering, systems planning.* New York: Wiley.

Matsuoka, R., M. Saito, M. Hara, and A. Otake. 1994. Design and development of high-density optical fiber slotted-core cable. In *Proceedings of the 43rd IWCS*, 22–26.

Miller, S. E., and A. G. Chynoweth. 1979. *Optical fiber telecommunications.* New York: Academic Press.

Miller, S. E., and I. P. Kaminow. 1988. *Optical fiber telecommunications II.* Boston: Academic Press.

Murata, H. 1988. *Handbook of optical fibers and cables.* New York: Marcel Dekker.

Obi, K., T. Takeda, T. Watanabe, K. Imamura, M. Fujita, and H. Tanaka. 1995. Developments of high density optical fiber cable consists of 0.3mm-thick fiber ribbons. In *Proceedings of the 44th IWCS*, 622–625.

Patel, P. D., and A. J. Panuska. 1990. Lightweight fiber optic cable. In *39th IWCS Proceedings*, 158–165.

Patel, P. D., M. R. Reynolds, M. D. Kinard, and A. J. Panuska. 1988. LXE — A fiber-optic cable sheath family with enhanced fiber access. In *37th IWCS Proceedings*, 72–78.

Tomita, S., M. Matsumoto, S. Nagasawa, and T. Tanifuji. 1993. Ultra-high density optical fiber cable with thin coated fibers and multi-fiber connectors. In *Proceedings of the 42nd IWCS*, 5–15.

Chapter 6 | Polarization Effects in Lightwave Systems

Craig D. Poole

EigenLight Corporation, Ocean, New Jersey

Jonathan Nagel

AT&T Laboratories–Research, Holmdel, New Jersey

6.1 Introduction

Polarization effects, although recognized in the earliest observations of optical fiber (Snitzer and Osterberg 1961), have historically played a minor role in the development of lightwave systems. The primary reason for this is that commercial optical receivers detect optical power rather than the optical field and thus are insensitive to polarization.

In recent years the importance of polarization in lightwave systems has grown as a result of two developments. First, the optical amplifier has dramatically increased the optical path lengths achievable with single-mode fiber and at the same time increased the number of optical elements that lightwaves encounter in a path. As a result, small effects such as polarization mode dispersion (PMD) and polarization-dependent loss (PDL) can accumulate in a span to the point where they become an important consideration for lightwave system developers.

The second reason that polarization effects have become important is that transmitter and receiver technologies have pushed the capacity of optical fiber to its limit, even in relatively short spans. This has occurred through dramatic increases in bit rates in digital systems and through the rapid advancement of analog transmission techniques in video distribution systems.

In this chapter we review the most important polarization phenomena affecting lightwave systems that use single-mode fiber as the transmission

OPTICAL FIBER TELECOMMUNICATIONS,
VOLUME IIIA

ISBN: 0-12-395170-4

medium. We begin in Section 6.2 with a review of the underlying mechanisms responsible for polarization effects in telecommunications fiber and provide definitions for some of the most common terms. In Section 6.3 we focus on PMD and its effects on digital and analog lightwave systems. A review of various techniques for measuring PMD is given in Section 6.4. Finally, in Section 6.5 we review polarization effects that are of particular concern to designers of lightwave systems employing long amplifier chains.

We limit the discussion in this chapter to linear optical effects. For discussions of nonlinear polarization effects, the reader is directed to Chapter 12 in Volume IIIA.

6.2 Origin of Polarization Effects in Fiber

In single-mode fiber, an optical wave of arbitrary polarization can be represented as the linear superposition of two orthogonally polarized HE_{11} modes. In ideal fiber, the two HE_{11} modes are indistinguishable (degenerate) in terms of their propagation properties owing to the cylindrical symmetry of the waveguide. This degeneracy, which gives some validity to the term *single mode*, is realized to a greater or lesser degree in real fibers depending on the manufacturing process and the extent to which external mechanical forces act on the fiber after manufacture.

Real fibers, as shown in Fig. 6.1, contain some amount of anisotropy owing to an accidental loss of circular symmetry. This loss occurs either through a noncircular waveguide geometry or a nonsymmetrical stress field in the glass. In either case, the loss of circular symmetry gives rise to two distinct HE_{11} polarization modes with distinct phase and group velocities. All polarization effects in single-mode fiber are a direct consequence of this accidental loss of degeneracy for the polarization modes.

6.2.1 BIREFRINGENCE MECHANISMS

Figure 6.2 illustrates the perturbations that act on single-mode fiber to produce distinct polarization modes. The perturbations are divided into two groups: intrinsic and extrinsic.

Intrinsic perturbations are accidentally introduced in the manufacturing process and are a permanent feature of the fiber. These include a noncircular core and nonsymmetrical stress fields in the glass around the core region. A noncircular core gives rise to *geometric birefringence*, whereas a nonsymmetrical stress field creates *stress birefringence* (Fig. 6.2a).

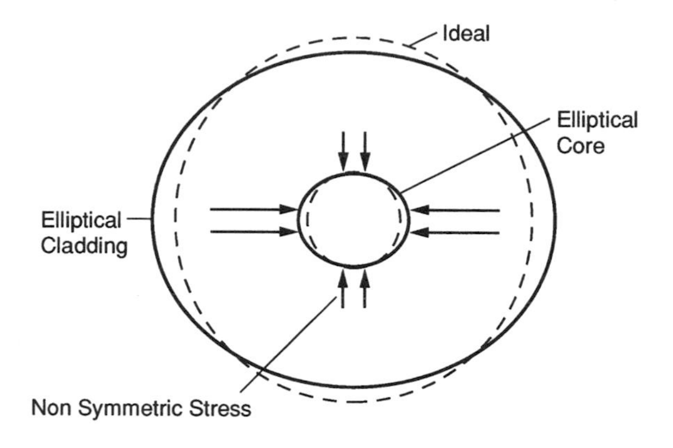

Fig. 6.1 Anatomy of a real fiber.

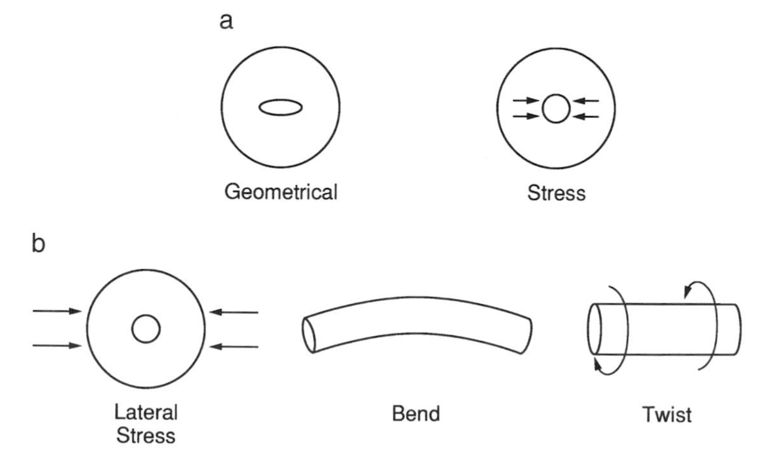

Fig. 6.2 (a) Intrinsic and (b) extrinsic mechanisms of fiber birefringence.

A noncircular fiber geometry can usually be traced to the preform from which the fiber is drawn. In telecommunications fiber, deviations of less than 1% in the circularity of the fiber core can translate into noticeable effects in lightwave systems. Consequently, even extremely small imperfections in the processing of the preform can have a significant impact on fiber performance. In the case of the modified chemical vapor deposition (MCVD) process, for example, nonideal symmetry in the starting tube or nonsymmetrical heating during collapse to form the preform is a potential cause of a loss of circular symmetry.

Stress birefringence caused by a nonsymmetrical stress field in the core region of a fiber typically arises in combination with a noncircular fiber geometry. The necessarily different chemical composition of the core relative to the cladding in single-mode fiber usually results in a slightly different thermal expansion coefficient for the two regions. This gives rise to radially directed stresses when the fiber is cooled after being drawn. In an ideal, circularly symmetrical fiber, these stress fields are symmetrical and thus do not cause anisotropy. However, if there is a noncircular shape to either the core or the cladding in the preform, the drawn fiber will have internal stress fields that are not circularly symmetrical.

Birefringence can also be created in a fiber when it is subjected to external forces in handling or cabling. Such extrinsic sources of birefringence, as shown in Fig. 6.2b, include lateral stress or pinching of the fiber, fiber bending, and fiber twisting. All three of these mechanisms are usually present to some extent in spooled and field-installed telecommunications fiber.

6.2.2 BIREFRINGENCE AND PMD: SHORT FIBERS

When considering polarization effects in lightwave systems, one is primarily interested in the behavior of fiber spans of at least several hundred meters. Over such lengths, the perturbations illustrated in Fig. 6.2, which are typically present by accident, vary both in magnitude and in orientation from position to position along the length of a span.

To understand the ways in which polarization effects arise in lightwave systems, it is best to start by considering a short section of fiber within a long fiber span. Although it is assumed that the perturbations acting on the fiber span vary along its length, the particular section of fiber to be considered is assumed to be short enough that any perturbations acting on it are constant over its entire length. Any single-mode fiber span can be modeled as a concatenation of such uniformly perturbed fiber sections.

When subjected to a uniform perturbation such as one of the perturbations shown in Fig. 6.2, a single-mode fiber becomes bimodal owing to a loss of degeneracy for the two HE_{11} modes. This can be expressed as a difference in the local propagation constants for the modes:

$$\beta_s - \beta_f = \frac{\omega n_s}{c} - \frac{\omega n_f}{c} = \frac{\omega \Delta n_{eff}}{c}, \tag{6.1}$$

where $\omega = 2\pi c/\lambda$ is the angular frequency of the light, c is the speed of light in vacuum, n_s and n_f are the effective indices of refraction for the slow and fast modes, and $\Delta n_{eff} = n_s - n_f$ is the differential index of refraction. By definition, $n_s > n_f$.

The difference in propagation constants expressed in Eq. (6.1) is referred to as *birefringence* and is often specified by the differential index, Δn_{eff}. Under normal conditions, Δn_{eff} is much smaller than the index difference between core and cladding in telecommunications fiber, the latter typically having a value of $\sim 3 \times 10^{-3}$, while values of Δn_{eff} range between 10^{-7} and 10^{-5}.

All the perturbations shown in Fig. 6.2, except fiber twist, create an orthogonal set of two linearly polarized waveguide modes having electric field vectors aligned with the symmetry axes of the fiber as shown at the bottom of Fig. 6.1. In the case of fiber twist, shear stresses in the fiber couple components of the optical field that are $\pi/2$ out of phase, which leads to left- and right-hand circularly polarized modes (Ulrich and Simon 1979; Frigo 1986). In general, the state of polarization of an arbitrary optical field can be represented by the vector sum of field components aligned with the two polarization modes.

Figure 6.3 illustrates how birefringence affects the polarization of a lightwave as it propagates down a fiber. The figure shows the special case of an input wave linearly polarized at 45° to the axes of a linearly birefringent fiber. In this situation both polarization modes are equally excited. Phase slippage between orthogonal components of the field causes the polarization to evolve in a cyclic fashion from the initial linear state through various elliptical states, returning after a length L_B to its original linear polarization. This cyclic evolution continues as the light travels down the fiber.

The characteristic length L_B for this oscillation is referred to as the *beat length* of the fiber and is often used as an alternative measure of fiber birefringence. The beat length is directly related to the differential index, Δn_{eff}, by

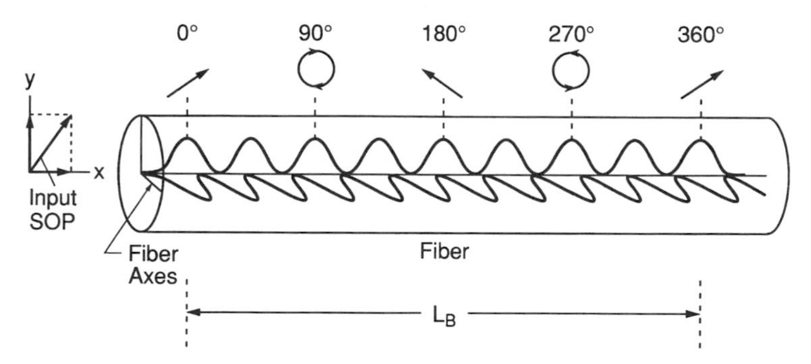

Fig. 6.3 Spatial evolution of polarization caused by uniform birefringence. SOP, state of polarization.

$$L_B = \frac{\lambda}{\Delta n_{eff}}. \tag{6.2}$$

Figure 6.3 illustrates one possible evolution of the polarization in a birefringent fiber. In general, the particular states of polarization that a lightwave cycles through are determined by both the state of polarization launched into a fiber and the orientation and type of fiber birefringence. In telecommunications applications these factors are typically arbitrary and unknown.

The differential phase velocity indicated by Eq. (6.1) is usually accompanied by a difference in the local group velocities for the two polarization modes, as illustrated in Fig. 6.4. This differential group velocity, which can limit the bandwidth of a fiber by broadening pulses, is described by a group-delay time per unit length between the two modes and is obtained by taking the frequency derivative of the propagation constants of Eq. (6.1):

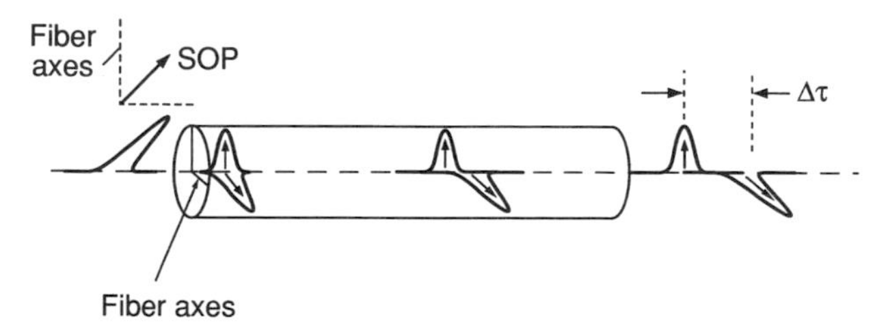

Fig. 6.4 Time-domain effect of polarization mode dispersion (PMD) in a short fiber.

$$\frac{\Delta\tau}{L} = \frac{d}{d\omega}(\beta_s - \beta_f) = \frac{\Delta n_{eff}}{c} - \frac{\omega d\Delta n_{eff}}{c \ d\omega}. \tag{6.3}$$

The quantity $\Delta\tau/L$ is referred to as the PMD of a fiber and is usually expressed in units of picoseconds per kilometer of fiber length. The linear length dependence of PMD applies to short fiber lengths where the birefringence can be assumed to be uniform. As is discussed in the next section, PMD in long fiber spans has a square root of length dependence and is usually quoted in units of picoseconds per square-root kilometer. To distinguish between the short- and long-length PMD parameters, the parameter defined by Eq. (6.3) is often referred to as the *short-length*, or *intrinsic*, PMD of a fiber.

The differential group delay, $\Delta\tau$, of Eq. (6.3) is the time-domain manifestation of PMD. There is also a frequency-domain manifestation that is a direct consequence of the frequency dependence of the differential phase velocities indicated in Eq. (6.3). In the frequency domain, PMD causes the state of polarization at the output of a fiber to vary with frequency (wavelength) for a fixed input polarization (Eickhoff, Yen, and Ulrich 1981). This occurs in a cyclic fashion, as shown in Fig. 6.5. The evolution of the state of polarization is the result of frequency-dependent phase slippage between orthogonal components of the output field. When displayed on the Poincaré sphere, the polarization at the output moves on a circle on the surface of the sphere as the optical frequency is varied, ultimately returning to the starting polarization after a characteristic frequency shift, $\Delta\omega_{cycle}$. This characteristic frequency is an alternative measure of PMD and is related to the differential delay time defined in Eq. (6.3) as follows:

$$\Delta\omega_{cycle} = \frac{2\pi}{\Delta\tau}. \tag{6.4}$$

Using Eq. (6.4) allows measurement of polarization evolution in the frequency domain to be directly related to the time-domain effects of PMD.

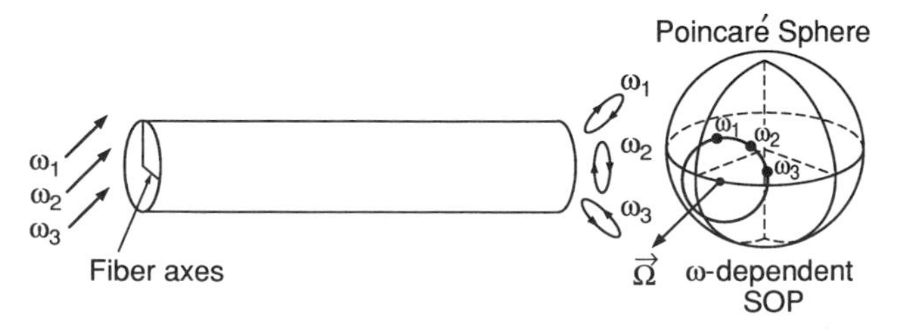

Fig. 6.5 Frequency-domain effect of PMD in a short fiber.

Table 6.1 **Effect of Perturbations: Single-Mode Fiber, λ = 1550 nm**

Perturbation	*Birefringence* (Δn_{eff})	*PMD* (*ps/km*)	*Scaling*
1% Core ellipticity	5×10^{-8}	1.5	Proportional to ellipticity
1 g-wt/cm Lateral stress	7×10^{-8}	0.23	Proportional to stress
10-cm Radius bend	5×10^{-8}	0.17	$1/R^2$
1 Turn/m twist	2.3×10^{-7}	~0	Proportional to twist rate

This relationship is the basis for a number of PMD measurement techniques, as discussed in Section 6.4.

Understanding the relationship between the perturbations illustrated in Fig. 6.2 and the magnitude of birefringence and PMD that results in a fiber has been the subject of much early research into optical fiber properties (for reviews of this subject, see Kaminow 1981; Sakai and Kimura 1981; Rashleigh 1983). Table 6.1 summarizes, by means of examples, the results of this work for the case of standard telecommunications fiber operating at 1550 nm. The magnitude of the perturbations illustrated are typical of what might be expected for field-deployed fiber.

6.2.3 POLARIZATION EFFECTS IN LONG FIBER SPANS

With the development of the erbium-doped fiber amplifier, optical path lengths in lightwave systems now range beyond 10,000 km. The numerical examples shown in Table 6.1 suggest that over such span lengths even the smallest perturbation on a fiber could create PMD-induced dispersion greater than 1 ns. Such dispersion would be a serious limitation to long-distance digital transmission. Indeed, early measurements of PMD in short pieces of fiber seemed to indicate that PMD might be an important limitation in spans as short as 100 km (Rashleigh and Ulrich 1978).

Fortunately, polarization effects do not accumulate in long fiber spans in a linear fashion. Instead, because of random variations in the perturbations along a fiber span, the effects of one section of a fiber span may either add to or subtract from the effects of another section. As a result, PMD in long fiber spans accumulates in a random-walk-like process that leads to a square root of length dependence and lower time dispersion than would be predicted from Table 6.1.

The lower dispersion in long spans comes at a price, however. Owing to the temperature dependence of many of the perturbations that act on a fiber, the transmission properties typically vary with ambient temperature. In practice, this manifests as a random, time-dependent drifting of the state of polarization at the output of a fiber or, in the case of PMD, as random fluctuations in the bandwidth of a fiber. To evaluate the polarization properties of long fiber spans, one is thus forced to adopt a statistical approach.

An important parameter for distinguishing between the short-length regime, where polarization effects are deterministic, and the long-length regime, where they become statistical is the correlation length, l_c. This parameter is also sometimes referred to as the *coupling length.* In the context of polarization-maintaining fibers, the reciprocal of l_c is typically used and is referred to as the *h parameter* (Kaminow 1981). Figure 6.6 shows how this parameter is defined.

One imagines a large population of uniformly birefringent fibers all subjected to the same random perturbations (in a statistical sense). One such fiber is illustrated in Fig. 6.6. Into each fiber in this population a lightwave is launched such that only one of the two local polarization modes is excited at the input.

As the lightwave propagates down the fiber, it initially remains in the starting polarization mode and thus retains its linear polarization state. Eventually, however, the state of polarization evolves away from the initial linear state as a result of power leaking over to the other polarization mode. This leakage of power occurs because of variations in the birefringence along the fiber caused by the random perturbations.

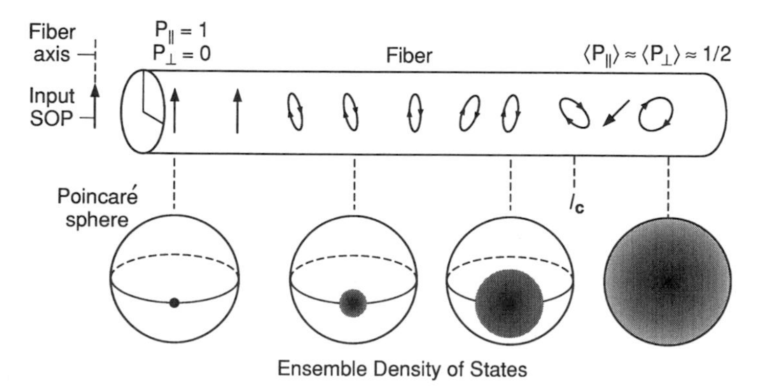

Fig. 6.6 Decorrelation of polarization in long fibers.

The particular state of polarization that would be observed at some distance from the input in any given fiber would depend on the details of the perturbations acting on the fiber and would be different for each fiber in the population under consideration. However, if one were to *average* the amount of optical power that had leaked to the orthogonal state over all the fibers in the population, one would find that this average power would grow with distance from the input until, at very large distances, the average power in the two polarization modes was approximately the same. This equalizing of the average power in the two modes reflects the fact that at large distances from the input, all polarization states are equally likely to be observed.

The loss of correlation between the input and output polarization states can be graphically depicted by plotting the states of polarization observed in the population of fibers as points on the surface of the Poincaré sphere. As shown in the lower part of Fig. 6.6, the states of polarization diffuse over the surface of the sphere with length, until at long lengths the surface of the sphere is uniformly covered.

The correlation length, l_c, is defined to be the length at which the average power in the orthogonal polarization mode, $P_\perp$, is within $1/e^2$ of the power in the starting mode $P_\|$ (Kaminow 1981) — i.e.,

$$\frac{\langle P_\|(l_c)\rangle - \langle P_\perp(l_c)\rangle}{P_{total}} = \frac{1}{e^2}. \tag{6.5}$$

This phenomenological definition of correlation length makes no assumption about the details of the perturbations acting on the fibers, only that they act on average the same on all fibers and at every position along the length of any one fiber.

Fibers whose length is short compared with l_c are considered to be in the short-length regime, in which transmission properties are free from statistical variation. Such fibers also exhibit PMD that grows linearly with length. Fibers whose length is long compared with l_c are considered to be in the long-length regime. These fibers show statistical variations in their polarization properties and a square root of length dependence for PMD. Figure 6.7 illustrates the different behavior of fibers in the two length regimes when physical parameters, such as fiber length, temperature, and wavelength, are varied.

Measuring the correlation length directly in telecommunications fiber is difficult owing to the need for a large population of statistically equivalent fiber. An approach that gets around this problem uses spectral averaging

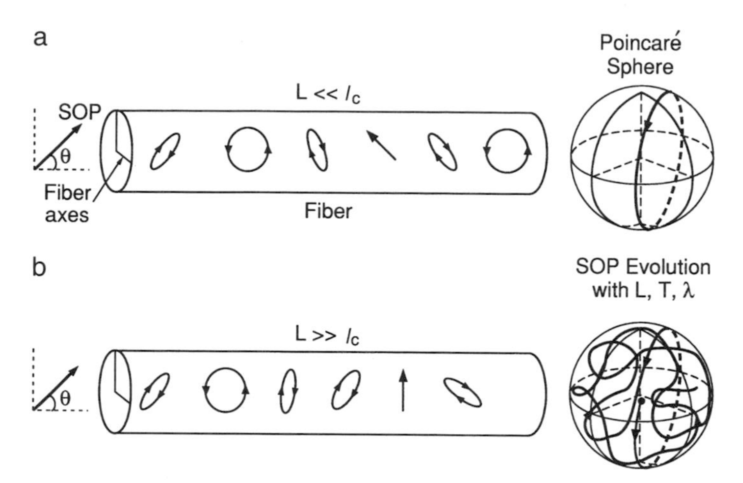

Fig. 6.7 Output polarization evolution for (a) short and (b) long fibers under varying length, temperature, or wavelength.

instead of ensemble averaging to measure l_c in a single fiber (Rashleigh *et al.* 1982). This approach has been applied to the measurement of l_c in conventional single-mode fiber (Burns, Moeller, and Chen 1983; Poole 1989; de Lignie, Nagel, and van Deventer 1994; Galtarossa *et al.* 1996). These measurements indicate that the correlation length is extremely sensitive to the way in which the fiber is deployed, with measured values ranging from less than 1m for fiber on a spool, up to more than 1 km for cabled fiber. This strong dependence of the correlation length on how the fiber is deployed is attributed to the differing magnitudes and spatial frequencies of the perturbations that act on fiber in different modes of deployment (Kaminow 1981; Wai and Menyuk 1994).

6.3 PMD in Systems

PMD in long fiber spans is a special case of the general problem of modal dispersion in a multimode waveguide subjected to random perturbations. In finding a solution to this problem, one adopts either a high-coherence model or a low-coherence model. The latter is usually referred to as the *coupled-power model* (Personick 1971). The most widely used high-coherence model, which was developed specifically for dealing with PMD in single-mode fiber, is the *principal states model* (Poole and Wagner 1986).

6.3.1 COUPLED-POWER MODEL

The problem is illustrated schematically in Fig. 6.8. The two modes of a two-mode waveguide are assumed to have different group velocities. A pulse launched into one of the two modes at the input encounters at some point in the waveguide a perturbation that causes a small amount of power to couple over to the other mode. In effect, the pulse is split into two pulses at that point. In the case of PMD, this splitting occurs because of a change in the local birefringence caused by one of the perturbing mechanisms shown in Fig. 6.2.

The two pulses continue down the waveguide, propagating with different velocities until they arrive at the next perturbation, where both pulses become split and thus form four pulses. This splitting of the pulses continues as more and more perturbations are encountered until there are a large number of small pulses propagating down the waveguide.

Because the relative distance traveled in the two modes is different for each pulse, the arrival times of the pulses at the output are different. The result is that the optical energy of the input pulse becomes dispersed in time at the output.

To this point, no assumption has been made regarding the relative coherence of the individual pulses at the output of the waveguide. However, to determine the net pulse shape at the output, one must decide whether to add the pulses coherently or incoherently. If one assumes that the pulses at the output are incoherent, the power at any instant in time is given by the sum of the *power* in the individual pulses. This low-coherence approach, which is the basis for the coupled-power model, predicts that in the long-length regime the net output pulse will be Gaussian in shape and broadened

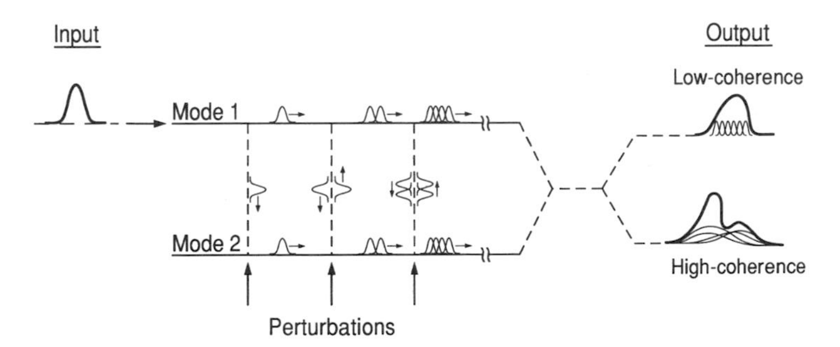

Fig. 6.8 Pulse propagation in a two-mode waveguide with random perturbations.

relative to the input pulse by an amount proportional to the square root of the waveguide length L and the differential group velocity ΔV of the two waveguide modes (Personick 1971):

$$\sigma_t = \frac{1}{2\Delta V}\sqrt{Ll_c}, \qquad L \gg l_c, \tag{6.6}$$

where σ_t is the root mean square (rms) broadening and l_c is the correlation, or coupling, length defined in the previous section.

The coupled-power model was first developed for multimode fibers and has proved useful in describing modal dispersion in these fibers when low-coherence sources such as light-emitting diodes (LEDs) are used (Marcuse 1972; Kawakami and Ikeda 1978). More recently the model has been applied to PMD in single-mode fiber (Gisin, Von der Weid, and Pellaux 1991). Its usefulness for describing PMD in lightwave systems is limited, however, owing to the relatively high coherence of the sources that are used and the small delay times typical of PMD.

When the source coherence is high, the small pulses at the output of the waveguide of Fig. 6.8 interfere coherently. As a result, the shape of the net pulse depends on the relative phase of the constituent pulses and will be extremely different from Gaussian, as shown in the lower right of Fig. 6.8. In this situation the coupled-power model can predict only the *average* pulse shape at the output of a waveguide. It cannot predict the actual pulse shape that might be observed in a given waveguide, or how the pulse shape might change when the state of polarization or the wavelength of the source is changed.

6.3.2 *PRINCIPAL STATES MODEL*

To address the problem of PMD in lightwave systems, a high-coherence model was developed based on the concept of *principal states of polarization* (Poole and Wagner 1986). The model assumes that the optical loss in the fiber does not depend on polarization and that the coherence time of the source is greater than the PMD-induced time shifts involved. For a digital lightwave system, this latter assumption is equivalent to assuming that the net time delay caused by PMD in the span is small compared with the bit period. The model thus addresses the regime in which most lightwave systems are expected to operate.

Under the stated assumptions, the transmission properties of any linear birefringent fiber can be represented by a frequency-dependent transmission matrix of the following form (Jones 1941):

$$\mathbf{T}(\omega) = e^{\alpha(\omega)}\begin{bmatrix} u_1(\omega) & u_2(\omega) \\ -u_2^*(\omega) & u_1^*(\omega) \end{bmatrix}, \tag{6.7}$$

where $\alpha(\omega)$, $u_1(\omega)$, and $u_2(\omega)$ are complex quantities, and the latter two satisfy the relation

$$|\, u_1 \,|^2 + |\, u_2 \,|^2 = 1. \tag{6.8}$$

The complex electric field vector $\mathbf{E}_2(\omega)$ of an optical wave at the output of a fiber is related to the input field vector $\mathbf{E}_1(\omega)$ by

$$\mathbf{E}_2(\omega) = \mathbf{T}(\omega)\mathbf{E}_1(\omega). \tag{6.9}$$

The form of $\mathbf{T}(\omega)$ shown in Eq. (6.7) comes directly from the assumption that the loss through the fiber does not vary with polarization.

The principal states model is based on the observation that for any transmission matrix $\mathbf{T}(\omega)$ there exists at every frequency an orthogonal pair of input *principal states* of polarization. These states have the property that when an input is aligned with one of them the corresponding *output* polarization is invariant to first order with changes in frequency.

An immediate consequence of the property of polarization invariance with frequency is that an optical pulse that is aligned with a principal state at the input of a fiber will emerge at the output with its spectral components all having the same state of polarization. This in turn implies that the only distortion on the pulse that can occur is a pure phase distortion, which, to first order, does not change the shape of the pulse but only shifts it in time. An optical pulse that is aligned with a principal state at the input of a fiber thus emerges both polarized and unchanged in shape to first order.

Because the principal states form an orthogonal pair, any input can be expressed as the vector sum of two components, each of which is aligned with a principal state. Thus, provided the source spectrum satisfies the narrow band assumption, the time-varying output electric field vector $\mathbf{E}_2(t)$ of a randomly birefringent fiber will have the general form (Poole and Giles 1988)

$$\mathbf{E}_2(t) = c_+\hat{\boldsymbol{\varepsilon}}_+\mathrm{E}_1(t + \tau_+) + c_-\hat{\boldsymbol{\varepsilon}}_-\mathrm{E}_1(t + \tau_-), \tag{6.10}$$

where $\mathrm{E}_1(t)$ is the time-varying input field, c_+ and c_- are complex weighting coefficients, and $\hat{\boldsymbol{\varepsilon}}_+$ and $\hat{\boldsymbol{\varepsilon}}_-$ are unit vectors specifying the output polarization states of the two components. The latter two states are referred to as the *output principal states*.

The difference in arrival time, $\Delta\tau = \tau_+ - \tau_-$, for the two component pulses on the right side of Eq. (6.10) gives rise to pulse broadening at the output of a fiber when energy is split between the two principal states at the input. In general, the amount of broadening depends on the power-splitting ratio between the pulses determined by the coefficients c_+ and c_-, as well as the differential delay, $\Delta\tau$.

The surprising and counterintuitive feature of the principal states model, when one compares Eq. (6.10) and Fig. 6.8, is that it predicts a simple splitting of an input pulse into two pulses at the output of a randomly perturbed waveguide as a consequence of the coherent superposition of an essentially *infinite* number of smaller pulses. This comes about because the phases of the constituent pulses shown in Fig. 6.8 are constrained by the requirement that the total output pulse energy must equal the input pulse energy in a lossless waveguide.

Equation (6.10) can be derived directly by expanding in frequency the transmission matrix $\mathbf{T}(\omega)$ and truncating the expansion after the first-order terms (Poole and Giles 1988). In this sense, Eq. (6.10) expresses the "first-order" effects of PMD. It can also be viewed as a generalization of the pulse bifurcation caused by PMD in short fibers (see Fig. 6.4). In short fibers the principal states simply correspond to the polarization modes of the fiber. In long fiber spans, however, these states are determined by the cumulative effects of the birefringence over the entire span. In either case the differential delay time, $\Delta\tau$, is related to the matrix elements of $\mathbf{T}(\omega)$ as follows (Poole and Wagner 1986):

$$\Delta\tau = 2\sqrt{|u_1'|^2 + |u_2'|^2}, \tag{6.11}$$

where the primes denote differentiation with respect to angular frequency.

Figure 6.9 shows the first time-domain observation confirming the principal states model. The figure shows three waveforms observed at the output of 10 km of spooled single-mode fiber. The waveforms marked $\hat{\boldsymbol{\varepsilon}}_-$ and $\hat{\boldsymbol{\varepsilon}}_+$ correspond to the input being aligned with the two principal states, whereas the intermediate waveform was obtained when the pulse power was equally split between the two principal states. The latter pulse was slightly broadened, as indicated by an approximately 10% reduction in the pulse height.

An important difference between the behavior of short fibers and long fiber spans is that the dispersion in long fiber spans is highly sensitive to ambient temperature changes. This can be thought of as arising from small changes in the relative phases of the constituent pulses shown in Fig. 6.8.

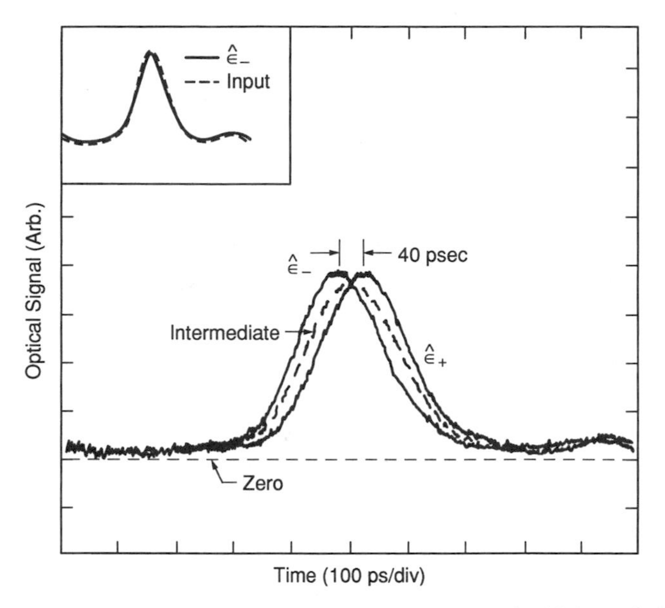

Fig. 6.9 Measured differential delay time due to PMD in 10 km of dispersion-shifted fiber. (Reproduced from Poole, C. D., and C. R. Giles. 1988. *Opt. Lett.* 13:155–157, with permission.)

As a result, the net dispersion and the principal states themselves change slowly with time. This effect is illustrated in Fig. 6.10, where the measured differential delay time between the principal states in the fiber of Fig. 6.9 is shown to change by a factor of 2 during a 24-h period.

The generalization of pulse bifurcation to long fiber spans made possible by the principal states model extends as well to the frequency domain. In the frequency domain, polarization dispersion manifests as a frequency-dependent state of polarization at the output of a fiber (Andresciani *et al.* 1987; Bergano, Poole, and Wagner 1987; Poole *et al.* 1988). Figure 6.11 shows the evolution of the polarization at the output of a 147-km-long undersea fiber cable over a narrow-wavelength window for three input polarizations. As in short fibers, the output polarization undergoes a rotation on the Poincaré sphere that can be represented by the following differential equation (Eickhoff, Yen, and Ulrich 1981; Andresciani *et al.* 1987; Poole *et al.* 1988):

$$\frac{d\hat{\mathbf{s}}}{d\omega} = \boldsymbol{\Omega} \otimes \hat{\mathbf{s}}, \tag{6.12}$$

where $\hat{\mathbf{s}}$ is the unit Stokes vector describing the output polarization state.

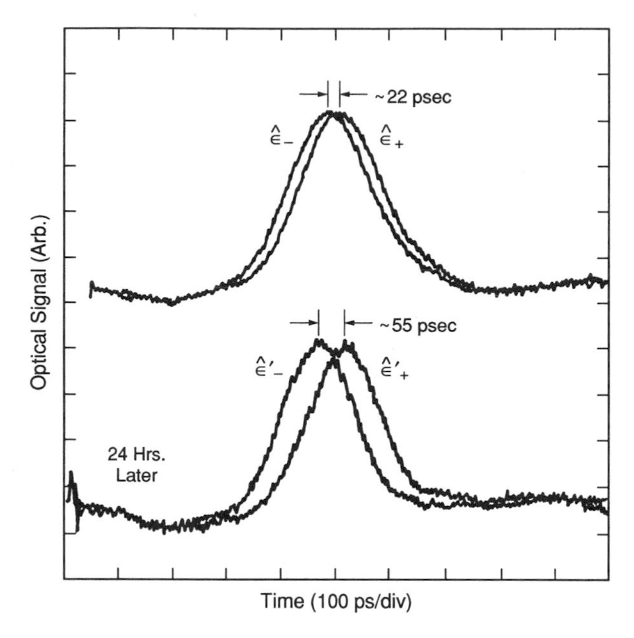

Fig. 6.10 Drift of differential delay time during a 24-h period.

The rotation vector, **Ω,** describes the rate of rotation and is usually referred to as the *dispersion vector*. The magnitude of this vector is equal to the differential delay time:

$$|\,\mathbf{\Omega}\,| = \Delta\tau. \tag{6.13}$$

The direction of the vector **Ω** defines an axis whose two intercepts with the surface of the Poincaré sphere correspond to the two principal states of polarization at the fiber output.

The polarization-dispersion vector provides the most compact representation of PMD and can be related to the microscopic birefringence in a fiber through the vector equation (Poole, Winters, and Nagel 1991)

$$\frac{\partial \mathbf{\Omega}}{\partial z} = \frac{\partial \mathbf{W}}{\partial \omega} + \mathbf{W} \otimes \mathbf{\Omega}, \tag{6.14}$$

where z represents the position along the fiber and **W** is a three-dimensional vector representing the local birefringence of the fiber (Eickhoff, Yen, and Ulrich 1981).

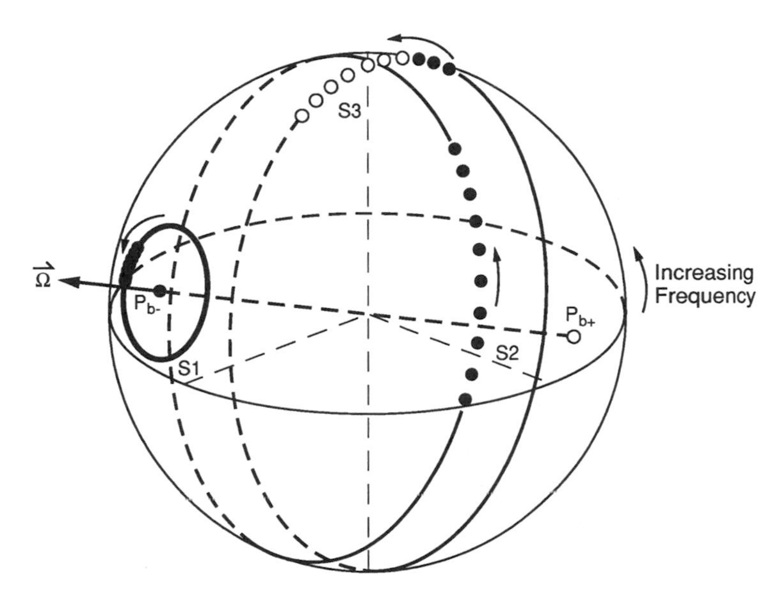

Fig. 6.11 Output polarization variation with optical frequency in 147-km-long cable for three input polarizations. The data points are separated by 2 GHz. (Reprinted from Poole, C. D., *et al.* 1988. *IEEE J. Lightwave Tech.* LT-6:1185–1190. Copyright © 1988 IEEE.)

6.3.3 HIGHER ORDER DISPERSION EFFECTS

The principal states model is based on a frequency expansion of the transmission matrix of a fiber. The attractiveness of the model is that it allows one to characterize the first-order effects of PMD both in the time domain and in the frequency domain by a single observable parameter, namely the differential delay time, $\Delta\tau$. However, the model need not be limited to first-order effects. In general, the differential delay time and the dispersion vector, $\mathbf{\Omega}$, are themselves frequency dependent and may vary over the bandwidth of a source. This frequency dependence is a manifestation of higher order dispersion that must be considered when the coherence time of the source becomes comparable to the differential delay time (Poole, Winters, and Nagel 1991; Betti *et al.* 1991).

Studies of higher order dispersion have thus far been limited to second-order effects (Poole and Giles 1988; Foschini and Poole 1991). In the frequency domain, second-order dispersion manifests as a linear frequency dependence of the dispersion vector. In the time domain, second-order dispersion manifests as a linear frequency dependence in the polarization vectors and the delay items τ_+ and τ_- of Eq. (6.10). The latter effect acts

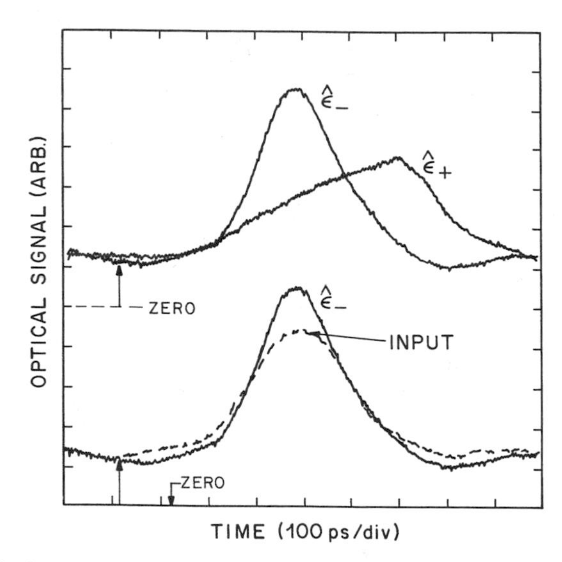

Fig. 6.12 Polarization-dependent pulse compression and broadening due to second-order PMD. The upper two pulses correspond to input polarization being aligned with two principal states. The lower two pulses show compression of the $\boldsymbol{\varepsilon}_-$ pulse relative to input. (Reproduced from Poole, C. D., and C. R. Giles. 1988. *Opt. Lett.* 13:155–157, with permission.)

as an effective chromatic dispersion that is opppsite in sign for the two principal states. For chirped pulses, this has the interesting consequence of leading to either compression or broadening of chirped pulses depending on the relative sign of the chirp and the dispersion (Poole and Giles 1988). This effect is illustrated in Fig. 6.12.

6.3.4 STATISTICAL TREATMENT OF PMD

Using the principal states model as a starting point, researchers of the statistical properties of PMD in long fiber spans have focused on the length dependence and probability density function for the differential delay time, $\Delta\tau$. The results of these studies have revealed that the average differential delay time has a square root of length dependence (Fig. 6.13) (Poole 1988; Curti *et al.* 1989; Tsubokawa and Ohashi 1991; Namihira and Wakabayashi 1991; Namihira, Kawazawa, and Wakabayashi 1992), and that the probability density function is Maxwellian (Fig. 6.14) (Curti *et al.* 1990; Poole, Winters, and Nagel 1991; De Angelis *et al.* 1992; Gisin *et al.* 1993). Both

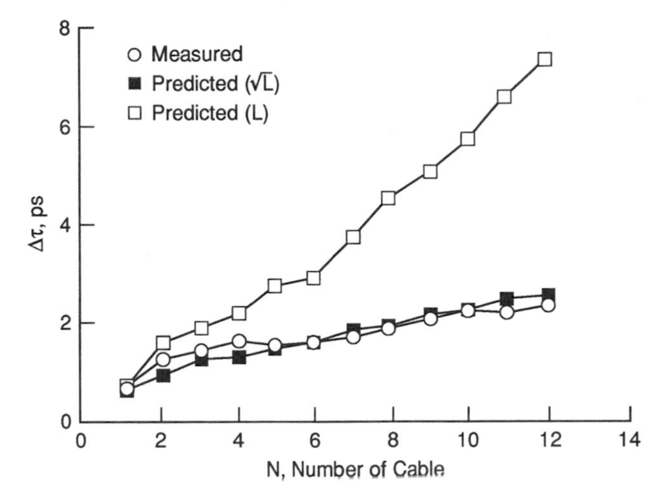

Fig. 6.13 Length dependence of PMD in a concatenated cable span (Reproduced from Curti *et al.* 1989. Concatenation of polarisation dispersion in single-mode fibres. *Electron. Lett.* 25:290–292, with permission of IEE.)

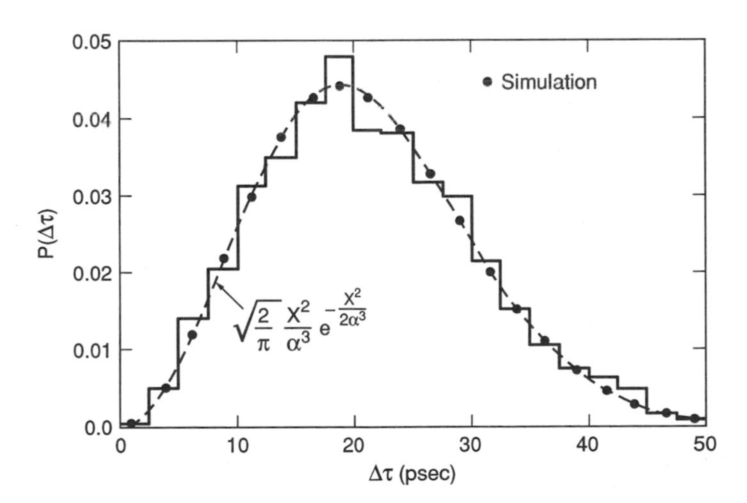

Fig. 6.14 Measured distribution of differential delay times in 10 km of spooled fiber subjected to varying temperature. The *dashed curve* is a Maxwellian fit. (Reproduced from Poole, C. D., J. H. Winters, and J. A. Nagel. 1991. *Opt. Lett.* 16:372–374, with permission.)

these results can be obtained theoretically by modeling the PMD using the three-dimensional dispersion vector, $\mathbf{\Omega}$, and allowing this vector to grow in a random-walk-like process along the fiber.

Alternatively, Eq. (6.14) can be solved explicitly while treating the local birefringence as a stochastic process (Foschini and Poole 1991). This approach allows the PMD of a long span to be related to the statistical properties of the local fiber birefringence. For example, a fiber with uniform birefringence $\Delta\beta$ subjected to random perturbing birefringences, has mean square differential delay time between the principal states in the long-length regime given by (Poole 1988; Foschini and Poole 1991)

$$\langle \Delta\tau^2 \rangle = \left(\frac{d\Delta\beta}{d\omega} \right)^2 L l_c, \qquad L \gg l_c, \tag{6.15}$$

where L is the fiber length and l_c is the correlation length defined previously. The frequency derivative of the birefringence in Eq. (6.15) is just the intrinsic PMD of the fiber (see Eq. [6.3]). This result has been generalized to the strong coupling limit as well (Menyuk and Wai 1994).

Comparison of Eqs. (6.15) and (6.6) shows that the principal states model and the coupled-power model results are related through a simple numeric constant:

$$\langle \Delta\tau^2 \rangle = 4\sigma_t^2, \qquad L \gg l_c. \tag{6.16}$$

As noted previously, the correlation length l_c in Eq. (6.15) is highly sensitive to mechanical stresses induced by spooling or cabling fiber. Because of the dependence of PMD on the correlation length, PMD is also dependent on such perturbations. As a result, a single fiber may show varying levels of PMD depending on whether it is spooled or cabled. Recent measurements indicate that cabling may tend to increase l_c and, thus, PMD levels by reducing the mechanical stresses relative to those on the spool (de Lignie, Nagel, and van Deventer 1994).

The solution of Eq. (6.14) has also revealed information about the statistics of higher order dispersion effects related to the frequency dependence of the dispersion vector. These results have proved useful in understanding PMD in analog systems where higher order dispersion is the primary source of system impairment (Poole and Darcie 1993).

6.3.5 IMPACT ON DIGITAL SYSTEMS

Digital lightwave systems rely on undistorted transmission of optical pulses through long lengths of fiber. Dispersive effects such as PMD cause pulse spreading and distortion and thus can lead to system penalties. As discussed already, PMD is a time-varying, stochastic effect; therefore, the system penalties are also time varying.

Figure 6.15 shows an experimental demonstration of PMD-induced variations in bit error rate performance in a digital lightwave system in which the transmission fiber was subjected to ambient temperature changes in a field environment. The correlation between rapid fluctuations in bit error rate and steep changes in ambient temperature during sunrise and sunset illustrates how performance variations related to PMD can be directly tied to environmental conditions.

A system designed with adequate margins for normal conditions may have unacceptable penalties under rare conditions of extremely high PMD. This complicates the design of lightwave systems because such rare events almost never occur under laboratory conditions and so cannot be characterized by measurements. Thus margin planning for digital systems, when PMD is a factor, relies to an unusual degree on theory and simulations.

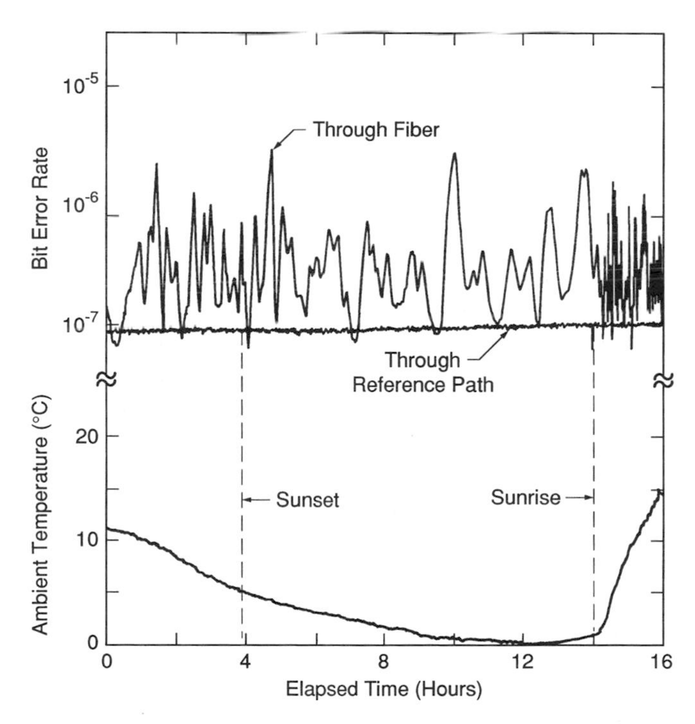

Fig. 6.15 Bit error rate fluctuations in a digital system caused by PMD and changing ambient temperature. (Reprinted from Poole, C. D., *et al.* 1991. *IEEE Phot. Tech. Lett.* 3:68–70. Copyright © 1991 IEEE.)

To obtain an estimate of the limitations imposed by PMD on digital systems, we assume that pulse bifurcation as described by Eq. (6.10) is the dominant mechanism for pulse broadening. Under small penalty conditions, the power penalty ε incurred by a non-return-to-zero (NRZ) digital system in which pulse energy is split into two components that are shifted in time relative to each other has the following form (Poole *et al.* 1991):

$$\varepsilon(\mathrm{dB}) \cong A \frac{\Delta\tau^2 \gamma(1 - \gamma)}{T^2}, \tag{6.17}$$

where $0 \leq \gamma \leq 1$ is the power-splitting ratio between the two components, T is the full width at half maximum of the optical pulse, and A is a dimensionless parameter that depends on the optical pulse shape and receiver filter characteristics. For NRZ pulses the bit rate is equal to the reciprocal pulse width, $B = 1/T$. Table 6.2 shows values of A for several pulse shapes determined by computer simulation using a fourth-order Bessel receiver.

Although Eq. (6.17) applies to only small penalty conditions, large penalty conditions have also been studied theoretically (Wagner and Elrefaie 1988; Winters *et al.* 1992; Iannone *et al.* 1993).

Equation (6.17) shows that the system penalty has a quadratic dependence on both the differential delay time and the bit rate. Note that the penalty goes to zero for the special case where all the power is in one of the component pulses (i.e., $\gamma = 0,1$), because no pulse broadening occurs in this case.

In real systems, the power penalty caused by PMD will vary in a random way owing to the random variation of the parameters γ and $\Delta\tau$. For the purpose of establishing a PMD "limit," it is stipulated that penalties in excess of 1 dB are unacceptable. By analogy with fading in radio systems, such penalty conditions are considered to be "outages." It is further stipu-

Table 6.2 **A Values for Various Pulse Shapes**

Pulse Shape	A
Gaussian	25
Raised cosine	22
Square	12
25% Rise–fall	15
Triangular	24

lated, somewhat arbitrarily, that the allowable probability for such an outage is less than 1 in 18,000. This probability corresponds to 4σ on a Gaussian distribution. In terms of fractional outage time, this probability translates to 30 min per year. It is important to note that in the following derivation no assumption is made about the duration of a single outage, only that the *cumulative* outage time must be less than 30 min per year.

The probability of observing a penalty greater than 1 dB is determined by the probability density function for ε. Assuming a Maxwellian distribution for $\Delta\tau$ and a uniform distribution of γ, and statistical independence for the two parameters, Eq. (6.17) leads to a simple exponential probability density function for the power penalty:

$$P(\varepsilon) = \eta e^{-\eta\varepsilon}, \tag{6.18}$$

where $\eta = 16T^2/A\pi\langle\Delta\tau\rangle^2$ and $<\Delta\tau>$ denotes the average differential delay.

The probability of observing a penalty greater than 1 dB is obtained by integrating Eq. (6.18) from 1 to infinity:

$$Prob\{\varepsilon \geq 1\} = \int_1^\infty \eta e^{-\eta\varepsilon} d\varepsilon = e^{-\eta}. \tag{6.19}$$

Figure 6.16 shows the distribution of power penalties obtained in a computer simulation of a digital system in which fiber PMD is modeled by a concatenation of 1000 uniformly birefringent fiber sections having randomly oriented fiber axes. Chirp-free Gaussian pulses were sent through the simulated PMD and detected by a receiver with a fourth-order Bessel-shape filter characteristic. The points in Fig. 6.16 show the logarithm of frequency of occurrence for penalties in 10,000 simulated fibers. The solid curve demonstrates the expected exponential dependence according to Eq. (6.18). Because the model used for PMD included all higher order dispersion effects, the agreement between the data in Fig. 6.16 and Eq. (6.18) demonstrates the validity of assuming that first-order effects are the dominant source of pulse broadening under small-penalty conditions.

To obtain a PMD "limit," the left-hand side of Eq. (6.19) is set equal to 1/18,000 in accordance with the 30 min per year "outage" criterion. This leads to the condition $\eta = 9.8$, or assuming Gaussian pulse shape with $A = 25$,

$$\left.\frac{\langle\Delta\tau\rangle}{T}\right|_{Limit} = 0.14. \tag{6.20}$$

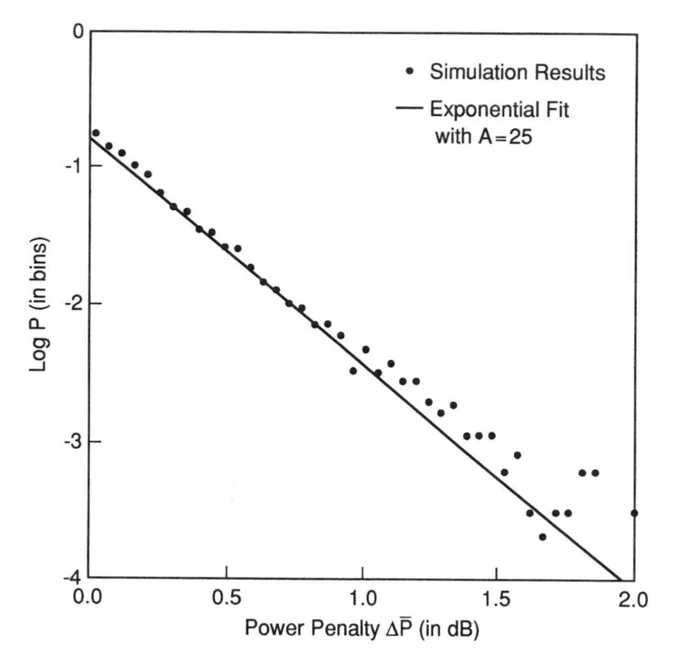

Fig. 6.16 Distribution of power penalties and exponential fit for a simulated digital system containing PMD.

Equation (6.20) indicates that for a digital system to avoid incurring a power penalty of 1 dB or greater for a fractional time of 30 min per year, the *average* differential delay time between the principal states must be less than 0.14 of the bit period.

To generate a bit-rate limit curve using Eq. (6.20), we make use of the length-normalized parameter $PMD = \langle \Delta\tau \rangle / \sqrt{L}$ and the bit rate $B = 1/T$ so that Eq. (6.20) becomes

$$B^2 L \approx \frac{0.020}{(PMD)^2}. \tag{6.21}$$

Figure 6.17 shows a plot of the PMD limit corresponding to Eq. (6.21). The limit curve shows, for example, that a system operating at a bit rate of 10 Gb/s over a span length of 100 km should have PMD of less than approximately 1.4 ps/$\sqrt{\text{km}}$. This corresponds to a total span PMD of 14 ps. For transoceanic systems where lengths can approach 10,000 km, PMD will have to be kept at less than 0.14 ps/$\sqrt{\text{km}}$ for the same bit rate.

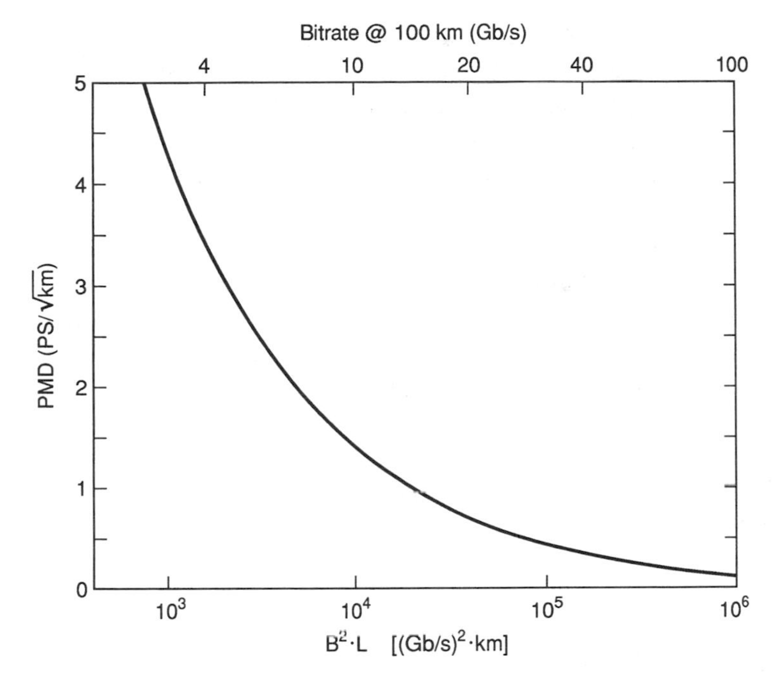

Fig. 6.17 PMD limit for a digital system using a chirp-free source.

It is important to note that the PMD limit curve shown in Fig. 6.17 is not a hard limit. A system operating at the PMD limit would operate normally for the majority of time. It is only during the rare PMD "outage" events that the system would experience significant performance degradation. In addition, systems using directly modulated laser sources will experience PMD impairment greater than that predicted for the simple model used here, owing to higher order dispersion effects that become more important as the spectral width of the modulated source increases (Poole *et al.* 1991).

6.3.6 IMPACT ON ANALOG SYSTEMS

The use of analog transmission techniques in lightwave systems, primarily in cable television (CATV) applications, has motivated studies on PMD-induced impairment in analog systems (Poole and Darcie 1993; Kikushima *et al.* 1994; Phillips *et al.* 1994; see Chapter 14 of Volume IIIA). Although span lengths and modulation frequencies are typically much less in analog systems than in digital systems, the stringent requirement on channel linearity in analog systems makes PMD as important a consideration for analog system designers as it is for digital system designers. As with digital systems, PMD-induced impairment in analog systems is characterized by random variations in system performance with time. The mechanisms by which PMD affects analog system performance, however, are significantly different from those affecting digital systems.

In analog systems, PMD causes impairment through several mechanisms, all of which require laser chirp. The most important of these mechanisms manifest as nonlinear distortion, which occurs when harmonics of the carrier frequency are unintentionally created in the analog channel. In CATV systems using subcarrier modulation techniques, harmonics are usually required to be 60 dB below the carrier to avoid noticeable degradation in the received signal.

The creation of harmonics in an analog system using directly modulated laser sources can be thought of as a two-step process. The first step is the conversion of the frequency-modulated (FM) signal emerging from the chirped laser source into an amplitude-modulated (AM) signal by means of some impairment. In the second step, the AM signal created by this FM-to-AM conversion beats with the carrier to create the undesirable harmonics. Sources of nonlinear distortion can thus be identified by identifying the mechanisms that create FM-to-AM conversion of an optical signal.

Figure 6.18 illustrates how FM-to-AM conversion can occur when PMD in a fiber span is combined with PDL. In analog systems one of the most common sources of PDL is from fiber splitters used at the output of fiber spans.

When light from a directly modulated laser source is passed through a fiber containing PMD, the frequency modulation of the light (i.e., chirp) is converted at the output to polarization modulation by the PMD in the fiber. This polarization modulation is then converted to amplitude modula-

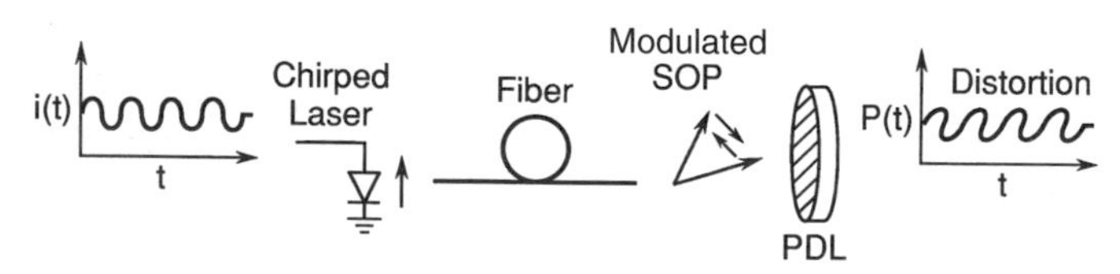

Fig. 6.18 Distortion caused by the combination of laser chirp, PMD, and polarization-dependent loss (PDL).

tion when the optical signal passes through a PDL element. The amplitude-modulated signal then beats with the carrier to create harmonics.

In long fiber spans, the average second-harmonic distortion that is created by this mechanism is related to the PMD in the span and the PDL as follows (Poole and Darcie 1993):

$$\langle \eta^{(2)} \rangle = \gamma^2 m_0^2 \pi \frac{\Delta T^2 \langle \Delta\tau \rangle^2}{192}, \qquad L \gg l_c, \tag{6.22}$$

where γ is the chirp parameter of the laser, m_0 is the modulation index, and ΔT is the PDL. In Eq. (6.22), $\eta^{(2)}$ is the ratio of RF power in the second harmonic to the RF power in the carrier and the product γm_0 is the optical frequency change caused by laser chirp.

A second mechanism that has been identified as leading to distortion in analog systems containing PMD is illustrated in Fig. 6.19. This mechanism relies on polarization-mode coupling to redistribute optical power between the modes (Nielsen 1982, 1983). As with the previous mechanism, frequency modulation at the source is converted to polarization modulation in the fiber. The modulating polarization causes the time delay through the fiber to also be modulated as a result of a redistribution of the optical power between the fast and slow axes. This restribution of power occurs at points where there is polarization-mode coupling caused by perturbing birefringences. Frequency modulation at the source is thus converted to a modulation of the time of flight through the fiber. This in turn causes the optical energy to become bunched in time and leads to amplitude modulation at the output. This effect is analogous to FM-to-AM conversion caused by chromatic dispersion in fibers (Meslener 1984; Chraplyvy *et al.* 1986; Phillips *et al.* 1991). The resulting amplitude-modulated signal beats with the carrier to create the undesirable harmonics.

In long fiber spans the average second-harmonic signal power created by this mechanism is given by the following equation (Poole and Darcie 1993):

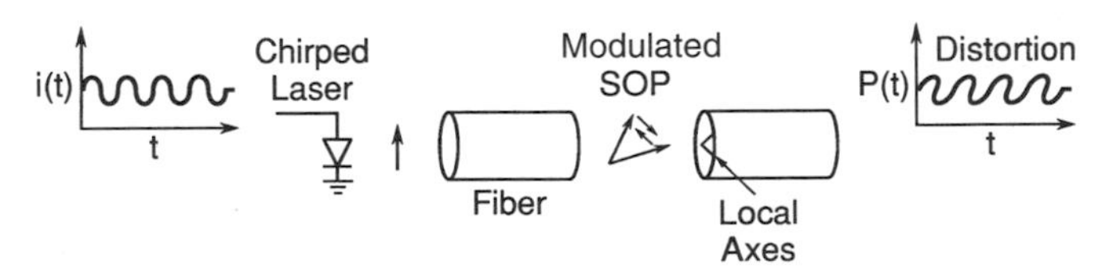

Fig. 6.19 Distortion caused by the combination of laser chirp, PMD, and polarization-mode coupling.

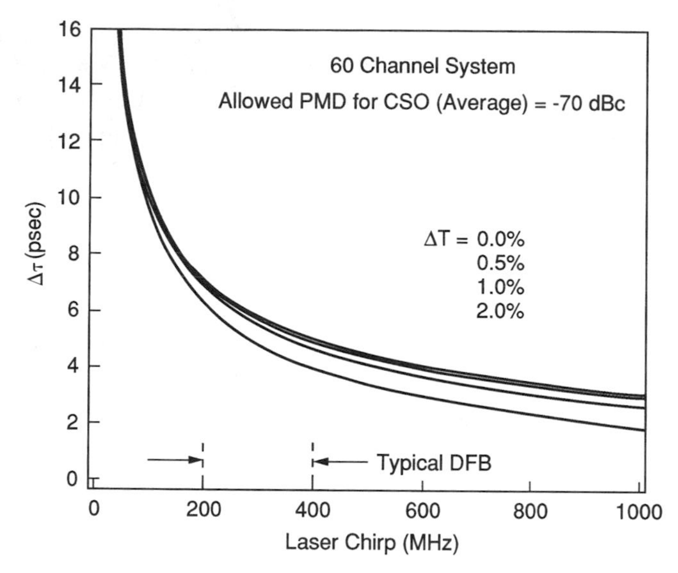

Fig. 6.20 PMD limits in a 60-channel analog system versus laser chirp for several levels of PDL. CSO, composite second-order (distortion); DFB, distributed feedback. (Reprinted from Poole, C. D., and T. E. Darcie. 1993. *IEEE J. Lightwave Tech.* LT-11:1749–1759. Copyright © 1993 IEEE.)

$$\langle \eta^{(2)} \rangle = \gamma^2 m_0^2 \omega_m^2 \pi^2 \frac{\langle \Delta\tau \rangle^4}{256}, \qquad L \gg l_c, \tag{6.23}$$

where ω_m is the carrier frequency.

Figure 6.20 shows the system requirements to achieve an average composite second-order distortion less than −70 dB for a 60-channel analog system based on Eqs. (6.22) and (6.23). The family of curves correspond to different levels of PDL. The figure shows that a 60-channel system should have an average span PMD less than about 7 ps in the absence of any PDL. A fiber splitter placed at the output of the span having 2% (0.1-dB) polarization-dependent transmission would lower this value to 4 ps.

6.3.7 *REDUCING THE EFFECTS OF PMD*

Strategies for reducing the effects of PMD in lightwave systems have taken on new importance as it has become evident that some embedded commercial fiber optic cables may have PMD values unacceptable for certain network upgrade scenarios (Gisin, Perny, and Passy 1991; Passy *et al.* 1991; de Lignie, Nagel, and van Deventer 1994; Galtarossa *et al.* 1996). In addition, recent studies have indicated the importance of cabling effects in producing elevated levels of PMD in cabled fiber relative to spooled fiber (Galtarossa and Schiano 1993; de Lignie, Nagel, and van Deventer 1994; Galtarossa *et al.* 1996).

Early strategies for reducing PMD in fiber focused on reducing the intrinsic PMD of the fiber by altering the manufacturing process (Norman *et al.* 1979; Chiang 1985a,b; Barlow, Ramskov-Hansen, and Payne 1981; Vengsarkar *et al.* 1993). This has led to exceptionally low and stable values of PMD in the new generation of single-mode fiber being manufactured (Judy 1994).

More recent efforts are just beginning to examine ways in which lightwave systems can be designed to better accommodate otherwise unacceptable levels of PMD and thus make full use of the existing embedded fiber base. Such strategies include reduction of intersymbol interference by electronic equalization in the receiver (Winters and Gitlin 1990; Winters and Santoro 1990; Winters and Kasturia 1992) and optical equalization techniques that employ automatic polarization control at the receiver or transmitter (Winters *et al.* 1992; Takahashi, Imai, and Aiki 1994).

6.4 Measurement of PMD

PMD in telecommunications fiber is measured using a variety of techniques. The majority of these techniques seek to measure, either in the time domain or in the frequency domain, the differential delay time, $\Delta\tau$, between the principal states of polarization.

Early measurements of PMD in single-mode fiber focused on short fiber samples. More recently, the focus has shifted to measuring fibers that fall in the long-length regime, because this is the regime of interest for telecommunications applications. Because of the statistical nature of PMD in the long-length regime, most measurement techniques are designed to measure the average or rms value of $\Delta\tau$ in a given fiber. For fibers in the long-length regime, these two quantities are related by (Curti *et al.* 1990)

$$\Delta\tau_{rms} = \sqrt{\langle\Delta\tau^2\rangle} = \sqrt{\frac{3\pi}{8}}\langle\Delta\tau\rangle. \tag{6.24}$$

The averages indicated in Eq. (6.24) are usually obtained by varying either source wavelength or fiber temperature to sweep the fiber through varying differential delays. The usual assumption is that varying wavelength provides the same average value as varying fiber temperature, and that these averages are the same as ensemble averages. Theoretical and experimental studies seem to support this assumption (Rashleigh *et al.* 1982; Gisin 1991; De Angelis *et al.* 1992; Menyuk and Wai 1994).

Table 6.3 summarizes the distinguishing characteristics of the most common techniques used for measuring PMD in telecommunications fiber. These techniques are described in more detail next.

6.4.1 PULSE DELAY MEASUREMENT

A simple, direct method to measure PMD is to launch short pulses into a fiber, vary the input polarization state, and measure the maximum differential time of flight using an oscilloscope triggered by the same clock source as the input pulses. This method of measurement was used to obtain the waveforms shown in Figs. 6.9 and 6.10 of Section 6.3.2.

Because short pulses are required to resolve small differential group delays, this method works best with large values of PMD. Although direct and simple, the method is complicated by the need to search the range of input polarization states to find the two principal states. Thus, this method is primarily used to determine discrete values of $\Delta\tau$ at specific wavelengths, rather than for measuring average dispersion.

6.4.2 INTERFEROMETRIC METHOD

One of the first techniques to be used for measuring PMD in fibers uses the classic Michaelson interferometer (Mochizuki, Namihira, and Wakabayashi 1981; Thevenaz *et al.* 1989; Gisin, Von der Weid, and Pellaux 1991; Thevenaz, Nikles, and Robert 1992; Namihira, Nakajima, and Kawazawa 1993). Figure 6.21 shows an example of an apparatus used for measuring PMD in which the beam splitter of the classic interferometer is replaced by a polarization-maintaining fiber coupler. In this setup broadband light is sent through the test fiber and then into the interferometer. A fringe pattern is generated at the detector by changing the position of a movable mirror in one of the interferometer arms.

The visibility of the fringe pattern as a function of mirror position provides information on the PMD of the test fiber. Fiber samples that fall in the short-length regime exhibit a fringe pattern that has two peaks corresponding to the two polarization modes of the fiber. The separation of these peaks is a direct measure of the differential delay time for the two polarization modes.

Table 6.3 **Comparison of PMD Measurement Techniques**

Technique	*Time or Frequency Domain*	*Source*	*Measures*	*Higher Order Dispersion*	*Measurement Range (ps)*
Pulse delay	Time	Chirp-free laser	$\Delta\tau$	No	~10–>1000
Interferometric	Time	Broadband	$<\Delta\tau>$	No	~0.002–100
RF response	Time	Chirp-free laser	$\Delta\tau$, $<\Delta\tau>$	No	~25–>1000
Poincaré sphere	Frequency	Tunable laser	$\Delta\tau$, $<\Delta\tau>$	Yes	~0.002–>1000
Jones matrix	Frequency	Tunable laser	$\Delta\tau$, $<\Delta\tau>$	Yes	~0.002–>1000
Fixed analyzer	Frequency	Broadband or tunable laser	$<\Delta\tau>$	No	~0.1–100

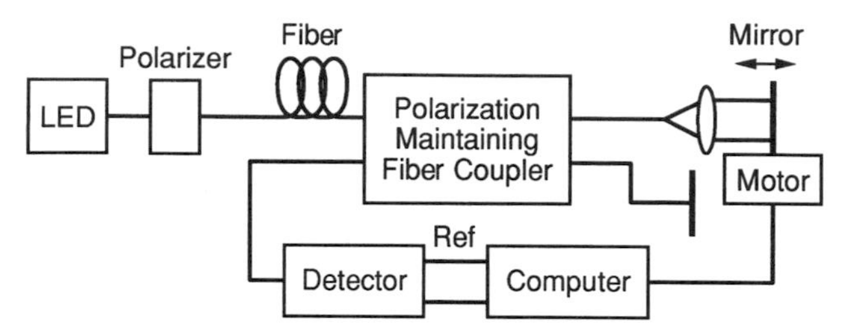

Fig. 6.21 Apparatus for interferometric measurement of PMD. LED, light-emitting diode (Reproduced from Gisin *et al.* 1991. Experimental comparison between two different methods for measuring polarisation mode dispersion in single-mode fibres. *Electron. Lett.* 27:2292–2293, with permission of IEE.)

Fibers that fall in the long-length regime produce a fringe pattern that contains a large number of peaks distributed about a central point (Gisin, Von der Weid, and Pellaux 1991). The large number of peaks results from polarization-mode coupling that causes the light to travel many paths through the fiber (see Fig. 6.8). Because of the broadband nature of the source, the distribution of peaks is interpreted according to the coupled-power model (see Section 6.3.1) and indicates the temporal spreading caused by PMD as given by Eq. (6.6) (Gisin, Von der Weid, and Pellaux 1991).

Although this technique does not measure the differential delay between the principal states directly, the average differential delay can be inferred by making use of Eq. (6.16) (Galtarossa *et al.* 1992; Gisin and Pellaux 1992; Matera and Someda 1992; Gisin *et al.* 1993). However, care must be taken to account for the dependence of the measured fringe pattern on the source spectrum (Heffner 1996).

The advantages of the interferometer technique include femtosecond temporal resolution, wide dynamic range, and good stability under changing fiber conditions.

6.4.3 RF SPECTRAL RESPONSE

In the RF spectral response method, the optical power from a single-frequency laser source is sinusoidally modulated using an external modulator and launched into the fiber under test (Bahsoun, Nagel, and Poole 1990). At the fiber output, the strength of the RF signal received by a photodiode is measured as a function of RF frequency using a network analyzer. The strength of the RF signal is a function of the relative time delay between the principal states of polarization, the relative power in the two states, and the RF frequency. For fixed input polarization, the RF spectrum shows a minimun when the differential delay between the principal states is equal to one-half the RF cycle. The frequency corresponding to this minimum is thus given by

$$f_{\min} = \frac{1}{2\Delta\tau}. \tag{6.25}$$

By identifying the frequency $f_{\min}$, one obtains a measure of the differential delay time.

Figure 6.22 shows a superposition of 50 response curves obtained in a single fiber that was subjected to varying temperature. The shifting in the minimum from curve to curve is indicative of a changing differential delay time. The average differential delay time is obtained by applying Eq. (6.25) to each curve and averaging.

The primary advantage of this method is the ability to measure large values of PMD using commercially available network analyzers. The disadvantage is that the depth of the minima that are observed depends on how

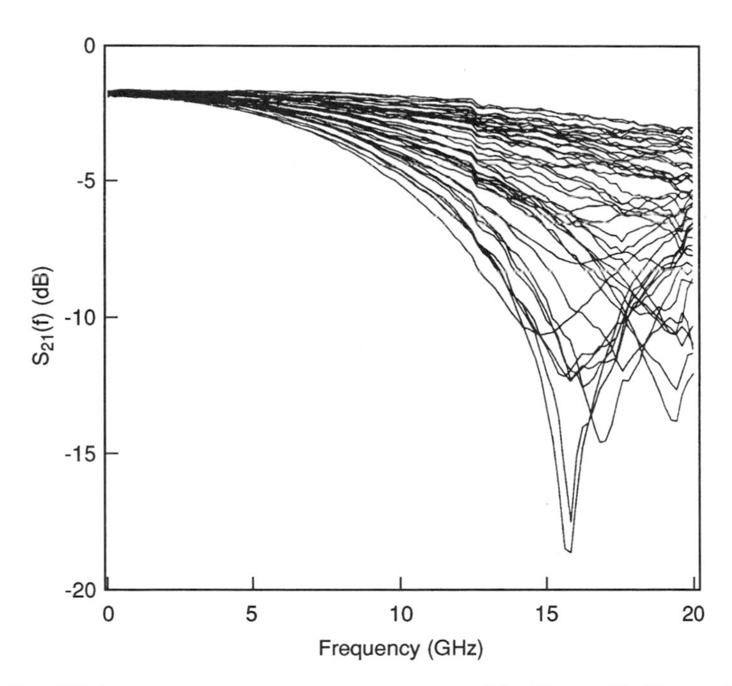

Fig. 6.22 RF frequency response curves measured in fiber with 23 ps of PMD (Bahsoun *et al.* 1990).

evenly the power is split between the two principal states. Under certain conditions the minimum may be too shallow for accurate identification. Another disadvantage of this method is that chromatic dispersion in the test fiber will also create minima in the response curve, which complicates the analysis when measurements are made away from the dispersion zero of the test fiber.

6.4.4 POINCARÉ SPHERE METHOD

The Poincaré sphere method measures PMD in the frequency domain by measuring the state of polarization at the output of a fiber as a function of frequency (wavelength) with a fixed input polarization (Andresciani *et al.* 1987; Bergano, Poole, and Wagner 1987; Poole *et al.* 1988). By differentiating the data with respect to frequency, and by making measurements for at least two input polarizations, one can use Eq. (6.12) to determine both the magnitude and the direction of the dispersion vector, Ω, at a given wavelength, and thus the differential delay $\Delta\tau = |\Omega|$.

This technique can be used to measure values of differential delay at specific wavelengths or to determine the average delay by scanning over a broad wavelength range and averaging. This technique has the advantage that the complete dispersion vector can be measured as a function of frequency, thus yielding information about the principal states and higher order dispersion. This method is accurate for both large and small values of PMD, its accuracy usually being limited by the stability of the fiber under test.

6.4.5 JONES MATRIX METHOD

The Jones matrix method is similar to the Poincaré sphere method in that the polarization at the output of a test fiber is measured as a function of optical frequency (Heffner 1992, 1993). The key difference between the two techniques is that the Jones matrix method uses a set of predetermined launch polarization states to determine the complete Jones matrix of the fiber at each frequency. This is usually accomplished by using a set of calibrated polarizers that are alternately placed in front of the source as shown in Fig. 6.23. By determining the Jones matrix of the test fiber at closely spaced frequency intervals, one can compute the frequency derivative and from this the differential delay time using Eq. (6.11). This method also yields the principal states of polarization for the test fiber at every frequency.

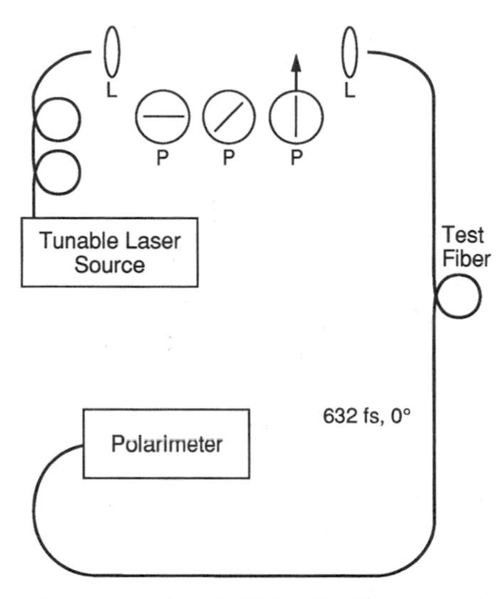

Fig. 6.23 Apparatus for measuring PMD by the Jones matrix method. (Reprinted from Heffner, B. L. 1993. *IEEE Phot. Tech. Lett.* 5:814–817. Copyright © 1993 IEEE.)

The Jones matrix method provides the most complete characterization of a fiber because the transmission properties are completely specified by the frequency-dependent Jones matrix. Thus the method offers the flexibility to measure PMD at specific wavelengths or, depending on the tunability of the source, to measure averages by sweeping the source wavelength over a suitable range. The primary drawbacks of this technique are the need for sophisticated algorithms for computing derivatives and the need to have a stable test fiber during the measurement. As with the Poincaré sphere technique, the latter drawback usually limits measurement accuracy.

6.4.6 FIXED ANALYZER METHOD

In the fixed analyzer method, light with a fixed polarization is transmitted through a test fiber and then through a polarizer (analyzer) placed at the output of the fiber, as shown in Fig. 6.24a (Poole 1989; Poole and Favin 1994). The normalized transmission through the analyzer is measured as a function of wavelength. In practice, this is accomplished using either a

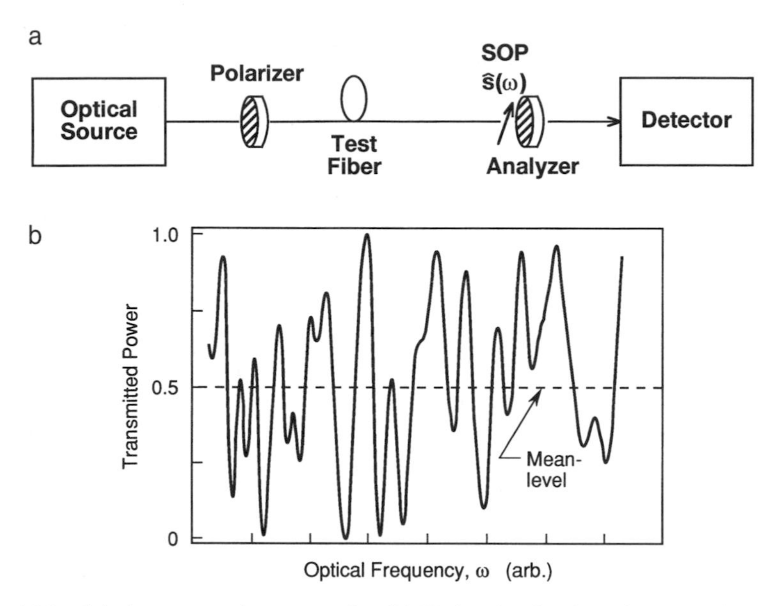

Fig. 6.24 (a) Apparatus for measuring PMD by the fixed analyzer method. (Reprinted from Poole, C. D., and D. L. Favin. 1994. *IEEE J. Lightwave Tech.* LT-12:917–929. Copyright © 1994 IEEE.) (b) Example of a transmission curve with the mean transmission level shown as a *dashed curve.*

broadband source and an optical spectrum analyzer as the detector or a narrow band, tunable laser source and power meter.

The transmitted spectrum (Fig. 6.24b) shows a series of peaks and valleys as a result of variation of the polarization incident on the analyzer caused by PMD in the test fiber. The larger the PMD, the more rapidly the output state of polarization changes with frequency, and thus the more densely spaced are the peaks and valleys in the spectrum.

The average PMD of the test fiber can be inferred from such transmission spectra by any one of several methods. The first of these is to count the number of times that the transmission curve crosses the mean transmission level (i.e., 0.5) per unit frequency interval. The PMD is related to this mean-level crossing density by (Poole and Favin 1994)

$$\langle \Delta\tau \rangle = k_1 \frac{N_m}{\Delta\omega}, \qquad N_m \longrightarrow \infty, \tag{6.26}$$

where N_m is the number of mean-level crossings in the frequency interval $\Delta\omega$, and k_1 is a constant. For fibers in the short-length regime, $k_1 = \pi$, and in the long-length regime $k_1 = 4$.

Another approach is to count the number of extrema (i.e., minima and maxima) per unit frequency interval. The PMD is related to this extrema density by (Poole and Favin 1994)

$$\langle \Delta \tau \rangle = k_2 \frac{N_e}{\Delta \omega}, \qquad N_e \longrightarrow \infty, \tag{6.27}$$

where N_e is the number of extrema in the frequency interval $\Delta\omega$. For fibers in the short-length regime, $k_2 = \pi$, and in the long-length regime $k_2 = 0.82\pi$.

In practice the number of extrema or mean-level crossings is finite, thus Eqs. (6.26) and (6.27) are used as approximations whose accuracy improves as the square root of the number of extrema or mean-level crossings observed.

By measuring both mean-level crossings and extrema densities, one can infer the length regime that the test fiber falls into, because $N_e/N_m = 1.55$ for $L \gg l_c$, and $N_e/N_m = 1$ for $L \ll l_c$.

Another approach to analyzing the data of Fig. 6.24b is to determine the Fourier transform of the transmission spectrum. The width of the spectrum is proportional to the PMD in the test fiber and is closely related to the autocorrelation obtained by the interferometric technique (Gisin, Passy, and Von der Weid 1994; Heffner 1994).

The advantage of the fixed analyzer technique is the simplicity both in the experimental apparatus and in the data analysis required. This technique is limited, however, to measuring the average PMD and requires a stable fiber during measurement.

6.5 Polarization Effects in Amplified Systems

The application of optical amplifiers in lightwave systems has greatly increased the number and variety of optical elements that a signal encounters in an optical path. In a transoceanic system, for example, an optical signal may pass through as many as 300 optical amplifiers, each containing several optical components. As a result, extremely small polarization effects associated with individual components can accumulate to produce noticeable performance degradation (see Chapter 10 of Volume IIIA).

Figure 6.25 shows an example of a lightwave system containing most of the elements common to amplified systems. PDL may be introduced by optical isolators (1), splitters (4), or optical filters (5), whereas PMD can be introduced by isolators (1), amplifiers (3), or the transmission fiber (2).

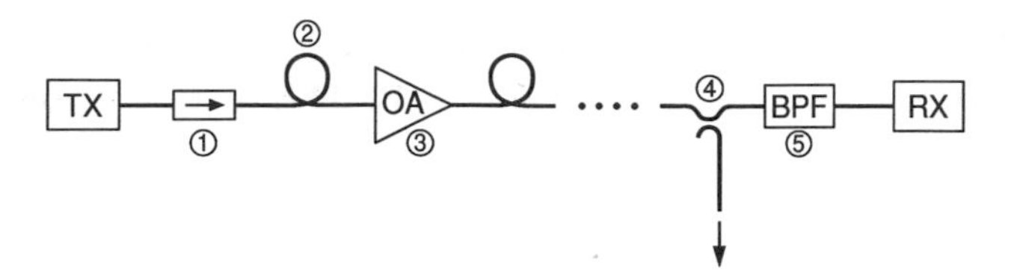

Fig. 6.25 Amplified system showing common sources of polarization effects. BPF, band-pass filter; OA, optical amplifier; Rx, receiver; Tx, transmittor.

Polarization effects unique to optical amplifiers (3) are introduced in the form of polarization-dependent gain and polarization hole burning. All these polarization effects share the trait that they depend on the state of polarization of light entering a component. Thus, in long amplified systems where the state of polarization is random and time varying, these effects lead to random, time-varying system performance (Yamamoto *et al.* 1989).

6.5.1 POLARIZATION-DEPENDENT LOSS

The signal-to-noise ratio at the output of amplified systems depends sensitively on the accumulation of amplified spontaneous emission (ASE) noise. Mechanisms that enhance the noise at the expense of signal power lead to system impairment through a reduction in the signal-to-noise ratio at the receiver.

One such mechanism involves PDL. For a linear optical component this is usually expressed as the ratio of maximum to minimum transmission in decibels:

$$PDL = 10 \log_{10} \left(\frac{T_{\max}}{T_{\min}} \right). \tag{6.28}$$

The polarization properties of any passive optical component or system can be modeled by an equivalent bulk-optic system comprising a single linear partial polarizer positioned between two polarization-transforming elements, as shown in Fig. 6.26 (Jones 1941; Hurwitz and Jones 1941). The polarization transformers induce an arbitrary transformation of the polarization without inducing loss, while the partial polarizer serves to attenuate one polarization more than the other. The polarization transformers exhibit the additional property that orthogonal polarizations at the input remain orthogonal at the output.

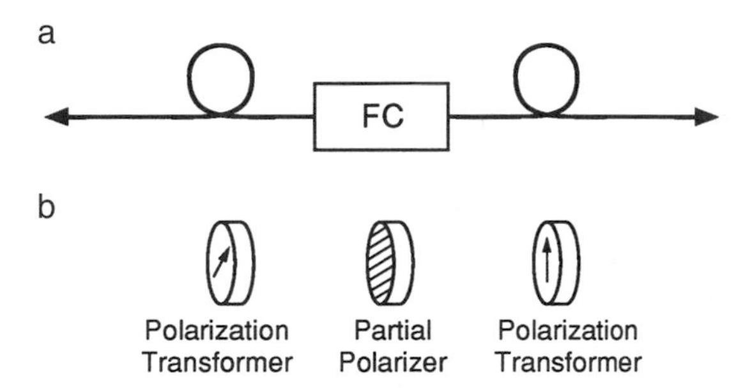

Fig. 6.26 Equivalent bulk-optic representation of a fiber optic component. (a) Fiber component (FC). (b) Equivalent optical system.

From the equivalent bulk-optic model of Fig. 6.26b one can conclude that all linear optical components have one input polarization state that produces the maximum transmission loss and an orthogonally polarized input state that produces the minimum transmission loss. Optical components that have a single input and a single output are characterized by a single value of PDL. Devices with multiple inputs and outputs, such as fused fiber couplers, are characterized by values of PDL for each optical path through the device.

An optical component containing PDL and embedded in a long amplified system causes the local signal power to fluctuate in response to slowly varying changes in the incident polarization state. These power fluctuations rarely manifest at the receiver because of the power limiting effect of the amplifiers downstream, which are usually operated in saturation. Instead, the power fluctuations are converted to fluctuations in the ASE noise produced by the amplifiers. The primary effect of PDL in an amplified system is to cause slowly varying changes in signal-to-noise levels at the receiver (Yamamoto *et al.* 1989; Bruyere and Audouin 1994; Lichtman 1995a,b). Such fluctuations are readily observable in transoceanic systems and are minimized by stringent control on the PDL of system components.

6.5.2 POLARIZATION HOLE BURNING AND POLARIZATION-DEPENDENT GAIN

Another mechanism that can lead to a reduced signal-to-noise ratio in long amplified systems involves polarization hole burning in erbium-doped fiber amplifiers (Taylor 1993; Mazurczyk and Zyskind 1994; see Chapter 2, in Volume IIIB). Polarization hole burning arises in an amplifier when a saturating signal causes selective deexcitation of erbium ions that are aligned with the polarization of the saturating signal. This causes the gain seen by unpolarized ASE noise accompanying the signal to be reduced for components of the noise parallel to the signal and enhanced for components orthogonal to the signal (see Fig. 6.27). The result is that the orthogonal noise component grows at the expense of the signal power, which leads to a reduction in the signal-to-noise ratio at the receiver (Bruyere and Audouin 1994; Lichtman 1995a).

In bulk material, polarization hole burning is maximum when the saturating signal has a linear polarization (Hall *et al.* 1983; Wysocki and Mazurczyk

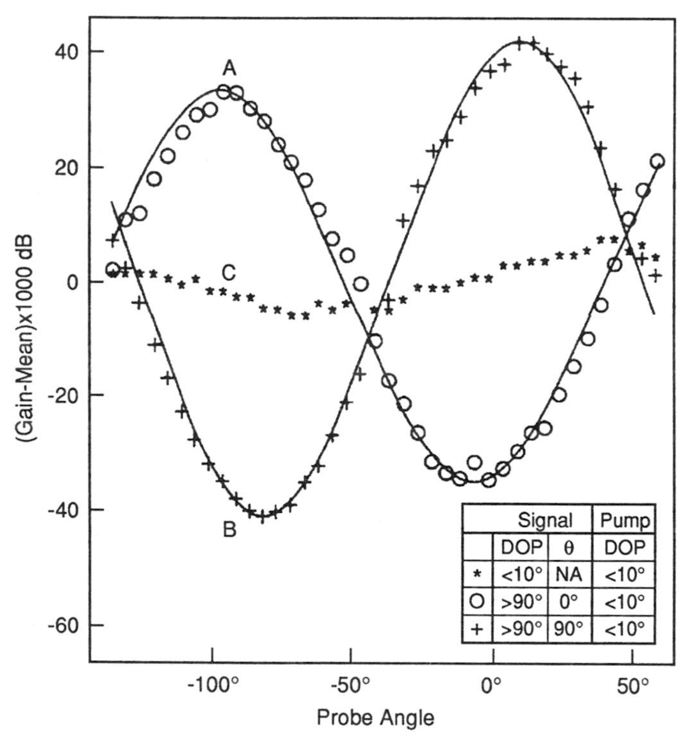

Fig. 6.27 Variation of gain with polarization for a weak probe and fixed saturating signal in an Erbium-doped fiber amplifier. Curves A and B correspond to different saturating signal polarizations. Curve C is for unpolarized saturating signal. (Reprinted from Mazurczyk, V. J., and J. L. Zyskind. 1994. *IEEE Phot. Tech. Lett.* 6:616–618. Copyright © 1994 IEEE.)

1994). In erbium-doped fiber amplifiers, polarization hole burning is much less sensitive to the polarization of the saturating signal owing to birefringence in the amplifying fiber (Mazurczyk and Poole 1994; Wysocki and Mazurczyk 1994, 1996). This birefringence, which is usually induced in packaging the erbium-doped fiber, causes the signal polarization to walk through all possible polarization states as it propagates from input to output. This spatial randomization of the polarization greatly reduces any dependence on the polarization of the saturating signal. As a result, polarization hole burning in amplified systems does not lead to large time-varying fluctuations in system performance but rather induces a constant impairment through a reduced signal-to-noise ratio (Lichtman 1995a).

Another effect that can cause orthogonal components of the noise to grow at the expense of the signal is polarization-dependent gain caused by the pump light used to excite the erbium ions in an amplifier (Mazurczyk and Zyskind 1994). Unlike polarization hole burning, the gain becomes anisotropic for both signal and noise because of the preferential excitation of erbium ions by the polarized pump light. The anisotropy in the gain causes the signal power to vary depending on the relative polarization of signal and pump in the amplifier. The effect on system performance is exactly analogous to that of PDL and can lead to fluctuating signal-to-noise levels at the receiver.

The effects of polarization hole burning and polarization-dependent gain have been shown to be greatly reduced by modulating the signal polarization (Bruyere *et al.* 1994) and pump polarizations (Mazurczyk and Zyskind 1994), respectively, at rates that exceed the response time of the amplifiers.

References

Andresciani, D., F. Curti, F. Matera, and B. Daino. 1987. Measurement of the group-delay difference between the principal states of polarization on a low-birefringence terrestrial fiber cable. *Opt. Lett.* 12:844–846.

Bahsoun, S., J. Nagel, and C. Poole. 1990. Measurement of temporal variations in fiber transfer characteristics to 20 GHz due to polarization-mode dispersion. In *Proceedings of the European Conference on Optical Communications, ECOC'90, Amsterdam,* 1003. Postdeadline paper.

Barlow, A. J., J. J. Ramskov-Hansen, and D. N. Payne. 1981. Birefringence and polarization mode-dispersion in spun single-mode fibers. *Appl. Opt.* 20:2962–2968.

Bergano, N. S., C. D. Poole, and R. E. Wagner. 1987. Investigation of polarization dispersion in long lengths of single-mode fiber using multilongitudinal mode lasers. *IEEE J. Lightwave Tech.* LT-5:1618–1622.

Betti, S., F. Curti, B. Daino, G. De Marchis, E. Iannone, and F. Matera. 1991. Evolution of the bandwidth of the principal states of polarization in single-mode fibers. *Opt. Lett.* 16:467–469.

Bruyere, F., and O. Audouin. 1994. Penalties in long-haul optical amplifier systems due to polarization dependent loss and gain. *IEEE Photon. Tech. Lett.* 6:654–656.

Bruyere, F., O. Audouin, V. Letellier, G. Bassier, and P. Marmier. 1994. Demonstration of an optimal polarization scrambler for long-haul optical amplifier systems. *IEEE Photon. Tech. Lett.* 6:1153–1155.

Burns, W. K., R. P. Moeller, and C. Chen. 1983. Depolarization in a single-mode optical fiber. *IEEE J. Lightwave Tech.* LT-1:44–49.

Chiang, K. S. 1985a. Conditions for obtaining zero polarisation-mode dispersion in elliptical-core fibres. *Electron. Lett.* 21:592–593.

Chiang, K. S. 1985b. Linearly birefringent fibres with zero polarisation-mode dispersion. *Electron. Lett.* 21:916–917.

Chraplyvy, A. R., R. W. Tkach, L. L. Buhl, and R. C. Alferness. 1986. Phase modulation to amplitude modulation conversion of CW laser light in optical fibres. *Electron. Lett.* 22:409–411.

Curti, F., B. Daino, G. De Marchis, and F. Matero. 1990. Statistical treatment of the evolution of the principal states of polarization in single-mode fibers. *IEEE J. Lightwave Tech.* LT-8:1162–1166.

Curti, F., B. Daino, Q. Mao, F. Matera, and C. G. Someda. 1989. Concatenation of polarisation dispersion in single-mode fibres. *Electron. Lett.* 25:290–292.

De Angelis, C., A. Galtarossa, G. Gianello, F. Matera, and M. Schiano. 1992. Time evolution of polarization mode dispersion in long terrestrial links. *IEEE J. Lightwave Tech.* LT-10:552–555.

de Lignie, M. C., H. G. J. Nagel, and M. O. van Deventer. 1994. Large polarization mode dispersion in fiber optic cables. *IEEE J. Lightwave Tech.* LT-12:1325–1329.

Eickhoff, W., Y. Yen, and R. Ulrich. 1981. Wavelength dependence of birefringence in single-mode fiber. *Appl. Opt.* 20:3428–3435.

Foschini, G. J., and C. D. Poole. 1991. Statistical theory of polarization dispersion in single mode fibers. *IEEE J. Lightwave Tech.* LT-9:1439–1456.

Frigo, N. J. 1986. A generalized geometrical representation of coupled mode theory. *IEEE J. Quantum Electron.* QE-22:2131–2140.

Galtarossa, A., G. Gianello, C. G. Someda, and M. Schiano. 1996. In-field comparison among polarization-mode dispersion measurement techniques. *IEEE J. Lightwave Tech.* LT-14:42–49.

Galtarossa, A., and M. Schiano. 1993. Polarization mode dispersion in high-density ribbon cables. In *Proceedings of the Optical Fibre Measurement Conference,* 181–184.

Galtarossa, A., M. Schiano, C. G. Someda, B. Daino, F. Matera, R. Zaninello, and F. Bergamin. 1991. Two different methods for measuring polarisation mode dispersion in singlemode fibres. *Electron. Lett.* 27:2292–2293.

Gisin, N. 1991. Solutions of the dynamical equation for polarization dispersion. *Opt. Commun.* 86:371–373.

Gisin, N., R. Passy, J. C. Bishoff, and B. Perny. 1993. Experimental investigations of the statistical properties of polarization mode dispersion in single mode fibers. *IEEE Photon. Tech. Lett.* 5:819–821.

Gisin, N., R. Passy, B. Perny, A. Galtarossa, C. Someda, F. Bergamin, M. Schiano, and F. Matera. 1991. Experimental comparison between two different methods for measuring polarisation mode dispersion in singlemode fibres. *Electron. Lett.* 27:2292–2293.

Gisin, N., R. Passy, and J. P. Von der Weid. 1994. Definitions and measurements of polarization mode dispersion: Interferometric versus fixed analyzer methods. *IEEE Photon. Tech. Lett.* 6:730–732.

Gisin, N., and J. P. Pellaux. 1992. Polarization mode dispersion: Time versus frequency domains. *Opt. Commun.* 89:316–323.

Gisin, N., B. Perny, and R. Passy. 1991. Polarization mode dispersion measurements in optical fibers, cables, installed terrestrial cables and fiber ribbons. In *Proceedings of the Optical Fibre Measurement Conference,* 85–88.

Gisin, N., J. Von der Weid, and J. Pellaux. 1991. Polarization mode dispersion of short and long single-mode fibers. *IEEE J. Lightwave Tech.* LT-9:821–827.

Hall, D. W., R. A. Haas, W. F. Krupke, and M. J. Weber. 1983. Spectral and polarization hole burning in neodymium glass lasers. *IEEE J. Quantum Electron.* QE-19:1704–1717.

Heffner, B. L. 1992. Automated measurement of polarization mode dispersion using Jones matrix eigenanalysis. *IEEE Photon. Tech. Lett.* 4:1066–1069.

Heffner, B. L. 1993. Accurate, automated measurement of differential group delay dispersion and principal state variation using Jones matrix eigenanalysis. *IEEE Photon. Tech. Lett.* 5:814–817.

Heffner, B. L. 1994. Single-mode propagation of mutual temporal coherence: Equivalence of time and frequency measurements of polarization-mode dispersion. *Opt. Lett.* 19:1104–1106.

Heffner, B. L. 1996. Influence of optical source characteristics on the measurement of polarization-mode dispersion of highly mode-coupled fibers. *Opt. Lett.* 21:113–115.

Hurwitz, H., and R. C. Jones. 1941. A new calculus for the treatment of optical systems II. *J. Opt. Soc. Am.* 31:493–499.

Iannone, E., F. Matera, A. Galtarossa, G. Gianello, and M. Schiano. 1993. Effect of polarization dispersion on the performance of IM–DD communication systems. *IEEE Photon. Tech. Lett.* 5:1247–1249.

Jones, R. C. 1941. A new calculus for the treatment of optical systems I. *J. Opt. Soc. Am.* 31:488–503.

Judy, A. F. 1994. Improved PMD stability in optical fibers and cables. In *Proceedings of the 43rd International Wire & Cable Symposium,* 658–664.

Kaminow, I. P. 1981. Polarization in optical fibers. *IEEE J. Quantum Electron.* QE-17:15–22.

Kawakami, S., and M. Ikeda. 1978. Transmission characteristics of a two-mode optical waveguide. *IEEE J. Quantum Electron.* QE-14:608–614.

Kikushima, K., K. Suto, H. Yoshinaga, and E. Yoneda. 1994. Polarization dependent distortion in AM-SCM video transmission systems. *IEEE J. Lightwave Tech.* LT-12:650–657

Lichtman, E. 1995a. Limitations imposed by polarization-dependent gain and loss on all-optical ultralong communication systems. *IEEE J. Lightwave Tech.* LT-13:906–913.

Lichtman, E. 1995b. Performance limitations imposed on all-optical ultralong lightwave systems at the zero-dispersion wavelength. *IEEE J. Lightwave Tech.* LT-13:898–905.

Marcuse, D. 1972. Pulse propagation in mutlimode dielectric waveguides. *Bell Syst. Tech. J.* 51:1199–1232.

Matera, F., and C. G. Someda. 1992. Concatenation of polarization dispersion in single-mode fibers: A Monte-Carlo simulation. *Opt. Commun.* 3:289–294.

Mazurczyk, V. J., and C. D. Poole. 1994. The effect of birefringence on polarization hole burning in erbium doped fiber amplifiers. In *Proceedings of the optical amplifiers topical meeting.* Paper THB3.

Mazurczyk, V. J., and J. L. Zyskind. 1994. Polarization dependent gain in erbium doped fiber amplifiers. *IEEE Photon. Tech. Lett.* 6:616–618.

Menyuk, C. R., and P. K. A. Wai. 1994. Polarization evolution and dispersion in fibers with spatially varying birefringence. *J. Opt. Soc. Am. B* 11:1288–1296.

Meslener, G. J. 1984. Chromatic dispersion induced distortion of modulated monochromatic light employing direct detection. *IEEE J. Quantum Electron.* QE20:1208–1216.

Mochizuki, K., Y. Namihira, and H. Wakabayashi. 1981. Polarisation mode dispersion measurements in long single mode fibres. *Electron. Lett.* 17:153–154.

Namihira, Y., T. Kawazawa, and H. Wakabayashi. 1992. Polarisation mode dispersion measurements in 1520 km EDFA system. *Electron. Lett.* 28:881–883.

Namihira, Y., K. Nakajima, and T. Kawazawa. 1993. Fully automated interferometric PMD measurements for active EDFAs, fiber optic devices and optical fibers. In *Proceedings of the Optical Fibre Measurement Conference,* 189–192.

Namihira, Y., and H. Wakabayashi. 1991. Fiber length dependence of polarization mode dispersion measurements in long-length optical fibers and installed optical submarine cables. *J. Opt. Commun.* 12:1–8.

Nielsen, C. J. 1982. Influence of polarization-mode coupling on the transmission bandwidth of single-mode fibers. *J. Opt. Soc. Am.* 72:1142–1146.

Nielsen, C. J. 1983. Impulse response of single-mode fibers with polarization-mode coupling. *J. Opt. Soc. Am.* 73:1603–1611.

Norman, S. R., D. N. Payne, M. J. Adams, and A. M. Smith. 1979. Fabrication of single-mode fibres exhibiting extremely low polarisation birefringence. *Electron. Lett.* 24:309–311.

Passy, R., B. Perny, A. Galtarossa, C. Someda, F. Bergamin, M. Schiano, and F. Matera. 1991. Relationship between polarisation dispersion measurements before and after installation of optical cables. *Electron. Lett.* 27:595–596.

Personick, S. D. 1971. Time dispersion in dielectric waveguides. *Bell Syst. Tech. J.* 50:843–859.

Phillips, M. R., G. E. Bodeep, X. Lu, and T. E. Darcie. 1994. 64-QAM BER measurements in an analog lightwave link with large polarization-mode dispersion. In *Proceedings of the Optical Fiber Communications Conference.* Paper WH6.

Phillips, M. R., T. E. Darcie, D. Marcuse, G. E. Bodeep, and N. J. Frigo. 1991. Nonlinear distortion generated by dispersive transmission of chirped intensity-modulated signals. *IEEE Photon. Tech. Lett.* 3:481–483.

Poole, C. D. 1988. Statistical treatment of polarization dispersion in single-mode fiber. *Opt. Lett.* 13:687–689.

Poole, C. D. 1989. Measurement of polarization-mode dispersion in single-mode fibers with random mode coupling. *Opt. Lett.* 14:523–525.

Poole, C. D., N. S. Bergano, R. E. Wagner, and H. J. Schulte. 1988. Polarization dispersion and principal states in a 147-km undersea lightwave cable. *IEEE J. Lightwave Tech.* LT-6:1185–1190.

Poole, C. D., and T. E. Darcie. 1993. Distortion related to polarization-mode dispersion in analog lightwave systems. *IEEE J. Lightwave Tech.* LT-11:1749–1759.

Poole, C. D., and D. L. Favin. 1994. Polarization-mode dispersion measurements based on transmission spectra through a polarizer. *IEEE J. Lightwave Tech.* LT-12:917–929.

Poole, C. D., and C. R. Giles. 1988. Polarization-dependent pulse compression and broadening due to polarization dispersion in dispersion-shifted fiber. *Opt. Lett.* 13:155–157.

Poole, C. D., R. W. Tkach, A. R. Chraplyvy, and D. A. Fishman. 1991. Fading in lightwave systems due to polarization-mode dispersion. *IEEE Photon. Tech. Lett.* 3:68–70.

Poole, C. D., and R. E. Wagner. 1986. Phenomenological approach to polarisation dispersion in long single-mode fibres. *Electron. Lett.* 22:1029–1030.

Poole, C. D., J. H. Winters, and J. A. Nagel. 1991. Dynamical equation for polarization dispersion. *Opt. Lett.* 16:372–374.

Rashleigh, S. C. 1983. Origins and control of polarization effects in single-mode fibers. *IEEE J. Lightwave Tech.* LT-1:312–331.

Rashleigh, S. C., W. K. Burns, R. P. Moeller, and R. Ulrich. 1982. Polarization holding in birefringent single mode fibers. *Opt. Lett.* 7:40–42.

Rashleigh, S. C., and R. Ulrich. 1978. Polarization mode dispersion in single-mode fibers. *Opt. Lett.* 3:60–62.

Sakai, J., and T. Kimura. 1981. Birefringence and polarization characteristics of single-mode optical fibers under elastic deformations. *IEEE J. Quantum Electron.* QE-17:1041–1051.

Snitzer, E., and H. Osterberg. 1961. Observed dielectric waveguide modes in the visible spectrum. *J. Opt. Soc. Am.* 51:499–505.

Takahashi, T., T. Imai, and M. Aiki. 1994. Automatic compensation technique for time-wise fluctuating polarisation mode dispersion in in-line amplifier systems. *Electron. Lett.* 30:348–349.

Taylor, M. G. 1993. Observation of new polarization dependence effect in long haul optically amplified system. In *Proceedings of the Optical Fiber Communications Conference.* Paper PD5.

Thevenaz, L., M. Nikles, and P. Robert. 1992. Interferometric loop method for polarization dispersion measurements. In *Proceedings of the Symposium on Optical Fiber Measurement.* 151–154.

Thevenaz, L., J. Pellaux, N. Gisin, and J. Von der Weid. 1989. Birefringence measurements in fibers without polarizer. *IEEE J. Lightwave Tech.* LT-7:1207–1212.

Tsubokawa, M., and M. Ohashi. 1991. A consideration of polarization dispersion determining from a Stokes parameter evaluation. *IEEE J. Lightwave Tech.* LT-9: 948–951.

Ulrich, R., and A. Simon. 1979. Polarization optics of twisted single-mode fibers. *Appl. Opt.* 18:2241–2251.

Vengsarkar, A. M., A. H. Moesle, L. G. Cohen, and W. L. Mammel. 1993. Polarization mode dispersion in dispersion shifted fibers: An exact analysis. In *Proceedings of the OFCC.* Paper ThJ7.

Wagner, R. E., and A. F. Elrefaie. 1988. Polarization dispersion limitations in lightwave systems. In *Proceedings of the Optical Fiber Communications Conference.* Paper TU16.

Wai, P. K. A., and C. R. Menyuk. 1994. Polarization decorrelation in optical fibers with randomly varying birefringence. *Opt. Lett.* 19:1517–1519.

Winters, J. H., and R. D. Gitlin. 1990. Electrical signal processing techniques in long-haul, fiber-optic systems. *IEEE Trans. Commun.* 38:1439–1453.

Winters, J. H., Z. Haas, M. A. Santoro, and A. H. Gnauck. 1992. Optical equalization of polarization dispersion. *Proc. Soc. Photo-Opt. Instr. Eng.* 1787:346–357.

Winters, J. H., and S. Kasturia. 1992. Adaptive nonlinear cancellation for high-speed fiber-optic systems. *IEEE J. Lightwave Tech.* LT-10:971–977.

Winters, J. H., and M. A. Santoro. 1990. Experimental equalization of polarization dispersion. *IEEE Photon. Tech. Lett.* 2:591–593.

Wysocki, P. F., and V. Mazurczyck. 1994. Polarization hole-burning in erbium-doped fiber amplifiers with birefringence. In *Proceedings of the optical amplifiers topical meeting.* Paper THB4.

Wysocki, P. F., and P. Mazurczyck. 1996. Polarization dependent gain in erbium-doped fiber amplifiers: Computer model and approximate formulas. *IEEE J. Lightwave Tech.* LT-14, 572–584.

Yamamoto, S., N. Edagawa, H. Taga, Y. Yoshida, and H. Wakabayashi. 1989. Observation of BER degradation due to fading in long distance optical amplifier system. *Electron. Lett.* 29:209–210.

Chapter 7 | Dispersion Compensation for Optical Fiber Systems

A. H. Gnauck
AT&T Laboratories–Research, Holmdel, New Jersey

R. M. Jopson
Lucent Technologies, Bell Laboratories, Holmdel, New Jersey

I. Introduction

Two developments in lightwave system design, one revolutionary and the other evolutionary, have stimulated the invention of many methods for compensating for optical fiber chromatic dispersion. The evolutionary development has been the inexorable increase in the bit rate used in long-haul lightwave transport. The recent history of this trend is displayed in Fig. 7.1. It can be seen that the bit rate achieved in laboratory hero experiments as well as that used in highly reliable commercial systems has been increasing exponentially. Since 1982, the speed of commercially available systems has doubled every 2.4 years. By the mid-1990s, 2.5-Gb/s systems were deployed in several national networks. It is likely that 10-Gb/s systems will be widely deployed before the advent of the 21st century.

The revolutionary development in lightwave systems has been the deployment of erbium-doped fiber amplifiers (EDFAs) (Desurvire 1994). These amplifiers facilitate the transmission of lightwave signals over long distances by providing periodic analog amplification rather than digital regeneration. Although EDFAs can reduce system cost and increase flexibility, these advantages are accompanied by a disadvantage: penalties can accumulate over the entire system length. Thus, the designer of a system containing EDFAs may have to accommodate the chromatic dispersion of hundreds or even thousands of kilometers of fiber and, depending on the type of fiber used, may be forced to employ dispersion compensation.

OPTICAL FIBER TELECOMMUNICATIONS,
VOLUME IIIA

ISBN: 0-12-395170-4

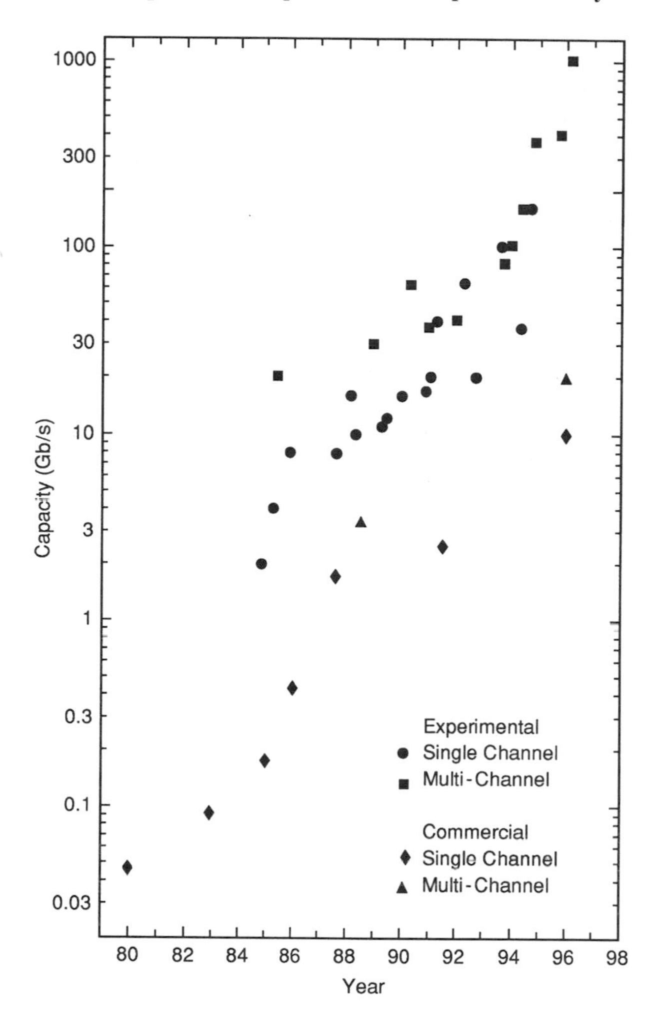

Fig. 7.1 Progress in lightwave transmission capacity for single- and multiple-channel (WDM) systems.

II. Lightwave Systems

Lightwave systems are discussed extensively elsewhere in this book, so only a brief description is provided here. In the absence of EDFA deployment, most 2.5-Gb/s long-haul fiber optic transmission links consist of a linear assemblage of regenerator spans, each about 40 km long. The transmitter end of a regenerator span contains a distributed feedback laser operating at a wavelength lying in either the 1.3-μm telecommunications window of transmission fiber or the window at 1.55 μm. The NRZ (non-return-to-zero) on–off modulation is provided by direct modulation, that is, by simply modulating the current used to drive the semiconductor laser. Direct amplitude modulation of these lasers is inherently accompanied by a large frequency modulation, or chirp, that increases with the bit rate. At multigigabit data rates, the chirp is typically about 0.5 nm (60–80 GHz). The receiver end of the regenerator span contains an APD (avalanche photodiode) detector followed by gain, clock recovery, and a decision circuit. The transmitter and the receiver are joined by approximately 40 km of single- (transverse) mode silica fiber. After each span of fiber, the signal is retimed and reshaped before being transmitted over the next span. This regeneration has the advantage of limiting the accumulation of signal impairments such as optical noise, linear distortion, and nonlinearity, but it is expensive to implement and maintain.

As mentioned previously, when EDFAs are employed, the regenerator spacing can be increased to hundreds or thousands of kilometers. In addition, the system must be operated in the gain window of the amplifier, approximately 1.53–1.56 μm. Much of the fiber that has been deployed in the national networks has been so-called conventional fiber with the zero in the chromatic dispersion located at a wavelength near 1300 nm. If a system using this fiber is upgraded to use EDFAs, the system chromatic dispersion will increase, not only because of the longer distances between regenerators but also because of the shift from operation at the zero-dispersion wavelength to a wavelength with significant dispersion (17 ps/km/nm). Even in new systems for which the fiber can be chosen to have any desired dispersion, compensation of dispersion may be required. There are three ways in which this may arise: (1) combating system nonlinearity requires the use of a nonzero dispersion that then must be compensated for, (2) a system using multiple wavelengths may require dispersion compensation of some channels because the transmission fiber will generally have significant dispersion for some channels, and (3) as single channel bit rates approach 1 Tb/s, it becomes desirable to compensate the wavelength dependence of the chromatic dispersion.

III. Chromatic Dispersion

The electric field of a linearly polarized lightwave signal propagating in a single-mode fiber can be described by

$$\vec{\mathbf{E}}(x,y,z,t) = \frac{\hat{\mathbf{n}}}{2}\left[E(x,y,z,t)e^{i[\beta(\omega)z-\omega t]} + C.C.\right], \tag{7.1}$$

where $\hat{\mathbf{n}}$ is a unit vector, $E(x,y,z,t)$ is a complex scalar that can vary along the propagation direction, z, and in time, t, and has some mode shape in the transverse dimensions, x and y. $\beta(\omega)$ is the propagation constant, ω is the angular frequency, and $C.C.$ denotes complex conjugate. It is well known that in the absence of fiber nonlinearity, pulse propagation and distortion can be estimated from the derivatives of $\beta(\omega)$ with respect to ω. $\beta(\omega)$ is expanded in a Taylor series about some desired frequency, ω_0:

$$\beta(\omega) = \sum_{n=0}^{\infty} \frac{1}{n!}\,\beta_n\,(\omega - \omega_0)^n, \tag{7.2}$$

where

$$\beta_n \equiv \left.\frac{\partial^n \beta}{\partial \omega^n}\right|_{\omega_0}. \tag{7.3}$$

The phase velocity is ω_0/β_0, whereas the group velocity, which is the speed of pulse or data propagation, is $1/\beta_1$. Chromatic dispersion, or more precisely group-velocity dispersion, in lightwave systems is caused by a variation in the group velocity in a fiber with changes in optical frequency. Since it can cause pulse spreading in a lightwave signal, chromatic dispersion can impair system performance. This is shown schematically in Fig. 7.2a. An isolated pulse, if for no reason other than its modulation, contains a spectrum of wavelengths. As it traverses the fiber, the shorter wavelength components of the pulse (shown darker) travel faster than the longer wavelength components (shown lighter). Thus the pulse broadens as it travels down the fiber and by the time it reaches the receiver, it may have spread over several bit periods and cause errors. The measure of chromatic dispersion used by the lightwave community is D, in units of ps/nm/km, which is the amount of broadening in picoseconds that would occur in a pulse with a bandwidth of 1 nm while propagating through 1 km of fiber. D is given by

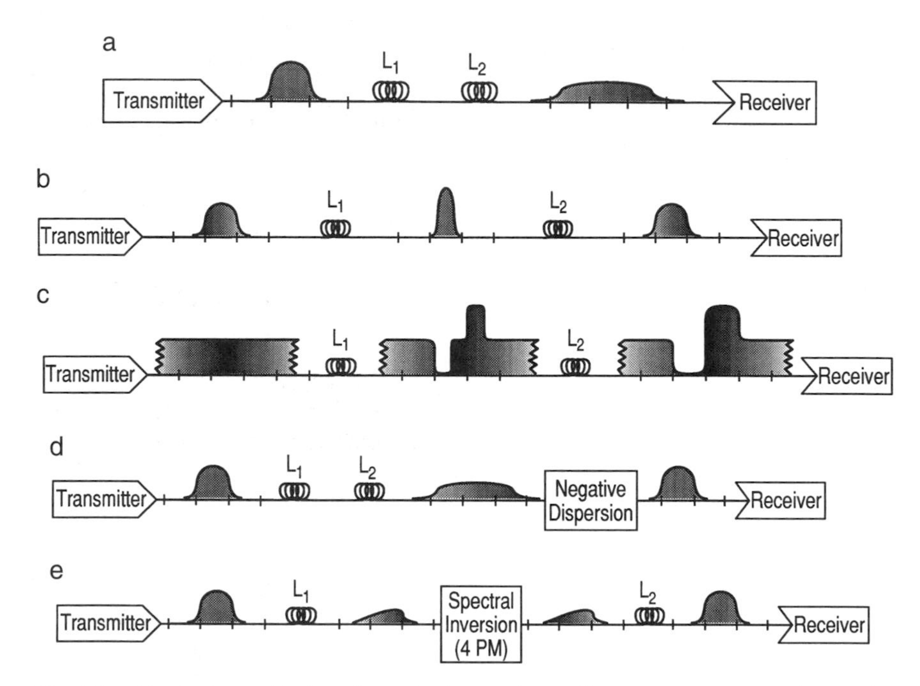

Fig. 7.2 Evolution of isolated marks for various dispersion-compensation techniques. Shorter-than-average wavelength components are shown darker, and longer-than-average wavelength components are shown ligher. (a) No dispersion compensation, (b) prechirped pulses, (c) dispersion-supported transmission using frequency-modulation, (d) negative dispersion at the end of a span, (e) midsystem spectral inversion.

$$D = \frac{d}{d\lambda}\frac{1}{v_g} = \frac{d^2\beta}{d\lambda d\omega} \approx -\frac{2\pi c}{\lambda^2}\beta_2, \tag{7.4}$$

where λ, v_g, β, ω, and c are the wavelength, group velocity, propagation constant, angular frequency, and speed of light, respectively. "Conventional" fiber, so-called because it was the first widely deployed single-mode fiber, contains a step-index waveguide with a zero in the dispersion near 1.31 μm. In the 1.55-μm window, D is about 17 ps/nm/km. For some purposes, a more convenient measure of chromatic dispersion is β_2, often expressed in units of ps^2/km. Because, at a wavelength of 1.55 μm, $2\pi c/\lambda^2 = 0.781$/ps/nm, at this wavelength, $D \approx -\beta_2$ when they are expressed in units of ps/nm/km and ps^2/km, respectively.

The dispersion of optical fiber can vary significantly over the EDFA gain bandwidth. In conventional fiber, the slope ($dD/d\lambda$) of D is 0.08 ps/

nm^2/km. Unfortunately this has been labeled *second-order dispersion*, a term generating confusion because its value contains the third-order derivative of β with respect to ω:

$$dD/d\lambda = \frac{2\pi c}{\lambda^3}\left[2\beta_2 - \frac{2\pi c}{\lambda}\beta_3\right]. \tag{7.5}$$

This wavelength dependence of the chromatic dispersion may be important in long wavelength-division multiplexed (WDM) systems because different wavelengths may need different dispersion compensation. The bit-error rates observed in 100 Gb/s and higher systems can be significantly affected by second-order dispersion, so there is now interest in compensation of the dispersion slope.

Chromatic dispersion places a limit on the maximum distance a signal can be transmitted without regenerating the original digital signal. This distance, the dispersion limit or dispersion length, can be estimated by determining the transmission distance at which a pulse has broadened by one bit interval. The estimated dispersion limit for a signal of width $\Delta\lambda$, is then given by

$$L_D = 1/(B\ D\ \Delta\lambda). \tag{7.6}$$

For a commercially available directly modulated 2.5-Gb/s system, we can use 2.5 Gb/s, 17 ps/nm/km, and 0.5 nm for B, D, and $\Delta\lambda$, respectively, to estimate a dispersion limit of 47 km. In practice, somewhat longer dispersion limits have been achieved by fabricating lasers with low chirp. However, in commercial systems not employing EDFAs, the loss budget limits regenerator spans to 40–50 km, so dispersion is not a big concern. However, to significantly increase the bit rate or to use an optically amplified system and increase regenerator span lengths in these systems, one must turn to low-chirp modulation of the optical signal. This can be accomplished by using an external modulator. When an external modulator is used, the optical bandwidth of an NRZ signal in frequency units can be approximately 1.2 times the bit rate, B. Converting to units of wavelength, we find, for externally modulated systems, that

$$L_D = \frac{6100}{B^2}\ \mathrm{km(Gb/s)^2} \tag{7.7}$$

for conventional fiber at 1.55 μm. The solid trace to the left in Fig. 7.3 shows L_D as a function of bit rate for a system employing an externally modulated source and conventional fiber. The dashed loss line is the

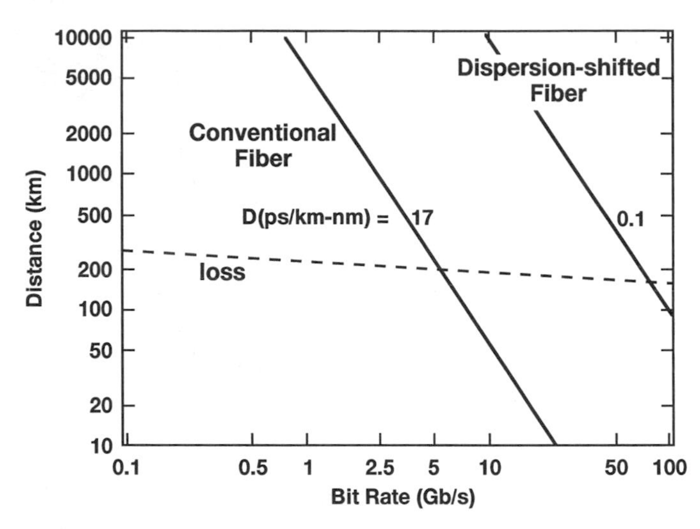

Fig. 7.3 Chromatic dispersion limit (km) as a function of bit rate (Gb/s) for externally modulated 1.55-μm NRZ on–off signals. Limits for conventional fiber (17 ps/km-nm) and dispersion-shifted fiber (0.1 ps/km-nm) are shown.

single-hop transmission-distance limit imposed by fiber loss under laboratory conditions when EDFAs are not employed. It can be seen that a 1.55-μm externally modulated system using conventional fiber has an estimated dispersion limit of nearly 1000 km at a bit rate of 2.5 Gb/s. At 10 Gb/s, this estimated limit is reduced to 61 km, in agreement with the 60–70 km observed in both practice and computer simulation (Elrefaie *et al.* 1988) for a 1-dB penalty. The results of our own simulation are shown in Fig. 7.4, and show that the dispersion penalty reaches 1 dB at 70 km and exceeds 10 dB at 150 km. The penalty is infinite at 160 km. To go farther or go faster, one can reduce the dispersion of the fiber, either by operating near the 1.31-μm dispersion zero of conventional fiber, or by modifying the dispersion of the fiber. Currently, the overwhelming advantages of erbium amplifiers and the low fiber loss at 1.55 μm virtually rule out operation in conventional fiber near 1.31 μm, but it is possible in new installations to use dispersion-shifted fiber (DSF). This fiber is designed such that the waveguide contribution to the chromatic dispersion shifts the dispersion zero to some desired wavelength in the 1.55-μm window. In most of the DSF designs, the slope in the dispersion remains near 0.08 ps/nm^2/km.

Even in these new installations, dispersion compensation may be required. For example, to minimize the effects of four-wave mixing in WDM

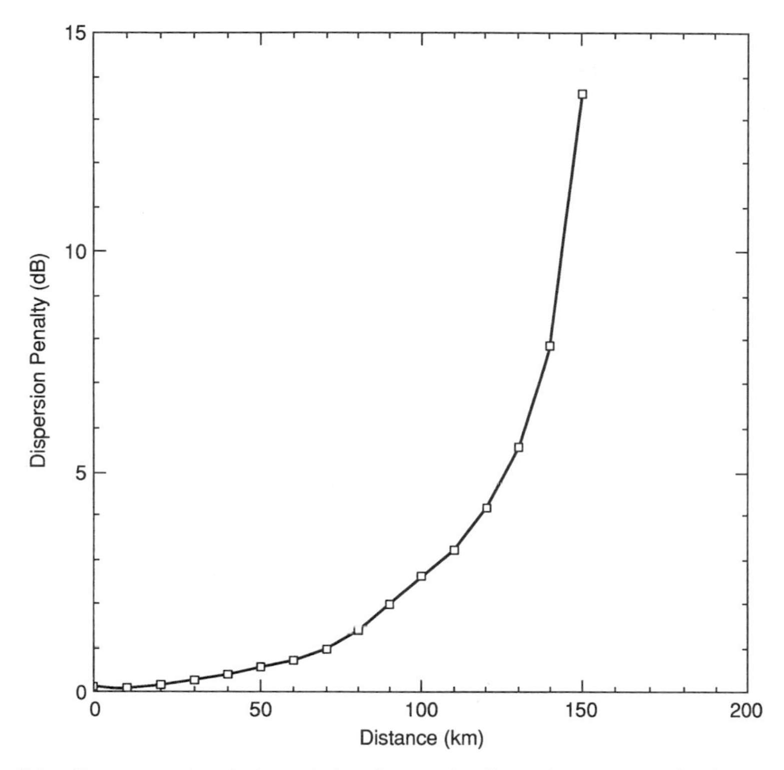

Fig. 7.4 Computer simulation of the chromatic-dispersion penalty (dB) as a function of distance (km) for externally modulated 1.55-μm 10-Gb/s NRZ on–off signals propagated through conventional fiber. The penalty reaches 1 dB at 70 km and is infinite at 160 km.

systems, non-zero-dispersion fiber may be used instead of DSF (Clark, Klein, and Peckham 1994). This fiber has a dispersion of 2–4 ps/nm/km or -2 to -4 ps/nm/km. Some form of dispersion compensation or "dispersion management" will then be distributed in the transmission line to keep the accumulated dispersion low. Alternatively, these systems may utilize dispersion compensation at the transmitter or receiver to ameliorate the effects of the total line dispersion.

Although the use of DSF or non-zero-dispersion fiber can enhance the performance of completely new systems greatly, systems must often be designed to use the existing conventional fiber network. During the late 1980s and early 1990s, many nations throughout the world invested in lightwave long-haul networks. If present trends continue, sometime during the twilight of the 20th century, this fiber deployment will reach 200 million kilometers. At first, these networks contained conventional fiber almost exclusively. Now, some DSF and non-zero-dispersion fiber is starting to be deployed, but the national networks still predominantly contain conventional fiber. Because these networks are an enormous investment, there is strong incentive to design systems that use conventional fiber.

The reasons for interest in chromatic dispersion compensation are now clear. Designers of lightwave systems want to use erbium amplifiers, which offer the advantages of higher signal power levels, WDM, and a many-fold increase in the distance signals can travel before detection and regeneration. The first two advantages allow nonlinear effects to play a role, a role that can be minimized by adding dispersion to the system, and this added dispersion must be compensated before detection. The third advantage places strict limits on the total amount of system dispersion that can be tolerated. The use of erbium amplifiers forces the use of the wavelengths near 1.55 μm, where the conventional fiber that has been deployed throughout the world has high dispersion, and once again we find that dispersion compensation is needed. Finally, as the cost of 10-Gb/s electronics and electrooptics decreases, there will be a desire to operate at higher bit rates. In the examples discussed in this chapter, we concentrate on systems that use the installed base of conventional single-mode fiber, with the understanding that the compensation techniques discussed can be applied to systems using other types of fiber.

IV. Dispersion Compensation

Myriad techniques for compensating chromatic dispersion have been demonstrated or proposed. Before we examine them, it is instructive to calculate the amount of pulse broadening occurring in a typical high-speed system. Consider a 1.55-μm, 10-Gb/s system with regenerators separated by 300 km of conventional fiber. With a fiber dispersion of 17 ps/nm/km, the net dispersion in a regenerator span is 5100 ps/nm. If chirp-free modulation is employed, an isolated mark requires 0.1 nm of optical bandwidth and it will broaden by 510 ps during passage through the 300-km regenerator span. Thus the pulse will occupy 5 or 6 bit slots at the end of the span, with the blue spectral components of the pulse being some 10 cm of fiber ahead of the red spectral components. (Despite our exclusive interest in the infrared region of the optical spectrum, we use "blue" to mean the shorter wavelength spectral components of a signal and "red" to denote the longer wavelength spectral components of the signal.) This large separation means that if optical components are employed to compensate the dispersion, they will have to be large. Such a device must provide 15 cm (in air) of differential path length between the spectral components of the signal and it must operate over a bandwidth of at least 0.1 nm centered on the signal wavelength. Even allowing for multipassing, the smallest of these components will have dimensions measured in centimeters and, as we shall see, some of them will be many kilometers long. If a directly modulated 10-Gb/s signal is transmitted over the 300-km regenerator span, the pulse will occupy 0.5 m or 26 bit slots at the end of the span! An optical device used to compensate dispersion for this signal must provide 77 cm (in air) of differential path difference over a bandwidth of 0.5 nm.

A. *SPECTRAL SHAPING AT THE TRANSMITTER*

It is possible to provide only a sample of the many methods used to compensate chromatic dispersion. It is not surprising, given the importance of the problem and the presence of three subsystems in a lightwave regenerator span, that attempts have been made to attack dispersion in the transmitter, in the fiber and in the receiver. We first describe modifications to the transmitter. Most transmitter techniques modify the spectrum of the data stream, often by "prechirping" (Koch and Alferness 1985). This is illustrated in Fig. 7.2b. The spreading of the bit into adjacent slots can be delayed by arranging for the light in the leading edge of the bit to be of longer-than-average wavelength and that in the trailing edge of the bit to be of shorter-than-average wavelength. By prechirping the pulse in this manner, long-wavelength light in the leading edge must pass through the entire pulse slot before it starts impinging on the trailing pulse and causing errors. Light in the trailing edge will behave similarly.

Various techniques have been used to prechirp the transmitted signal. One simple method for generating a chirped bit stream is to add phase modulation through the use of an unbalanced Mach–Zehnder amplitude modulator (Koyama and Iga 1988; Schiess and Carlden 1994). This technique has demonstrated modest compensation of dispersion (Gnauck, Korotky, *et al.* 1991). Another method is to frequency modulate (FM) the laser to provide chirp in the optical signal entering an external modulator. The laser drive signal can be as simple as a properly phased sinusoid at a frequency equal to the bit rate. Transmission of 10-Gb/s data over 100 km of fiber (2-dB dispersion penalty) has been demonstrated in this way (Patel *et al.* 1992). A third technique takes advantage of the transmission fiber nonlinearity. At modest power levels of several milliwatts, self-phase modulation (see Chapter 8 in Volume IIIA) produces a shift to longer wavelength at the beginning of an optical pulse, and a shift to shorter wavelength at the trailing edge. This effect has been shown to provide compensation approximately equivalent to the previously described prechirping techniques (Ogata *et al.* 1992). At a launched power of 14 mW, 10-Gb/s, 100-km transmission (1-dB penalty) has been achieved (Gnauck, Tkach, and Mazurczyk 1993).

One of the most successful transmitter techniques for dispersion compensation, referred to as *dispersion-supported transmission* (Wedding, Franz, and Junginger 1994), has used FM alone. In this scheme, illustrated in Fig. 7.2c, the transmitter generates an FM signal with the modulation tailored to the span length so that the dispersion in the span converts the FM at the transmitter to amplitude modulation at the receiver. With this technique, span lengths in excess of 200 km have been achieved at 10 Gb/s with a 6-dB receiver-sensitivity penalty compared with baseline. Drawbacks to dispersion-supported transmission include having to tailor the laser drive current to the span length, the requirement for a laser with good broadband FM response, and the need to decode a three-level optical signal at the receiver.

Combinations of several techniques have also been investigated. For example, computer simulation shows that 120-km transmission is feasible using an unbalanced Mach–Zehnder modulator in conjunction with self-phase modulation (Gnauck, Tkach, and Mazurczyk 1993). An integrated electroabsorption modulator–DFB laser reached 107 km at 10 Gb/s with a slight improvement in sensitivity over back-to-back measurements when the launched power was raised to +15 dBm to take advantage of self-phase modulation (Kuindersma *et al.* 1993). Transmission over 300 km may be possible through a combination of dispersion-supported transmission and self-phase modulation (Kurtzke and Gnauck 1993a) and, with the addition of residual amplitude modulation, this might be extended to 350 km (Rasmussen *et al.* 1995). Another technique that can be employed is to lower the optical extinction ratio, provided that the associated penalty can be tolerated. In on–off keying, the optical signal is unipolar. As the extinction ratio is reduced, the signal begins to resemble a bipolar signal. The effect of dispersion on the ones and zeros becomes more symmetrical as light from one can coherently interact with light from the other. Although the dispersion limit is largely unchanged for an unchirped signal, the effect is

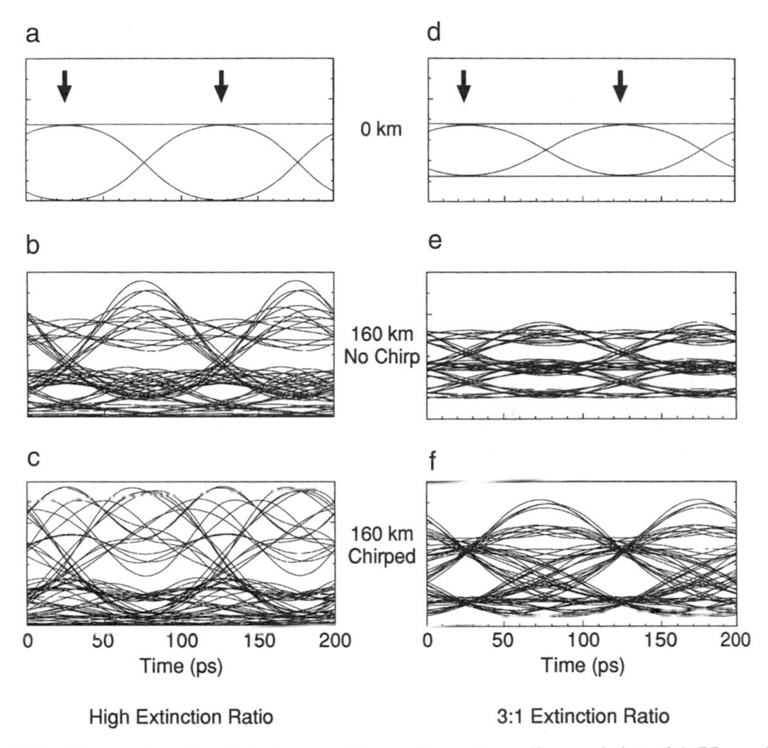

Fig. 7.5 Computer-simulated eye patterns for externally modulated 1.55-μm NRZ signals before and after propagation through 160 km of conventional fiber. A high-extinction-ratio signal is compared with a low (3 : 1)-extinction-ratio signal, both with and without chirp. *Arrows* show the points at which decisions are made.

dramatic when combined with chirp. Figure 7.5 shows a computer simulation of 10-Gb/s transmission using an adjustable-chirp Mach–Zehnder modulator running at either high or low (3 : 1) extinction ratio. The eye patterns in this figure are generated by overlaying portions of a bit stream that have been delayed by an integer number of bit periods. The arrows indicate the decision points. As mentioned previously in this section, the eye of a high-extinction, unchirped signal is completely closed after 160-km transmission, as shown in (b). In fact, the eyes of a high-extinction chirped signal (c), or a low-extinction unchirped signal (e), are also closed. However, the chirped, low-extinction-ratio eye exhibits a dispersion penalty of only 1 dB (f). An unbalanced Mach–Zehnder modulator, lowered extinction ratio, and self-phase modulation have been combined to obtain 200-km transmission at 10 Gb/s with a penalty relative to baseline of less than 2 dB (Jørgensen 1994). A low-chirp integrated electroabsorption modulator–DFB laser run with an extinction ratio of 5 dB has demonstrated 10-Gb/s, 130-km transmission with no dispersion penalty (Park *et al.* 1996).

Through the transmitter-based compensation techniques discussed thus far, the onset of chromatic-dispersion impairments can be delayed. Unfortunately, they cannot be delayed forever. This is because the modulation requires a minimum bandwidth for the optical signal and because the light from a bit is initially confined to a single bit period. Once the outlying spectral components in the pulse disperse one clock period, they will start to cause errors. Thus prechirping techniques have extended the dispersion limit by a factor of 3 and this is probably about as far as they will go. The limitations of the technique have been attacked in several ways. One method is a proposal to widen the transmitted pulses to 1.5 clock periods, chirp them, multiplex them into a single (overlapping) bit stream, and then use dispersion to narrow the pulses into a single bit period (Schiess 1994). Another method is to filter the modulating signal to reduce its bandwidth while generating controlled intersymbol interference. In one implementation, this results in the generation of a narrower bandwidth duobinary optical signal that can propagate further before incurring a large dispersion penalty (May, Solheim, and Conradi 1994). Transmission over 138 km of fiber at 10 Gb/s (2.5-dB dispersion penalty) has been demonstrated in this way (Gu *et al.* 1994). Conventional duobinary transmission involves a three-level optical signal that increases signal-to-noise requirements and requires decoding at the receiver. Recent duobinary proposals use a Mach–Zehnder-modulator transmitter driven by a duobinary electrical signal whose magnitude is twice the switching voltage of the modulator. This produces a conventional optical binary signal, but with phase reversals in the electric field (Price and Le Mercier 1995; Yonenaga *et al.* 1995). This modulation scheme also results in an increased threshold for stimulated Brillouin scattering. Recent experiments have demonstrated 210-km transmission with a 1.7-dB penalty (Price *et al.* 1995), and 164-km transmission with a 1.8-dB penalty (Kuwano, Yonenaga, and Iwashita 1995). Although there is a limit to the dispersion that can be compensated for by transmitter techniques, they may become very important at 10 Gb/s because they are easily implemented and because 200 km of dispersion compensation is sufficient for a large fraction of the world's point-to-point links.

B. DATA RECOVERY AT THE RECEIVER

Receiver-based compensation techniques can be used with either coherent-detection receivers or direct-detection receivers. A coherent receiver mixes the incoming signal with a local oscillator, thereby shifting any phase and amplitude fluctuations on the optical carrier to a carrier at an electronic frequency. Then linear dispersion compensation can be performed on the electronic carrier. One 8-Gb/s demonstration (Takachio, Norimatsu, and Iwashita 1992) used heterodyne detection and a 31.5-cm-long microwave-stripline dispersion compensator operating over an IF band of 6–18 GHz. It compensated for 188 km of "conventional" fiber.

In a direct-detection receivers, the optical signal is simply coupled to a photodiode and converted to a current proportional to the square the optical electric field. This detection process loses information about the phase of the dispersed signal so it is not surprising that linear equalization of a dispersed signal cannot compensate much dispersion. However, it has been recognized that nonlinear equalization can be of benefit. The signal distortion caused by chromatic dispersion is constant and predictable, and although an individual bit may be spread over many bit periods, the detected waveform is changed by its presence. Two methods of processing the detected waveform have been proposed and modeled (Winters and Gitlin 1990). In one method, the threshold of the decision circuit is varied depending on preceding bits. In another, the decision about a given bit is made by analyzing the analog waveform for a band of clock periods surrounding the bit in question. Modeling suggests that these techniques, particularly the latter one, will be able to compensate many dispersion lengths of fiber, but confirmation awaits experimental verification.

In contrast to single-bit transmitter-based techniques, receiver-based techniques for dispersion compensation can, in principle, compensate many dispersion lengths of fiber. This results from the availability of many bit slots of received signal for processing. However, network designers are reluctant to employ coherent detection in long-haul systems, so the heterodyne technique has a dim future. Because (1) the direct-detection compensator requires logic operating at the bit rate, (2) the complexity of the logic increases as 2^N where N is the number of bits over which a pulse has been broadened, and (3) the required signal-to-noise ratio is proportional to N, the direct-detection compensator will probably be limited to compensation of several dispersion lengths of fiber.

C. THE OPTICAL REGIME

In contrast to the electrical processing used in transmitter- and receiver-based compensation techniques, those operating on the fiber span employ optical processing. Figure 7.2d illustrates the use of optical compensation at the end of a regenerator span. After the pulse has been broadened by the positive dispersion in a span, it passes through an equal amount of negative dispersion and the original pulse shape is restored before detection. Figure 7.6a shows the pulse broadening experienced by a 1.5-μm, 10-Gb/s, externally modulated signal over a 600-km span of conventional fiber, in the absence of dispersion compensation. Figure 7.6b shows the pulse broadening that would occur if compensating negative dispersion were added at the beginning of the span (dashed line) the end of the span (solid line), or distributed every 100 km along the span (dotted line). The latter design is an attractive architecture, not only because it can reduce nonlinearity impairments, but also because it allows the dispersion-compensation components to be included in the optical-amplifier repeaters. With proper design, the optical attenuation of the compensation components can then be mitigated without an increase in the component count.

1. Interferometers

Many optical techniques for dispersion compensation have been demonstrated. They can be classified into three categories: interferometers, negative-dispersion fibers, and phase-conjugation techniques. Interferometers compensate dispersion by providing wavelength-dependent paths of different lengths for different spectral components of the signal. The Gires-Tournois (GT) interferometer is an etalon filter in which the first mirror is partially reflective and the second mirror is 100% reflective. Therefore, all the incident light is reflected, but there is a wavelength-dependent differential group delay near the cavity resonance. This characteristic can be

Fig. 7.6 Pulse broadening experienced by a 1.5-μm 10-Gb/s, externally modulated signal over a 600-km span of conventional fiber for various dispersion compensation techniques. (a) No compensation, (b) negative dispersion at the beginning (*dashed*) or end (*solid*) of a span or distributed every 100 km (*dotted*), (c) midsystem spectral inversion, (d) spectral inversion not at the middle of a system, (e) midsystem spectral inversion in a multiwavelength system. The horizontal lines show the tolerable amount of broadening for an initially transform-limited spectrum.

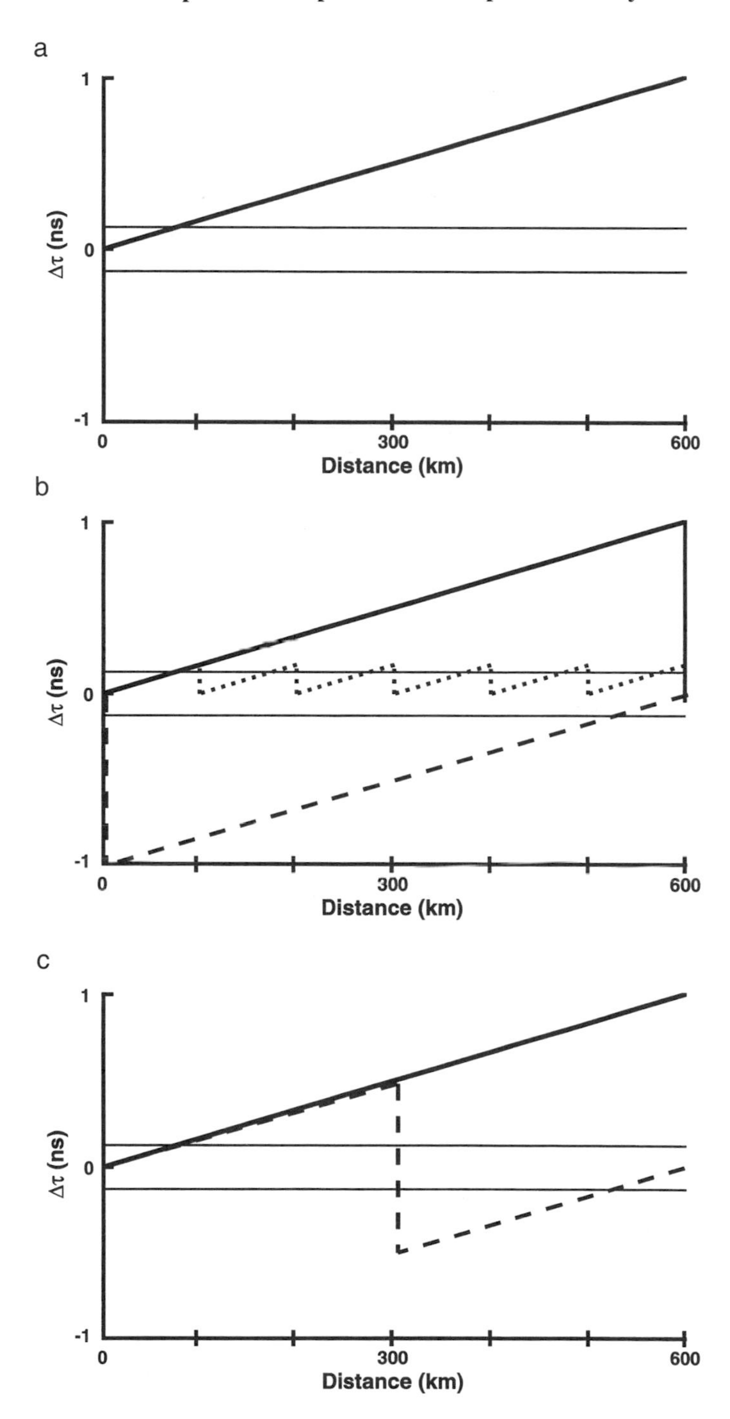
a
1
0
-1
Δτ (ns)
0
300
600
Distance (km)
b
1
0
-1
Δτ (ns)
0
300
600
Distance (km)
c
1
0
-1
Δτ (ns)
0
300
600
Distance (km)

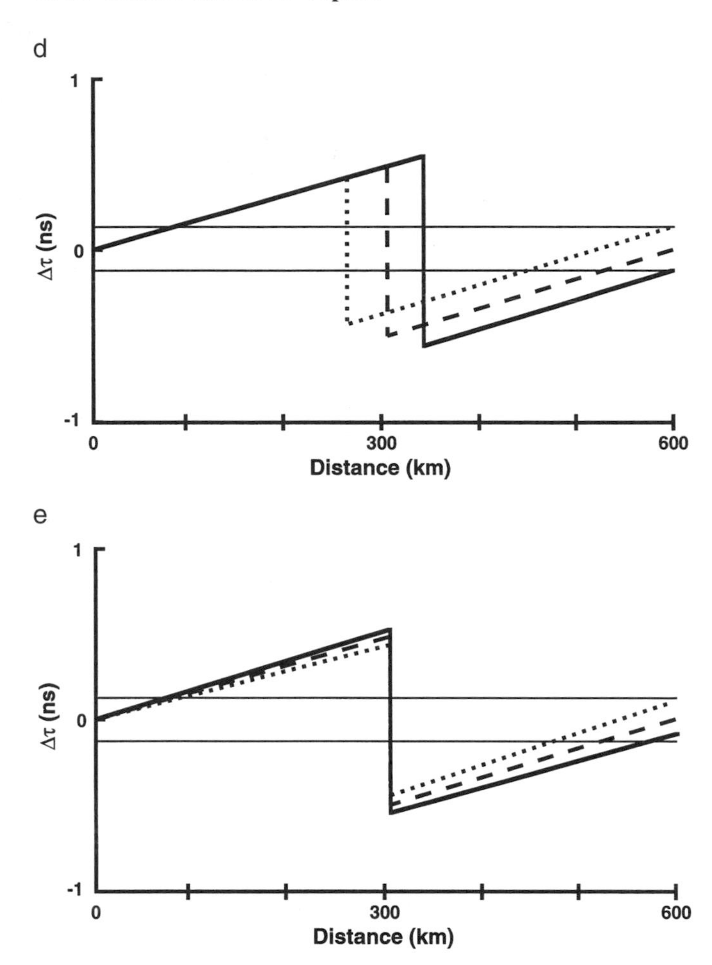

Fig. 7.6 *(Continued)*

used for dispersion compensation (Cimini, Greenstein, and Saleh 1990; Gnauck, Giles, *et al.* 1991). The mirror spacing can be adjusted to tune the resonance to the signal wavelength, and the compensated signal is recovered through an optical coupler or circulator. The dispersion of a GT interferometer is periodic. Thus, by matching the periodicity to the channel spacing in a WDM system, all channels might be simultaneously compensated. Recently, a GT interferometer reached a peak dispersion value of 2000 ps/nm (Garthe *et al.* 1996). It was used in a 10-Gb/s experiment in which 160-km transmission was achieved with only a 0.2-dB dispersion penalty.

Another promising interferometric device is a silica-on-silicon planar circuit containing cascaded Mach–Zehnder interferometers. In a Mach–Zehnder interferometer, the light is split into two, generally unequal, length paths and then recombined, in this case, by a 2×2 combiner. The distribution of light in the two output ports will depend on the relative phase delay provided by the two arms. In the planar dispersion-compensation circuit, these interferometers are cascaded with the path lengths adjusted so that the blue light travels mostly in the longer arms while the red light travels mostly in the shorter arms. A 5- by 8-cm circuit containing five cascaded interferometers has provided a dispersion of 836 ps/nm over a 10-GHz bandwidth with a loss of 3.5 dB (Takiguchi, Okamoto, and Moriwaki 1994). This is sufficient to compensate for 50 km of conventional fiber. The device is polarization dependent and suffers from the disadvantage of a narrow bandwidth of operation and a relatively limited compensation capability, deficiencies common to many interferometric devices. However, the phase in one arm of each interferometer is thermally tunable, so the device should be broadly tunable. In addition, the periodicity of the cascaded Mach–Zehnder device may allow all channels in a WDM system to be simultaneously compensated.

The interferometric device that is currently attracting the most attention is the chirped fiber Bragg grating. Its principle of operation is easily understood. A grating is written down the length of an optical fiber by periodically changing the fiber refractive index. Light in the fiber with a wavelength of twice the grating period is reflected. In a dispersion compensator, the grating period is reduced linearly down the length of the device (i.e., the grating is "chirped"). The blue light is therefore reflected at a point farther into the device than the red light and is thus delayed relative to the red light. These devices can be compact. A 5-cm-long grating, in principle, can compensate the 300-km, 10-Gb/s, externally modulated system used in the previous example. Long (multi-cm) gratings are difficult to make because submicron tolerances must be maintained over the length of the device. Most fiber grating compensators have the disadvantage that the compensated signal is retroreflected, so an optical circulator must be employed to separate the input from the output. They also have a narrow bandwidth of operation, and unlike other interferometric devices, the dispersion does not have a useful periodicity. However, the gratings offer low loss to nonresonant light passing through them, so multiwavelength operation can be obtained by putting several gratings in series down the fiber, each centered on a different wavelength. Recently, chirped gratings as long as 10 cm have been made. One of these provided −8000 ps/nm of dispersion over a bandwidth of 9 GHz (Loh *et al.* 1996). In a 10-Gb/s experiment using a duobinary transmitter and optical-amplifier repeaters, 650 km was spanned (1.4-dB dispersion penalty) by inserting this grating near the middle of the link. Also, an 11-cm unchirped grating has now been reported that operates in transmission mode, thus eliminating the need for an optical circulator (Eggleton *et al.* 1996). Preliminary results using this device have demonstrated 10-Gb/s, 100-km transmission with a 2-dB dispersion penalty. The suitability of chirped fiber gratings for compensation is advancing more rapidly than that of any other compensation technique. It is possible that in the future they will be the method of choice.

2. Negative-Dispersion Fiber

The method of choice at the present time is single-mode negative-dispersion fiber, generally referred to as dispersion-compensating fiber (DCF). Negative dispersion is generally achieved by guiding the mode very weakly so that small increases in wavelength are accompanied by relatively large changes in mode size. This puts more of the mode into the cladding where the lower refractive index provides an increase in the speed of propagation. Unfortunately, as the guiding is reduced, the fiber attenuation and bending loss increase. The optimal singly cladded fiber designs (designs containing one core and one cladding region) have a dispersion of around −100 ps/nm/km and an attenuation of 0.35 dB/km. In these fibers, the slope in the wavelength dependence of the dispersion has the same sign as that of conventional fiber (positive), so the fibers will cancel dispersion completely only at one wavelength. However, here, in contrast to the interferometric devices, the compensation mismatch changes weakly with wavelength, so the constraints imposed by this incomplete compensation are not very severe. In principle, a 10-Gb/s externally modulated, 30-nm-wide WDM system will have a dispersion limit of about 700 km when compensated with this fiber. This limit has already begun to assert itself in experimental systems. A 16-channel, 10-Gb/s, 10-nm-wide system has been reported in which 1000-km transmission was achieved using DCF (Oda *et al.* 1995). Applying the final compensating-fiber length on a channel-by-channel basis reduced the variation in receiver sensitivity from 6.6 to 3.3 dB.

By adding more cladding layers, a fiber designer can obtain better control over the mode-expansion-induced changes in the propagation constant and achieve broadband compensation by designing a fiber with a negative slope to its dispersion versus wavelength curve (Vengsarkar and Reed 1993; Onishi *et al.* 1994). A DCF of this type was used in a recent WDM experiment that demonstrated the transmission of eight 20-Gb/s channels over 232 km of conventional fiber (Tkach *et al.* 1995). The effective dispersion slope was reduced from 0.08 to 0.02 ps/nm^2/km, and no individual-channel dispersion compensation was necessary. This indicates the enhanced potential for future WDM upgrades when using negative-slope DCF. (Note that because the dispersion limit decreases as the minus two power of the bit rate, this demonstration, in principle, suffered from the same dispersion penalty as the aforementioned 10-Gb/s, 1000-km demonstration.)

A commonly used figure of merit (FOM) for compensating fibers is the absolute value of the dispersion divided by the attenuation (in dB/km), usually expressed in units of ps/nm/dB. The designer should employ caution when using this FOM because a very high FOM can be obtained by using very weak waveguiding. This performance "improvement" comes with a few problems; among them is a very high bend loss sensitivity (Hawtoff, Berkey, and Antos 1996). When using this FOM to compare different fibers, one must strive to measure the FOM using the conditions under which the fibers will be used. The best reported FOMs in fiber with demonstrated practicality are around 300 ps/nm/dB, with dispersion near −100 ps/nm/km (Vengsarker *et al.* 1994; Akasaka *et al.* 1996; Onishi *et al.* 1996). By dividing the FOM by the "FOM" of the transmission fiber, one obtains the number of decibels of transmission loss that will be compensated for by 1 dB of DCF loss. For conventional fiber, we have 17 ps/nm/km ÷ 0.2 dB/km = 85 ps/nm/dB. Thus a high-FOM compensating fiber will increase each span loss by about 30%. Because loss budgets will usually not absorb such a large loss, an amplifier is usually placed between the span and the compensating fiber. The high cost of this gain stage can be mitigated somewhat in system designs that require two-stage or three-stage amplifiers because, in that case, the DCF can be placed between gain stages (Delavaux *et al.* 1994). The traditional FOM is useful for comparing component lengths and hence costs, but it does not provide a reliable measure of the impact of the DCF on system performance. The additional loss provided by the DCF reduces the signal-to-noise ratio of the system and nonlinearity in the DCF increases signal distortion. A different figure of merit, F_{NL}, has been described that provides means to compare the system performance of various DCF designs (Forghieri, Tkach, and Chraplyvy 1996). The results depend somewhat on the dominant nonlinearity in the system (Brillouin scattering, self-phase modulation, cross-phase modulation, or Raman gain).

F_{NL} provides a prescription for optimally balancing the input powers to the transmission fiber and the DCF and also provides an estimate of the system penalty incurred by the use of DCF rather than a lossless, completely linear, method of dispersion compensation. Optimal use of typical DCF with a FOM of 200, a dispersion of -100 ps/nm/km and a nonlinear coefficient, γ, that is four times that of conventional fiber will incur a signal-to-noise-ratio penalty of 2–3 dB. The nonlinear coefficient, γ, is defined in Chapter 8 of Volume IIIA. In general, for constant FOM, F_{NL} increases for larger (more negative) values of dispersion. DCF offers the advantages of low polarization dependence, broadband compensation, and passivity. Unfortunately, a fiber length that is a significant fraction ($\frac{1}{4}$ to $\frac{1}{5}$) of the span length is needed to compensate a span. The required lengths of fiber are expensive and the additional amplifier gain that they necessitate may be expensive. A 300-km span might require some 50 km of compensating fiber with an attendant loss of 18 dB. Despite this disadvantage, the fiber is commercially available and is likely to be the first compensation technique to be widely deployed.

Another fiber dispersion-compensating technique uses a higher-order (LP_{11}) fiber mode (Poole, Wiesenfeld, and DiGiovanni 1993). This mode can provide much more negative dispersion per given length than fundamental (LP_{01}) mode and it can compensate a broad bandwidth as well. Fiber that compensates 35 times its own length had been demonstrated. This technique requires mode conversion between the fundamental mode and the LP_{11} mode. Although this has been demonstrated to have high efficiency, it is a barrier to field deployment. The higher-order-mode technique currently suffers a higher loss than the fundamental-mode technique. Demonstrated figures of merit are around 100. In principle, the LP_{11} mode should not exhibit high loss, so it remains to be seen whether the two-mode technique will become practical.

3. Spectral Inversion

All the methods of dispersion compensation discussed previously require varying amounts of compensation, depending on the length of the span. Midsystem spectral inversion (MSSI) is unique in that a single component compensates any span, regardless of length. Although the concept of MSSI was proposed as long ago as 1979 (Yariv, Fekete, and Pepper 1979), it was demonstrated in fiber optic transmission systems only in 1993 (Watanabe, Naito, and Chikama 1993; Jopson, Gnauck, and Derosier 1993). This technique is illustrated in Fig. 7.2e. The signal is dispersed in the first half of the span resulting in distorted pulse shapes with the blue light leading the red light. This dispersed signal is then phase conjugated. (For practical reasons, it usually is also shifted in frequency by several times its bandwidth, but the frequency shift is not needed for dispersion compensation.) The phase conjugation reverses or inverts the optical spectrum of the signal so that red becomes blue and blue becomes red. As shown in Fig. 7.2e, the shape of a pulse remains the same, but now the leading edge is red and the trailing edge is blue. Now the dispersion in the second half of the span reshapes the pulse so that, if the dispersion before the phase conjugator matches the dispersion after it, then the original pulse shape will be restored at the end of the span. The change in pulse broadening is shown in Fig. 7.6c. It can be seen that the phase conjugator near the center of the span reverses the broadening and that this allows the broadening to be zero at the receiver.

By phase conjugation, we mean a process in which a modulated signal,

$$E(x,y,z,t) = E(x,y,z,t)\ e^{i[\beta(\omega)z-\omega t]}, \tag{7.8}$$

is converted to

$$E'(x,y,z,t) = E^*(x,y,z,t)\ e^{i[\beta(2\omega'-\omega)z-(2\omega'-\omega)t]}. \tag{7.9}$$

Here, $E(x,y,z,t)$ is an amplitude that varies along the propagation direction, z, and in time, t, and has some mode shape in the transverse dimensions, x and y. $\beta(\omega)$ is the propagation constant and ω denotes angular frequency. ω' is some constant frequency determined by the properties of the phase conjugator and system constraints on the converted frequency. Because system constraints generally require that both the signal and converted signal be in the EDFA bandwidth, ω' is usually near ω. Both ω and ω' are usually present at the output of the phase conjugator; therefore it is convenient to facilitate filtering by offsetting ω' somewhat from ω. The critical point is that variations in ω or equivalently, temporal phase variations in the input signal, appear with the opposite sign in the converted signal. Thus, for example, some feature that appears in the longer-wavelength side in the input signal spectrum will be in the shorter wavelength side of the output spectrum. An input pulse with positive chirp will have negative chirp in the conjugate signal.

A more formal understanding of MSSI can be obtained by assuming that compared with the optical carrier, the envelope of the electric field varies slowly in time, t, and propagation distance, z. Consider a system containing a length, L_1, of fiber, a phase conjugator, and another length, L_2, of fiber. Assume that all signals are initially in the same state of polarization and that this polarization evolves identically for all signals as they traverse the system. The (scalar) electric field can then be written as a carrier propagating at some frequency, with an envelope, $A(z,t)$, and a transverse mode shape, $F(x,y)$, that is constant in time and propagation distance. With no loss of generality, the "carrier" frequency can be set to ω', the inversion frequency of the phase conjugator (see Eq. [7.9]):

$$\vec{\mathbf{E}}(x,y,z,t) = \frac{\hat{\mathbf{n}}}{2}\left[F(x,y)\ A(z,t)\ e^{i[\beta_0(\omega')z-\omega' t]} + C.C.\right]. \tag{7.10}$$

$C.C.$ denotes complex conjugate. Note that β_0 is evaluated at ω'. The units of $A(z,t)$ are chosen to be the square root of power. $A(z,t)$, which is generally complex, contains any frequency, phase, or amplitude modulation, chirp, noise, frequency offset, or other deviations of the electric field from a constant-amplitude carrier. Thus a modulated carrier at a frequency $\omega' + \Omega$ can be described in Eq. (7.10) with

$$A(z,t) = C(z,t)\ e^{-i\Omega t}, \tag{7.11}$$

where $C(z,t)$ now contains the modulation and the second factor contains the frequency offset. This description can be recursive in the sense that $C(z,t)$ may contain further frequency offsets or perhaps a comb of channels of different offsets, all relative to $\omega' + \Omega$.

The propagation equation for the envelope can be obtained (Agrawal 1995) by assuming small nonlinearity and loss and treating dispersion perturbatively:

$$\frac{\partial A}{\partial z} + \sum_{n=1}^{\infty} \frac{i^{n-1}}{n!}\beta_n \frac{\partial^n A}{\partial t^n} + \frac{\alpha}{2}A = i\gamma\,|\,A\,|^2 A. \tag{7.12}$$

Here, β_n is described in Eqs. (7.2) and (7.3):

$$\beta_n = \left.\frac{\partial^n \beta}{\partial \omega^n}\right|_{\omega'}. \tag{7.13}$$

Once again, with no loss in generality, ω' can be used, this time as the center frequency for expansion of $\beta(\omega)$. The terms in the summation account for the group velocity and its successive orders of dispersion. The term containing α, the (power) loss coefficient of the fiber, causes fiber attenuation. We have included on the right side of the equation, the third-order nonlinear term because this derivation will also shed some light on the use of MSSI to cancel fiber nonlinearity (Pepper and Yariv 1980). The nonlinear coefficient, γ, is discussed in Chapter 8 of Volume IIIA. Now define a new amplitude, B, for which the attenuation is removed,

$$B(z,t) = A(z,t)\, e^{\alpha z/2}, \tag{7.14}$$

and a retarded time,

$$\tau = t - z\beta_1, \tag{7.15}$$

so that our reference frame is moving at the group velocity, $v_g(\omega')$. Combining Eqs. (7.12), (7.14), and (7.15) allows the first-order time derivative to be eliminated:

$$\frac{\partial B}{\partial z} + \sum_{n=2}^{\infty} \frac{i^{n-1}}{n!}\, \beta_n\, \frac{\partial^n B}{\partial \tau^n} = ie^{-\alpha z}\gamma \mid B \mid^2 B. \tag{7.16}$$

The description can be simplified without loss of generality by using a coordinate system with $z = 0$ in the middle of the system at the location of the conjugator. The conjugator is assumed to have negligible length. This assumption is not always valid, but the major problems arising from nonnegligible conjugator length can be mitigated (Gnauck, Jopson, and Derosier 1995). Assume that $B_1(z,\tau)$ is the solution to Eq. (7.16) for $z \leq 0$ given some input $B(-L_1,\tau)$. The signal at the middle of the system, $B_1(0,\tau)$ is conjugated as described in Eq. (7.9) and then coupled into the second half of the system. Define $B_2(z,\tau)$ to be the signal in the second part of the system. Thus for $z > 0$, $B_2(z,\tau)$, will be a solution of Eq. (7.16) with the input boundary condition, $B_2(0,\tau) = B_1^*(0,\tau)$. Inserting $B_2(z,\tau)$ into Eq. (7.16):

$$\frac{\partial B_2}{\partial z} + \sum_{n=2}^{\infty} \frac{i^{n-1}}{n!}\, \beta_n\, \frac{\partial^n B_2}{\partial \tau^n} = ie^{-\alpha z}\gamma \mid B_2 \mid^2 B_2. \tag{7.17}$$

Take the conjugate and substituting $z' = z$ yields:

$$\frac{\partial B_2^*}{\partial z'} + \sum_{n=2}^{\infty} \frac{(-1)^n i^{n-1}}{n!}\, \beta_n\, \frac{\partial^n B_2^*}{\partial \tau^n} = ie^{\alpha z'}\gamma \mid B_2 \mid^2 B_2^*. \tag{7.18}$$

Under the conditions that even orders of dispersion can be neglected ($\beta_n \approx 0$ for odd n) and negligible fiber nonlinearity (small $\gamma \mid B_2 \mid^2$), this is just Eq. (7.16). $B_1(z',\tau)$ and $B_2^*(z,\tau)$ have the same boundary conditions at $z = 0$, $z' = 0$, and, under the above conditions, are solutions to the same equation. Therefore they are equal: $B_2(z,\tau) = B_1^*(-z,\tau)$. For $L_2 = L_1$,

$B_2(L_2,\tau) = B_1^*(-L_1,\tau)$, which is the conjugate of the input signal. After (square-law) detection, the output pulse will be identical to the input pulse.

If β_3 is nonzero (the usual case) then its influence will accumulate over the entire length of the span. The system will then behave like an equivalent length of dispersion-shifted fiber having the same β_3 and a dispersion zero at ω'. For the usual case of a signal with a spectrum not centered around ω', the β_3-induced distortion can be reduced by moving the conjugator so that $L_2 \neq L_1$. This can be understood intuitively by realizing that as a consequence of nonvanishing β_3, and for spectra not centered on ω', the magnitude of the dispersion differs in the two parts of the system. By adjusting the position of the conjugator, the integrated dispersion in the two "halves" can be equalized for any one wavelength. This can be understood formally by using Eq. (7.11) to express the signal. From Eq. (7.12), the propagation equation for $C(z,t)$ is:

$$\frac{\partial C}{\partial z} + \sum_{n=1}^{\infty} \frac{i^{n-1}}{n!} \beta_n \sum_{k=0}^{n} \frac{n!}{k!\,(n-k)!} (-i\Omega)^{n-k} \frac{\partial^k C}{\partial t^k} + \frac{\alpha}{2} C = i\gamma \mid C \mid^2 C. \tag{7.19}$$

The zeroth-order derivative of a function is understood to be the function itself. The terms containing each order of time derivative can be collected and the argument proceeds in a manner similar to that presented previously. The retarded frame now follows the group velocity at the signal frequency ($\omega' + \Omega$ in the first half and $\omega' - \Omega$ in the second half). Eq. (7.15) becomes

$$\tau = t - z \left[\sum_{n=1}^{\infty} \frac{\beta_n}{(n-1)!} (\pm\,\Omega)^{n-1} \right]. \tag{7.20}$$

The upper and lower signs are used in the first half and the second half of the system, respectively. As before, the transformations to the retarded frame removes the first-order time derivatives from the equation, and now conjugation is used to cancel the even-order time derivatives as before. However, to accomplish this, it is necessary to scale z by the ratio of the dispersions in the system evaluated at $\omega' + \Omega$ in the first half and $\omega' - \Omega$ in the second half. Hence, even orders of dispersion can be cancelled exactly at one spectral component, $\omega' + \Omega$, of the input signal.

In addition to showing that MSSI can compensate dispersion, Eq. (7.16) shows that MSSI can, in principle, compensate some fiber nonlinearities under the conditions of vanishing dispersion of even order and vanishing fiber loss. The latter condition is stronger than necessary because it arises from the assumption implicit in Eq. (7.14) that the transmission fiber provides exponentially decreasing power. A weaker condition that still provides nonlinearity compensation is that the power variation after the conjugator be a mirror image of the variation before the conjugator. Thus in the absence of even-order dispersion, nonlinearity can still be compensated for in a system containing exponential loss on one side of the conjugator if that loss is mirrored by exponential gain on the other side of the conjugator.

Theoretical studies of using MSSI for compensation generally do not consider polarization mode dispersion (PMD). One would expect that the amount of PMD encountered in practical systems will not significantly impair MSSI dispersion compensation, but will limit MSSI nonlinearity compensation to short (Watanabe and Chikama 1994; Gnauck *et al.* 1995) or to narrow-bandwidth (Kikuchi and Lorattanasane 1993; Watanabe, Chikama *et al.* 1993; Kurtzke and Gnauck 1993b; Pieper *et al.*, 1994) systems.

An intuitive understanding of the use of MSSI for dispersion compensation can be obtained by calculating the dispersion experienced by the signal as it traverses a system. Consider the evolution of the frequency, ω, dependence of the time-indepenent portion of the phase, $\Theta(\omega,z)$, of a lightwave signal as it propagates in the z direction through a system consisting of a length of fiber, L_1, providing a propagation phase shift, a phase conjugator where the accumulated phase is negated, and a second length of fiber, L_2, which adds more propagation delay. As before, the coordinant system is defined so that $z = 0$ at the conjugator. At the beginning of the system, the phase is

$$\Theta(\omega, -L_1) = \theta(\omega), \tag{7.21}$$

where $\theta(\omega)$ is the phase of the input field, $E(x,y,z,t)$ at frequency ω. After propagation through the fibers and the phase conjugator, the phase of the signal component originally at frequency ω is

$$\Theta(\omega,L_2) = -\theta(\omega) - \beta(\omega) L_1 + \beta(2\omega' - \omega) L_2. \tag{7.22}$$

Here the first argument of Θ shows that we are following the phase of the frequency component originally at frequency ω even though the component actually has a frequency of $2\omega' - \omega$ after the conjugator.

Methods for phase conjugating signals in system demonstrations of MSSI include both degenerate and nondegenerate four-wave mixing between the signal and a high-power pump or pumps using either a semiconductor laser (Iannone, Gnauck, and Prucnal 1994), a semiconductor amplifier (Tatham, Sherlock, and Westbrook 1993), or DSF (Watanabe, Naito, and Chikama 1993; Jopson, Gnauck, and Derosier 1993) as the nonlinear element. Three-wave mixing in quasi-phase-matched waveguide devices is another promising method for phase conjugating a signal (Xu *et al.* 1993; Yoo *et al.* 1996). Polarization-independent phase conjugation (Jopson and Tench 1993; Inoue *et al.* 1993) has been used to compensate 560 km of conventional fiber in a 2-channel WDM system operating at 10 Gb/s (Gnauck *et al.* 1994). Eye patterns from this experiment are shown in Fig. 7.7. Without the dispersion compensation provided by phase conjugation, a bit-error-rate

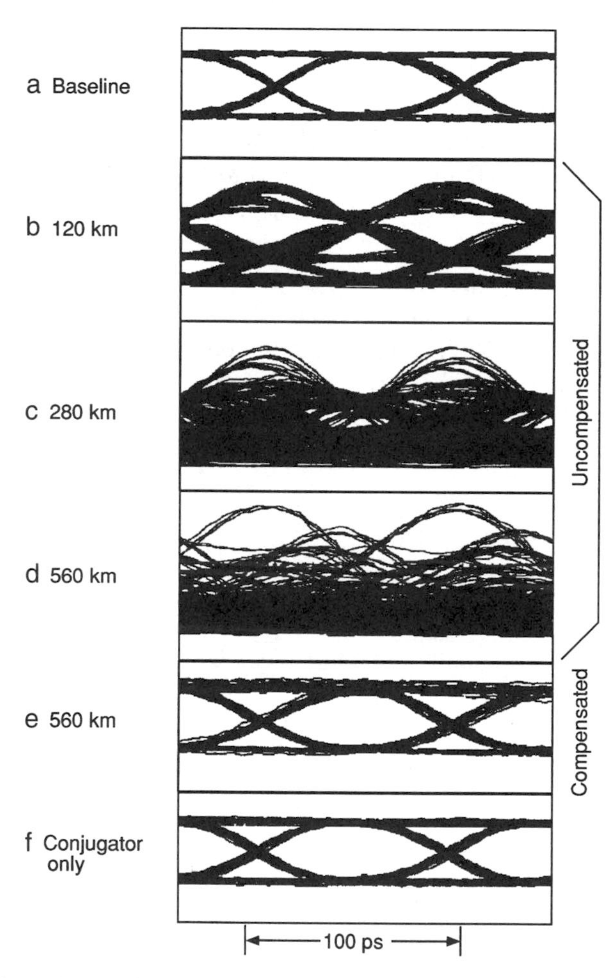

Fig. 7.7 Eye patterns from a 10-Gb/s MSSI experiment. (a) Baseline response. (b)–(d) Uncompensated response at 120, 280, and 560 km. (e) The response at 560 km using MSSI. (f) The baseline response using phase conjugation without transmission fiber.

floor was encountered at 120 km. With MSSI, the dispersion penalty at 560 km was less than 2 dB. Other high bit-rate and short-pulse experiments have included 40-Gb/s transmission over 202 km of fiber (Ellis *et al.* 1995), and transmission of 10-ps pulses over 318 km of fiber (Røyset *et al.* 1995).

It can be seen from Fig. 7.6c that the optimal position for the phase conjugator is not the physical center of the span. This is a consequence of the frequency shift that accompanies the phase conjugation. Because the signal has a different (in Fig. 7.6c, longer) wavelength in the second half of the system than in the first half, the dispersion in the fiber is different and it therefore requires different fiber lengths to balance the dispersions on either side of the conjugator. Figure 7.6d shows the results of not balancing the dispersions before and after the conjugator. The dispersion in the second section over- (under-) compensates the conjugated dispersion in the first section when the conjugator is too close to the transmitter (receiver). The tolerable error in the placement of the conjugator can be determined by geometric construction on Fig. 7.6d, to be plus or minus one-half the dispersion limit (e.g., ±35 km for 10 Gb/s). Figure 7.6e shows the broadening that would occur in various wavelength channels of a WDM system unless each channel is conjugated individually at its optimal location. The location of the conjugator can be perfect for only one wavelength. Channels having a longer than optimal wavelength in the first section will experience more broadening than the optimal wavelength in the first section. After conjugation, they will have a shorter than optimal wavelength and will experience less broadening than the conjugated optimal wavelength. Thus, channels having longer than optimal wavelength will be undercompensated and those with shorter than optimal wavelength will be overcompensated. This limits the regeneration span length of an MSSI-compensated, 10 Gb/s, 30-nm-wide WDM system to about 1000 km. One disadvantage of MSSI is its complexity. It requires one or two single-frequency pump lasers, several narrow band optical filters and, if DSF is used as a conjugator, at least one high-power optical amplifier. It offers the advantage that it can be implemented with commercially available components and that once assembled, it can compensate any span length.

V. The Next Step

One of the next steps in chromatic dispersion compensation will be the commercialization of the more promising techniques. This has already taken place for negative-dispersion fiber. Dispersion compensation will start to be incorporated into other components such as optical amplifiers. The use of multiple gain stages in amplifier modules will allow the compensation to be placed between the gain stages where it will neither materially affect the signal-to-noise ratio nor the output power of the amplifier module. The desired distribution of dispersion compensation in a high-speed link is currently being studied. The viability of many of the techniques discussed here depends on the results of these studies.

Given the large variety of compensation techniques with vastly different capabilities and properties, it is not surprising that dispersion compensation is attracting the attention of an ITU-T (International Telecommunications Union—Telecommunication Standardization Sector) standards committee (Question 25 of Working Party 15/4). At this early stage, they are investigating DCF and some transmitter-based dispersion compensation techniques. As other techniques mature, they will probably be considered as well.

VI. Summary

Dispersion compensation will be required in optically amplified, long-haul 10-Gb/s systems using conventional fiber. Many compensation techniques have been demonstrated and they exhibit a variety of different and often complimentary properties. Transmitter compensation techniques are the most easily implemented but provide a limited amount of compensation. The most commercially advanced technique is negative-dispersion fiber. Chirped Bragg gratings are advancing rapidly but will always be hampered by their narrow bandwidth. The adoption of any particular techniques for use in a high-speed network will depend on the constraints imposed by the, as yet, undefined network architecture.

References

Agrawal, G. P. 1995. *Nonlinear fiber optics,* 2nd edition, p. 42. New York: Academic Press.

Akasaka, Y., R. Sugizaki, A. Umeda, and T. Kamiya. 1996. High-dispersion-compensation ability and low nonlinearity of W-shaped DCF. In *Technical Digest Optical Fiber Communications Conference, San Jose,* 201–202.

Cimini, L. J., Jr., L. J. Greenstein, and A. A. M. Saleh. 1990. Optical equalization for high-bit-rate fiber-optic communications. *IEEE Photon. Tech. Lett.* 2:200–202.

Clark, L., A. A. Klein, and D. W. Peckham. 1994. Impact of fiber selection and nonlinear behavior on network upgrade strategies for optically amplified long interoffice routes. In *Technical Digest NFOEC'94, San Diego,* 231.

Delavaux, J-M. P., J. A. Nagel, K. Ogawa, and D. DiGiovanni. 1994. COBRA: Compensating optical balanced reflective amplifier. In *Technical Digest European Conference on Optical Communications, Firenze,* 5–9. Postdeadline papers.

Desurvire, E. 1994. *Erbium-doped fiber amplifiers.* New York: Wiley.

Eggleton, B. J., T. Stephens, P. A. Krug, G. Dhosi, Z. Brodzeli, and F. Ouellette. 1996. Dispersion compensation over 100 km at 10 Gbit/s using a fiber grating in transmission. In *Technical Digest Optical Fiber Communications Conference, San Jose.* Postdeadline paper PD-5.

Ellis, A. D., M. C. Tatham, D. A. O. Davies, D. Nesset, D. G. Moodie, and G. Sherlock. 1995. 40 Gbit/s transmission over 202 km of standard fiber using midspan spectral inversion. *Electron. Lett.* 31:299–301.

Elrefaie, A. F., R. E. Wagner, D. A. Atlas, and D. G. Daut. 1988. Chromatic dispersion limitations in coherent lightwave transmission systems. *IEEE J. Lightwave Tech.* 6:704–709.

Forghieri, F., R. W. Tkach, and A. R. Chraplyvy. 1996. Dispersion-compensating fiber: Is there merit in the figure of merit? In *Technical Digest Optical Fiber Communication Conference, San Jose.* Paper ThM5.

Garthe, D., J. Ip, P. Colbourne, R. E. Epworth, W. S. Lee, and A. Hadjifotiou. 1996. Low-loss dispersion equaliser operable over the entire erbium window. *Electron. Lett.* 32:371–373.

Gnauck, A. H., S. K. Korotky, J. J. Veselka, J. Nagel, C. T. Kemmerer, W. J. Minford, and D. T. Moser. 1991. Dispersion penalty reduction using an optical modulator with adjustable chirp. *IEEE Photon. Tech. Lett.* 3:916–918.

Gnauck, A. H., C. R. Giles, L. J. Cimini, Jr., J. Stone, L. W. Stulz, S. K. Korotky, and J. J. Veselka. 1991. 8-Gb/s-130 km transmission experiment using Er-doped fiber preamplifier and optical dispersion equalization. *IEEE Photon. Tech. Lett.* 3:1147–1149.

Gnauck, A. H., R. W. Tkach, and M. Mazurczyk. 1993. Interplay of chirp and self phase modulation in dispersion-limited optical transmission systems. In *Technical Digest European Conference on Optical Communications, Montreux,* 105–108.

Gnauck, A. H., R. M. Jopson, P. P. Iannone, and R. M. Derosier. 1994. Transmission of two wavelength-multiplexed 10 Gbit/s channels over 560 km of dispersive fibre. *Electron. Lett.* 30:727–728.

Gnauck, A. H., R. M. Jopson, and R. M. Derosier. 1995. Compensating the compensator: A demonstration of nonlinearity cancellation in a WDM system. *IEEE Photon. Tech. Lett.* 7:582–584.

Gu, X., S. J. Pycock, D. M. Spirit, A. D. Ellis, and C. J. Anderson. 1994. 10 Gbit/s, 138 km uncompensated duobinary transmission over installed standard fibre. *Electron. Lett.* 30:1953–1954.

Hawtoff, D. W., G. E. Berkey, and A. J. Antos. 1996. High figure of merit dispersion compensating fiber. In *Technical Digest Optical Fiber Communications Conference, San Jose.* Postdeadline paper PD-6

Iannone, P. P., A. H. Gnauck, and P. R. Prucnal. 1994. Dispersion-compensated 333-km 10-Gb/s transmission using mid-span spectral inversion in an injection-locked InGaAsP V-groove laser. *IEEE Photon. Tech. Lett.* 6:1046–1049.

Inoue, K., T. Hasegawa, K. Oda, and H. Toba. 1993. Multichannel frequency conversion experiment using fibre four-wave mixing. *Electron. Lett.* 29:1708–1710.

Jopson, R. M., A. H. Gnauck, and R. M. Derosier. 1993. Compensation of fibre chromatic dispersion by spectral inversion. *Electron. Lett.* 29:576–578.

Jopson, R. M., and R. E. Tench. 1993. Polarisation-independent phase conjugation of lightwave signals. *Electron. Lett.* 29:2216–2217.

Jørgensen, B. F. 1994. Unrepeatered transmission at 10 Gbit/s over 204 km standard fiber. In *Technical Digest European Conference on Optical Communication, Firenze,* 685–688.

Kikuchi, K., and C. Lorattanasane. 1993. Compensation for pulse waveform distortion in ultra-long distance optical communication systems by using nonlinear optical phase conjugator. In *Technical Digest Optical Amplifiers and Their Applications Conference, Yokohama,* 22–25.

Koch, T. L., and R. C. Alferness. 1985. Dispersion compensation by active predistorted signal synthesis. *IEEE J. Lightwave Tech.* LT-3:800–805.

Koyama, F., and K. Iga. 1988. Frequency chirping in external modulators. *IEEE J. Lightwave Tech.* 6:87–93.

Kuindersma, P. I., P. P. G. Mols, G. L. A. v. d. Hofstad, G. Cuypers, M. Tomesen, T. van Dongen, and J. J. M. Binsma. 1993. Non-linear dispersion compensation: Repeaterless transmission of 10 Gb/s NRZ over 107 km standard fibre with an EA-MOD/DFB module. In *Technical Digest European Conference on Optical Communications, Montreux,* 89–92.

Kurtzke, C., and A. H. Gnauck. 1993a. Operating principle of in-line amplified dispersion-supported transmission. *Electron. Lett.* 29:1969–1971.

Kurtzke, C., and A. H. Gnauck. 1993b. How to increase capacity beyond 200 Tbit/s-km without solitons. In *Technical Digest European Conference on Optical Communications, Montreux.* Postdeadline paper ThC12.12.

Kuwano, S., K. Yonenaga, and K. Iwashita. 1995. 10-Gbit/s repeaterless transmission experiment of optical duobinary modulated signal. *Electron. Lett.* 31:1359–1361.

Loh, W. H., R. I. Laming, A. D. Ellis, and D. Atkinson. 1996. Dispersion compensated 10 Gbit/s transmission over 700 km of standard single mode fiber with 10 cm chirped fiber grating and duobinary transmitter. In *Technical Digest Optical Fiber Communications Conference, San Jose.* Postdeadline paper PD-30.

May, G, A. Solheim, and J. Conradi. 1994. Extended 10 Gb/s fiber transmission distance at 1538 nm using a duobinary receiver. *IEEE Photon. Tech. Lett.* 6:648–650.

Oda, K., M. Fukutoku, T. Kukui, T. Kitoh, and H. Toba. 1995. 16-channel × 10-Gbit/s optical FDM transmission over 1000 km conventional single-mode fiber employing dispersion compensating fiber and gain equalization. In *Technical Digest Optical Fiber Communications Conference, San Diego.* Postdeadline paper PD-22.

Ogata, T., S. Nakaya, Y. Aoki, T. Saito, and N. Henmi. 1992. Long-distance, repeaterless transmission utilizing stimulated Brillouin scattering suppression and dispersion compensation. In *Technical Digest Fourth Optoelectronics Conference, Makuhari Messe,* 104–105.

Onishi, M., C. Fukuda, H. Kanamori, and M. Nishimura. 1994. High NA double-cladding dispersion compensating fiber for WDM systems. In *Technical Digest European Conference on Optical Communications, Firenze,* 681–684.

Onishi, M., H. Kanamori, T. Kato, and M. Nishimura. 1996. Optimization of dispersion compensating fibers considering self-phase modulation suppression. In *Technical Digest Optical Fiber Communications Conference, San Diego,* 200–201.

Park, Y. K., T. Nguyen, P. A. Morton, J. E. Johnson, O. Mizuhara, J. Jeong, L. D. Tzeng, P. Yeates, T. Fullowan, P. F. Sciortino, A. M. Sergent, and W. T. Tsang. 1996. Dispersion penalty free transmission over 130 km standard fiber using a 1.55 μm, 10 Gb/s integrated EA/DFB laser with low extinction ratio and negative chirp. *IEEE Photon. Tech. Lett.* 8:1255–1257.

Patel, B. L., E. M. Kimber, M. A. Gibbon, E. J. Thrush, D. J. Moule, A. Hadjifotiou, and J. G. Farrington. 1992. Transmission at 10 Gb/s over 100 km using a high performance electroabsorption modulator and the direct prechirping technique. In *Technical Digest European Conference on Optical Communications, Berlin,* 859–862.

Pepper, D. M., and A. Yariv. 1980. Compensation for phase distortions in nonlinear media by phase conjugation. *Opt. Lett.* 5:59–60.

Pieper, W., C. Kurtzke, R. Schnabel, D. Breuer, R. Ludwig, K. Petermann, and H. G. Weber. 1994. Nonlinearity-insensitive standard-fibre transmission based on optical-phase conjugation in a semiconductor-laser amplifier. *Electron. Lett.* 30:724–726.

Poole, C. D., J. M. Wiesenfeld, and D. J. DiGiovanni. 1993. Elliptical-core dual-mode fiber dispersion compensator. *Photon. Tech. Lett.* 5:194–197.

Price, A. J., and N. Le Mercier. 1995. Reduced bandwidth optical digital intensity modulation with improved chromatic dispersion tolerance. *Electron. Lett.* 31:58–59.

Price, A. J., L. Pierre, R. Uhel, and V. Havard. 1995. 210 km repeaterless 10 Gb/s transmission experiment through nondispersion-shifted fiber using partial response scheme. *IEEE Photon. Tech. Lett.* 7:1219–1221.

Rasmussen, C. J., B. F. Jørgensen, R. J. S. Pedersen, and F. Ebskamp. 1995. Optimum amplitude- and frequency-modulation in an optical communication system based on dispersion supported transmission. *Electron. Lett.* 31:746–747.

Røyset, A., S. Y. Set, I. A. Goncharenko, and R. I. Laming. 1995. Transmission of <10 ps pulses over 318 km standard fibers using midspan spectral inversion. In *Technical Digest European Conference on Optical Communications, Brussels,* 577–580.

Schiess, M. 1994. Extension of the dispersion limit by pulse shaping and profiting of fibre nonlinearities. In *Technical Digest European Conference on Optical Communications, Firenze,* 423–426.

Schiess M., and H. Carlden. 1994. Evaluation of the chirp parameter of a Mach–Zehnder intensity modulator. *Electron. Lett.* 30:1524–1525.

Takachio, N., S. Norimatsu, and K. Iwashita. 1992. Optical PSK synchronous heterodyne detection experiment using fiber chromatic dispersion equalization. *IEEE Photon. Tech. Lett.* 4:278–280.

Takiguchi, K., K. Okamoto, and K. Moriwaki. 1994. Dispersion compensation using a planar lightwave circuit optical equalizer. *IEEE Photon. Tech. Lett.* 6:561–564.

Tatham, M. C., G. Sherlock, and L. D. Westbrook. 1993. Compensation of fibre chromatic dispersion by mid-way spectral inversion in a semiconductor laser amplifier. In *Technical Digest European Conference on Optical Communications, Montreux.* Postdeadline paper ThP12.3.

Tkach, R. W., R. M. Derosier, F. Forghieri, A. H. Gnauck, A. M. Vengsarkar, D. W. Peckham, J. L. Zyskind, J. W. Sulhoff, and A. R. Chraplyvy. 1995. Transmission of eight 20-Gb/s channels over 232 km of conventional single-mode fiber. *IEEE Photon. Tech. Lett.* 7:1369–1371.

Vengsarkar, A. M., and W. A. Reed. 1993. Dispersion-compensating single-mode fibers: Efficient designs for first- and second-order compensation. *Opt. Lett.* 18:924–926.

Vengsarkar, A. M., A. E. Miller, M. Haner, A. H. Gnauck, W. A. Reed, and K. L. Walker. 1994. Fundamental-mode dispersion-compensating fibers: Design considerations and experiments. In *Technical Digest Optical Fiber Communications Conference, San Jose,* 225–227.

Watanabe, S., T. Naito, and T. Chikama. 1993. Compensation of chromatic dispersion in a single-mode fiber by optical phase conjugation. *IEEE Photon. Tech. Lett.* 5:92–95.

Watanabe, S., T. Chikama, G. Ishikawa, T. Terahara, and H. Kuwahara. 1993. Compensation of pulse shape distortion due to chromatic dispersion and Kerr effect by optical phase conjugation. *IEEE Photon. Tech. Lett.* 5:1241–1243.

Watanabe, S., and T. Chikama. 1994. Cancellation of four-wave mixing in multichannel fibre transmission by midway optical phase conjugation. *Electron. Lett.* 30:1156–1157.

Wedding, B., B. Franz, and B. Junginger. 1994. 10-Gb/s optical transmission up to 253 km via standard single-mode fiber using the method of dispersion-supported transmission. *IEEE J. Lightwave Tech.* 12:1720–1727.

Winters, J. H., and R. D. Gitlin. 1990. Electrical signal processing techniques in long-haul fiber-optic systems. *IEEE Trans. Commun.* 38:1439–1453.

Xu, C. Q., H. Okayama, K. Shinozaki, and M. Kawahara. 1993. Wavelength conversions at 1.5 μm by difference frequency generation in periodically domain-inverted LiNb O_3 channel waveguides. *Appl. Phys. Lett.* 63:1170–1172.

Yariv, A., D. Fekete, and D. M. Pepper. 1979. Compensation for channel dispersion by nonlinear optical phase conjugation. *Opt. Lett.* 4:52–54.

Yonenaga, K., S. Kuwano, S. Norimatsu, and N. Shibata. 1995. Optical duobinary transmission system with no receiver sensitivity degradation. *Electron. Lett.* 31:302–304.

Yoo, S. J. B., C. Caneau, R. Bhat, and M. A. Koza. 1996. Transparent wavelength conversion by difference frequency generation in AlGaAs waveguides. In *Technical Digest Optical Fiber Communications Conferences, San Jose.* Paper WG7.

Chapter 8 | Fiber Nonlinearities and Their Impact on Transmission Systems

F. Forghieri

AT&T Laboratories–Research, Holmdel, New Jersey

R. W. Tkach

AT&T Laboratories–Research, Holmdel, New Jersey

A. R. Chraplyvy

Lucent Technologies, Bell Laboratories, Holmdel, New Jersey

I. Introduction

Advances in 1.5-μm amplifier technology have reshaped the lightwave communications landscape. Erbium-doped fiber amplifiers have fundamentally altered two aspects of lightwave systems. Regenerator spacings can be increased from the typical 40 km to distances spanning oceans. In addition, wavelength-division multiplexing (WDM) is not only practical but, in many cases, a more economical method of increasing capacity than time-division multiplexing (TDM). However, implementation of long unregenerated spans and WDM gives rise to new problems, namely optical nonlinearities. The references [1–164] at the end of this chapter are a sampling of the rich literature associated with these effects and their impact on non-return-to-zero (NRZ) lightwave systems. Soliton transmission systems are considered in Chapter 12 in Volume IIIA. Interactions between propagating light signals and the transmission medium can lead to interference, distortion, or excess attenuation of the optical signals.

The nonlinearities in silica fibers can be classified into two categories: stimulated scattering (Raman and Brillouin), and effects arising from the nonlinear index of refraction. Stimulated scattering is manifested as intensity-dependent gain or loss, while the nonlinear index gives rise to an intensity-dependent phase of the optical field.

OPTICAL FIBER TELECOMMUNICATIONS,
VOLUME IIIA

ISBN: 0-12-395170-4

In stimulated Brillouin scattering (SBS), a strong wave traveling in one direction provides narrow band gain, with a linewidth on the order of 20 MHz, for light propagating in the opposite direction. In general, this effect does not couple channels in a wavelength multiplexed system. The system impact of SBS arises when signal power is transferred in a backward direction, due to amplification of spontaneous scattering, leading to depletion of the forward traveling signal. The threshold for this effect is typically a few mW for an unmodulated wave. Stimulated Raman scattering (SRS) is a similar but much weaker effect. The main differences are that SRS can occur in either the forward or backward direction, and the gain bandwidth for SRS is on the order of 12 THz or 100 nm at 1.5 μm. Thus, SRS can couple WDM channels and give rise to cross talk. Cross talk begins to be significant for 10 1-mW channels with 1-nm spacing in a multi-Mm length system. The intensity-dependent refractive index of silica gives rise to three effects: self-phase modulation (SPM), where fluctuations in the signal power give rise to modulation of the signal phase and lead to broadening of the spectrum; cross-phase modulation (CPM), in which intensity fluctuations in one channel propagating in the fiber modulate the phase of all other channels; and four-photon mixing (FPM), where the beating between two channels at their difference frequency modulates the signal phase at that frequency, generating new tones as sidebands. Modulation instability (MI) can be described as FPM phase matched by SPM, leading to exponential gain for FPM products. The ways in which these effects impair lightwave system performance and the constraints they impose on advanced system design are described. More important, techniques for reducing the impact of optical nonlinearities are discussed.

The effects of nonlinear interactions between two copolarized signals in a fiber can be expressed by the change in the electric field of one of the signals caused by the other after propagating through a distance dz. In general,

$$E_1(z + dz) = E_1(z) \exp\left[(-\alpha/2 + ik)\, dz + gP_2(z)dz/2A_e\right], \quad (8.1)$$

where α is the loss coefficient of the fiber, g is the frequency-dependent gain coefficient of the nonlinear process, $P_2(z)$ is the injected power of the other signal, and A_e is the effective area of the fiber. The effective area can be evaluated by overlap integrals [65, 139, 141], but it is easier to simply note that, in general, conventional unshifted ($\lambda_0 = 1.3$ μm) fibers have effective areas of 80 μm^2, dispersion-shifted and non-zero-dispersion fibers have 55-μm^2 effective areas. The effective areas of dispersion-compensating fibers (DCF) are typically much smaller, on the order of 20 μm^2.

The gain coefficient g represents the strength and nature of the nonlinearities. For example, if g is a real quantity, the nonlinearity will produce optical gain or loss such as in SRS and SBS. If g is imaginary, the nonlinearity produces phase modulation and leads to SPM, CPM, and FPM. In some cases, for example, in the absence of nonlinear signal depletion and pulse distortion, Eq. (8.1) can be integrated over the length of the fiber

$$E_1(L) = E_1(0) \exp [(-\alpha/2 + ik) L + gP_2(0)L_e/2A_e], \qquad (8.2)$$

where L_e is the effective nonlinear length of the fiber that accounts for the fiber loss and is given by

$$L_e = \frac{1 - e^{-\alpha L}}{\alpha}. \qquad (8.3)$$

There are two interesting limits for L_e. For $\alpha L \ll 1$, $L_e \approx L$, the actual fiber length; for $\alpha L \gg 1$, $L \approx 1/\alpha$. For typical fibers, the $1/\alpha$ length is about 20 km.

In more complex situations, for example, in the presence of many signals or in the case of signal depletion due to nonlinearities or even modest pulse distortions due to dispersion, Eq. (8.1) cannot be easily integrated. In these situations, the evolution of signals can only be determined by solving the nonlinear Schrödinger equation for the entire propagating electric field. This equation can be solved analytically for solitons (see Chapter 12 in Volume IIIA). In other cases, computer simulations using the split-step method [2] can be used. In this chapter, such simulations are used for several illustrative examples. These simulations begin by creating an electric field for each channel, including laser linewidth as a random walk of the phase, which is externally modulated. The modulator is typically modeled as a Mach–Zehnder modulator with variable chirp. The electric fields of the channels are then summed to give a net electic field input to the transmission fiber. Propagation through the fiber is simulated by solving the nonlinear Schrödinger equation with a split-step algorithm. To keep calculation error within acceptable limits, the step size is limited by a maximum allowable nonlinear phase shift as well as a maximum phase shift due to chromatic dispersion between the furthest separated signal components. At each amplifier site, the signal is amplified so that the signal energy is restored to that at input. Some simulations incorporate detailed models of erbium-doped amplifiers, others assume flat gain. At the output of the system, after a final amplification, the channels are filtered optically with a Bessel filter of variable order with a bandwidth of several times the bit rate, and then electrically filtered at base band. Output bit patterns, eye diagrams, and power spectra can then be plotted.

The parameters used in the various simulations presented in this chapter are as follows. Typically the chirp parameter of the modulator was set to zero. The step size was calculated to keep the phase shift due to nonlinearities below 3 mrad in a single step. In any case, the maximum step size was limited to 0.1 km. Tenth-order optical Bessel filters and twentieth-order electrical Bessel filters (with bandwidth of 0.7 times the bit rate) were used. Typically 64-bit patterns were used in the simulations.

II. Stimulated Brillouin Scattering

Brillouin scattering is the interaction of light with sound waves in matter. Sound waves in glass, for example, cause a variation in the index of refraction corresponding to the density variations of the wave. Light can be diffracted by these index gratings if the Bragg condition is met. In single-mode optical fibers, which have a well-defined axis of propagation, the only possible diffraction from these moving gratings corresponds to reflection in the backward direction. Since the grating is formed by a wave moving at the speed of sound, the reflected wave experiences a Doppler shift, given by:

$$\nu_B = \frac{2nV_s}{\lambda},$$

where n is the index of refraction, V_s is the speed of sound in the glass, and λ is the wavelength of the light. This frequency shift is to lower frequencies for sound waves traveling in the same direction as the incident light (called Stokes scattering), and to higher frequencies for counterpropagating sound waves (called anti-Stokes scattering). The bandwidth of this process is determined by the acoustic attenuation of the glass, which limits the spatial extent of the grating formed by sound waves. For silica fibers at 1500 nm, this results in a bandwidth of about 20 mHz. For sources with linewidths larger than this, the efficiency of the Brillouin process is reduced. If there is significant optical power at the frequency of the downshifted light, the interference pattern between the incident light and downshifted light can create sound waves, which in turn scatter more of the incident light. SBS occurs when the incident light is of sufficiently high intensity that the energy added to the sound waves by Stokes scattering significantly increases their amplitude, thus significantly increasing the probability of scattering more of the incident light. It is interesting to note that the same process cannot occur for anti-Stokes scattering since energy is removed from the sound waves in that process. This leads to an exponential gain for light at the stokes frequency. For a fiber of length L, with a pump power P_p injected at $z = 0$, and a Stokes power of P_s, injected at $z = L$,

$$P_s(0) = P_s(L) \exp\left(-\alpha L + \frac{g_B P_P L_e}{2A_e}\right), \tag{8.4}$$

where $g_B = 4 \times 10^{-9}$ cm/W is the Brillouin gain coefficient. The factor of 2 in the denominator accounts for averaging over polarization. The SBS threshold is defined as the input power at which the scattered power grows as large as the input power (in the undepleted pump approximation). The threshold power [32, 133] is given by

$$P_B{}^{th} = \frac{42A_e}{g_B L_e} \cdot \left(1 + \frac{\Delta\nu_s}{\Delta\nu_B}\right), \tag{8.5}$$

where $\Delta\nu_s$ is the linewidth of the source, and $\Delta\nu_B$ is the Brillouin linewidth. This equation shows that the threshold power is increased as the source linewidth increases beyond the Brillouin linewidth [139]. Similarly, the threshold increases if the speed of sound in the glass is nonuniform, leading to a broadening of the bandwidth of the Brillouin process in the fiber, and a concomitant reduction in the peak value of the gain coefficient, g_B.

Figure 8.1 shows the behavior of the power transmitted through an optical fiber as a function of the launched power for a narrow linewidth source [102]. Also shown is the backscattered power. The transmitted power grows linearly with input power for low input powers but becomes constant for high input powers. In this example, no more than 3 mW can be transmitted. Unlike other nonlinear effects in lightwave systems, the threshold for observing SBS does not decrease in a long amplified system. Because practical optical amplifiers contain one or more optical isolators, the backscattered light from the Brillouin process does not accumulate from amplifier span to amplifier span.

Clearly, lightwave systems will be impaired if power intended to be transmitted is instead reflected, since above threshold the loss of the fiber appears to grow linearly with input power. One might imagine scenarios where this power-limiting effect could be useful in regulating systems; however, this limiting behavior is accompanied by a dramatic increase in intensity noise, as can be seen in Fig. 8.2. This figure shows the carrier-to-

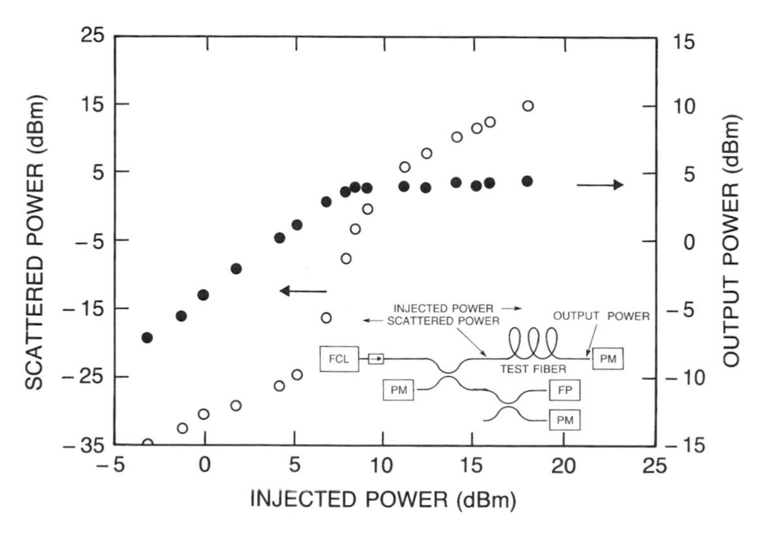

Fig. 8.1 Plot of transmitted and scattered power for a 13-km length of dispersion-shifted fiber. The experimental diagram is shown as an inset. FCL, F-center laser; FP, Fabry–Perot equation; PM, power meter.

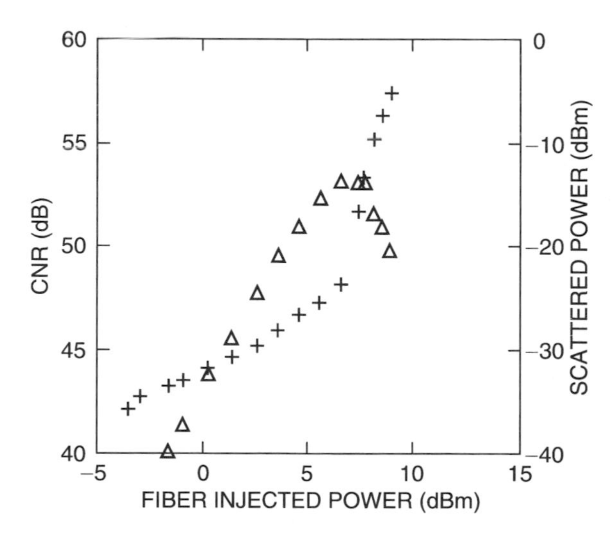

Fig. 8.2 Plot of scattered power (*crosses*) and CNR (*triangles*) for the same fiber as in Fig. 8.1, when subcarrier-modulated laser light is injected (AM-VSB format).

noise ratio (CNR) for a subcarrier-modulated laser transmitted over the same fiber as in Fig. 8.1 [103]. At low power, the CNR is determined by shot noise and increases with launched power. However, just as the scattered power begins to rise, the noise performance degrades. Similar degradation in performance is seen in digital systems [40]. If reflections are present in the fiber plant, the large gains provided by the Brillouin process can result in laser oscillation at powers less than the threshold quoted previously. In fact, experimental measurements of the SBS threshold require extreme care to eliminate all reflections approaching −33 dB, which is the reflection level provided by Rayleigh scattering.

Thus, from a system point of view, it is essential that no optical signal in a fiber should exceed the SBS threshold. In WDM systems, the signals need only remain individually below threshold since there is no interaction among the channels. As noted previously, the threshold observed in a system may be higher than that given by Eq. (8.5) if either the source linewidth is broader than the natural Brillouin linewidth of 20 MHz, or if fiber properties result in a broadening of the Brillouin linewidth for the fiber used. Sources modulated at high bit rates typically have bandwidths larger than the natural Brillouin linewidth, and hence reduced Brillouin gain. The SBS threshold for an unmodulated, narrow-linewidth source has been measured to be about 6 dBm for long conventional fibers [102]. For narrow linewidth sources, modulated with zero chirp, at bit rates higher than 100 Mb/s, the threshold is due to the remaining carrier component of the signal, which carries half the average power. When an on–off keyed signal exceeds the SBS threshold, it is this carrier component that is backscattered. This leads to distortion of the transmitted signal as observed by Kawakami *et al.* [86].

The relevant source characteristic for determining the SBS threshold is the linewidth of the carrier component of the modulated signal spectrum. Under ideal conditions, this carrier component retains the linewidth of the source laser and carries half the average power, which gives a 3-dB increase in the threshold power, to 9 dBm. However, most sources will not have such narrow 3-dB widths under operating conditions. If directly modulated lasers are used, or even lasers with integrated electroabsorption modulators, the 3-dB width under modulation is likely to be larger than the 20-MHz Brillouin linewidth. For a source whose 3-dB linewidth under modulation is 100 MHz, the threshold would be raised by another 7 dB to 16 dBm.

While these power levels are high enough for many applications, higher powers can be needed, for example, in repeaterless systems [59]. The SBS threshold can be further increased by artificially broadening either the source linewidth [40, 59, 139] or the Brillouin bandwidth of the fiber [102, 132]. An alternative method is to suppress the carrier component of the modulated signal, for example using the duobinary modulation scheme proposed by Kuwano *et al.* [93].

Dithering of the transmitter laser frequency provides a simple way of broadening the source linewidth and is effective, provided the dither frequency is high enough. In practice, this means that the dither frequency should be on the order of the transit time of the light through an effective length of the fiber — roughly, tens of kHz. However, as the injected power is increased, the relevant length is decreased due to the extremely large Brillouin gain. The dither frequency should scale as the ratio of the injected power to the SBS threshold. For injected powers as high as 1 W, the dither frequency should approach 10 MHz. The amplitude of the frequency dither for this high power should exceed 5 GHz. Although it may appear that dithering can suppress SBS without limit, the excess bandwidth resulting from the dither will eventually result in penalties from filtering or chromatic dispersion. A similar Brillouin suppression technique uses phase modulation of the source [59].

Concatenation of fibers with different Brillouin shifts leads to a fiber with a broadened Brillouin linewidth. This broadening can increase the SBS threshold. However, it is important to note that the most important part of the fiber for SBS is the first absorption length after the input. This is roughly 20 km; thus any segmentation of the fiber to significantly suppress SBS will have to involve lengths of the order of 1 km [102]. SBS can also be suppressed by simply adding isolators to the line [147]. Again, the isolated sections must be short near the fiber input. The fiber's Brillouin linewidth can also be broadened by strain applied in cabling, or varying the draw conditions when the fiber is manufactured [163].

This abundance of suppression techniques results in system limitations being set by other nonlinearities, in general.

III. Self-Phase Modulation

The index of refraction of silica is weakly intensity dependent, so that the index has the form [135]

$$n = n_0 + n_2 \frac{P}{A_e}. \tag{8.6}$$

The coefficient n_2, for silica fibers, is 2.6×10^{-20} m^2/W [90]. This number takes into account averaging of the polarization state of the light as it travels down the fiber. The nonlinear contribution to the index of refraction results in a phase change for light propagating in a fiber of

$$\Phi_{NL} = \gamma P L_e, \tag{8.7}$$

where we have defined the nonlinear coefficient

$$\gamma = \frac{2\pi n_2}{\lambda A_e}. \tag{8.8}$$

This phase change becomes significant (roughly $\pi/2$) when the power times the net effective length of the system reaches 1 mW-Mm or 1 W-km. The first set of units is appropriate for long amplified systems and the second for repeaterless systems. Although the nonlinearity of the index is small, the lengths and powers that have been made possible by the use of the optical amplifiers guarantee that the nonlinear index will play a role in future systems.

SPM occurs when an intensity-modulated signal travels through an optical fiber. Due to the nonlinear index of refraction, the peak of a pulse travels more slowly (or more rigorously, accumulates phase more quickly) than the wings. This results in the wavelength being stretched on the leading edge of the pulse and compressed on the trailing edge. Thus, the leading edge of the pulse acquires a "red shift" and the trailing edge acquires a "blue shift." The signal then is broadened in the frequency domain by an amount

$$\Delta B = \gamma L_e \frac{dP}{dt}, \tag{8.9}$$

where dP/dt is the time derivative of the pulse power. This is the frequency shift experienced by the transition portion of the pulse. This broadening may result in penalties due to filter bandwidth or pulse distortion arising from chromatic dispersion.

In the "normal" dispersion regime, where the chromatic dispersion is negative, corresponding to decreasing group delay with wavelength, the leading edge of a pulse, which is red-shifted by nonlinearity, travels more quickly, and moves away from the center of the pulse. The trailing edge, which has been blue-shifted, travels more slowly and also moves away from the center of the pulse. So the pulse is broadened by the combined effects of nonlinearity and dispersion. In the "anomalous" dispersion regime, corresponding to increasing group delay with wavelength and positive chromatic dispersion, the red-shifted leading edge, travels more slowly, and moves toward the center of the pulse. Similarly, the trailing edge of the pulse, which has been blue-shifted, travels more quickly, and also moves toward the center of the pulse. Thus, the pulse narrows.

This pulse narrowing in the anomalous dispersion regime suggests that SPM can be used to compensate for dispersion. The ultimate example of this is soliton transmission (see Chapter 12 in Volume IIIA); however, even with conventional NRZ pulses, there is some compensation. Figure 8.3 shows simulations of the performance of a 10-Gb/s system in conventional fiber with 17 ps/nm-km chromatic dispersion (anomalous) for 10, 20, and 40 mW average power, as well as the linear case (obtained by setting the nonlinear index to zero in the simulation). The linear case shows a 1-dB penalty at roughly 60 km, while when nonlinearity is taken into account, the penalty decreases becoming effectively zero at 40 mW average power.

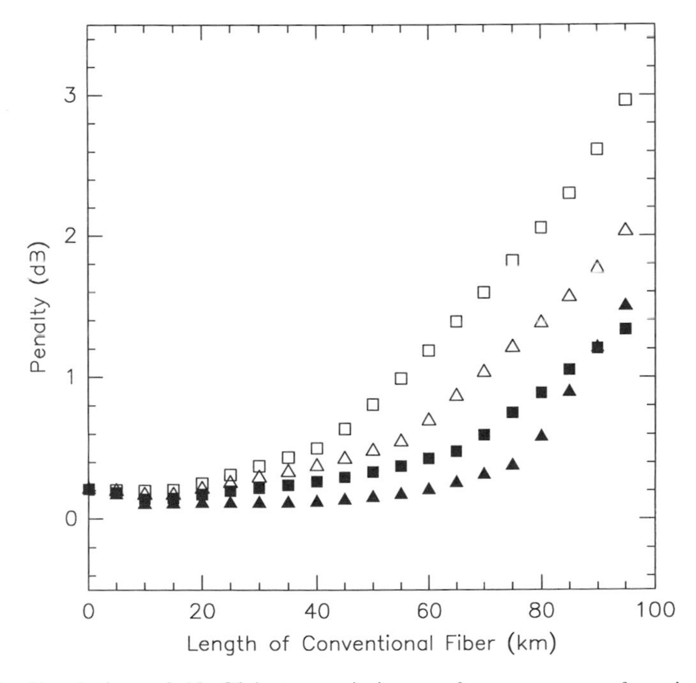

Fig. 8.3 Simulation of 10 Gb/s transmission performance as a function of the length of conventional fiber used. Results are shown for linear transmission (*open squares*), 10 mW average power (*open triangles*), 20 mW (*filled squares*), and 40 mW (*filled triangles*).

Note that the curve for 40 mW crosses that for 20 mW at 85 km. It is typically seen that higher powers exhibit a more sudden degradation in performance as the length is increased.

In systems with dispersion compensation, this competition between dispersion and nonlinearity leads to the use of an amount of compensation that is less than would be expected on linear grounds. Figure 8.4 shows the penatly for 10-GB/s transmission over 170 km of conventional fiber with an average power of 40 mW, as a function of the amount of dispersion compensation. The horizontal axis is labeled with the length of the transmission fiber that is compensated. In other words, at the right end of the axis the compensation is complete, and at the left end there is a residual dispersion corresponding to 80 km of conventional fiber. In the nonlinear case, there is clearly an optimum range of compensation that is less than full compensation. The penalty in the nonlinear case is lower than in the linear case over most of the plot, but rises steeply near full compensation because of the broadening of the signal spectrum arising from SPM.

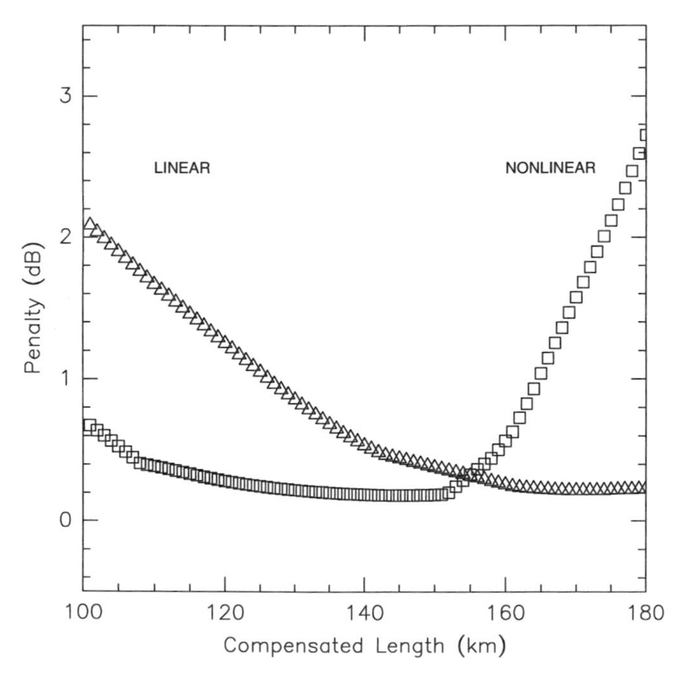

Fig. 8.4 Simulation of 10 Gb/s transmission over 180 km as a function of the degree of dispersion compensation; 180-km compensated length corresponds to full compensation. Linear transmission (*triangles*), nonlinear transmission (*squares*).

More extensive investigation of these effects for multiple amplified spans, including some analytical modeling, can be found in the paper by Kikuchi and Sasaki [88]. Nuyts and Park [116] have performed extensive simulations of a 360-km three-span system, determining the optimal amount of dispersion compensation in each span.

The picture presented so far has neglected any nonlinearity in the compensation. If dispersion compensating fiber is used, the small core areas and relatively high losses of these fibers lead to significant nonlinearity. Kikuchi and Sasaki [89] and Miyata *et al.* [113] add this consideration. Forghieri *et al.* [47] showed that the traditional figure of merit (ps/nm-dB) for DCF is insufficient when nonlinearity is considered. In particular, for equal ps/nm-dB, higher dispersion is preferable.

Another approach to lessen the impact of SPM is spectral inversion. This subject is examined in detail in Chapter 7 in Volume IIIA.

IV. Cross-Phase Modulation

Cross-phase modulation is another way in which intensity fluctuations affect the phase of a signal [81, 139]. In this case, the responsible intensity fluctuations arise from the modulation of other channels present in a WDM system. The frequency shift caused by intensity variations in another channel is given by

$$\Delta B = 2\gamma L_e \frac{dP}{dt}, \tag{8.10}$$

where dP/dt is the time derivative of the power in the interfering pulse. This is analogous to the expression given for SPM (see Eq. [8.9]) except for a factor of 2. This factor of 2 arises from the counting of terms in the expansion of the nonlinear polarization. Thus one is lead to assume that CPM will be a more severe limitation than SPM for WDM systems because the effect is twice as large for each interfering channel and there are presumably many other channels to generate this interference. However, chromatic dispersion plays a very significant role in the system impact of CPM. As in the case of SPM, there is no impact of the spectral broadening introduced by CPM if there is no chromatic dispersion to transform this spectral broadening into pulse broadening. However, in addition, since CPM is an interaction between distinct channels, the presence of chromatic dispersion also means that pulses in interfering channels will not, in general, remain superimposed on the pulses in the channel of interest. This tends to lessen the impact of CPM. This is most easily seen if we consider two isolated pulses in different channels propagating in a lossless fiber. As the leading edge of the interfering pulse is superimposed on the center of the pulse of interest, the signal pulse receives a red shift due to the rising intensity of the interfering pulse. As the pulses walk through each other, and the trailing edge of the interfering pulse is superimposed, the signal pulse receives a blue shift. In fact, the net frequency shift experienced by a particular part of the signal pulse is proportional to the integral of the derivative of the intensity of the interfering pulse. Thus, when the pulses have completely walked through each other, there is no residual effect.

Figure 8.5 shows simulation results that demonstrate this effect. The simulation considers two 400-ps pulses separated by 1 nm propagating in lossless fiber with a chromatic dispersion equal to that of conventional single-mode fiber (16 ps/nm-km). At the beginning of the fiber, the pulses have a center-to-center separation of 800 ps and the figure shows optical spectra at the input (0 km) and every 25 km through the fiber. The signal pulse has a 2-mW peak power so that SPM is negligible in this short fiber, while the interfering pulse has a peak power of 20 mW so that the effects of CPM on the signal pulse should be large. At the input (0 km) the pulse spectra have the expected form arising from the square input pulse shape. At 25 km into the fiber, we can see the effect of SPM on the interfering pulse broadening the spectrum. The effect of CPM on the signal pulse is beginning to be visible and new signals generated by FPM can also be seen. At 50 km, the pulses are perfectly overlapped. It is interesting to note that the broadening of the signal due to CPM is smaller than the broadening of the interfering pulse arising from SPM. Absent dispersion, one would expect the CPM effect to be larger owing to the factor of 2 in Eq. (8.10); however, the fact that the edge of the interfering pulse moves through the signal means that its dwell time on any particular portion of the signal pulse is diminished. At 75 km, the falling edge of the interfering pulse begins to pass through the signal pulse. The broadening of the signal pulse is now less than at 50 km since the portion of the interfering pulse with falling intensity produces a frequency shift of opposite sign to that previously produced by the rising edge. A similar mechanism is also responsible for the diminution of the signals due to FPM [106]. At 100 km, the pulses have completely passed through each other, and the effects of CPM have completely vanished, while the effect of SPM on the interfering pulse remains.

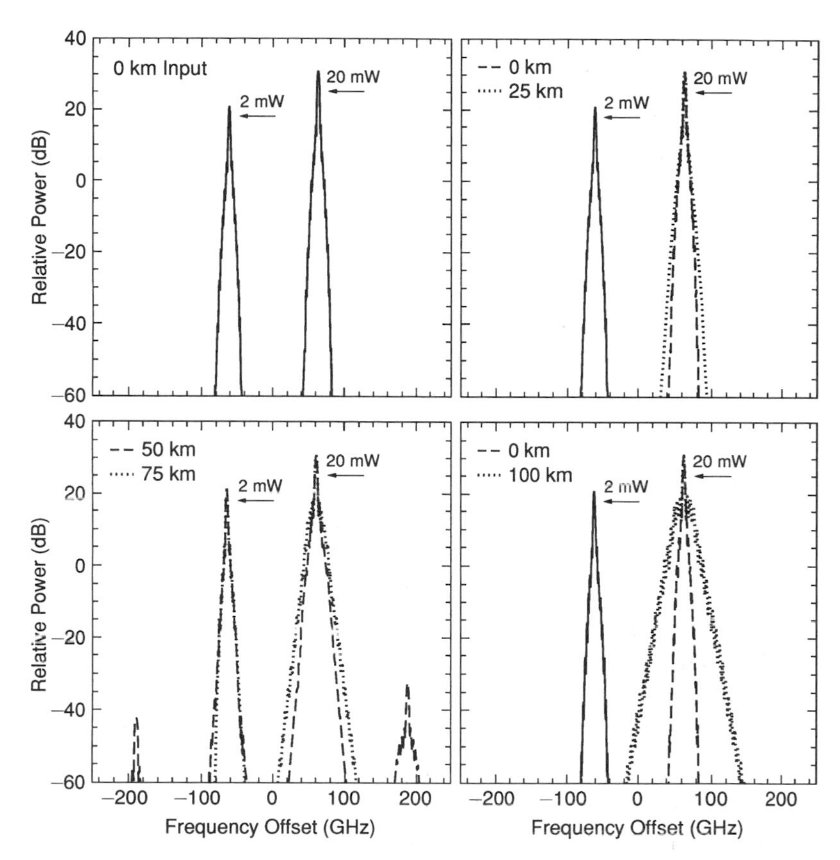

Fig. 8.5 Simulation of two 400-ps pulses propagating over 100 km of lossless fiber with chromatic dispersion of 16 ps/nm-km and a channel spacing of 1 nm. The signal pulse has a peak power of 2 mW, while the interfering pulse has a peak power of 20 mW. The pulses are separated by 800 ps (center to center) at the input (0 km), begin to overlap at 25 km, are perfectly superimposed at 50 km, and are separated again at 100 km.

This illustrates that dispersion can result in a lessening of the effects of CPM, at least in a lossless fiber. Figure 8.6 shows results from a somewhat less idealized case [108]. Three channels, each with an average power of 5 mW, are launched into conventional fiber with chromatic dispersion of 16 ps/nm-km. The horizontal axis is the channel spacing and plotted on the vertical axis is the ratio of the mean height of the "ones" level to the standard deviation of the "ones" level. The performance of a single-channel system is given by the dotted line. The multichannel performance is substan-

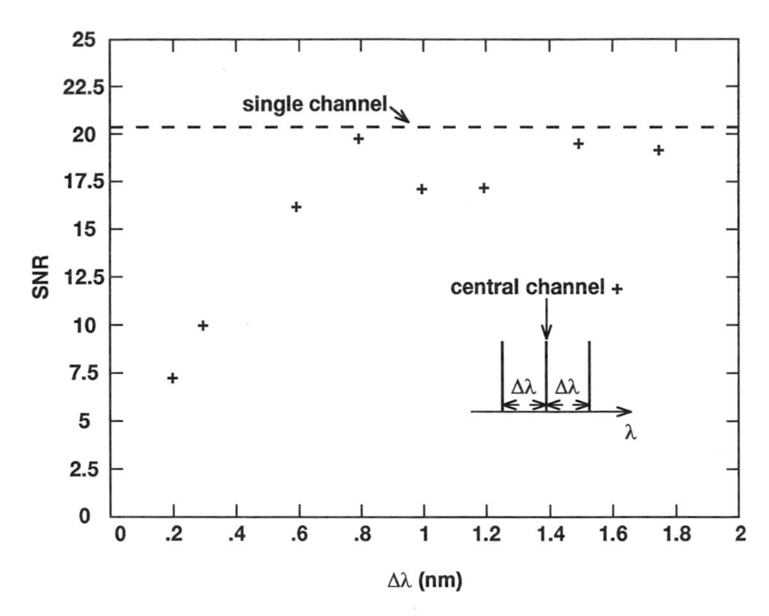

Fig. 8.6 Plot of "SNR," defined as the mean height of the ones level divided by the standard deviation of the ones level, versus channel spacing for a three-channel system at 2.5 Gb/s operating over three 120-km spans of conventional fiber with 16 ps/nm-km chromatic dispersion.

tially degraded relative to single channel for channel spacings of a few tenths of a nm, but improves to roughly the single-channel value for channel spacings larger than 0.8 nm.

Chiang *et al.* [20] have performed calculations and experiments using sinusoidal modulation as well as on–off keying, and have observed a decrease in the degree of phase modulation when the walk-off parameter (product of dispersion and channel spacing) is large. They calculate the frequency-dependent transfer function of modulation from the intensity of one channel to the phase of another and find that the bandwidth of this interaction is limited by dispersion. This transfer function exhibits resonances in the case of multiple amplifier spans, arising from the spatial periodicity of the system. They also show that too-frequent dispersion compensation can result in the loss of the suppression of CPM by dispersion.

Some insight into the dependence of CPM effects on dispersion and channel spacing can be obtained from a simple analysis. We assume that there is sufficient dispersion to ensure that bits in adjacent channels slip past each other in a propagation distance less than the absorption length of the fiber. In this regime, using the results of Fig. 8.5, we consider only the effects of "incomplete" collisions, since complete pulse collisions have little effect. By "incomplete" we mean that an edge of an interfering pulse has passed only part way through the signal. The frequency shift induced on a portion of the signal by such an incomplete collision is given, in analogy with Eq. (8.10) by

$$\Delta B = 2\gamma \frac{\varepsilon T}{D\Delta\lambda} \cdot \frac{P}{\varepsilon T}, \tag{8.11}$$

where γ is the nonlinear coefficient, T is the bit interval, ε is the fraction of the bit interval required for the transition from zero to one, D is the chromatic dispersion, $\Delta\lambda$ is the channel spacing, and P is the interfering pulse power. This heuristic equation is derived under the assumption that loss may be neglected during the collision. We have also assumed that the transition from zero to one is made with a linear slope, but this is not essential, in fact the transition time cancels in Eq. (8.11). To consider the effect of this frequency shift on a system we calculate the time-delay displacement of the frequency-shifted edge of a pulse with respect to the unshifted edge; in other words, the broadening (or compression) of the pulse.

$$\delta t = DL\delta\lambda = \lambda^2 DL \frac{\Delta B}{c}, \tag{8.12}$$

where L is the total system length, and $\delta\lambda$ is the wavelength separation corresponding to ΔB. Substituting for ΔB,

$$\delta t = 2\frac{\gamma P L \lambda^2}{\Delta\lambda c} = 2\frac{\gamma P L}{\Delta f}, \tag{8.13}$$

with Δf representing the channel spacing in frequency. Note that the chromatic dispersion has cancelled in the derivation of Eq. (8.13). This occurs since the magnitude of the broadening ΔB decreases as D increases, but the impact of that broadening increases with D. This temporal broadening corresponds to that arising from a single incomplete collision near the input to the system. In a system with 3 channels, like the one in Fig. 8.6, it is possible that both interfering channels give rise to an incomplete collision on the same bit. This would increase the broadening by a factor of 2. For a system with more channels, the worst case broadening would be given by a sum where the more distant channels would contribute progressively less due to the factor $\Delta\lambda$ in the denominator in Eq. (8.13). For a 10-channel system, this results in the 2 in Eq. (8.13) being replaced by 8. Typically, for acceptable transmission performance, this broadening of the pulse, δt must be less than $T/4$. Using this, Eq. (8.13) gives

$$\frac{\Delta f}{R} > 16\gamma PL, \tag{8.14}$$

where R is the bit rate and the factor 16 relates to a 3-channel system. For 2.5 Gb/s at 5 mW/channel and a total length of 360 km (the parameters used in Fig. 8.6), this results in the condition that the channel spacing should exceed 80 GHz, in astonishing agreement with the figure. The impact of CPM can of course be lessened by dispersion management techniques, which maintain high local chromatic dispersion, and the associated reduction in CPM-induced broadening, while through the use of concatenated spans of fiber of opposite signs of dispersion achieve low overall chromatic dispersion.

V. Four-Photon Mixing

Four-photon mixing (FPM) is a third-order nonlinearity, which is analogous to intermodulation distortion in electrical systems. Like SPM and CPM, FPM is generated by the intensity-dependent refractive index of silica; however, its impact on the performance of a WDM system is completely different. In FPM, the beating between two channels of a WDM system at their difference frequency modulates the phase of one of the channels at that frequency, generating new tones as sidebands. When three waves of frequencies f_i, f_j, and f_k ($k \neq i, j$) interact through the third-order electric susceptibility of the optical fiber, they generate a wave of frequency [65]

$$f_{ijk} = f_i + f_j - f_k. \tag{8.15}$$

Thus, three copropagating waves give rise, by FPM, to nine new optical waves [25]. In a WDM system, this happens for every choice of three channel waves with $k \neq i, j$; therefore, even if the system has only 10 channels, hundreds of new components are generated by FPM. In WDM systems with equally spaced channels, all the produce terms generated by FPM in the bandwidth of the system fall at the channel frequencies, giving rise to cross talk [67, 100, 127, 154]. In the following, the worst case of copolarized signals is considered. A general treatment including polarization effects can be found in [72].

Assuming that the input signals are not depleted by the generation of mixing products, the peak power of the mixing product is given by (in MKS)

$$P_{ijk} = \left(\frac{D_{ijk}}{3}\gamma L_e\right)^2 P_i P_j P_k e^{-\alpha L}\eta, \tag{8.16}$$

where $D_{ijk} = 3$ for two-tone products and 6 for three-tone products, γ is the nonlinear coefficient, defined in Eq. (8.8), L_e is the fiber effective length and P_i, P_j, and P_k are the input peak powers of the channels. The efficiency η is given by [130]:

$$\eta = \frac{\alpha^2}{\alpha^2 + \Delta\beta^2}\left(1 + \frac{4e^{-\alpha L}\sin^2(\Delta\beta L/2)}{(1 - e^{-\alpha L})^2}\right). \tag{8.17}$$

The quantity $\Delta\beta$ is the difference of the propagation constants of the various waves, due to dispersion, given by [70]

$$\Delta\beta = \beta_i + \beta_j - \beta_k - \beta_{ijk} = \frac{2\pi\lambda^2}{c}(f_i - f_k)(f_j - f_k)$$

$$\left[D(\lambda) - \frac{\lambda^2}{c}\left(\frac{f_i + f_j}{2} - f\right)\frac{dD}{d\lambda}\right], \tag{8.18}$$

where the dispersion D and its slope are computed at the generic wavelength λ and $f = c/\lambda$. Equation (8.17) has been extended to multiple amplified spans [73, 126] and is given in Section X. Also extension to the case of nonuniform chromatic dispersion, applicable to both dispersion managed systems and systems with randomly varying disperion, is given in [78].

High-speed WDM systems require simultaneously high launched power and low dispersion. This greatly enhances the efficiency of FPM generation, making FPM the dominant nonlinear effect in these systems [148]. For sufficiently low fiber chromatic dispersion, $\Delta\beta \approx 0$ and $\eta \approx 1$. If in addition, all channels have the same input peak power P, the ratio of the generated power P_{ijk} to the power of the channel at the receiver can be written as

$$\frac{P_{ijk}}{P(L)} = \left(\frac{D_{ijk}}{3}\right)^2 (\gamma P L_e)^2. \tag{8.19}$$

Note that this ratio scales with P^2. For the case of a three-tone product ($D_{ijk} = 6$), assuming $L_e = 22$ km, $A_e = 55\ \mu\text{m}^2$, the mixing product to transmitted channel power ratio is about $0.01 \times P^2$ [mW^2].

Figure 8.7 shows a measured spectrum at the output of a 25-km dispersion-shifted fiber (DSF) ($\lambda_0 = 1547.3$ nm) when three 3-mW signals are launched at the input. Nine new frequencies are observable with a maximum peak ratio of 0.01 to the input signals. For N channels the number of mixing products generated is

$$M = \frac{1}{2}(N^3 - N^2), \tag{8.20}$$

so that for eight channels 224 products are generated. The top trace in Fig. 8.8 shows the effect on the bit pattern of the central channel in the experiment of Fig. 8.7 when the three channels are equally spaced. Clearly, even at these modest power levels and short system lengths, there is substantial impairment. Uniform spacing of the input channel frequencies causes the large mixing product near 1545 nm to fall at the same frequency as the central channel, and the resulting beating gives rise to the width of the "one" level. Thus, although the product itself has only 1% of the power of the channel, it results in a ± 1 dB fluctuation in the received power. Figure 8.9 shows a histogram of the "one" level illustrating that the impairment in this case is a power penalty, since the distribution has well-defined maximum and minimum values.

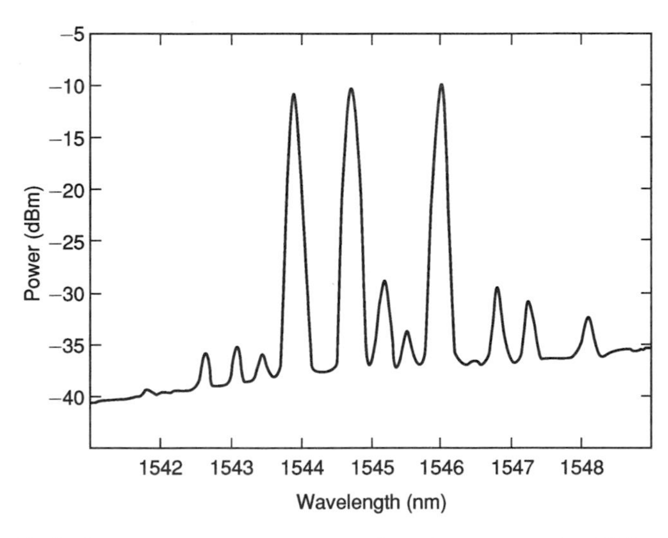

Fig. 8.7 Optical power spectrum measured at the output of a 25-km length of dispersion-shifted fiber ($D = -0.2$ ps/nm-km at the central channel) when three 3-mW channels are launched.

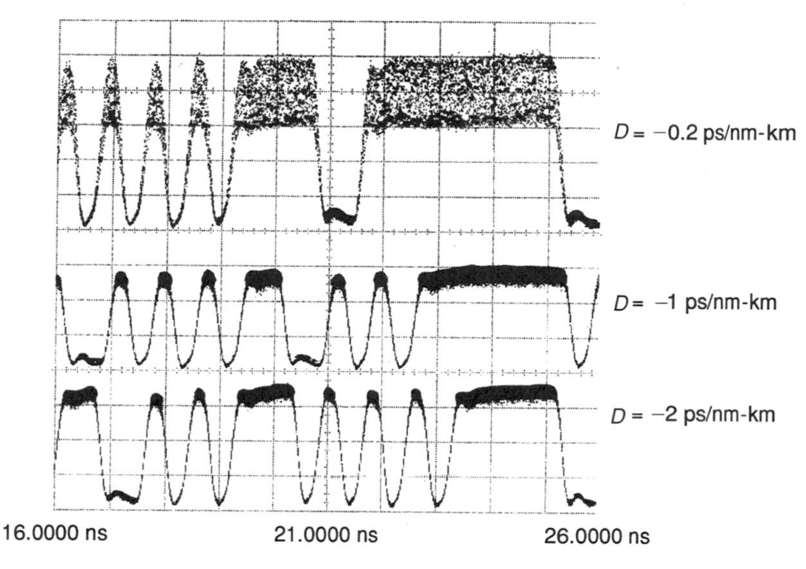

Fig. 8.8 Bit patterns observed for the central channel when the three channels are equally spaced so that the largest mixing product falls on the central channel, shown for three different values of fiber chromatic dispersion. Launched power was 3 mW/channel.

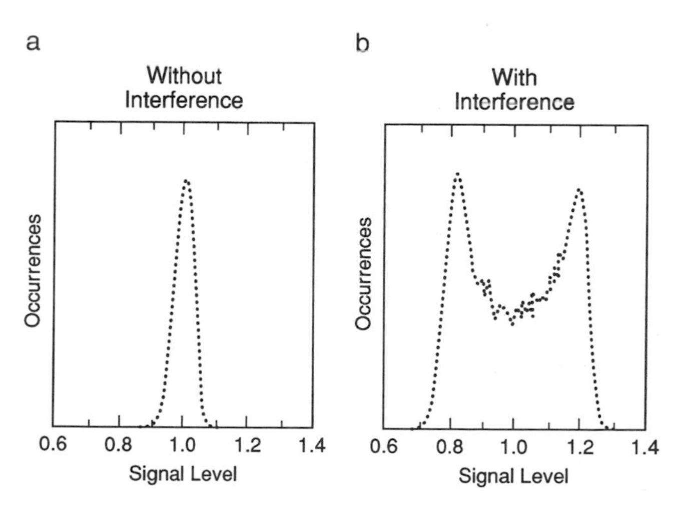

Fig. 8.9 Histograms taken at the sampling point of the marks both (a) without and (b) with the outside channels present for the $D = -0.2$ ps/nm-km case.

The evaluation of the performance degradation due to FPM in WDM systems is rather complicated. Although for the simple case of a single mixing product generated at a channel frequency, the impairment can be clearly viewed as a power penalty, as shown in Fig. 8.9, when several mixing products are generated at the same frequency, the total interference strongly depends on the relative phases of the mixing products. A power-penalty approach may not be suitable, since the probability of all mixing products aligning in counterphase with the channel would be small. When a large number of mixing products of comparable size are generated at a channel frequency, the central limit theorem can be invoked and the ensemble of the mixing products generated at the same frequency can be treated as Gaussian noise with power equal to the sum of the powers of the single mixing products [77, 79]. The effect of FPM on the performance of an amplified WDM system can in this case be viewed as an optical SNR degradation.

The intermediate cases of a few mixing products at each frequency, or mixing products with unequal powers, can be analytically treated, including the effect of modulation statistics, assuming that the mixing products have independent phases, uniformly distributed between 0 and 2π [45]. In presence of FPM, the power on channel n at the output of the fiber can be found from Eq. (8.16) to be approximately given by

$$P_n(L) = Pe^{-\alpha L}\left[1 + 2\gamma P L_e \sum_{\mathcal{A}_n} \sqrt{\eta_{ijk}} \left(\frac{D_{ijk}}{3}\right) \cos\left(\phi_{ijk}\right)\right], \qquad (8.21)$$

where $\mathcal{A}_n = \{(i,j,k) \mid i,j,k \in 1 \ldots N,\ k \neq i,j,\ k = i + j - n\}$ is the set of all the triplets of channel numbers that generate a mixing product on channel n, N is the number of channels, η_{ijk} is the efficiency, and ϕ_{ijk} is the phase of the mixing product relative to the channel and depends on the phases of the generating channels. The coherent nature of mixing interference is apparent in this equation. The optical power of channel n at the receiver $P_n(L)$ is increased or decreased by mixing interference, depending on the actual values of the ϕ_{ijk}. Figure 8.9b depicts the above relation for the simple case of a single mixing product. Since the probability density function (pdf) of the random variable $X = \cos(\phi)$, if ϕ is uniformly distributed between 0 and 2π, is given by

$$f_X(x) = \frac{1}{\pi\sqrt{1 - x^2}}, \qquad (8.22)$$

its characteristic function is $\Phi_X(\omega) = \mathcal{F}(f_X(x)) = J_0(\omega)$, where $\mathcal{F}$ indicates the Fourier transform operator and J_0 is the Bessel function of 0th order. The effect of mixing is the sum of several contributions with the functional form of X and can therefore be computed in the transform domain. The pdf of the relative power at the receiver $f_n(p)$ is then given by

$$f_n(p - 1) = \mathcal{F}^{-1}\left\{\prod_{\mathcal{A}_n} K_{ijk} J_0(K_{ijk}\omega)\right\}, \tag{8.23}$$

where p is the optical power at the receiver normalized to one and $K_{ijk} = (2/3)\sqrt{\eta_{ijk}} \mathrm{D}_{ijk} \gamma L_e P$. To include the effect of modulation, it is assumed that "0"s and "1"s are equally likely and that bits transmitted on different channels are independent. The effect of dispersion is neglected, and it will be assumed that the modulation in all channels is synchronous so that each bit in one channel interacts with a single whole bit in each other channel. When modulation is present, the number (and type) of mixing products created on each channel depends on the particular modulation pattern being transmitted on the other channels. This can be viewed as a quasi-cw problem in which only some of the N channels are present. In this case, the pdf of the relative power at the receiver $f_{mod}(p)$ can be found, by the total probability theorem, by adding the conditional probability densities of the received power for each of the possible modulation patterns (which are all equally likely and have probability $1/2^N$):

$$f_{mod}(p) = \frac{1}{2^N} \sum_{\mathcal{P}} f_n(p \mid \mathcal{P}), \tag{8.24}$$

where $\mathcal{P}$ is the modulation pattern, $f_n(p \mid \mathcal{P})$ is the pdf of the relative received power when the modulation pattern $\mathcal{P}$ is transmitted given by

$$f_n(p - 1 \mid \mathcal{P}) = \mathcal{F}^{-1}\left\{\prod_{\mathcal{A}_n \mid \mathcal{P}} K_{ijk} J_0(K_{ijk}\omega)\right\}, \tag{8.25}$$

where $\mathcal{A}_n \mid \mathcal{P}$ is the set of active mixing products generated when $\mathcal{P}$ is the modulation pattern.

An eight-channel experiment has been performed to confirm the validity of this model [45]. Eight cw external-cavity lasers with an equal spacing of 250 GHz were multiplexed and transmitted over 25 km of DSF with loss $\alpha =$ 0.23 dB/km and zero-dispersion wavelength $\lambda_0 = 1547.25$, which coincided with the wavelength of the second channel, numbering channels in order of increasing wavelength. At the receiver, the second channel was extracted with a tunable filter and a histogram of the received power was computed using 1 million independent samples. The result is shown in Fig. 8.10 by the circles. The dashed lines show the pdf derived with the model. In the theory, a shift of 0.1 nm of λ_0 has been included to fit the experimental data. With this shift, the agreement with the cw experimental data is excellent. Monte Carlo simulations have also been performed to study the validity of the hypothesis of independent phases, and the result, shown by the solid line, is in remarkable agreement with both model and experiment. The dashed-dotted line shows the prediction that would be provided by the Gaussian approximation in which the mixing products are merely added in power. The large discrepancy results in the overestimation of performance degradation [45].

Since FPM is the dominant effect for WDM systems, it has received considerable attention and several methods have been proposed to suppress it. Using orthogonally polarized signals [69, 101] can suppress FPM, due to its strong polarization dependence [72]. However, polarization-mode dispersion changes the relative polarizations and introduces slowly varying fading. A differential delay from PMD of 10 ps serves to evolve the relative polarization states of two signals separated by 100 GHz through a full cycle. Thus, even modest PMD serves to make the initial states of polarization

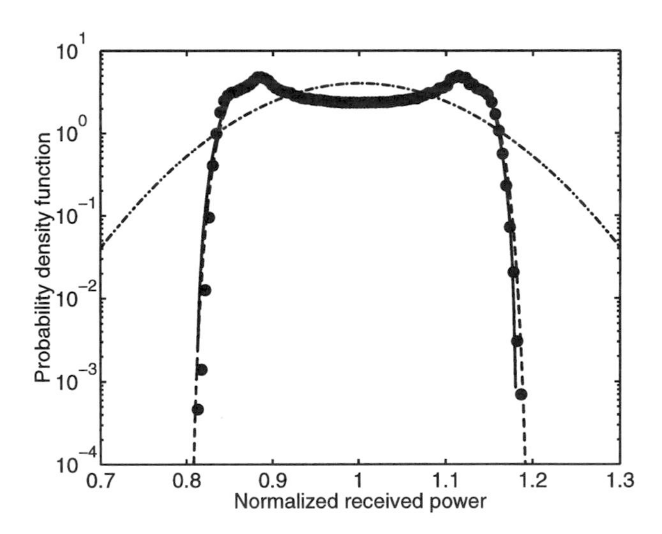

Fig. 8.10 Probability density function of the received power of the second of eight WDM channels. *Circles* show experimental results, the *dashed line* shows the result of the model with independent mixing phases, the *solid line* is obtained with Monte Carlo simulations, and the *dashed-dotted line* is for the Gaussian approximation.

of channels in a long WDM system irrelevant. Other schemes have been proposed based either on specialized modulation of the transmitted light [71, 75] or using devices introduced into the transmission line [74, 146]. Increasing the channel spacing can be used to reduce the efficiency of generation of mixing products [31] but the improvement comes at the expense of optical bandwidth, and when DSF is employed, care has to be taken to avoid positioning channels at λ_0 or equidistant from λ_0 in any fiber span. With non-zero-dispersion fiber, the approximation $\eta \propto 1/\Delta f^4$ holds, therefore doubling the bandwidth suppresses FPM by 12 dB and (see Eq. [8.19]) allows an increase in power per channel by 6 dB. Spectral inversion can also in principle be used to suppress FPM [54, 109], although the phase- and polarization-sensitive nature of FPM make it a very challenging task to effectively utilize this technique in a long system. This subject is further discussed in Chapter 7 in Volume IIIA.

In the following, two schemes that allow strong suppression of FPM impairment are described in more detail: unequal spacing of the channel frequencies and management of the fiber dispersion. These methods allow the use of conventional transmitters and require no additional devices in the transmission line.

FPM degradation mainly arises from the coherent interference between mixing waves and signal at the detector. If the channels are arranged so that no mixing products fall on any of the channels, this coherent cross talk can be avoided. A method of determining such channel arrangements has been described [41, 44]. Since typical WDM systems use channel spacings large compared to the bandwidth of the signals, there is an opportunity to use this "wasted bandwidth" to place mixing products away from the channels. Note that in systems with channels equally spaced in *wavelength,* the frequency spacing will not be uniform. However, the unequal frequency spacing inherent in equal *wavelength* spacing is not sufficient to prevent interference. The difference in frequency spacing, and hence the offset of the mixing product from the channel, must be at least twice the bit rate to avoid interference [41, 58]. For the 3-channel case of Fig. 8.8, the channel spacing would need to be 4 nm for the difference in frequency spacing, arising from equal wavelength spacing, to be 2.5 GHz. The requirement that no mixing product should fall on any channel was shown to be equivalent to requiring that the difference between any two channel frequencies be unique [41]. For reasonable numbers of channels, such an arrangement can be found by an exhaustive computer search. Figure 8.11 shows the number of mixing products at various frequencies for equal and unequal spacing

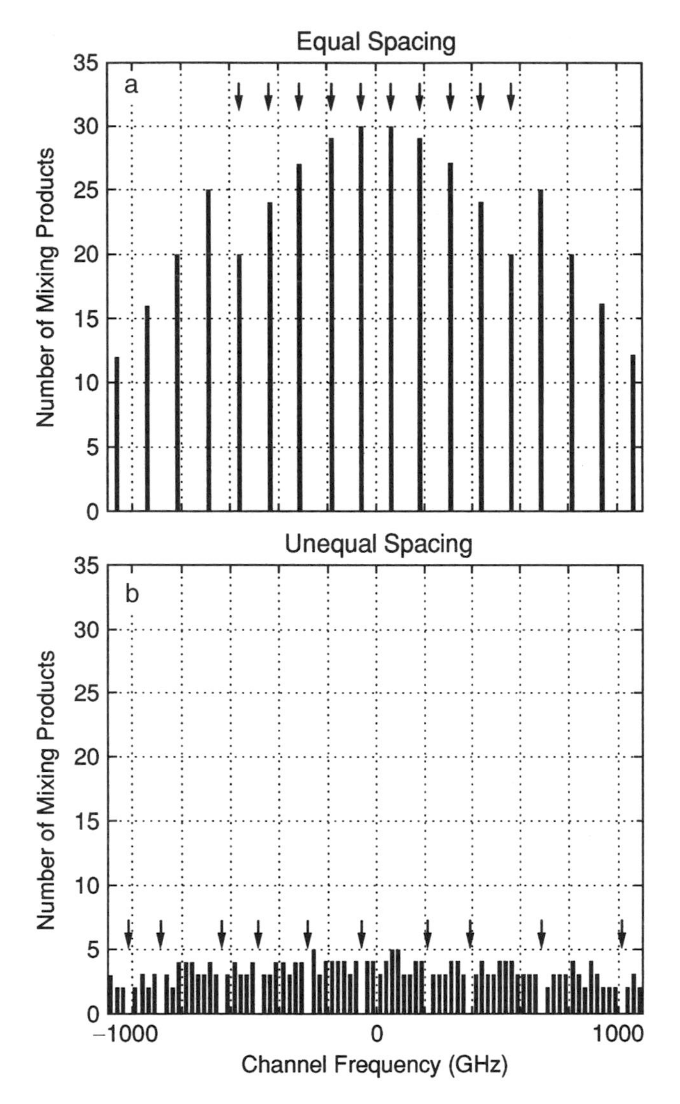

Fig. 8.11 Graphs showing the number of mixing products falling at various frequencies for a 10-channel system. (a) Equally spaced channels. (b) Unequally spaced channels. *Arrows* indicate the channel frequencies.

for a 10-channel system, where the mixing products are offset from the channels by at least 25 GHz and the channels are separated by at least 1 nm. Figure 8.12 shows a simulation for this case; the improvement in

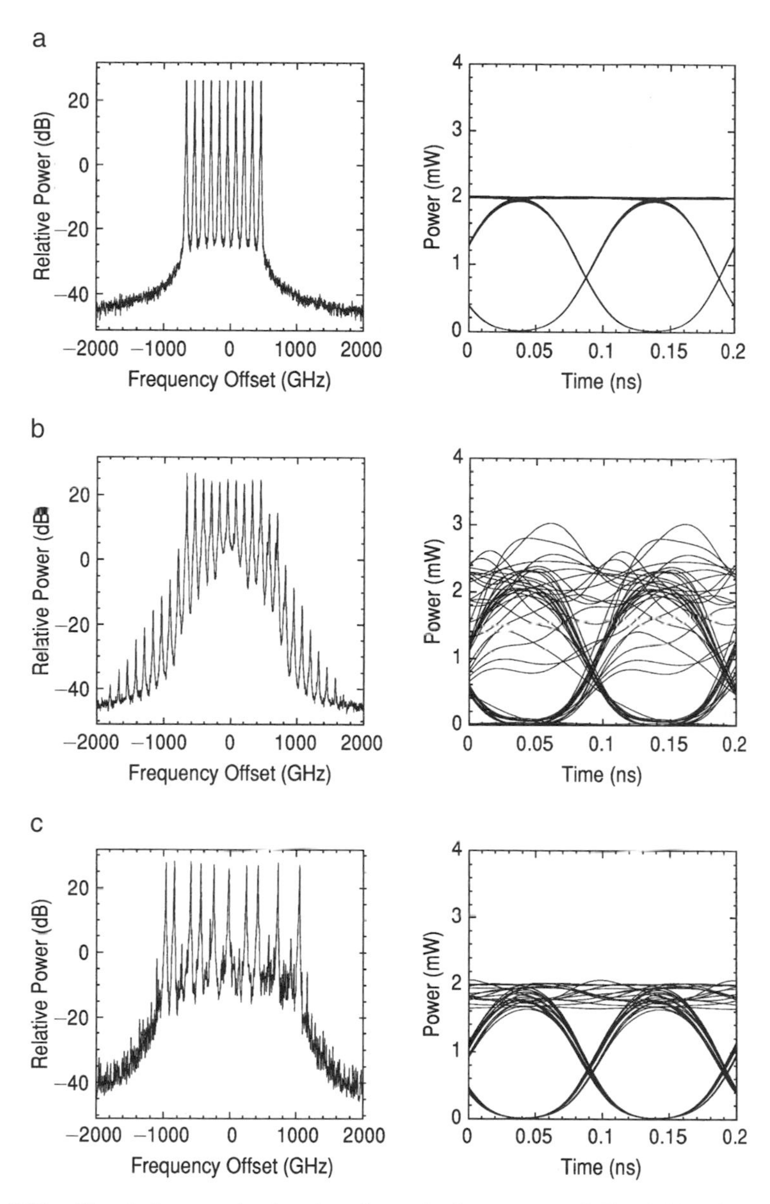

Fig. 8.12 Simulation results for the channel placements of Fig. 8.11. The system consists of 500 km of fiber with the dispersion zero located at zero frequency offset. Optical amplifiers are spaced every 50 km. The channels are modulated at 10 Gb/s and are launched into each span of fiber with 1 mW average power. (a) System input power spectrum and eye diagram. (b) System output, equally spaced channels. (c) System output, unequally spaced channels.

performance is obvious. The use of unequal channel spacing may require a bandwidth expansion, which depends on system parameters and is usually less than a factor of two for systems of up to 10 channels [44]. Also, the use of unequal spacing poses additional demands on the frequency stabilization of the laser sources. A rough estimate of the performance improvement achievable with unequal channel spacing can be given by considering the generation of one mixing product. For an equally spaced system, a mixing product with a power 20 dB lower than the signal is sufficient to generate a 20% eye closure and, therefore, produce a 1-dB penalty, as seen previously, due to the coherent interference between mixing wave and signal at the detector. With unequal spacing, the only effect of FPM is depletion of the channel power. In this case, to produce a 1-dB penalty, the mixing wave must have a power equal to 20% the power of the channel; therefore, its power must be 7 dB lower than the signal power and 13 dB larger than in the equal-spacing case. Since the relative mixing power grows as the square of the signal power (Eq. [8.19]), unequal channel spacing can be expected to allow an increase in the transmitted power of about 6.5 dB for the same penalty.

An experiment demonstrating this improved performance [42] is diagrammed in Fig. 8.13. Eight 10-Gb/s channels were launched into 137 km of DSF. The zero-dispersion wavelength of the initial 40 km of the line

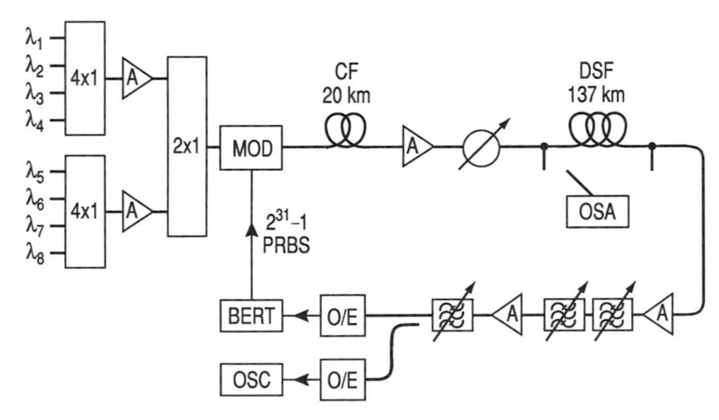

Fig. 8.13 Experimental diagram for a 137-km repeaterless transmission experiment with eight 10 Gb/s channels demonstrating the benefits of unequal channel spacing. The 20-km fiber after the modulator was conventional step-index fiber with 16 ps/nm-km chromatic dispersion and was used to decorrelate the bit patterns of the various wavelength channels prior to launching into the transmission fiber.

(the portion of the fiber relevant for the generation of mixing products) fell midway between the second and third channels. The remaining spools of fiber had slowly decreasing zero-dispersion wavelengths. Input and output spectra are shown in Fig. 8.14 for both equal and unequal spacing. The total occupied bandwidth was the same. Figure 8.15 shows the bit error ratio as a function of launched power for both cases. Note that the equally spaced case has a minimum BER of 10^{-6}, while the unequally spaced case operates essentially error free from 2 dBm to 6 dBm input powers. The degradation setting in at 7 dBm arises from depletion of the channel power by the mixing products. Thus, the unequally spaced system allows a 6 to 7 dB increase in launched power, but the generation of mixing products eventually degrades system performance.

Unequal channel spacing is a key technique for the WDM upgrade of already deployed fiber communication systems with DSF. It can also be

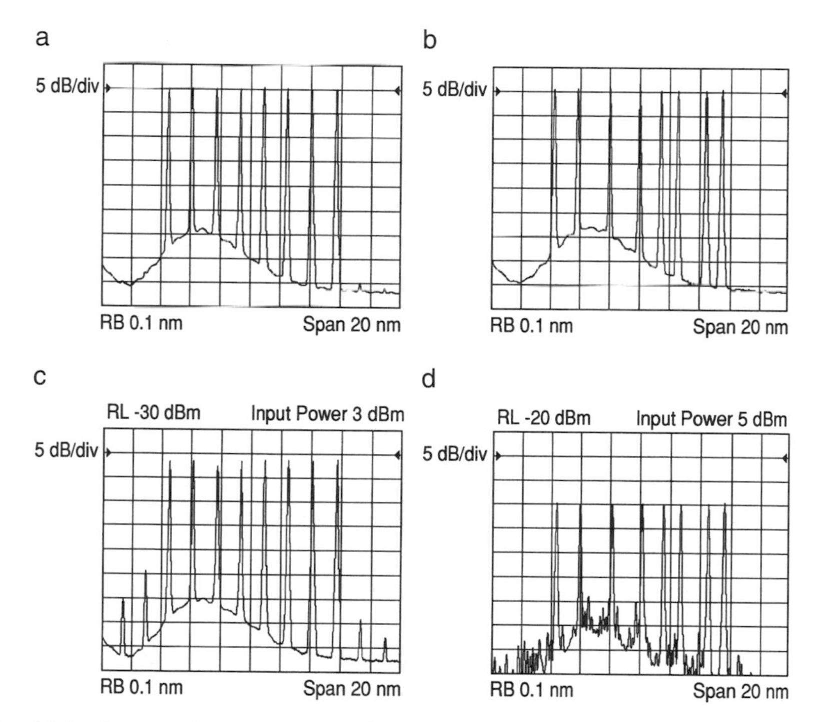

Fig. 8.14 Input and output spectra for the experiment of Fig. 8.13 for both equal and unequal spacing. (a) Equal spacing, input. (b) Unequal spacing, input. (c) Equal spacing, output. (d) Unequal spacing, output.

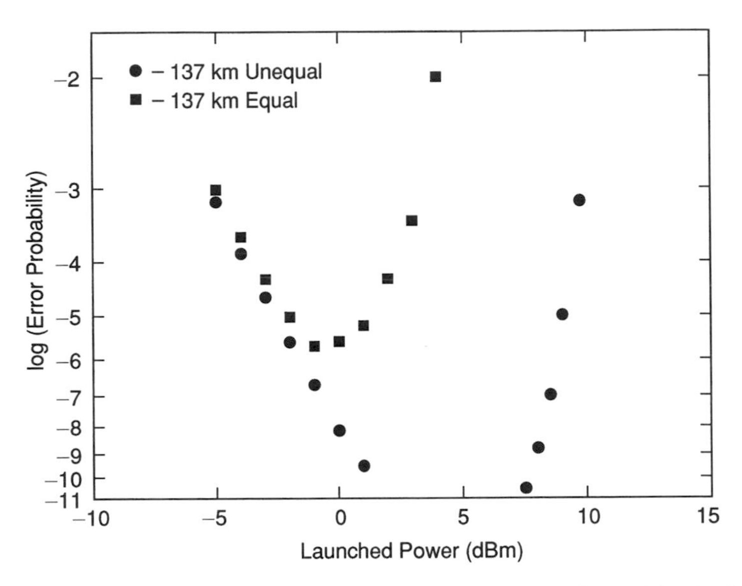

Fig. 8.15 Bit error ratio curves versus launched power per channel for both equal and unequal spacing.

compatible with a standardization of the optical spectrum in which the allowed channel frequecies are equally spaced, since an unequally spaced system can be designed by selecting the channel slots from the standard set [44]. Unequal spacing is even compatible with integrated optics. Unequally spaced waveguide grating multiplexers have been built [119] and demonstrated in a 10-channel experiment where add/drop of unequally spaced channels was performed [48].

While unequal spacing ameliorates the effects of mixing products on the performance of a WDM system, it is also possible to prevent the generation of those products in the first place by avoiding phase matching. Figure 8.16 illustrates the effect of fiber chromatic dispersion on the mixing impairment. The figure shows the ratio of generated power to transmitted power for a two-tone product as a function of channel spacing for three different values of chromatic dispersion. Note that the top curve, which corresponds to a case where one of the two tones is coincident with the fiber zero-dispersion frequency, and the mixing product farthest from the zero-dispersion frequency is considered, maintains high efficiency for channel spacings as large as 1 nm. Of course, the product on the other side of the channels is phase matched for all channel spacings in this case [70]. As can be seen, quite small values of dispersion (1–2 ps/nm-km) can have a dramatic effect. This

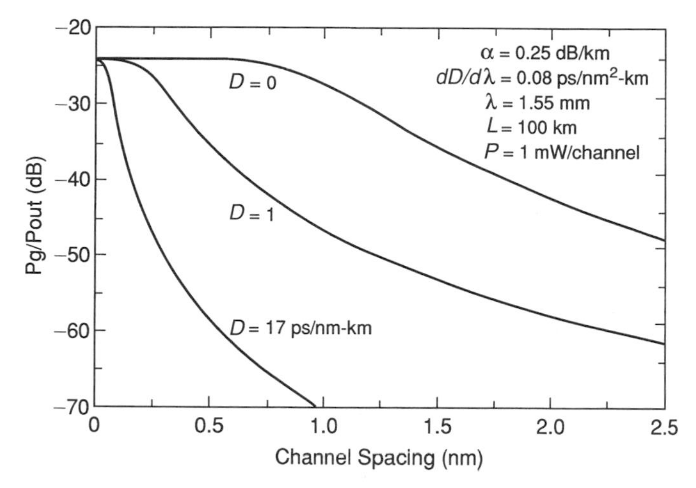

Fig. 8.16 Plot of the ratio of generated mixing product power to transmitted channel power versus channel spacing for two equal power channels. Curves are shown for three different values of fiber chromatic dispersion.

is also shown by the second and third bit patterns in Fig. 8.8. With $D = 1$ ps/nm-km and 1-nm channel spacing, FPM is suppressed by 20 dB; therefore (see Eq. [8.19]), the channel power can be increased by 10 dB. An example for a four-channel system is shown in Fig. 8.17. At the output of 25 km of DSF with λ_0 within the channels' bandwidth, the mixing products are clearly noticeable. With 50 km of non-zero-dispersion fiber ($D = 2.5$ ps/nm-km), instead, there is no evidence of mixing products.

To avoid a dispersion penalty in a high-speed system, the requirement is that the total chromatic dispersion be low. If however, the local chromatic dispersion of the fibers can be controlled, one may retain low total chromatic dispersion while avoiding low dispersion in any fiber segment. Dispersion management is discussed in detail in the next section. Dispersion management and unequal channel spacing are perfectly compatible and can be combined to achieve a FPM mixing suppression stronger than that provided by the two techniques separately.

VI. Dispersion Management

Fiber chromatic dispersion plays a schizophrenic role in high-speed WDM transmission systems. On the one hand, it is the bane of short optical pulses; on the other hand, it is a useful tool in suppressing FPM. This led to the

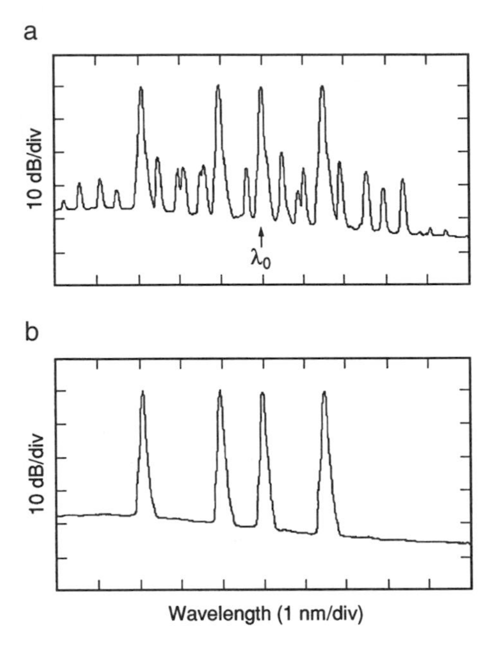

Fig. 8.17 Four WDM channels, each with average power 3 dBm, transmitted through (a) 25 km of dispersion-shifted fiber with λ_0 within the channels' bandwidth, and (b) 50 km of non-zero-dispersion fiber (D = 2.5 ps/nm-km). The channels are unequally spaced to check the presence of mixing products.

natural evolution of dispersion management [16, 18, 27–29, 53, 55, 61, 62, 89, 92, 98, 113, 116, 117, 122, 148, 149]. Dispersion management ensures that no fiber in the transmission path has a dispersion-zero wavelength close to the signal wavelengths but that the total accumulated dispersion for the signals between transmitter and receiver is near zero or at least smaller than the maximum (1-dB penalty) allowable accumulated dispersion for the system bit rate given by

$$DL[\text{ps/nm}] < 10^5/R^2, \tag{8.26}$$

where R is the bit rate in Gb/s.

Dispersion management can be accomplished in several different ways. Dispersion can be managed solely in the transmission fiber. Figure 8.18 shows the dispersion map for a 300-km long transmission experiment [53] with 50-km amplifier spacing. The adjacent fiber spans have dispersion magnitudes of 2.5 ps/nm-km but with span-by-span sign reversals. As ex-

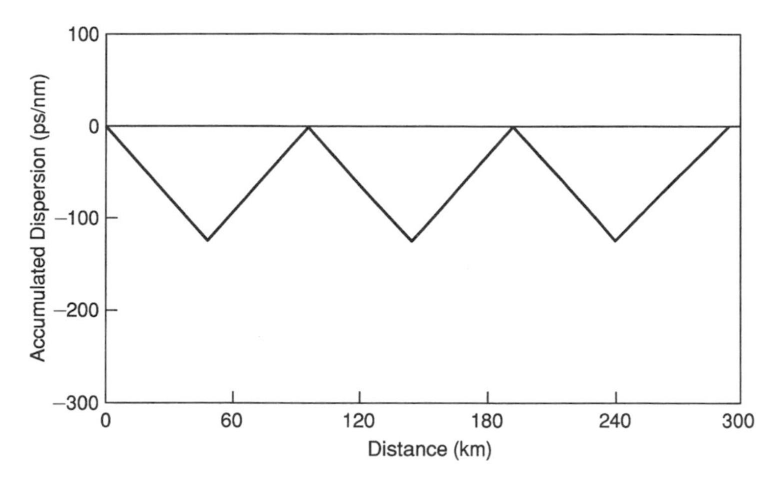

Fig. 8.18 An example of dispersion management. The overall accumulated dispersion is zero, yet no individual fiber has near-zero dispersion [53].

pected, there was little FPM (< −30 dB) and the accumulated system dispersion is nearly zero. One cautionary note should be discussed. In ultra-high bit-rate systems or in very long systems, the dispersion map in Fig. 8.18 is an oversimplification. In fact the slope of dispersion must be taken into account. For the eight-channel experiment, Fig. 8.18 represents the dispersion map for the central wavelength. Figure 8.19 shows the dispersion maps for the shortest-wavelength and longest-wavelength channels as well.

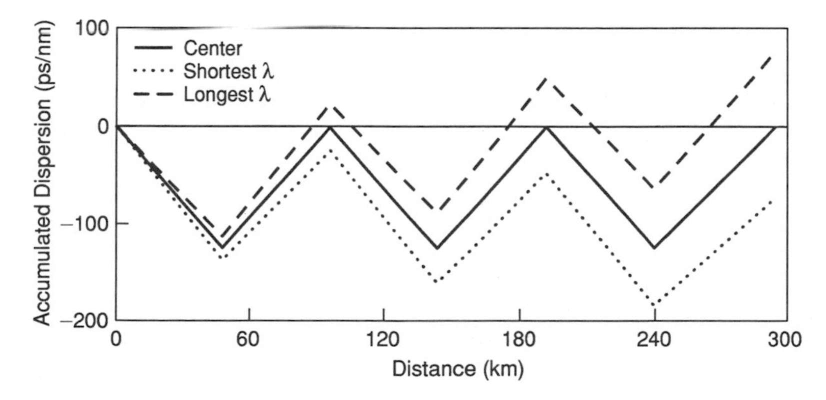

Fig. 8.19 Dispersion map for shortest and longest wavelength channels in addition to the dispersion map in Fig. 8.18. The spread in accumulated dispersion arises from the positive slope of dispersion in all the fibers comprising the transmission span.

Higher-order dispersion causes a spread in accumulated dispersion for various channels. In this case, the accumulated dispersion for the outermost channels was less than the maximum allowable dispersion for 20-Gb/s signals (250 ps/nm), but in longer systems or in systems with larger occupied bandwidths higher-order dispersion effects cannot be ignored.

Dispersion management in the transmission fiber does not have to be accomplished on a span-by-span basis. For example, in undersea WDM transmission experiments [16], accumulated dispersion is returned to near zero with a period of about 1000 km. About 900 km of non-zero-dispersion fiber with dispersion of about -2 ps/nm-km is followed by 100 km of conventional unshifted fiber with $\lambda_0 = 1.3\ \mu$m. In these experiments this 1000-km dispersion pattern was repeated six times. In transoceanic WDM systems with bit rates of 5 Gb/s or larger, the walkoff due to the slope of dispersion is significant. In such cases, the excess accumulated dispersion can be trimmed out on a channel-by-channel basis at the receiver [16].

Dispersion can also be managed in the transmission fiber on a "per span" basis [27]. In this 8×10 Gb/s experiment, the accumulated dispersion of each span was nearly zero. The spans consisted of about 8 km of unshifted fiber followed by 48 km of non-zero-dispersion fiber with dispersion equal to -2.5 ps/nm-km. FWM suppression in this configuration is as effective as in the span-by-span case. In general, efficient dispersion management does not necessarily require that the accumulated dispersion be approximately zero at the end of the transmission system, but only that it be below the allowable 1-dB limit. Figure 8.20 shows the dispersion map for a 16×2.5 Gb/s 1420-km transmission experiment [28]. Although the accumulated dispersion at the end of the system is large, it is still comfortably less than 16,000 ps/nm, the dispersion limit for 2.5 Gb/s signals.

The previous examples of dispersion management are useful for new fiber installations, but are not generally applicable to the enormous installed base of unshifted fiber. Compensation of the dispersion of conventional unshifted fibers at 1550 nm requires fibers with large negative dispersions. Over the last few years, a new type of fiber called dispersion-compensating fiber (DCF) [152] has been introduced. Because DCF has large negative dispersions ($D < -70$ ps/nm-km), it is ideal for compensating unshifted fibers. Some DCF not only have large negative dispersion but also negative slopes of dispersion [152]. Such compensating fibers can reduce the spread in accumulated dispersion in a WDM system due to dispersion slope. Because the attenuation coefficient of DCF is rather large (>0.4 dB/km), these fibers are typically incorporated in the center stage of tandem amplifi-

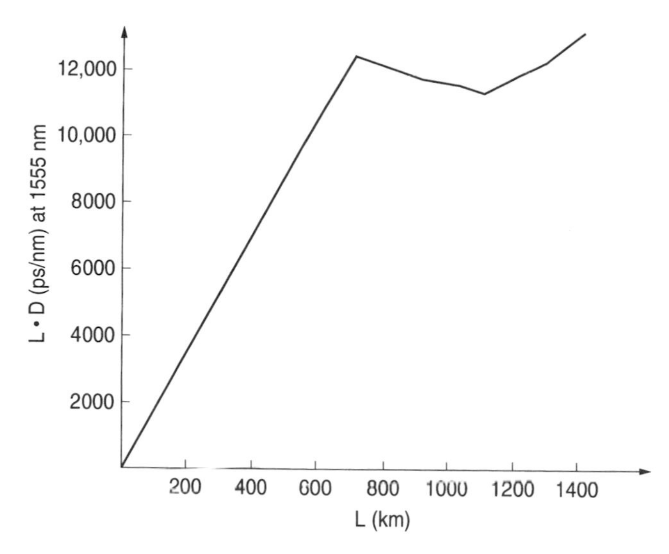

Fig. 8.20 Dispersion map in which accumulated dispersion does not return to zero at the end of the system [28].

ers. Figure 8.21 shows the dispersion map used in a 8 × 20 Gb/s experiment over 232 km of unshifted fiber using DCF with negative dispersion slope [149]. It should be pointed out that the dispersion length for 20-Gb/s signals in unshifted fiber is 15 km. Notice the absence of significant spread in the accumulated dispersions of the various channels. Had the DCF not had negative dispersion slope, the dispersion map would be given by Fig. 8.22. The dispersion spread in this case is prominent and the longest wavelength channel would be near the dispersion limit. DCF with negative slope allows much longer transmission distances before dispersion trimming of individual channels is required. Dispersion compensation using DCF have been demonstrated in various experiments [18, 117, 122, 149]. It has already been used in field experiments [18] and was used in one of the three recently reported Terabit/s-aggregate-capacity experiments [122]. In relatively short (≈100 km) systems using multiple 20-Gb/s channels, the most trivial type of dispersion management can be used. The entire fiber span can be made up of one type of non-zero-dispersion fiber. This fiber reduces chromatic dispersion effects to acceptable levels, while at the same time suppressing the effects of FPM. This was demonstrated in another of the Terabit/s transmission experiments [55].

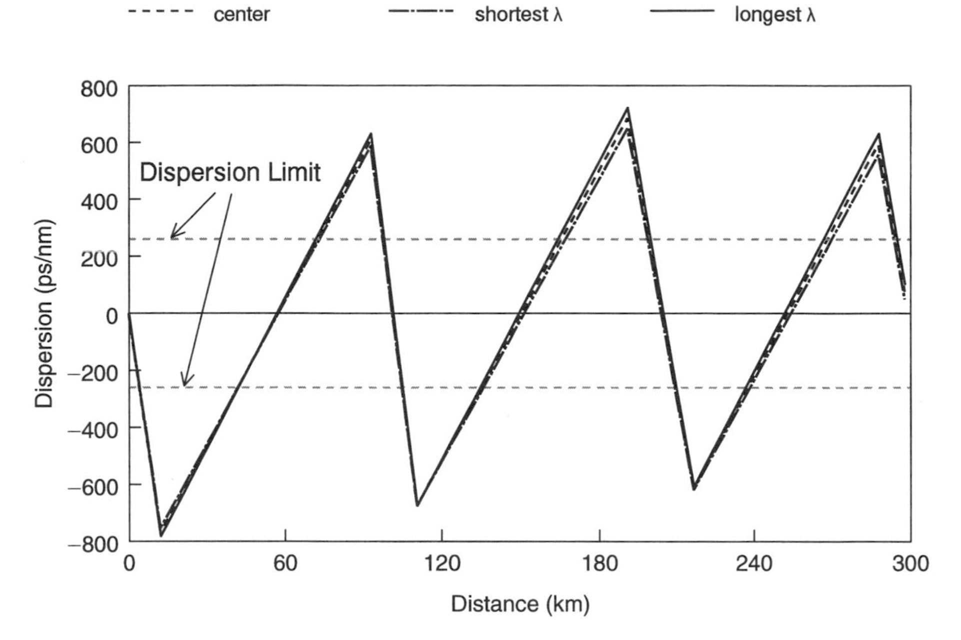

Fig. 8.21 Dispersion map for 232 km of conventional fiber compensated by DCF fiber with negative dispersion slope [149].

As was previously discussed, when using DCF optimum system performance might be achieved by undercompensating the dispersion of the previous span. One of the difficulties in implementing DCF is their susceptibility to nonlinearities due to their small core areas [47]. This precludes using DCF in high-power environments.

Computer simulations are useful in determining the effectiveness of various dispersion schemes before experimental implementation. Although there are few restrictions on dispersion management schemes, one such restriction is that the length scale over which the dispersion averages to zero cannot be too small [148]. Computer simulations proved to be particularly useful in this situation because experiments would have been inordinately costly and time consuming. A four-channel 10-Gb/s-per-channel system over 360 km with 120-km amplifier spacing and 200-GHz channel spacing was modeled. Positive- and negative-dispersion fiber with equal lengths and dispersions of 2 ps/nm-km and −2 ps/nm-km were used as transmission fibers. Figure 8.23 shows the fractional eye opening of each of the four channels as a function of the length of the fiber segments with constant dispersion, which was half the period of the alternation. For

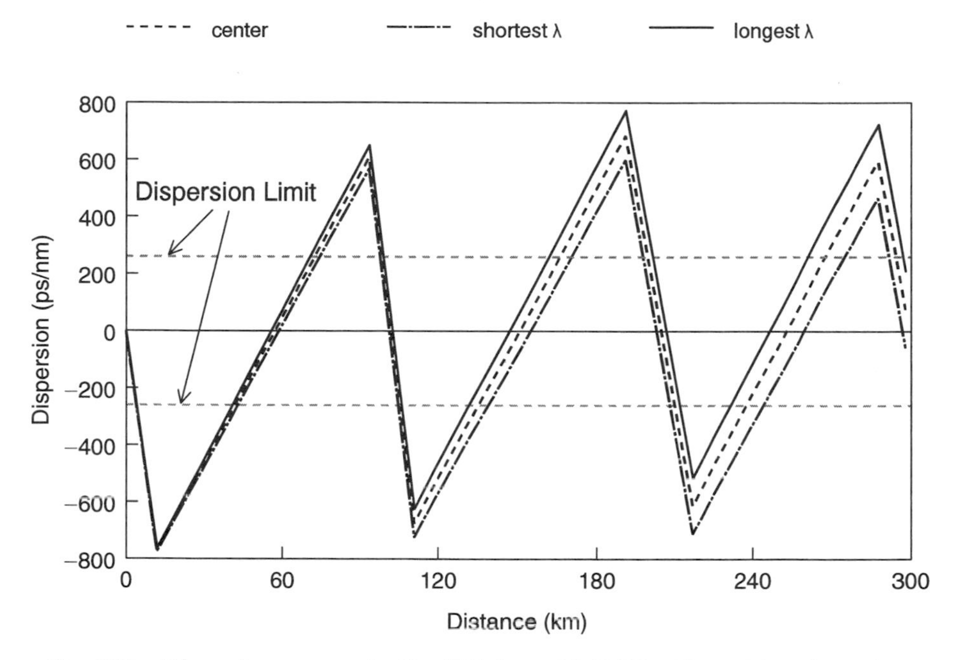

Fig. 8.22 Dispersion map as in Fig. 8.21 but with DCF with positive dispersion slope. Note the spread in accumulated dispersion after 232 km.

extremely short segment length, the fiber behaved as if it were a uniform fiber with the average, in this case, zero dispersion. As the segment length was increased, FPM was suppressed. The characteristic length required to achieve low penalties is the length over which the phase mismatch accumulated for the mixing products becomes much greater than 2π. For two-tone products with channels separated by 200 GHz, the "2π length" is 1.5 km. This explains the general shape of Fig. 8.23.

VII. Modulation Instability

Modulation instability (MI) [1–3, 9, 12, 50, 110, 125, 141, 144] can be phenomenologically described in two different ways. In the time domain, it can be viewed as pulse breakup or soliton formation [2]. In the frequency domain, MI can be described in terms of parametric gain or FPM phase matched by SPM producing exponential gain for the mixing products [2, 141]. This section follows the frequency-domain approach.

For simplicity, this discussion of MI is initially restricted to lossless fibers and to the case of one strong input signal (pump) and one or more weak

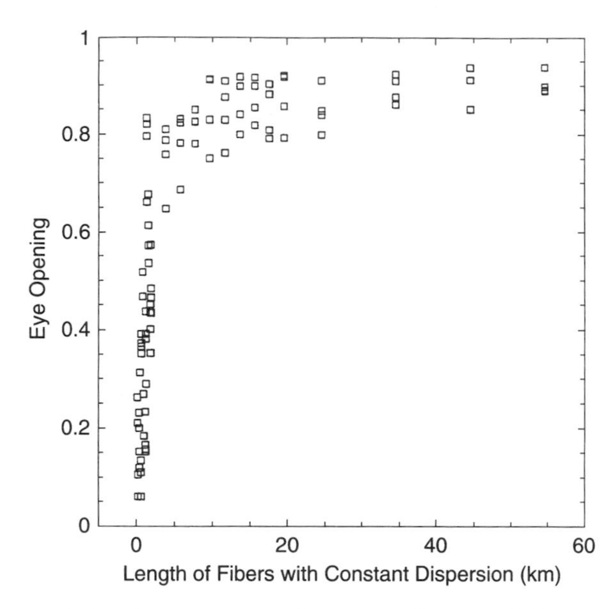

Fig. 8.23 Plot of fractional eye opening versus length of fibers with constant dispersion when the dispersion alternates between + 2 ps/nm-km and −2 ps/nm-km. This simulation was performed for a four-channel system with 200-GHz channel spacing, and an average input power of 6 dBm per channel [148].

signals, called Stokes signals if their wavelengths are longer than the pump wavelength, or anti-Stokes signals for wavelengths shorter than the pump wavelength. Even if only one strong signal is injected into the fiber, it is typically accompanied by amplified spontaneous emission noise, which acts as "seed" Stokes and anti-Stokes power. This discussion can be generalized to the case in which all signals are of arbitrary magnitude [141]. The effects of fiber loss are discussed later.

Equation (8.16) describes the generation of FPM mixing products for low-input powers. The general expression (valid for all powers in the absence of depletion) for the Stokes or anti-Stokes powers generated in fibers [141] is

$$\frac{P_a(L)}{P_s(0)} = (\gamma P_p L)^2 \frac{\sinh^2(g_{MI}L/2)}{(g_{MI}L/2)^2}, \tag{8.27}$$

where $P_s(0)$ is the injected Stokes power at the fiber input, $P_a(L)$ is the generated anti-Stokes power at the fiber output assuming no injected anti-

Stokes light, P_p is the input peak pump power, and g_{MI} is the gain coefficient given by

$$g_{MI} = 2[-(\Delta\beta/2)^2 - \Delta\beta\gamma P_p]^{1/2}, \tag{8.28}$$

where $\Delta\beta$, given in Eq. (8.18), is the difference in propagation constants of the pump, Stokes, and anti-Stokes waves, that is

$$\Delta\beta = 2\beta_p - \beta_a - \beta_s = -\frac{2\pi\lambda^2}{c}D\Delta f^2, \tag{8.29}$$

where Δf is the frequency difference between pump and Stokes frequencies. If $\Delta\beta$ is positive the gain coefficient is an imaginary quantity and the usual FPM as previously described takes place. If $\Delta\beta$ is negative and P is sufficiently large, g_{MI} will be real and amplification of Stokes and anti-Stokes waves will occur according to Eq. (8.27). Substituting into Eq. (8.28) gives an expression for the gain coefficient

$$g_{MI} = 2\left[-\left(\frac{\pi\lambda^2}{c}D\Delta f^2\right)^2 + \frac{2\pi\lambda^2}{c}D\Delta f^2\gamma P_p\right]^{1/2}. \tag{8.30}$$

The gain coefficient is real for positive D (anomalous dispersion region, i.e., for wavelengths longer than the dispersion-zero wavelength) in the frequency range

$$0 < \Delta f < \left(\frac{2c\gamma P_p}{\pi\lambda^2 D}\right)^{1/2}. \tag{8.31}$$

The maximum gain $g_{\max} = 2\gamma P_p$ occurs at the pump/Stokes frequency separation at which $\Delta\beta = -2\gamma P_p$:

$$\Delta f_{\max} = \left(\frac{c\gamma P_p}{\pi\lambda^2 D}\right)^{1/2}. \tag{8.32}$$

Figure 8.24 plots the gain coefficient as a function of pump/Stokes separation for a conventional unshifted fiber ($D = 16$ ps/nm-km or $\beta_2 = -20$ ps^2/km) at 1550 nm assuming $\gamma = 1.3$ W^{-1} km^{-1}. Figure 8.25 shows a plot (solid curve) of Eq. (8.27) for $g_{MI} = g_{\max}$ for a 20-km lossless fiber with 80-μm^2 effective area. The dashed curve shows the growth of perfectly phase-matched FPM in the absence of parametric gain. The onset of parametric gain for this fiber length occurs at pump powers of several tens of mW.

Thus far, fiber loss has been neglected. As the power decreases along a fiber, the parametric gain coefficient g_{MI} decreases as does $\Delta f_{\max}$ [3]. One

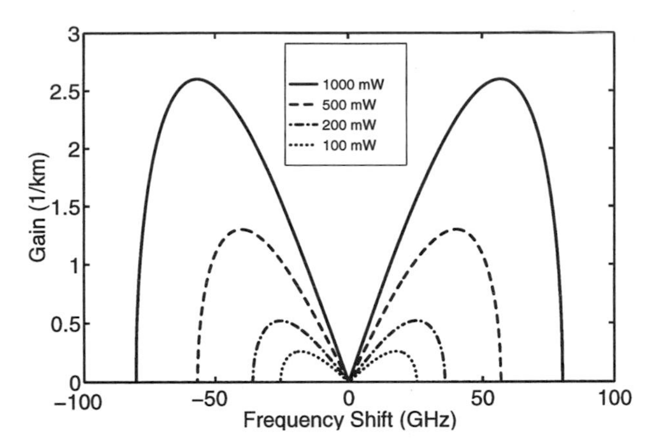

Fig. 8.24 Modulation instability gain coefficient versus frequency for conventional unshifted fibers with $\gamma = 1.3\ W^{-1}km^{-1}$. Note that the frequency at which g_{max} occurs increases with power according to Eq. (32) [2].

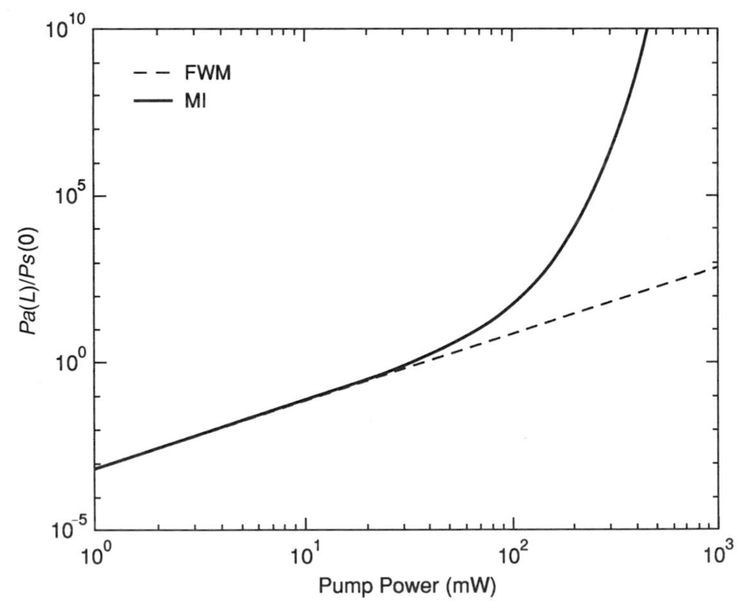

Fig. 8.25 Plot of Eq. (27) (*solid line*) compared to the linear approximation (*dashed line*).

way to determine the effects of MI taking into account fiber loss is through simulations. The results of two simulations and comparison with experiments comprise the remainder of this section.

Modulation instability using cw sources can be readily observed using an amplified source of about 100 mW. An experiment was performed using a DFB laser that was frequency dithered to suppress SBS and amplified to +19 dBm (80 mW). The ASE from the optical amplifier serves as the "seed" to be amplified, i.e., the ASE can be thought of as injected Stokes and anti-Stokes light that is then amplified by the strong pump wave. The amplified light was transmitted through fibers with 0.16 ps/nm-km dispersion. Figure 8.26 shows three spectra. The dashed line shows the spectrum of the amplified DFB laser. The dotted line shows the output from a 16-km long fiber. The output from this fiber was amplified to +19 dBm and passed through a 14-km long fiber. The output is shown by the solid curve. These results can be simulated using the split-step method to integrate the nonlinear Schrödinger equation. Figure 8.27 shows the output from a single 15-km span and from two 15-km spans separated by an amplifier. Aside from the amplifier ASE gain shape in the experimental figure, which was not included in the simulations, the agreement is quite good. Quasi-cw measurements of MI noise amplification were also per-

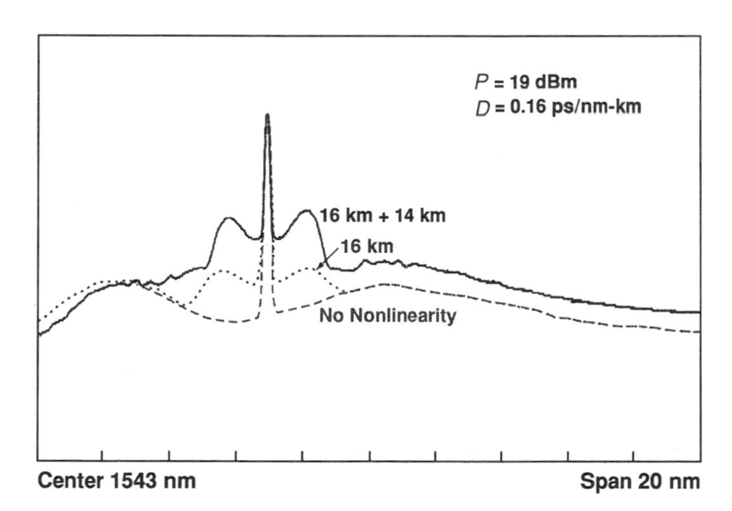

Fig. 8.26 Experimental results when +19 dBm (80 mW) cw light was transmitted through a 16-km long fiber (*dotted curve*), then amplified again to +19 dBm and transmitted through 14 km of fiber (*solid curve*). The input spectrum is given by the *dashed curve*.

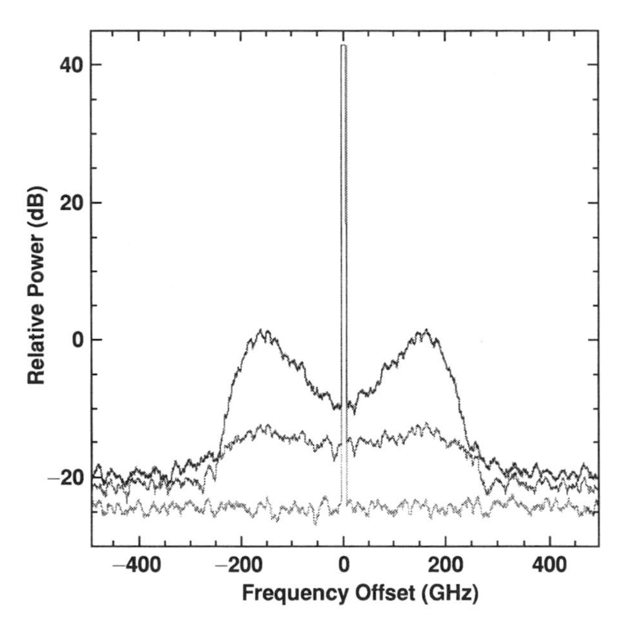

Fig. 8.27 Computer simulation of the experimental results shown in Fig. 8.26. The bottom trace is the input spectrum, the middle trace is the spectrum through 15 km, and the top trace is through 15 + 15 km. The amplifiers were assumed to have a flat gain spectrum in these simulations.

formed in 20-Gb/s transmission experiments [125]. The noise amplification was measured in the middle of a long string of NRZ "ones." In this type of measurement, the ASE amplified by MI appears to degrade systems using anomalous-dispersion fiber when compared to systems using normal-dispersion fiber. However, system degradation is not solely a function of noise but also of dispersive effects such as pulse shaping and SPM, as can be seen in the following experiments.

MI can be beneficial to transmission systems because of the pulse shaping it provides. Experiments and computer simulations show that at least in single-span systems positive dispersion provides superior system performance. A 10-Gb/s NRZ signal was transmitted through two nominally identical 145-km long fibers except with opposite signs of dispersion. One fiber had +330 ps/nm dispersion, the other −330 ps/nm. Figure 8.28a shows the 10-Gb/s eye diagram at the input (+14.5 dBm average power), Fig. 8.28b shows the output eye diagram from the positive-dispersion fiber, and Fig. 8.28c shows the output from the negative-dispersion fiber. Results

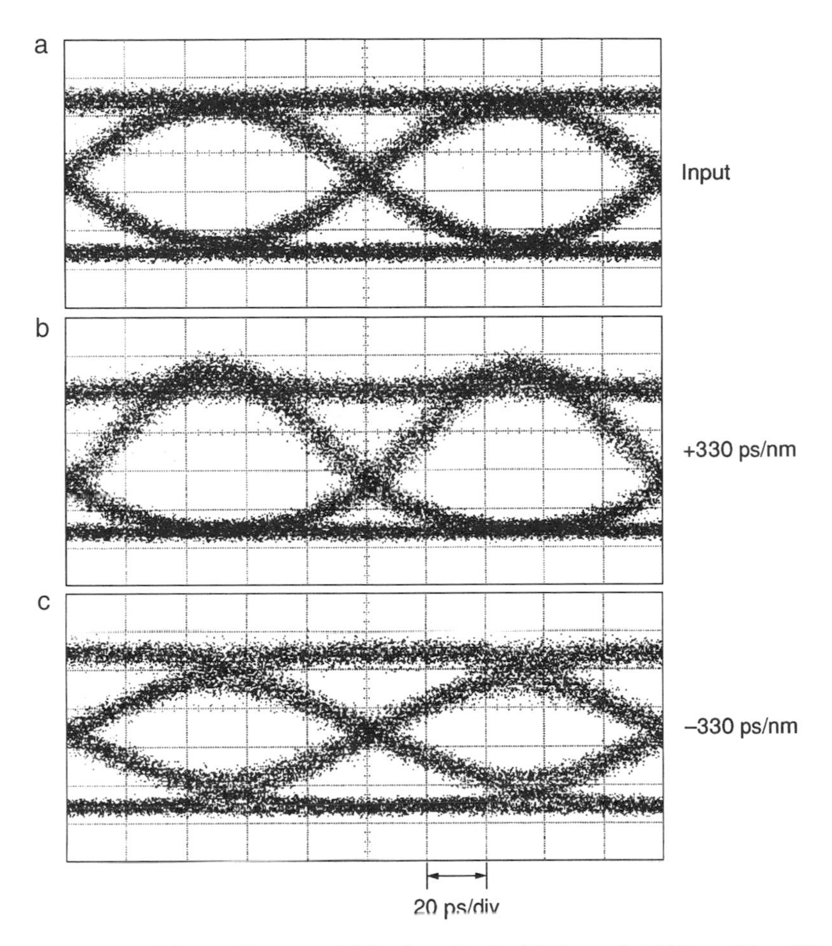

Fig. 8.28 10-Gb/s eye diagrams (a) back-to-back; (b) through 145 km of "positive" non-zero-dispersion fiber; and (c) through 145 km of "negative" non-zero-dispersion fiber. The average input power was +14.5 dBm.

of computer simulations, shown in Fig. 8.29a–c, using the experimental parameters (but with no amplifier noise) agree well with the experimental results. The eye diagram for negative dispersion is significantly closed. The penalty in the negative-dispersion fiber arises from pulse broadening due to SPM. However, in the case of positive dispersion the peaking and the lower crossing levels produce a larger eye opening. Obviously, the system penalty through the negative-dispersion fiber is larger than through the positive-dispersion fiber. Comparison of penalties is quantified in Fig. 8.30,

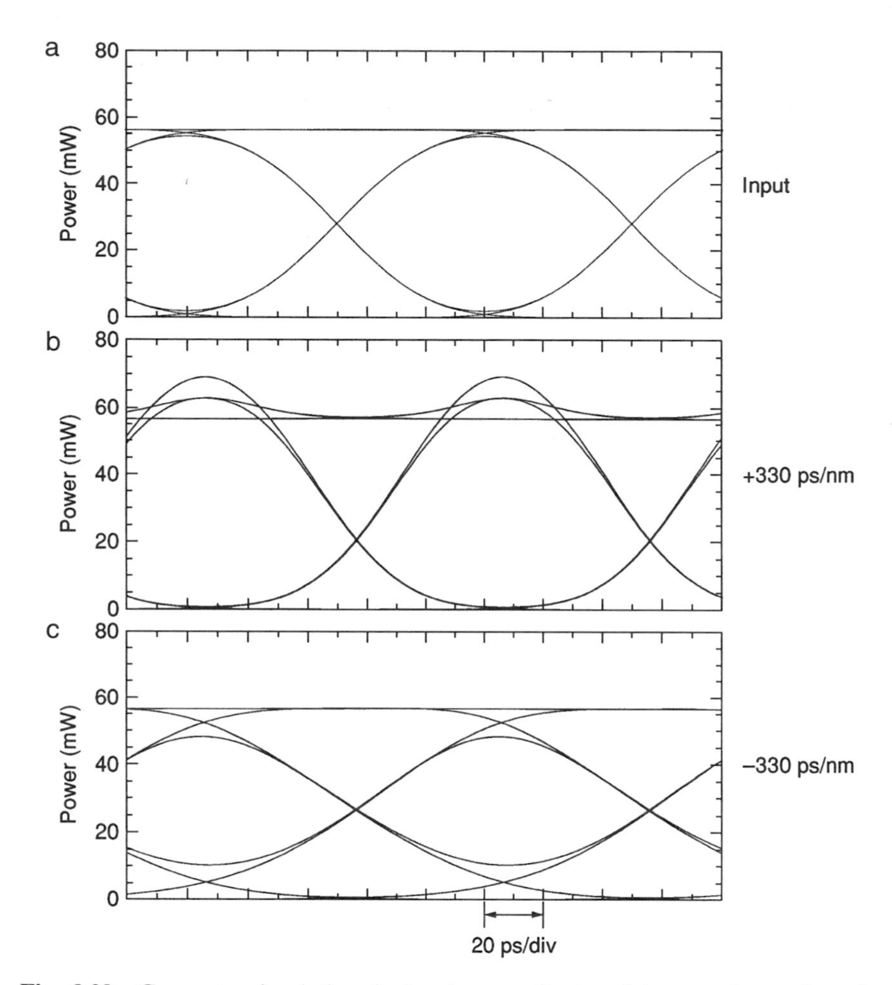

Fig. 8.29 Computer simulations in the absence of noise of the experimental results in Fig. 8.28.

which shows the experimental power penalty at 10^{-9} bit-error rate as a function of average launched power for the two fibers. For all input powers, the positive dispersion fiber provides superior performance, even exhibiting negative penalties over the power range between +7 dBm and +17 dBm. In the previously mentioned 20-Gb/s experiments [125], inspection of the bit patterns show similar behavior, i.e., in normal-dispersion fiber edges of bits spill over into adjacent "zeros" due to SPM, whereas in the anomalous-dispersion case the edges of bits become steeper.

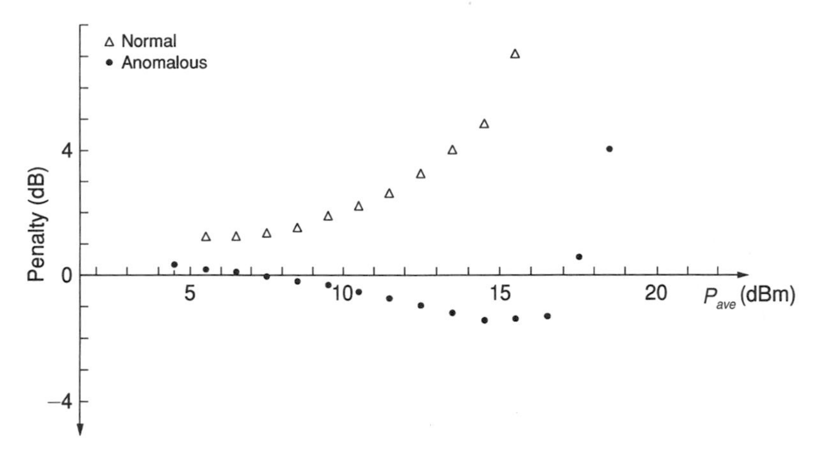

Fig. 8.30 Power penalty (at 10^{-9} BER) versus average launched power for "negative" non-zero-dispersion fiber (*triangles*) and "positive" non-zero-disperison fiber (*circles*).

VIII. Stimulated Raman Scattering

Stimulated Raman scattering (SRS) is a nonlinear parametric interaction between light and molecular vibrations. Light launched in an optical fiber is partially scattered and downshifted in frequency. The change in optical frequency corresponds to the molecular-vibrational frequency. SRS is similar to SBS, but can occur in either the forward or backward direction. The Raman gain coefficient is about three orders of magnitude smaller than the Brillouin gain coefficient, so in a single-channel system the SRS threshold is about three orders of magnitude larger than the SBS threshold given in Eq. (8.5), and is of the order of 1 W [2, 25, 133, 139]. However, the gain bandwidth for SRS, on the order of 12 THz or 100 nm, is much larger than that for SBS (see Section II). Thus, SRS can couple different channels in a WDM system and give rise to cross talk.

Figure 8.31 shows the Raman gain in fused silica fiber (at 1.5 μm). Due to SRS, in a WDM system, signals at longer wavelength are amplified by shorter-wavelength signals, which leads to degradation of the shorter wavelength signals. SRS couples channels separated in wavelength by up to 120 nm, and the interaction is stronger between channels further apart from each other. In the following calculations, the gain will be assumed to increase linearly with the wavelength separation (assuming that the total bandwidth is less than 120 nm) [23].

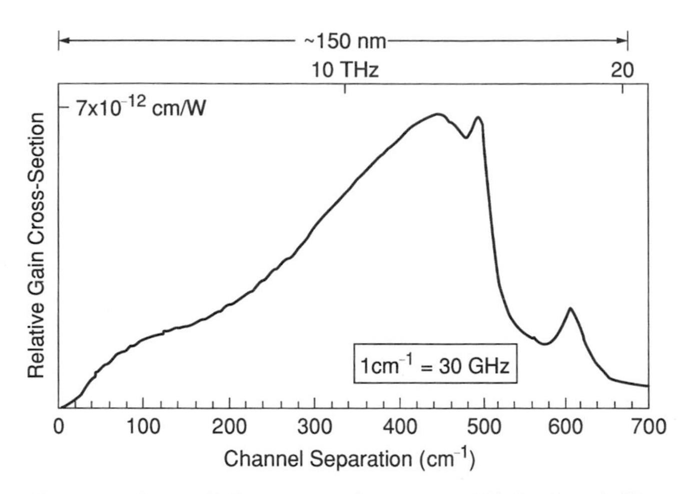

Fig. 8.31 Raman gain coefficient versus frequency shift for fused silica at a pump wavelength of 1.5 μm. 1 cm^{-1} = 30 GHz.

In regenerated systems, SRS causes a power penalty in the short-wavelength channels [21, 60, 64, 150], as shown in Fig. 8.32 for a simple two-channel case. If dispersion is neglected and looking at the worst case

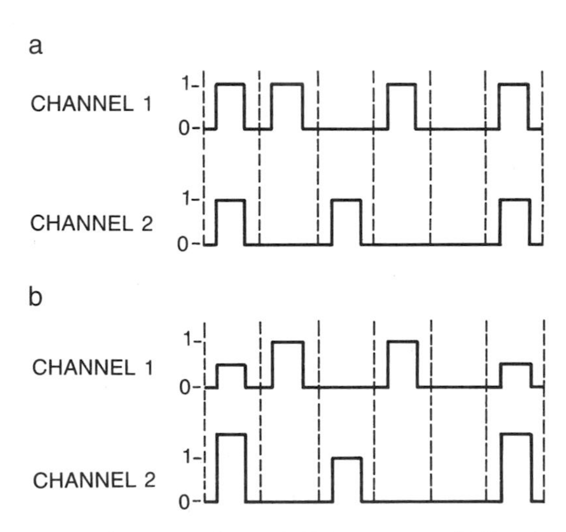

Fig. 8.32 Bit patterns in a two-channel WDM system (RZ pulses are shown for clarity) at (a) fiber input and (b) fiber output with SRS interaction between the channels ($\lambda_1 < \lambda_2$).

of marks being transmitted on all channels, one can obtain the very simple result that the product of total power and total bandwidth of an unrepeatered WDM system cannot exceed 500 GHz-W to guarantee a penalty for the shortest wavelength channel lower than 1 dB [25].

In amplified systems, the effects of SRS are more subtle. For single-channel systems, SRS can accumulate over multiple amplified spans; however, most of the scattered power is generated at wavelengths several tens of nanometers larger than the signal wavelength, as can be seen from Fig. 8.31, and is filtered by the optical amplifiers. Therefore, the single-channel SRS threshold decreases very slowly with the number of amplifiers. In WDM systems, the degradation is in the form of reduced SNR in the short-wavelength channels. Because the noise is added periodically over the entire length of a system, it experiences less Raman loss than the signal. For small degradations, the fractional depletion of the noise is half the fractional depletion of the signal (e.g., a 1-dB signal depletion corresponds to a 0.5-dB SNR degradation) [26]. Therefore, a system that, in the absence of SRS, operated at a particular required BER would no longer meet the required SNR and BER in the presence of SRS degradation. For amplified systems, and in the worst case of marks interacting on all channels, the requirement to ensure a SNR degradation of less than 0.5 dB in the worst channel is that the product of total power, total bandwidth, and total effective length of the system should not exceed 10 THz-mW-Mm [26]. This limit is detailed in Fig. 8.33, where the maximum allowable number of channels is given versus the total length of an amplified system with ideal amplifiers (3-dB noise figure), 0.2 dB/km fiber loss, 2.5 Gb/s per channel, 0.5-nm channel spacing, 10-GHz receiver optical bandwidth, optical SNR = 9 (for average power), corresponding to BER = 10^{-14} and for four values of amplifier spacings, 25, 50, 100, and 150 km.

In WDM systems with several channels, however, the probability of marks being simultaneously transmitted on all channels is very low. The fractional power lost by the shortest-wavelength channel (the most degraded one) is actually a random variable, since it depends on the bits transmitted in the other channels. For more than 10 channels and with zero-dispersion fiber, the depletion is well approximated by a truncated Gaussian with average value that depends only on the average transmitted power and is given, for the shortest-wavelength channel, by [43]

$$\eta_0 = \frac{N(N-1)}{4}K, \tag{8.33}$$

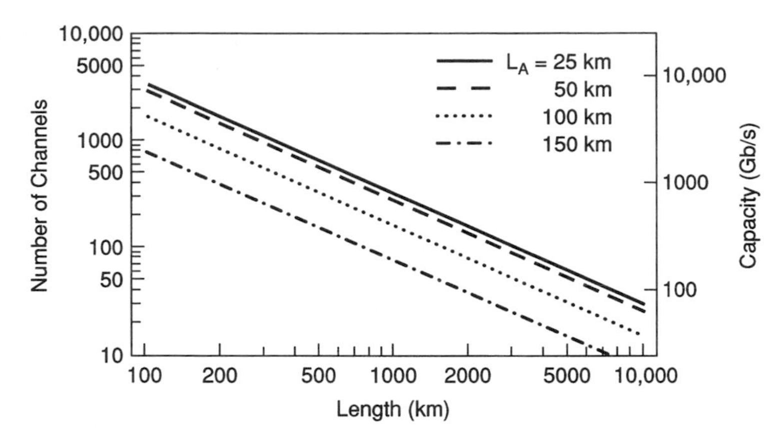

Fig. 8.33 The maximum number of 2.5 Gb/s channels and total capacity versus system length for four different amplifier spacings, L_A, assuming ideal amplifiers and in the worst case assumption of marks interacting on all channels. System parameters are given in the text.

and with variance

$$\sigma_0^2 = \frac{N(N-1)(2N-1)}{24}K^2, \tag{8.34}$$

where N is the number of channels, K is the depletion due to a neighboring channel, expressed by

$$K = \frac{P\Delta f \gamma_p L_e}{3 \times 10^{13} A_e}, \tag{8.35}$$

P is the peak power of each channel, Δf the channel spacing (in Hz), γ_p the Raman gain coefficient, and 1.5×10^{13} Hz is the region over which the Raman gain is assumed to increase linearly [23]. A factor of 2 is also included to account for averaging over polarization, as seen in Section II for SBS.

In the limit of infinitely many channels, with constant total power and bandwidth, the probability density function of the depletion converges to a delta function and the effect of SRS becomes completely deterministic. In this limit, the depletion can be in principle exactly compensated by using a filter after each in-line amplifier that provides an attenuation linearly varying in wavelength and with the correct slope to undo the SRS depletion [46]. This postamplifier wavelength-dependent loss does not introduce any SNR degradation. Furthermore, no new components are needed if optical filters are already present in the erbium-doped fiber amplifiers to compensate their gainshape. It is sufficient to slightly change the filter design to include the SRS gain tilt, since for the most part the natural gainshape of the amplifiers also has a positive slope. Compensation of the average depletion due to SRS without using filters can be also achieved by constraining the WDM channels in the region of the erbium-doped fiber amplifiers where the slope of the gainshape is negative, above the peak around 1560 nm [164]. Since the slope of the gainshape of the amplifiers is linear and negative only over a small bandwidth region, in this scheme the accuracy of the compensation has to be traded off with system bandwidth.

With a finite number of channels, only the average depletion can be suppressed by the filters. The residual statistical fluctuations of the depletion (and, therefore, of the channel power) cause a SNR degradation [46], which is computed later in this section. This degradation is a measure of the increase in optical SNR (as measured on a spectrum analyzer) required to achieve a given BER. Ideally, midspan spectral inversion (described in Chapter 7 in Volume IIIA) could completely eliminate SRS cross talk.

The effect of fiber chromatic dispersion, neglecting pulse shape distortion, is to introduce a walk-off between pulses transmitted in different channels [35], causing more (independent) bits to interact; therefore, its effect is similar to an increase in the number of channels. With large chromatic dispersion, and on each amplified fiber span, the number of bits of adjacent channels that have appreciable SRS interaction over each amplified span is approximately given by

$$N_b = \frac{L_e}{L_w}, \tag{8.36}$$

where

$$L_w = \frac{1}{RD\Delta\lambda} \tag{8.37}$$

is the walk-off length, R is the bit rate, D the fiber chromatic dispersion, and $\Delta\lambda$ the channel separation. The total number of bits that interact with each bit of the shortest-wavelength channel can be approximated with

$$N_t = \frac{L_e}{L_w}\frac{N(N-1)}{2}. \tag{8.38}$$

The total depletion caused on each bit of the shortest-wavelength channel by a bit in a different channel is given by

$$D_1 = \frac{L_w}{L_e} K, \tag{8.39}$$

and is the same for all channels. The reason is that the SRS gain increases linearly with channel frequency separation, while the interaction length decreases linearly with channel frequency separation, if the walk-off length is less than the effective length. With large chromatic dispersion ($L_w < L_e$), the average depletion is given by

$$\eta_D = N_t \frac{D_1}{2} = \frac{N(N-1)}{4} K = \eta_0, \tag{8.40}$$

while the variance of the depletion is

$$\sigma_D^2 = N_t \frac{D_1^2}{4} = \frac{N(N-1)}{8} \frac{L_w}{L_e} K^2. \tag{8.41}$$

A more accurate analysis including the decreased depletion introduced by bits encountered later in the fiber shows that σ_D^2 given in Eq. (8.41) should be divided by two, so that in one amplifier span

$$\frac{\sigma_D}{\eta_D} \approx \frac{1}{N} \sqrt{\frac{L_w}{L_e}}. \tag{8.42}$$

Figure 8.34 shows the relative fluctuation of the shortest-wavelength channel depletion (the ratio between standard deviation and mean) versus number of channels for a WDM system with 0.5-nm channel spacing, 10 Gb/s per channel, 50-km fiber length, and 0.2 dB/km fiber loss. The dashed line shows σ_0/η_0 (fiber with zero-chromatic dispersion); the solid line shows σ_D/η_D for conventional fiber with dispersion of 16 ps/nm-km. Chromatic dispersion is very effective in reducing the fluctuation of SRS depletion even for WDM systems with small number of channels. The truncated Gaussian that approximates the probability distribution function of power depletion is essentially a full Gaussian distribution when the ratio σ/η is less than 0.1, since the tails reach below 10^{-20}.

In lightwave systems with in-line amplifiers, the average of the depletion η_D increases linearly with the number of amplifiers N_A, while the increase of the standard deviation σ_D depends on the dispersion map and on the dispersion compensation technique (see Section VI and Chapter 7 in Volume IIIA). If all the dispersion is compensated at the end, due to walk-

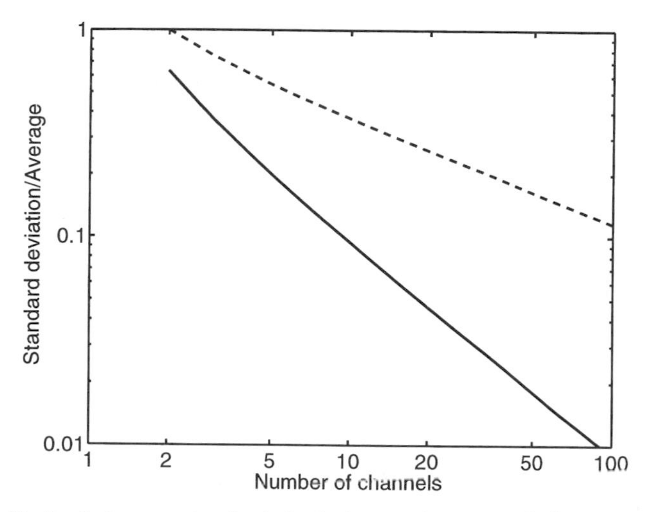

Fig. 8.34 Ratio between standard deviation and mean of shortest wavelength channel power depletion versus number of channels for zero-chromatic dispersion (*dashed line*) and conventional fiber with dispersion of 16 ps/nm-km (*solid line*). The other parameters are 0.5-nm channel spacing; 10 Gb/s bit rate per channel; 50-km fiber length; and 0.2 dB/km fiber loss.

off, new independent pulses interact with a given pulse on each amplified span, and the variance of the depletion σ_D^2 increases linearly with N_A. On the other hand, if dispersion-compensating fiber is used at each amplifier, the relative position of pulses in different channels is restored each time they are launched into a new fiber span, and the same pulses keep interacting. In this case, the standard deviation of the depletion σ_D increases linearly with N_A, and the SRS induced SNR degradation grows more rapidly. Thus, to reduce this degradation, the period of dispersion compensation of the dispersion map should be as large as possible. Alternatively, dispersion compensation could be provided on a channel-by-channel basis (e.g., with fiber gratings). In this case, dispersion is compensated without the negative side effect of pulse realignment, and σ_D^2 increases linearly with N_A independent of the dispersion map. In the examples that follow, σ_D^2 is assumed to increase linearly with N_A.

The worst case analysis discussed previously gives the safe region of operation of amplified systems with few channels and low dispersion. The worst case depletion is twice the average depletion η_0 [43], so the SNR degradation introduced by SRS is given by

$$\text{PEN}_{\omega c}[dB] = 10 \log_{10}(1 + \eta_0). \quad (8.43)$$

When the fiber chromatic dispersion and/or the number of channels are sufficiently large to assume a full Gaussian distribution for the power depletion ($\sigma/\eta < 0.1$), the SNR degradation becomes

$$\text{PEN}_D[dB] = 10 \log_{10}(1 + \eta_D/2 + Q^2\sigma_D^2), \quad (8.44)$$

where the Q-factor [15] depends on the required BER. Using filters, as mentioned previously, the deterministic part η_D of the SRS cross talk can in principle be completely suppressed; therefore, with filters, the SNR degradation becomes

$$\text{PEN}_{fil}[dB] = 10 \log_{10}(1 + Q^2\sigma_D^2). \quad (8.45)$$

Figure 8.35 shows SNR degradation versus number of channels for a 5000-km long WDM system with 50-km amplifier spacing and 10 Gb/s bit rate per channel. In the calculations, realistic system parameters have been assumed: 6-dB amplifier noise figure, 0.25 dB/km fiber loss, 80-μm^2 effective area for conventional fiber and 55 μm^2 for dispersion-shifted fiber, 0.5-nm channel spacing, 40-GHz receiver optical bandwidth, and 6 dB of margin at the receiver (SNR = 36). The dotted line is the result of the worst case

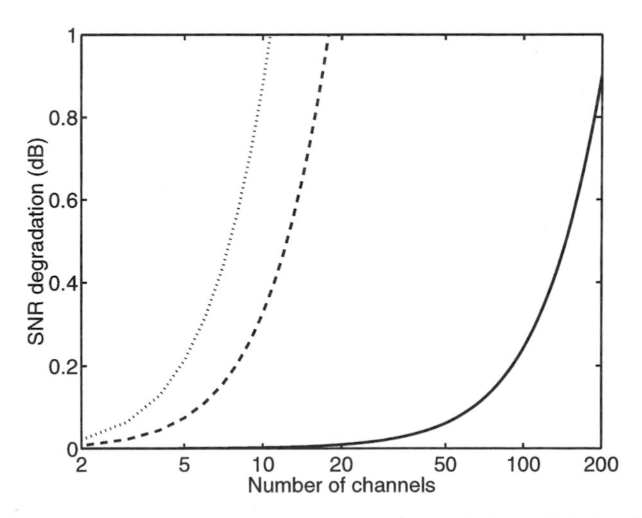

Fig. 8.35 SNR degradations versus number of channels for a 5000-km long WDM system obtained with worst case analysis and zero-dispersion fiber (*dotted line*), statistical analysis and conventional fiber (*dashed line*) power equalization at amplifiers, and conventional fiber (*solid line*). System parameters are given in the text.

analysis [26], which, for dispersion shifted fiber, is very close to the limit obtained with a statistical approach when power equalization is not used. The dashed line shows the improvement due to the chromatic dispersion of conventional fiber (partially due to the lower effective area). The solid line corresponds to a system with conventional fiber and proper filters placed after each amplifier, and is obtained, as detailed previously, under the assumption that bits in different channels are uncorrelated at each amplifier. The power equalization provided by the filters allows the number of channels to be increased tenfold.

Figure 8.36 shows the number of channels at which a 0.5-dB SNR degradation occurs versus system length, for the same system parameters of Fig. 8.35. The dotted line is the result of the worst case analysis (Eq. [8.43]). The dashed line shows the result of the statistical analysis (Eq. [8.44]). The solid line shows the improvement achievable with power equalization (Eq. [8.45]), with the same assumptions as in Fig. 8.35. There is a substantial increase in the maximum number of channels when the deterministic component in the SRS cross talk is suppressed with the power equalization technique. An excessive depletion over each amplifier span would require

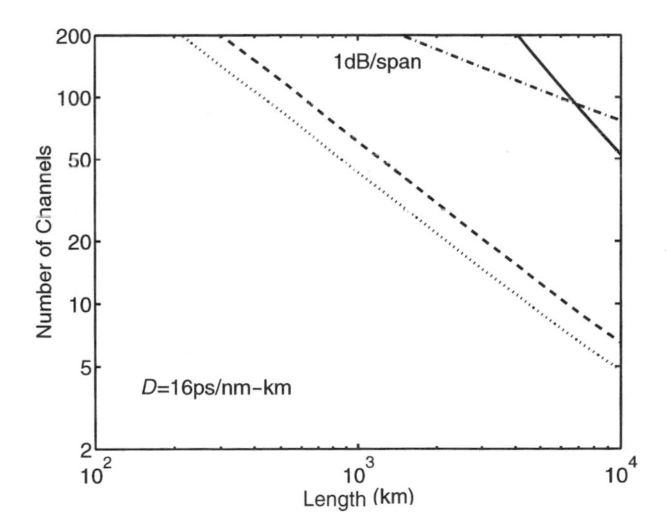

Fig. 8.36 Number of channels versus system length for a 0.5-dB SNR degradation with conventional fiber. The *dotted line* is for the worst case analysis, the *dashed line* is for the statistical analysis, and the *solid line* shows the effect of power equalization. The *dashed-dotted line* shows the 1 dB/span limit. System parameters are given in the text.

the use of filters with nonlinear shape or positioned along the fiber within each amplified span, and therefore should be in practice avoided. Also, the theory relies on the small gain linearization of SRS; therefore, the average depletion between filters should not exceed 1 dB. The dashed-dotted line shows the maximum number of channels achievable while still maintaining the average depletion of the worst channel below 1 dB over each amplified span. The improvement introduced by power equalization depends on system parameters; using lower receiver margin or better amplifiers rapidly increases the benefit of using power equalization.

SRS cross talk is effectively suppressed by the combined use of large fiber chromatic dispersion, to transform the cross talk into a quasideterministic effect, and power equalization, to suppress the deterministic part. Dispersion management is the key technique in designing WDM lightwave systems, since it facilitates simultaneous suppression of XPM (see Section IV), FPM (see Section VI), and SRS cross talk. This is further detailed in the next section with an example.

IX. Scaling Nonlinearities

In this section, the issue of scaling of nonlinearities is addressed. The linear and nonlinear effects that determine the BER of a WDM point-to-point transmission system with intensity modulation and direct detection are compared for two systems with different granularity and same total capacity of 100 Gb/s, S_1 with 10 channels at 10 Gb/s, and S_2 with 20 channels at 5 Gb/s. It is found that the impact of optical fiber nonlinearities is lower in the second system.

The two systems are assumed to have the same length and amplifier spacing and to use the same optical amplifiers, so total output power, total bandwidth and noise figure are the same, hence power per channel P and channel spacing Δf in S_2 are half those in S_1, as shown in Table 8.1. The fibers used in the two systems are assumed to be the same except for chromatic dispersion. Therefore fiber loss, effective length, effective area, and gain coefficients for nonlinear effects are the same in the two systems. Both systems are assumed to take advantage of dispersion management to balance the degradation introduced by chromatic dispersion and nonlinear effects.

The key element in this comparison is the choice of the dispersion map for the two systems. It is assumed here that S_1 utilizes the best possible

Table 8.1 **Two Systems with Total Capacity of 100 Gb/s**

	S_1 10 × 10 Gb/s	S_2 20 × 5 Gb/s
Number of channels	10	20
Bit rate per channel	10 Gb/s	5 Gb/s
Power per channel	P_1	$P_2 = P_1/2$
Channel spacing	1 nm	0.5 nm
Dispersion map	$D_1(l)$	$D_2(l) = 4D_1(l)$

dispersion map, the one that yields the lowest BER for the given system parameters (length, power, channel plan, amplifiers). Since the relation between the optimum dispersion map of S_1 and S_2 is in general unknown, it will be arbitrarily assumed here that the map for S_2 corresponds to the map for S_1 with a chromatic dispersion four times larger at any point. This choice of the dispersion map for S_2 is justified, at least from a linear point of view, by the fact that linear distortions due to chromatic dispersion are inversely proportional to the square of the bit rate R [2]. Clearly the choice of the dispersion map for S_2 could be suboptimal; therefore, it is expected that the performance of S_2 might improve with a more careful selection of its dispersion map. If S_2 can be proved to perform better than S_1 even with the suboptimal choice, all the more reason it will perform better with an optimal choice of its dispersion map.

In the absence of fiber nonlinearities, S_1 and S_2 with the previous choices are equivalent. They have the same overall loss and gain, the optical signal-to-noise ratio (SNR) at the receiver is the same and the pulse evolution due to the chromatic dispersion, D, determined by the product $R^2 D$, is the same. Also the walk-off length, defined as the length of fiber over which two bits of adjacent channels completely walk through each other due to dispersion, is the same since $R D \Delta f$ is the same.

When fiber nonlinearities are considered, one might expect that S_2 will be less affected than S_1 by the "single-channel"nonlinearities, SBS and SPM, since its power per channel is lower and more affected than S_1 by the "multi-channel" nonlinearities, CPM, FPM, and SRS, since its number of channels is larger.

For SBS (see Section II), the requirement is the power on each channel at the output of each amplifier be below threshold to avoid SBS-induced depletion. The SBS threshold is the same in the two systems; however, the power per channel is smaller in S_2, therefore S_2 is more robust to SBS than S_1. The effect of SPM can be estimated by looking at the spectral broadening ΔB induced on each signal, proportional to the product $P\ R$, as discussed in Section III. The degradation introduced by spectral broadening on each fiber span is proportional to $R\ \Delta\ B\ D$ [63], and is lower for S_2. Therefore, in agreement with the previous intuitive argument, S_2 is less affected than S_1 by "single-channel" nonlinearities.

For CPM, the spectrum of each channel is mostly broadened only by its immediate neighboring channels, since pulses belonging to channels that are widely separated pass through each other very rapidly without affecting each other, as discussed in Section IV. This is true provided that the dispersion is not too low, as must be the case for both S_1 and S_2 to achieve a sufficient FPM suppression, as discussed in Section VI. Since the walk-off length of adjacent channels is the same in the two systems, the number of relevant neighboring channels will also be the same. The CPM spectral broadening introduced by each immediate neighboring channel is proportional to $P\ R$; therefore, in analogy to the SPM case, the effect of each immediate neighboring channel is weaker in S_2. Hence, S_2 is less affected than S_1 by CPM.

The impact of a FPM wave generated at a channel frequency in a WDM system depends on its power relative to the channel power at the receiver and is proportional to the square of the channel powers and to FPM efficiency (see Eq. [8.16]). In a WDM system with equally spaced channels, the number of mixing products that are generated on each channel grows with the square of the number of channels. However, mixing products generated by channels beyond the near neighbors have rapidly decreasing efficiency, so the number of "relevant" mixing products generated on each channel is actually the same in S_1 and S_2. Hence, to compare the impact of mixing on S_1 and S_2, it is enough to compare the magnitude of the largest mixing products (and the magnitude of all "relevant" mixing products, suitably paired). If the best dispersion map for S_1 does not include fiber with very low dispersion, as is very likely, it can be shown that the efficiency of mixing products generated by the same set of channels in S_1 and S_2 is the same, since the increase in efficiency in S_2 due to the closer channel spacing is perfectly balanced by the reduced efficiency due to the increase in the fiber chromatic dispersion. Since the efficiency of the paired mixing products is the same in the two systems, but the power per channel is lower in S_2, the impact of FPM is smaller for S_2.

If the impact of SRS is assessed by considering the worst case SNR degradation of the shortest wavelength channel, assuming that all channels are transmitting marks and neglecting dispersion, as discussed in the previous section, the simple result is obtained that the effect is proportional to the product of total power and total bandwidth of the WDM system. Since S_1 and S_2 have same total power and bandwidth (they use the same amplifiers) the SRS effect in this worst case analysis is the same in the two systems. On the other end, it was shown in the previous section that in the presence of dispersion and with a large number of channels, SNR degradation of the different bits in the same channel tends to be the same, since each bit interacts with approximately the same number of marks and spaces in the other channels. When the number of bits that interact is sufficiently large, the SRS cross talk becomes a deterministic effect and the impact of SRS can be largely suppressed by channel pre-equalization. For SRS cross-talk suppression, what really matters is the number of bits that interact. The larger the number of bits, the better the suppression. Since the walk-off length between adjacent channels is the same in the two systems, the number of interacting bits is larger in S_2 because of the larger number of channels. So the worst case SRS cross talk is the same for S_1 and S_2, but in S_2 the SRS cross talk can be more effectively suppressed with power pre-equalization, due to improved averaging.

In conclusion, the intuition that "multi-channel" nonlinearities would affect S_2 more than S_1 because of the larger number of channels is wrong. CPM and FPM do not get worse, under the present assumptions, with an increase in the number of channels, and the lower power per channel of S_2 allows a reduction of these effects. For SRS, the increase in the number of channels is actually beneficial, under the present scaling assumptions, since it facilitates SRS cross-talk suppression.

A key role in the scaling of all the nonlinear effects is played by the scaling of the dispersion map. However, as mentioned previously, optimizing the dispersion map can only further improve the performance of S_2. Therefore, from a fiber nonlinearity point of view, 20 5 Gb/s channels is a better choice than 10 Gb/s channels to achieve a 100 Gb/s aggregate capacity. The conclusion that increasing the granularity is better from a transmission point of view is quite general, but the reasoning used here cannot be extended arbitrarily to, for example, compare one channel at 100 Gb/s with 100 channels at 1 Gb/s. A fundamental assumption here is indeed that the best dispersion map of S_1 does not include any segment of fiber with small dispersion. This is reasonable for a 10-channel 10 Gb/s WDM systems, but it is not for a single-channel system at 100 Gb/s. Furthermore, if the scaling factor is large, the assumed scaling of the dispersion map becomes unrealistic.

Table 8.2 **List of Symbols**[a]

Symbol	*Typical Value*	*Physical Meaning*
A_e	80 μm^2	Fiber effective area
c	3×10^8 m/s	Light speed in vacuum
D	16 ps/nm-km	Fiber chromatic dispersion
$\frac{dD}{d\lambda}$	0.08 ps/nm^2-km	Dispersion slope (derivative of $D(\lambda)$)
d_{12}	[s/m]	Walk-off parameter
D_{ijk}	3 or 6	FPM degeneracy factor
f	194 THz	Optical frequency
g_B	4×10^{-11} m/W	Brillouin gain coefficient
g_{MI}	[1/m]	Modulation instability gain coefficient
g_R	7×10^{-14} m/W	Raman gain coefficient
L	[m]	Fiber length
L_e	20.7 km	Fiber effective length
L_w	[m]	Walk-off length
N		Number of channels
N_A		Number of amplified spans
n	1.45	Fiber refractive index
n_2	2.6×10^{-20} m^2/W	Nonlinear index of refraction
P_B^{th}	4 mW	Brillouin scattering threshold power
P_R^{th}	1.8 W	Raman scattering threshold power
P	[W]	Optical power
R	[bit/s]	Bit rate
T	[s]	Bit time
α	0.21 dB/km	Fiber loss coefficient
β	[1/m]	Mode propagation constant
β_2	-20.4 ps^2/km	Group-velocity dispersion
β_3	0.16 ps^3/km	Second-order dispersion (derivative of $\beta_2(\omega)$)
$\Delta\beta$	[1/m]	Four-wave mixing phase mismatch
γ	1.3 W^{-1} km^{-1}	Nonlinearity coefficient
η		Four-wave mixing efficiency
λ	1550 nm	Light wavelength in vacuum
λ_0	1300 nm	Zero-dispersion wavelength
$\Delta\lambda$	[nm]	Wavelength difference
χ_{1111}	3.5×10^{-15} [esu]	Third-order nonlinear susceptibility
ω	1220 Trad/s	Angular optical frequency

[a] Fiber values are for conventional silica single-mode fiber.

X. Formulas and Symbols (see Table 8.2)

Relations between fiber chromatic dispersion and group-velocity dispersion

$$D = -\frac{2\pi c}{\lambda^2}\beta_2 \qquad \frac{dD}{d\lambda} = \frac{4\pi c}{\lambda^3}\beta_2 + \left(\frac{2\pi c}{\lambda^2}\right)^2\beta_3 \qquad \beta_3 = \left(\frac{\lambda^2}{2\pi c}\right)^2\left(\frac{2}{\lambda}D + \frac{dD}{d\lambda}\right)$$

Walkoff parameter and walkoff length

$$d_{12} = \int_{\lambda 1}^{\lambda 2} \frac{dD}{d\lambda}(\lambda - \lambda_0)d\lambda \qquad L_w = \frac{1}{R\,D\,\Delta\,\lambda}$$

Effective length and loss

$$L_e = \frac{1 - e^{-\alpha L}}{\alpha} \qquad \alpha = \log\left(10^{\alpha_{dB/10}}\right)$$

Definition of nonlinear coefficient γ and relation between χ_{1111} and η_2

$$\gamma = \frac{2\pi\eta_2}{\lambda A_e} \qquad \chi_{1111}\,[\text{esu}] = \frac{c\eta^2}{480\pi^2}\,\eta_2\,[\text{m}^2/\text{W}]$$

SBS threshold

$$\frac{g_B P_{\text{B}}^{\text{th}} L_e}{2A_e} \approx 21$$

SRS threshold

$$\frac{g_B P_{\text{R}}^{\text{th}} L_e}{2A_e} \approx 16$$

FPM

$$P_{ijk} = \left(\frac{D_{ijk}}{3}\gamma\, L_e\right)^2 P_i P_j P_k e^{-\alpha L}\eta \qquad [\text{MKS}]$$

$$P_{ijk} = \frac{1024\pi^6}{\eta^4\lambda^2 c^2}\left(\frac{D_{ijk}\chi_{1111}L_e}{A_e}\right)^2 P_i P_j P_k e^{-\alpha L}\eta \qquad [\text{esu}]$$

$$\eta = \frac{\alpha^2}{\alpha^2 + \Delta\beta^2}\left(1 + \frac{4e^{-\alpha L}\sin^2(\Delta\beta\, L/2)}{(1 - e^{-\alpha L})^2}\right)\cdot\left(\frac{\sin(N_A\,\Delta\beta\, L/2)}{\sin(\Delta\beta\, L/2)}\right)^2$$

$$\Delta\beta = \beta_i + \beta_j - \beta_k - \beta_{ijk} = \frac{2\pi\lambda^2}{c}(f_i - f_k)(f_j - f_k)$$
$$\left[D(\lambda) - \frac{\lambda^2}{c}\left(\frac{f_i + f_j}{2} - f\right)\frac{dD}{d\lambda}\right]$$
$$= -\frac{2\pi\lambda^4}{c^2}\frac{dD}{d\lambda}\left(\frac{f_i + f_j}{2} - f_0\right)(f_i - f_k)(f_j - f_k)$$
$$= \frac{2\pi c}{\lambda^2}\frac{dD}{d\lambda}\left(\frac{\lambda_i + \lambda_j}{2} - \lambda_0\right)(\lambda_i - \lambda_k)(\lambda_j - \lambda_k)$$

Relative magnitude of a phase-matched three-tone product in a WDM system

$$\frac{P_{ijk}}{P(L)} = \left(\frac{D_{ijk}}{3}\right)^2 (\gamma P\, L_e)^2 \approx 0.01\, P^2\ [\mathrm{mW}^2]$$

References

1. Agrawal, G. P. 1987. Modulation instability induced by cross-phase modulation. *Phys. Rev. Lett.* 59:880.
2. Agrawal, G. P. 1989. *Nonlinear fiber optics.* New York: Academic Press.
3. Anderson D., and M. Lisak. 1984. Modulational instability of coherent optical-fiber transmission signals. *Opt. Lett.* 9:468.
4. Aoki, Y., S. Kishida, and K. Washio. 1986. Stable cw backward Raman amplification in optical fibers by stimulated Brillouin scattering suppression. *Appl. Opt.* 25:1056.
5. Aoki, Y., K. Mito, and K. Tajima. 1987. Observation of stimulated Brillouin scattering in single-mode fibers with single-frequency laser-diode pumping. *Opt. Quantum Electron.* 19:141.
6. Aoki, Y., K. Tajima, and I. Mito. 1988. Input power limits of single-mode optical fibers due to stimulated Brillouin scattering in optical communications systems. *J. Lightwave Tech.* 6:710.
7. Aoki, Y. 1988. Properties of fiber Raman amplifiers and their applicability to digital optical communication systems. *J. Lightwave Tech.* 6:1225.
8. Aoki, Y. 1989. Fiber Raman amplifier properties for applications to long-distance optical applications. *Opt. Quantum Electron.* 21:89.
9. Artiglia, M., E. Ciaramella, and B. Sordo. 1995. Using modulation instability to determine Kerr coefficient in optical fibers. *Electron. Lett.* 31:1012.
10. Auyeung, J., and A. Yariv. 1978. Spontaneous and stimulated Raman scattering in long low-loss fibers. *IEEE J. Quantum Electron.* 4:347.

11. Azuma, Y., N. Shibata, T. Horiguchi, and M. Tateda. 1995. Wavelength dependence of Brillouin gain spectra for single-mode optical fibers. *Electron. Lett.* 24:250.
12. Bar-Joseph, I., A. A. Friesem, R. G. Waarts, and H. H. Yaffe. 1986. Parametric interaction of a modulated wave in a single-mode fiber. *Opt. Lett.* 11:534.
13. Bergano, N. S., J. Aspell, C. R. Davidson, P. R. Trischitta, B. M. Nyman, and F. W. Kerfoot. 1991. Bit error rate measurement of 14,000 km 5 Gb/s fiber-amplifier transmission system using circulating loop. *Electron. Lett.* 27:1889.
14. Bergano, N. S. 1993. Undersea lightwave transmission systems using Er-doped fiber amplifiers. *Opt. Photon. News* 4:8.
15. Bergano, N. S., and C. R. Davidson. 1995. Circulating loop transmission experiments for the study of long-haul transmission systems using erbium-doped fiber amplifiers. *J. Lightwave Tech.* 13:879.
16. Bergano, N. S., and C. R. Davidson. 1996. Wavelength division multiplexing in long-haul transmission systems. *J. Lightwave Tech.* 14:1229.
16. Botincau, J., and R. H. Stolen. 1982. Effect of polarization on spectral broadening in optical fibers. *J. Opt. Soc. Am.* 72:1592.
18. Chen, C. D., J. M. P. Delavaux, B. W. Hakki, O. Mizuhara, T. V. Nguyen, R. J. Nuyts, K. Ogawa, Y. K. Park, R. E. Tench, and P. D. Yeates. 1994. Field demonstration of 10 Gb/s–360 km transmission through embedded standard (non-DSF) fiber cables. *Electron. Lett.* 30:1139.
19. Chiang, T.-K., N. Kagi, T. K. Fong, and M. E. Marhic. 1994. Cross-phase modulation in dispersive fibers: theoretical and experimental investigation of the impact of modulation frequency. *IEEE Photon. Tech. Lett.* 6:733.
20. Chiang, T.-K., N. Kagi, M. E. Marhic, and L. G. Kazovsky. 1996. Cross-phase modulation in fiber links with multiple optical amplifiers and dispersion compensators. *J. Lightwave Tech.* 14:249.
21. Chraplyvy, A. R., and P. S. Henry. 1983. Performance degradation due to stimulated Raman scattering in wavelength-division-multiplexed optical-fiber systems. *Electron. Lett.* 19:641.
22. Chraplyvy, A. R., D. Marcuse, and P. S. Henry. 1984. Carrier-induced phase noise in angle-modulated optical-fiber systems. *J. Lightwave Tech.* 22:6.
23. Chraplyvy, A. R. 1984. Optical power limits in multichannel wavelength-division-multiplexed systems due to stimulated Raman scattering. *Electron. Lett.* 20:58.
24. Chraplyvy, A. R., and J. Stone. 1984. Measurement of crossphase modulation in coherent wavelength-division multiplexing using injection lasers. *Electron. Lett.* 20:996.
25. Chraplyvy, A. R. 1990. Limitations on lightwave communications imposed by optical-fiber nonlinearities. *J. Lightwave Tech.* 8:1548.
26. Chraplyvy, A. R., and R. W. Tkach. 1993. What is the actual capacity of single-mode fibers in amplified lightwave systems? *IEEE Photon. Tech. Lett.* 5:666.

27. Chraplyvy, A. R., A. H. Gnauck, T. W. Tkach, and R. M. Derosier. 1993. 8×10 Gb/s transmission through 280 km of dispersion-managed fiber. *IEEE Photon. Tech. Lett.* 5:1233.
28. Chraplyvy, A. R., J. M. Delavaux, R. M. Derosier, G. A. Ferguson, D. A. Fishman, C. R. Giles, J. A. Nagel, B. B. Nyman, J. W. Sulhoff, R. E. Tench, R. W. Tkach, and J. L. Zyskind. 1994. 1420-km transmission of sixteen 2.5-Gb/s channels using silica-fiber-based EDFA repeaters. *IEEE Photon. Tech. Lett.* 6:1371.
29. Chraplyvy, A. R., A. H. Gnauck, R. W. Tkach, and R. M. Derosier. 1995. One-third terabit/s transmission through 150 km of dispersion-managed fiber. *IEEE Photon Tech. Lett.* 7:98.
30. Cleland, D. A., A. D. Ellis, and C. H. F. Sturrock. 1992. Precise modeling of four wave mixing products over 400 km of step-index fiber. *Electron. Lett.* 28:1171.
31. Clesca, B., S. Artigaud, L. Pierre, and J.-P. Thiery. 1995. 8×10 Gbit/s repeaterless transmission over 150 km of dispersion-shifted fiber using fluoride-based amplifiers. In *Proc. OAA '95, Davos, Switzerland.* Paper SaB5-1, 220.
32. Cotter, D. 1982. Observation of stimulated Brillouin scattering in low-loss silica fiber at 1.3 μm. *Electron. Lett.* 18:495.
33. Cotter, D. 1983. Stimulated Brillouin scattering in monomode optical fiber. *J. Opt. Comm.* 4:10.
34. Cotter, D. 1983. Optical nonlinearity in fibers: A new factor in systems design. *Br. Telecom. Tech. J.* 1:17.
35. Cotter, D., and A. M. Hill. 1984. Stimulated Raman crosstalk in optical transmission: effects of group velocity dispersion. *Electron. Lett.* 20:185.
36. Cotter, D. 1987. Fiber nonlinearities in optical communications. *Opt. Quantum Electron.* 19:1.
37. Davey, S. T., D. L. Williams, B. J. Ainslie, W. J. M. Rothwell, and B. Wakefield. 1989. Optical gain spectrum of GeO_2-SiO_2 Raman fiber amplifiers. *IEEE Proc.* 136:301.
38. Dougherty, D. J., F. X. Kartner, H. A. Haus, and E. P. Ippen. 1995. Measurement of the Raman gain spectrum of optical fibers. *Opt. Lett.* 20:31.
39. Dziedzic, J. M., R. H. Stolen, and A. Ashkin. 1981. Optical Kerr effect in long fibers. *Appl. Opt.* 20:1403.
40. Fishman D. A., and J. A. Nagel. 1993. Degradations due to stimulated Brillouin scattering in multigigabit intensity-modulated fiber-optic systems. *J. Lightwave Tech.* 11:1721.
41. Forghieri, F., R. W. Tkach, A. R. Chraplyvy, and D. Marcuse. 1994. Reduction of four-wave mixing crosstalk in WDM systems using unequally spaced channels. *IEEE Photon. Tech. Lett.* 6:754.
42. Forghieri, F., A. H. Gnauck, R. W. Tkach, A. R. Chraplyvy, and R. M. Derosier. 1994. Repeaterless transmission of eight channels at 10 Gb/s over

137 km (11 Tb/s-km) of dispersion-shifted fiber using unequal channel spacing. *IEEE Photon. Tech. Lett.* 6:1374.
43. Forghieri, F., R. W. Tkach, and A. R. Chraplyvy. 1995. Effect of modulation statistics on Raman crosstalk in WDM systems. *IEEE Photon. Tech. Lett.* 7:101.
44. Forghieri, F., R. W. Tkach, and A. R. Chraplyvy. 1995. WDM systems with unequally spaced channels. *J. Lightwave Tech.* 13:889.
45. Forghieri, F., R. W. Tkach, and A. R. Chraplyvy. 1995. Statistics of four-wave mixing crosstalk. In *Proc. NLGW '95, Dana Point, CA,* Paper NSaD1, 256.
46. Forghieri, F., R. W. Tkach, and A. R. Chraplyvy. 1995. Suppression of Raman crosstalk in WDM systems. In *Proc. OAA '95, Davos, Switzerland,* Paper SaB3, 212.
47. Forghieri, F., R. W. Tkach, A. R. Chraplyvy, and A. M. Vengsarkar. 1996. Dispersion compensating fiber: is there merit in the figure of merit? In *Proc. OFC '96, San Jose, CA,* Paper ThM5, 255.
48. Fukui, M., K. Oda, H. Toba, K. Okamoto, and M. Ishii. 1995. 10 channel × 10 Gbit/s WDM add/drop multiplex/transmission experiment over 240 km of dispersion-shifted fiber employing unequally-spaced arrayed-waveguide-grating ADM filter with fold-back configuration. *Electron. Lett.* 31:1757.
49. Garth, S. J., and C. Pask. 1986. Four-photon mixing and dispersion in single-mode fibers. *Opt. Lett.* 11:380.
50. Garth, S. J. 1988. Phase matching the stimulated four-photon mixing process on single-mode fibers operating in the 1.55-μm region. *Opt. Lett.* 13:1117.
51. Gnauck, A. H., R. M. Jopson, and R. M. Derosier. 1993. 10-Gb/s 360-km transmission over dispersive fiber using midsystem spectral inversion. *IEEE Photon. Tech. Lett.* 5:663.
52. Gnauck, A. H., and R. W. Tkach. 1995. Interplay of chirp and self phase modulation in dispersion-limited optical transmission systems. In *Proc. 19th Eur. Conf. on Opt. Comm., Montreux.* Paper TuC4.4, 105.
53. Gnauck, A. H., A. R. Chraplyvy, R. W. Tkach, and R. M. Derosier. 1994. 160 Gbit/s (8 × 20 Gbit/s WDM 300 km transmission with 50 km amplifier spacing and span-by-span dispersion reversal. *Electron. Lett.* 30:1241.
54. Gnauck, A. H., R. M. Jopson, and R. M. Derosier. 1995. Compensating the compensator: a demonstration of nonlinearity cancellation in a WDM system. *IEEE Photon. Tech. Lett.* 7:582.
55. Gnauck, A. H., A. R. Chraplyvy, R. W. Tkach, J. L. Zyskind, J. W. Sulhoff, A. J. Lucero, Y. Sun, R. M. Jopson, F. Forghieri, R. M. Derosier, C. Wolf, and A. R. McCormick. 1996. One terabit/s transmission experiment. In *Proc. OFC '96, San Jose, CA,* Paper PD20-1.
56. Gordon, J. P., and L. F. Mollenauer. 1991. Effects of fiber nonlinearities and amplifier spacing on ultra-long distance transmission. *J. Lightwave Tech.* 9:170–173.

57. Hamaide, J. P., P. Emplit, and J. M. Gabriagues. 1990. Limitations in long haul IM/DD optical fiber systems caused by chromatic dispersion and nonlinear Kerr effect. *Electron. Lett.* 26:1451.
58. Hamazumi, Y., M. Koga, and K. Sato. 1996. Beat induced crosstalk reduction against wavelength difference between signal and four-wave mixing lights in unequal channel spacing WDM transmission. *IEEE Photon. Tech. Lett.* 8:718.
59. Hansen, P. B., L. E. Eskildsen, S. G. Grubb, A. M. Vengsarkar, S. K. Korotky, T. A. Strasser, J. E. Alphonsus, J. J. Veselka, D. J. DiGiovanni, D. W. Peckham, E. C. Beck, D. Truxal, W. Y. Cheung, S. G. Kosinski, D. S. Gasper, P. F. Wysocki, J. R. Simpson, and V. L. da Silva. 1995. 529 km unrepeatered transmission at 2.488 Gbit/s using dispersion compensation, forward error correction, and remote post- and pre-amplifiers pumped by diode-pumped Raman lasers. *Electron. Lett.* 31:1460.
60. Hegarty, J., N. A. Olsson, and M. McGlashan-Powell. 1985. Measurement of the Raman crosstalk at 1.5 μm in a wavelength-division-multiplexed transmission system. *Electron. Lett.* 21:395.
61. Henmi, N., T. Saito, and S. Nakaya. 1993. An arrangement of transmission-fiber dispersions for increasing the spacing between optical amplifiers in lumped repeater systems. *IEEE Photon. Tech. Lett.* 5:1337.
62. Henmi, N., Y. Aoki, T. Ogata, and T. Saito. 1993. A new design arrangement of transmission fiber dispersion for suppressing nonlinear degradation in long-distance optical transmission systems with optical repeater amplifiers. *J. Lightwave Tech.* 11:1615.
63. Henry, P. S. 1985. Lightwave primer. *IEEE J. Quantum Electron.* 21:1862.
64. Hill, A. M., D. Cotter, and I. Wright. 1984. Nonlinear crosstalk due to stimulated Raman scattering in a two-channel wavelength-division-multiplexed system. *Electron. Lett.* 20:247.
65. Hill, K. O., D. C. Johnson, B. S. Kawasaki, and R. I. MacDonald. 1978. CW three-wave mixing in single-mode optical fibers. *J. Appl. Phys.* 49:5098.
66. Iannone, E., F. Matera, and M. Settembre. 1994. Performance of very long intensity-modulated direct-detection optical links with low chromatic dispersion. *Electron. Lett.* 30:588.
67. Ikeda, M. 1981. Spectral power handling capability caused by stimulated Raman scattering effect in silica optical fibers. *Opt. Comm.* 37:388.
68. Inoue, K., and H. Toba. 1991. Error-rate degradation due to fiber four-wave mixing in four-channel FSK direct-detection transmission. *IEEE Photon. Tech. Lett.* 3:77.
69. Inoue, K. 1991. Arrangement of orthogonal polarized signals for suppressing fiber four-wave mixing in optical multichannel transmission systems. *IEEE Photon. Tech. Lett.* 3:560.
70. Inoue, K. 1992. Four-wave mixing in an optical fiber in the zero-dispersion wavelength region. *J. Lightwave Tech.* 11:1553.

71. Inoue, K. 1992. Reduction of fiber four-wave mixing influence using frequency modulation in multichannel IM/DD transmission. *IEEE Photon. Tech. Lett.* 4:1301.
72. Inoue, K. 1992. Polarization effect on four-wave mixing efficiency in a single-mode fiber. *IEEE J. Quantum Electron.* 28:883.
73. Inoue, K. 1992. Phase-mismatching characteristic of four-wave mixing in fiber lines with multistage optical amplifiers. *Opt. Lett.* 17:801.
74. Inoue, K. 1993. Suppression technique for fiber four-wave mixing using optical multi-/demultiplexers and a delay line. *J. Lightwave Tech.* 11:455.
75. Inoue, K. 1993. Fiber four-wave mixing suppression using two incoherent polarized lights. *J. Lightwave Tech.* 11:2116.
76. Inoue, K. 1994. Experimental study on channel crosstalk due to fiber four-wave mixing around the zero-dispersion wavelength. *J. Lightwave Tech.* 12:1023.
77. Inoue, K., K. Nakanishi, K. Oda, and H. Toba. 1994. Crosstalk and power penalty due to fiber four-wave mixing in multichannel transmissions. *J. Lightwave Tech.*12:1423.
78. Inoue, K., and H. Toba. 1995. Fiber four-wave mixing in multi-amplifier systems with nonuniform chromatic dispersion. *J. Lightwave Tech.* 13:88.
79. Inoue, K. 1995. A simple expression for optical FDM network scale considering fiber four-wave mixing and optical amplifier noise. *J. Lightwave Tech.* 13:856.
80. Ippen, E. P., and R. H. Stolen. 1972. Stimulated Brillouin scattering in optical fibers. *Appl. Phys. Lett.* 21:539.
81. Islam, M. N., L. F. Mollenauer, R. H. Stolen, J. R. Simpson, and H. T. Shang. 1987. Cross-phase modulation in optical fibers. *Opt. Lett.* 12:625.
82. Jopson, R. M., A. H. Gnauck, and R. M. Derosier. 1993. Compensation of fiber chromatic dispersion by spectral inversion. *Electron. Lett.* 29:576.
83. Jorgensen, B. F., R. J. S. Pedersen, and C. G. Joergensen. 1994. Self-phase modulation induced transmission penalty reduction in a 5 Gbit/s FM/AM conversion system experiment over 205 km of standard fiber. *IEEE Photon. Tech. Lett.* 6:279.
84. Kagi, N., T.-K. Chiang, T. K. Fong, M. E. Marhic, and L. G. Kazovsky. 1994. Frequency dependence of cross phase modulation in amplified optical fiber links. *Electron. Lett.* 30:1878.
85. Kakui, M., T. Kato, T. Kashiwada, and K. Nakazato. 1995. 2.4 Gbit/s repeaterless transmission over 306 km non-dispersion-shifted fiber using directly modulated DFB-LD and dispersion-compensating fiber. *Electron. Lett.* 31:51.
86. Kawakami, H., Y. Miyamoto, T. Kataoka, and K. Hagimoto. 1994. Overmodulation of intensity modulated signals due to stimulated Brillouin scattering. *Electron. Lett* 30:1507.
87. Kikuchi, K. 1993. Enhancement of optical-amplifier noise by nonlinear refractive index and group-velocity dispersion of optical fibers. *IEEE Photon. Tech. Lett.* 5:221–223.

88. Kikuchi, N., and S. Sasaki. 1995. Analytical evaluation technique of self-phase-modulation effect on the performance of cascaded optical amplifier systems. *J. Lightwave Tech.* 13:868.
89. Kikuchi, N., and S. Sasaki. 1996. Fiber nonlinearity in dispersion-compensated conventional fiber transmission. *Electron. Lett.* 32:570.
90. Kim, K. S., W. A. Reed, R. H. Stolen, and K. W. Quoi. 1994. Measurement of the non-linear index of silica core and dispersion-shifted fibers. *Opt. Lett.* 19:257.
91. Kurtzke, C., and A. Gnauck. 1993. Opcrating principle of in-line amplified dispersion-supported transmission. *Electron. Lett.* 29:1969.
92. Kurtzke, C. 1993. Suppression of fiber nonlinearities by appropriate dispersion management. *IEEE Photon. Tech. Lett.* 5:1250.
93. Kuwano, S., K. Yonenaga, and K. Iwashita. 1995. 10 Gbit/s repeaterless transmission experiment of optical duobinary modulated signal. *Electron. Lett.* 31:1359.
94. Li, A., C. J. Mahon, Z. Wang, G. Jacobsen, and E. Bodtker. 1995. Experimental confirmation of crosstalk due to stimulated Raman scattering in WDM AM-VSB CATV transmission systems. *Electron. Lett.* 31:1538.
95. Lichtman, E., A. A. Friesem, and R. G. Waarts. 1987. Exact solution of four-wave mixing of copropagating light beams in a Kerr medium. *J. Opt. Soc. Am.* 4:1801.
96. Lichtman E., R. G. Waarts, and A. A. Friesem. 1989. Stimulated Brillouin scattering excited by a modulated pump wave in single-mode fibers. *J. Lightwave Tech.* 7:171.
97. Lichtman, E., A. A. Friesem, S. Tang, and R. G. Waarts. 1991. Nonlinear Kerr interactions between modulated waves propagating in a single-mode fiber. *J. Lightwave Tech.* 9:422.
98. Lichtman, E., and S. G. Evangelides. 1994. Reduction of the nonlinear impairment in ultralong lightwave systems by tailoring the fiber dispersion. *Electron. Lett.* 30:346.
99. Lichtman, E. 1995. Performance limitations imposed on all-optical ultralong lightwave systems at the zero-dispersion wavelength. *J. Lightwave Tech.* 13:898.
100. Maeda, M. W., W. B. Sessa, W. I. Way, A. Yi-Yan, L. Curtis, R. Spicer, and R. I. Laming. 1990. The effect of four-wave mixing in fibers on optical frequency-division multiplexed systems. *J. Lightwave Tech.* 8:1402.
101. Mahon, C. J., L. Olofsson, E. Bodtker, and G. Jacobsen. 1996. Polarization allocation schemes for minimizing fiber four-wave mixing crosstalk in wavelength division multiplexed optical communication systems. *IEEE Photon. Tech. Lett.* 8:575.
102. Mao, X. P., R. W. Tkach, A. R. Chraplyvy, R. M. Jopson, and R. M. Derosier. 1992. Stimulated Brillouin threshold dependence on fiber type and uniformity. *IEEE Photon. Tech. Lett.* 4:66.

103. Mao, X. P., G. E. Bodeep, R. W. Tkach, A. R. Chraplyvy, T. E. Darcie, and R. M. Derosier. 1992. Brillouin scattering in externally modulated lightwave AM-VSB CATV transmission systems. *IEEE Photon. Tech. Lett.* 4:287.
104. Marcuse, D. 1991. Single-channel operation in very long nonlinear fibers with optical amplifiers at zero dispersion. *J. Lightwave Tech.* 9:356.
105. Marcuse, D. 1991. Bit-error rate of lightwave systems at the zero-dispersion wavelength. *J. Lightwave Tech.* 9:1330–1334.
106. Marcus, D., A. R. Chraplyvy, and R. W. Tkach. 1991. Effect of fiber nonlinearity on long-distance transmission. *J. Lightwave Tech.* 9:121.
107. Marcuse, D. 1992. RMS width of pulses in nonlinear dispersive fibers. *J. Lightwave Tech.* 10:17.
108. Marcuse, D., A. R. Chraplyvy, and R. W. Tkach. 1994. Dependence of cross-phase modulation on channel number in fiber WDM systems. *J. Lightwave Tech.* 12:885.
109. Marhic, M. E., N. Kagi, T. K. Chiang, and L. G. Kazovsky. 1995. Cancellation of third-order nonlinear effects in amplified fiber links by dispersion compensation, phase conjugation, and alternating dispersion. *Opt. Lett.* 20:863.
110. Matera, F., A. Mecozzi, M. Romagnoli, and M. Settembre. 1993. Sideband instability induced by periodic power variation in long-distance fiber links. *Opt. Lett.* 18:1499.
111. Mecozzi, A. 1994. Long-distance transmission at zero dispersion: combined effect of the Kerr nonlinearity and the noise of the in-line amplifiers. *J. Opt. Soc. Am.* 11:462.
112. Miyamoto, Y., T. Kataoka, A. Sano, and K. Hagimoto. 1994. 10 Gbit/s, 280 km nonrepeatered transmission with suppression of modulation instability. *Electron. Lett.* 30:797.
113. Miyata, H., H. Onaka, K. Otsuka, and T. Chikama. 1995. Dispersion compensation design for 10-Gb/s 16-wave WDM transmission system over standard single-mode fiber. In *Proc. 21st Eur. Conf. on Opt. Comm., Brussels,* Paper Mo.A.4.3, 63.
114. Morioka, T., H. Takara, S. Kawanishi, O. Kamatani, K. Takiguchi, K. Uchiyama, M. Saruwatari, H. Takahashi, M. Yamada, T. Kanamori, and H. Ono. 1996. 100 Gbit/s × 10 channel OTDM/WDM transmission using a single supercontinuum WDM source. In *Proc. OFC '96, San Jose, CA,* Paper PD21-1.
115. Naka, A., and S. Saito. 1992. Fiber transmission distance determined by eye opening degradation due to self-phase modulation and group-velocity dispersion. *Electron. Lett.* 28:2221.
116. Nuyts, R. J., and Y. K. Park. 1997. Dispersion equalization of a 10 Gb/s repeatered transmission system using dispersion compensating fibers. *J. Lightwave Technol.* (in press).
117. Oda, K. M. Fukutoka, M. Fukui, T. Kitoh, and H. Toba. 1995. 16-channel × 10-Gbit/s optical FDM transmission over a 1000-km conventional single-mode

fiber employing dispersion compensating fiber and gain equalization. In *Proc. OFC '95, San Diego, CA,* Paper PD22-1.

118. Ohmori, Y., Y. Sasaki, and T. Edahiro. 1981. Fiber-length dependence of critical power for stimulated Raman scattering. *Electron. Lett.* 17:593.
119. Okamoto, K., M. Ishii, Y. Hibino, Y. Ohmori, and H. Toba. 1995. Fabrication of unequal channel spacing arrayed-waveguide grating multiplexer modules. *Electron. Lett.* 31:1464.
120. Olsson, N. A., and J. P. VanderZiel. 1986. Fiber Brillouin amplifier with electronically controlled bandwidth. *Electron. Lett.* 22:488.
121. Olsson, N. A., and J. Hegarty. 1986. Noise properties of a Raman amplifier. *J. Lightwave Tech.* LT-4:396.
122. Onaka, H., H. Miyata, G. Ishikawa, K. Otsuka, H. Ooi, Y. Kai, S. Kinoshita, M. Seino, H. Nishimoto, and T. Chikama. 1996. 1.1 Tb/s WDM transmission over a 150 km 1.3 μm zero-dispersion single-mode fiber. In *Proc. OFC '96, San Jose, CA,* Paper PD19-1.
123. Otani, T., K. Goto, H. Abe, M. Tanaka, H. Yamaoto, and H. Wakabayashi. 1995. 5.3 Gbit/s 11,300 km data transmission using actual submarine cables and repeaters. *Electron. Lett.* 31:380.
124. Pare, C., A. Villeneuve, P.-A. Belanger, and N. J. Doran. 1996. Compensating for dispersion and the nonlinear Kerr effect without phase conjugation. *Opt. Lett.* 21:459.
125. Saunders, R. A., D. Garthe, B. L. Patel, W. S. Lee, and R. E. Epworth. 1995. Observation of parametric noise amplification owing to modulation instability in anomalous dispersion regime. *Electron. Lett.* 31:1088.
126. Schadt, D. G. 1991. Effect of amplifier spacing on four-wave mixing in multichannel coherent communications. *Electron. Lett.* 27:1805.
127. Schadt, D. G., and T. D. Stephens. 1992. Power limitations due to four-wave mixing effects in frequency division multiplexed coherent systems, using cascaded optical amplifiers. *J. Lightwave Tech.* 10:1715.
128. Schiess, M. 1995. Impact of different pulse shapes in nonlinear fibers. *J. Opt. Commun.* 16:168–172.
129. Shibata, N., R. P. Braun, and R. G. Waarts. 1987. Brillouin-gain spectra for single-mode fibers having pure-silica, GeO_2-doped, and P_2O_5-doped cores. *Opt. Lett.* 12:269.
130. Shibata, N., R. P. Braun, and R. G. Waarts. 1987. Phase-mismatch dependence of efficiency of wave generation through four-wave mixing in a single-mode optical fiber. *IEEE J. Quantum Electron.* QE-23:1205.
131. Shibata, N., Y. Nakano, Y. Azuma, and M. Tateda. 1988. Experimental verification of efficiency of wave generation through four-wave mixing in low-loss dispersion-shifted single-mode optical fiber. *Electron. Lett.* 24:1528.
132. Shiraki, K., M. Ohashi, and M. Tateda. 1995. Suppression of stimulated Brillouin scattering in a fiber by changing the core radius. *Electron. Lett.* 31:668.

133. Smith, R. G. 1972. Optical power handling capacity of low-loss optical fibers as determined by stimulated Raman and Brillouin scattering. *Appl. Opt.* 11:2489.
134. Stolen, R. H., and E. P. Ippen. 1973. Raman gain in glass optical waveguides. *Appl. Phys. Lett.* 22:276.
135. Stolen, R. H., and A. Ashkin. 1973. Optical Kerr effect in glass waveguide. *Appl. Phys. Lett.* 22:294.
136. Stolen, R. H., J. E. Bjorkholm, and A. Ashkin. 1974. Phase-matched three-wave mixing in silica fiber optical waveguides. *Appl. Phys. Lett.* 24:308.
137. Stolen, R. H. 1975. Phase-matched stimulated four-photon mixing in silica-fiber waveguides. *IEEE J. Quantum Electron.* QE-11:100.
138. Stolen, R. H., and C. Lin. 1978. Self-phase modulation in silica optical fibers. *Phys. Rev. A.* 17:1448.
139. Stolen, R. H. 1979. Nonlinear properties of optical fibers. In *Optical Fiber Telecommunications,* eds., S. E. Miller and A. G. Chynoweth. New York: Academic Press, p. 130.
140. Stolen, R. H. 1980. Nonlinearity in fiber transmission. In *Proc. IEEE* 68:1232.
141. Stolen, R. H., and J. E. Bjorkholm. 1982. Parametric amplification and frequency conversion in optical fibers. *IEEE J. Quantum Electron.* QE-18:1062.
142. Stolen, R. H., and C. Lee. 1984. Development of the stimulated Raman spectrum in single mode silica fibers. *J. Opt. Soc. Am. B.* 1:652.
143. Sugie, T. 1995. Maximum repeaterless transmission of lightwave systems imposed by stimulated Brillouin scattering in fibers. *Opt. Quantum Electron.* 27:643.
144. Tai, K., A. Hasegawa, and A. Tomita. 1986. Observation of modulational instability in optical fibers. *Phys. Rev. Lett.* 56:135.
145. Tajima, K. 1986. Self-amplitude modulation in PSK coherent optical transmission systems. *J. Lightwave Tech.* 4:900.
146. Takahashi, H., and K. Inoue. 1995. Cancellation of four-wave mixing by use of phase shift in a dispersive fiber inserted into a zero-dispersion transmission line. *Opt. Lett.* 20:860.
147. Takushima, Y., and T. Okoshi. 1992. Suppression of stimulated Brillouin scattering using optical isolators. *Electron. Lett.* 28:1155.
148. Tkach, R. W., A. R. Chraplyvy, F. Forghieri, A. H. Gnauck, and R. M. Derosier. 1995. Four-photon mixing and high-speed WDM systems. *J. Lightwave Tech.* 13:841.
149. Tkach, R. W., R. M. Derosier, F. Forghieri, A. H. Gnauck, A. M. Vengsarkar, D. W. Peckham, J. L. Zyskind, J. W. Sulhoff, and A. R. Chraplyvy. 1995. Transmission of eight 20-Gb/s channels over 232 km of conventional single-mode fiber. *IEEE Photon. Tech. Lett.* 7:1369.
150. Tomita, A. 1983. Crosstalk caused by stimulated Raman scattering in single-mode wavelength-division multiplexed systems. *Opt. Lett.* 8:412.

151. Uesugi, N., M. Ikeda, and Y. Sasaki. 1981. Maximum single-frequency input power in a long optical fiber determined by stimulated Brillouin scattering. *Electron. Lett.* 17:379.
152. Vengsarkar, A. M., and W. A. Reed. 1993. Dispersion-compensating single-mode fibers: efficient designs for first- and second-order compensation. *Opt. Lett.* 18:924.
153. Waarts, R. G. 1985. Crosstalk due to stimulated Brillouin scattering in monomode fiber. *Electron. Lett.* 21:1114.
154. Waarts, R. G., and R. P. Braun. 1986. System limitations due to four-wave mixing in single-mode optical fibers. *Electron. Lett.* 22:873.
155. Waarts, R. G., A. A. Friesem, and Y. Hefetz. 1988. Frequency-modulated to amplitude-modulated signal conversion by a Brillouin-induced phase change in single-mode fibers. *Opt. Lett.* 13:152.
156. Waarts, R. G., A. A. Friesem, E. Lichtman, H. H. Yaffe, and R. P. Braun. 1990. Nonlinear effects in coherent multichannel transmission through optical fibers. In *Proc. IEEE* 78:1344.
157. Walker, G. R., D. M. Spirit, P. J. Chidgey, E. G. Bryant, and C. R. Batchellor. 1992. Effect of fiber dispersion on four-wave mixing in multichannel coherent optical transmission system. *Electron. Lett.* 28:989.
158. Wang, Z., A. Li, C. J. Mahon, G. Jacobsen, and E. Bodtker. 1995. Performance limitation imposed by stimulated Raman scattering in optical WDM SCM video distribution systems. *IEEE Photon. Tech. Lett.* 7:1492.
159. Wang, Z., E. Bodtker, and G. Jacobsen. 1995. Effects of cross-phase modulation in wavelength-multiplexed SCM video transmission systems. *Electron. Lett.* 18:1591.
160. Watanabe, S., and M. Shirasaki. 1996. Exact compensation for both chromatic dispersion and Kerr effect in a transmission fiber using optical phase conjugation. *J. Lightwave Tech.* 14:243.
161. Yaffe, H. H., A. A. Friesem, E. Lichtman, and R. G. Waarts. 1987. Multiple-wave generation due to four-wave mixing in a single-mode fiber. *Electron. Lett.* 23:42.
162. Yonenaga, K., S. Kuwano, S. Norimatsu, and N. Shibata. 1995. Optical duobinary transmission system with no receiver sensitivity degradation. *Electron. Lett.* 31:302.
163. Yoshizawa, N. 1993. Stimulated Brillouin scattering suppression by means of applying strain distribution to fiber with cabling. *J. Lightwave Tech.* 11:1518.
164. Zou, X. Y., S.-M. Hwang, and A. E. Willner. 1996. Compensation of Raman scattering and EDFA's nonuniform gain in ultralong-distance WDM links. *IEEE Photon. Tech. Lett.* 8:139.

Chapter 9 | Terrestrial Amplified Lightwave System Design

Chungpeng Fan
Lucent Technologies, Bell Laboratories, Holmdel, New Jersey

J. P. Kunz
Lucent Technologies, Bell Laboratories, Holmdel, New Jersey

I. Introduction

This chapter discusses the building blocks, their associated technologies, and related issues of terrestrial lightwave systems.

A. *HISTORICAL BACKGROUND*

Since the advent of low-loss optic fiber [1-1] and the first field experiment of optical trunk equipment [1-2], rapid advances in lightwave research have contributed to dramatic progress in the expansion and improvement of terrestrial telecommunications networks [1-3, 1-4]. The widespread deployment of high-capacity lightwave systems enabled the service providers to complete their network digitization programs and establish their information superhighways to enter into the new era of competition in advanced services. Since the appearance of *Optical Fiber Telecommunications II* in 1988, revolutionary changes have occurred in many research fronts and significantly affected the evolution of the terrestrial transport network. For an in-depth understanding of the individual technology subjects, such as erbium-doped fiber amplifiers (EDFAs), wavelength-division multiplexing (WDM), narrow-width laser sources, fiber nonlinearity, and so forth, refer to other chapters of this book. In the following, a brief review is given on the impact of these new lightwave technologies on the terrestrial long-haul network evolution.

OPTICAL FIBER TELECOMMUNICATIONS,
VOLUME IIIA

ISBN: 0-12-395170-4

B. LIGHTWAVE NETWORK EVOLUTION

Stimulated by the successful commercial applications in U.S. and global markets abroad, suppliers of fiber optic and lightwave systems have recently grown their products through many phases of successive improvement and upgrade. An early consensus was reached by the international community in establishing comprehensive standards on common interfaces, protocols, and a signaling hierarchy, such as the synchronous optical network (SONET) [1-5], the synchronous digital hierarchy (SDH), and the asynchronous transfer mode (ATM), and, henceforth, moved the industry toward an open and leveled environment for competition. In the meantime, the service providers are driven by their concern for reducing cost to support price competition in long-distance services. They have adopted and exploited technology innovations, as described in other chapters of this book, such as Chapter 4 of Volume IIIA and Chapters 3, 7, and 8 of Volume IIIB, to enhance the capacity and quality of their lightwave networks by extending the repeater span to interoffice distances (~100 km) and increasing the data rate beyond the multigigabit per second level.

Another prevailing factor that exerts important influence over the evolution of terrestrial lightwave transport has been the realization by the network operators of the vulnerability of fiber routes to cable cuts; such vulnerability presents an ever-increasing economic risk due to the large concentration of traffic in a single strand of optic fiber. This focused concern by the industry on network reliability resulted in the rapid development of standards and products for self-healing SONET rings, which subsequently led to their introduction into the U.S. long-haul networks by many interexchange carriers [1-6].

With the emergence of fiber amplifiers offering wide-gain bandwidth for simultaneous amplification of multiple wavelengths, and the availability of narrow-linewidth laser sources and passive optical components, the earlier vision for applying optical WDM techniques as a networking approach to unlocking the vast information carrying capacity of the fiber medium is now gradually becoming a reality [1-7].

In summary, three successive stages of progress have been made in the evolution of the terrestrial lightwave network. The first stage, from 1977 to 1987, was the period of fiber optic transmission. Improvement of optical fiber as a transmission medium advanced from multimode to single mode, and the lightwave terminal increased its data speed by orders of magnitude from megabit to multigigabit per second. Applications were primarily point-to-point to replace radio or coaxial links.

This was followed by the second stage (1987–1996), the SONET networking period, when the industry adopted international standards and all vendors and users introduced standardized equipment to replace their earlier proprietary systems. These SONET products were first deployed by regional telephone companies for interoffice connections with protection access and diverse routing, and then were embraced by long-distance carriers in the form of self-healing rings for both short- and long-haul transmission. During this period, the new technologies of EDFAs and WDM were used in a limited fashion to reduce or eliminate the intermediate electronic regenerators and to increase the capacity for the optical transmission link [1-8]. The underlying networking technique was offered by SDH–SONET through electronic processing of signals at the network nodes.

In anticipation of the societal demand for high-capacity broadband services and encouraged by the concurrent advances in photonic component technologies and high-speed transmission, new research programs on optical networking architecture based on WDM are currently being pursued worldwide. This third stage of network evolution, building an optical layer for the transport infrastructure, has been initiated by the use of WDM in multipoint add–drop of wavelength channels along the optical transmission path [1-9, 1-10]. The vision of a transparent, scalable and flexible, all-optical network, although still remote, will become much more appealing and real with the appearance of novel devices such as the wavelength changer–router, tunable sources, and optical cross-connects.

Throughout these three stages, a common principle guiding the technology development for the new-generation lightwave systems has become evident [1-11]: it is to offer competitive advantages of improved performance and reduced cost to network operators by maximizing the reuse of embedded fiber in response to business needs for flexible capacity expansion. This will be achieved through the innovation and application of a platform of technology products built to industry standards and designed to simplify the infrastructure network to make it more reliable and valuable to the customers.

II. Architecture and Building Blocks of Lightwave Networks

The present and next-generation lightwave networks take advantage of three network features that were not available in previous networks: optical amplifiers, dense WDM, and SDH–SONET ring configurations. The following sections describe these new elements and how they are used in network applications.

A. *OPTICAL AMPLIFIERS*

The introduction of the optical amplifier has had a significant effect on lightwave network equipment design. An optical amplifier is, generically, any component that uses optical fiber as the amplification medium. In an optical amplifier, the optical signal is not converted to an electrical signal during amplification. Rather, the optical signal is amplified by passing through the fiber within the amplifier unit. There are several good references that describe optical amplifiers [2-1, 2-2], including Chapter 2 in Volume IIIB of this book. A simplified explanation of how optical amplifiers work is as follows: The input optical signal passes through a special optical fiber within the amplifier. This special fiber is also driven (pumped) with a signal from another optical source (laser). The special fiber has the characteristic that the optical power from the pump source is transferred to the input optical signal. Therefore, as a consequence of passing through the amplifier, the input optical signal is amplified. These optical amplifiers can amplify the optical signals passing through them to very large output power levels. The advantages that optical amplifiers provide are as follows:

(1) Greatly extended distances between terminals
(2) Great simplicity with resultant very low failure rates (high reliability)
(3) No jitter-accumulation problems

The advantages of the optical amplifier do, however, come with a price. That price is the maintenance subsystem requirements. In traditional regenerated systems, it was fairly easy for the regenerator to have access to the digital data stream that was being regenerated. Consequently, it was easy both to monitor the data stream and to make use of predefined locations in the data stream to send maintenance and control information along with the data. Because the optical amplifier never converts the optical signal to an electrical signal, no such easy access is available. A different method to maintain such systems must be produced. The resulting system features are described in Section III.B.5.

B. DENSE WDM

Two basic parameters have had a large influence on how telecommunications networks are built:

(1) The cost of installing new optical fiber is so high that almost any other method of capacity expansion is preferred.

(2) The needs for network capacity have continually exceeded the current network capacity.

These two facts explain why WDM has been popular in optical fiber networks. It is a way of increasing the capacity on an optical fiber that is already available. To realize this increase, two components are required: an optical combiner to put multiple signals of different wavelengths onto the same fiber and an optical filter to separate the signals.

Optical combiner technology has been available for some time. However, the optical filter technology has been more difficult to obtain. This technology has recently seen great advances in filter accuracy and stability. Both factors influence how close the optical signals can be spaced and still allow the optical filter to separate them again. Several applications have been installed in networks that combine and filter optical signals that operate at 1.3- and 1.5-μm wavelengths. A more recent technology has allowed two wavelengths in the 1.5-μm band to be combined onto a single fiber. With the latest technology, it is possible to put eight or more optical signals in the 1.5-μm region together on the same optical fiber.

C. SDH–SONET RING NETWORKS

The most common type of modern telecommunications network currently being installed is the SDH-SONET ring. SDH-SONET is a method of data transport and a set of requirements for terminal equipment. This standard was created to allow service providers to have a common type of network and network equipment [2-3]. The advantage of this would be to allow for a multivendor network equipment environment and, it was hoped, a lower cost network solution. The service provider acceptance of SDH-SONET has been such that SDH-SONET equipment has become dominant in the marketplace.

Bellcore document GR-253 [2-3] describes the generic requirements for equipment that operates in a SONET network. GR-253 describes the SONET signal format, the processing required for overhead bytes in that format, and the physical characteristics of the signals that flow through

SONET equipment. The standard SONET frame structure is shown in Fig. 9.1.

The SONET standard defines bytes in the SONET overhead stream that allow terminal equipment to operate in an add–drop mode (Fig. 9.2). In this mode of operation, a string of add–drop terminal sites can be configured such that a small piece of the total bandwidth of the data signal is accessed at any one add–drop site. For example, in Fig. 9.3 the line bandwidth is 2.5 Gb/s. However, at site *A* only one digital signal, level 3 (DS-3) of data is added–dropped from the line signal. The standard allows for an add–drop string to be configured in a loop where the end of the add–drop string is connected to the start of the string. This configuration is termed a *ring*.

The main advantage of SONET rings over strings of add–drop terminals is that the ring configuration allows for fast protection after fiber cuts in the network. As the telecommunications networks have evolved to a fiber-based network, the incidence of network outage caused by the fiber optical cable being cut has become a serious network problem. With the structure of a ring, it is always ensured that there are two paths into any ring node. In the case of a fiber cut on a ring, the ring protection switching will reroute traffic away from the cut fiber to travel the other way around the ring to the destination. This ring protection switch will occur in 60 ms or less. The SONET standard includes a very complete description of how a SONET ring switch is controlled [2-4].

1. Path-Switched Rings

A path-switched ring is shown in Fig. 9.4. This ring consists of add–drop nodes configured as shown. For a path-switched ring, data that enter the ring at point *A* and exit the ring at point *Z* would traverse the ring into two directions — from *A* to *Z* clockwise around the ring and from *A* to *Z* counterclockwise. The terminal at node *Z* receives both signals and uses path-level signal monitors to determine which of the signals was better and should be used. A path-switched ring always creates two dedicated paths for every data signal that enters the ring. These two signals always travel around the ring in different directions to arrive at the destination.

The advantage of path-switched rings is that protection switching is relatively simple. The terminal at the exit point of the signal makes all the decisions about which direction of transmission should be used (termed *tail-end switch*). The head-end terminal has the job of sending the data in

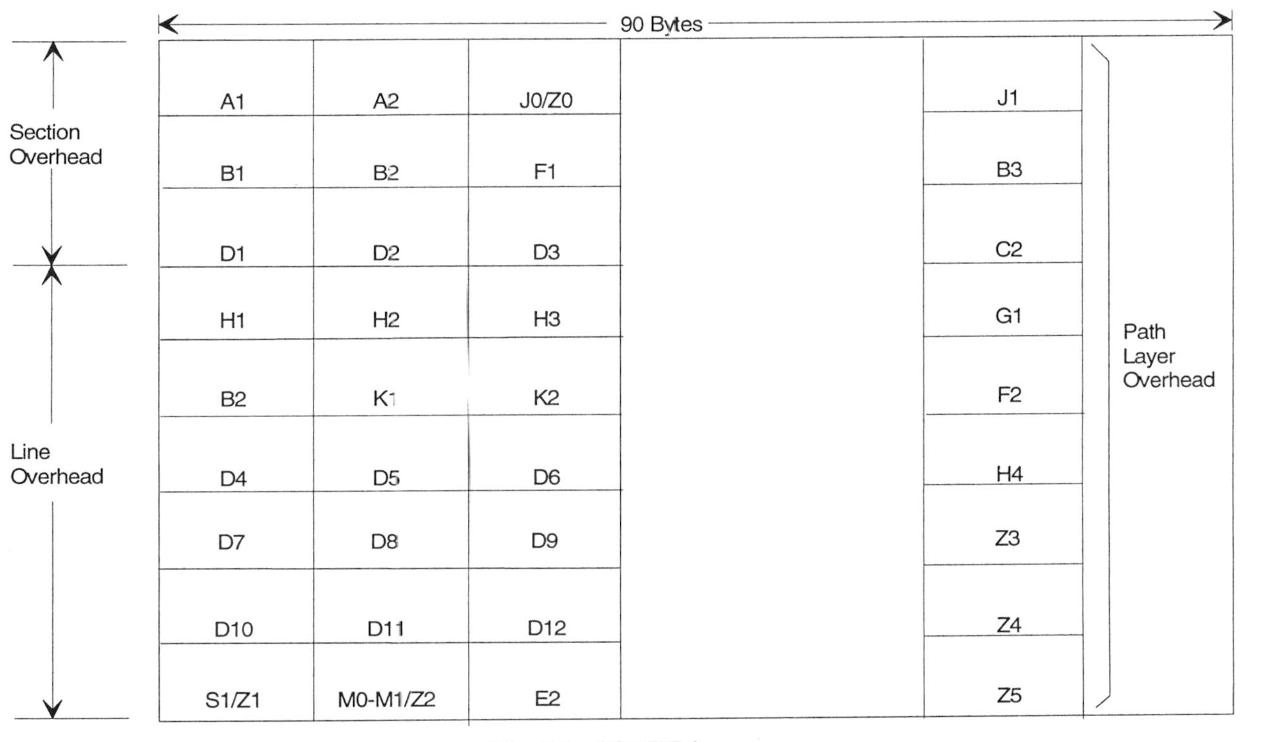

Fig. 9.1 SONET frame.

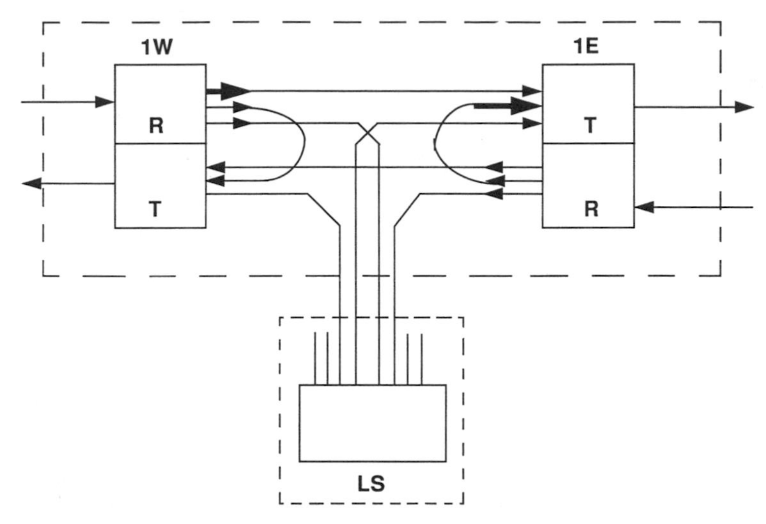

Fig. 9.2 Generic add–drop ring node (two-fiber ring). LS, low speed; R, receive; T, transmit.

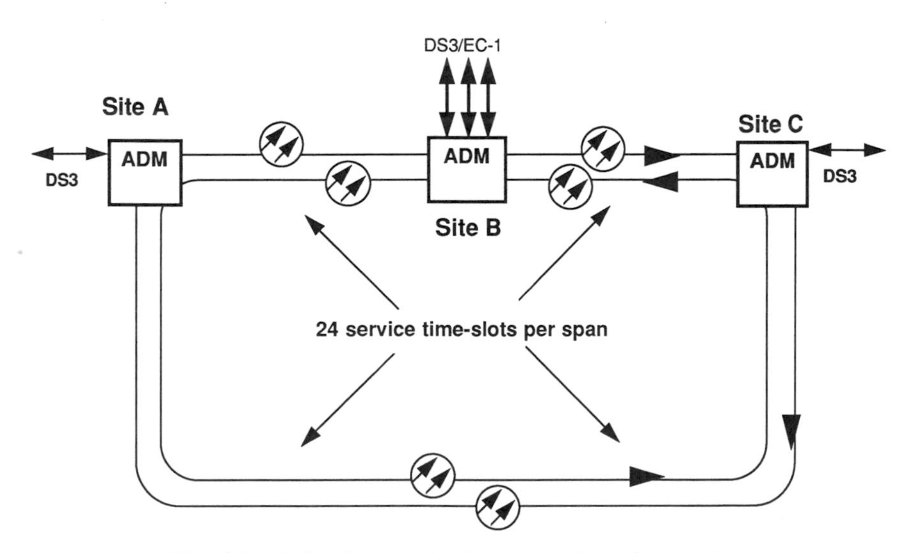

Fig. 9.3 Add–drop multiplexers configured as a ring.

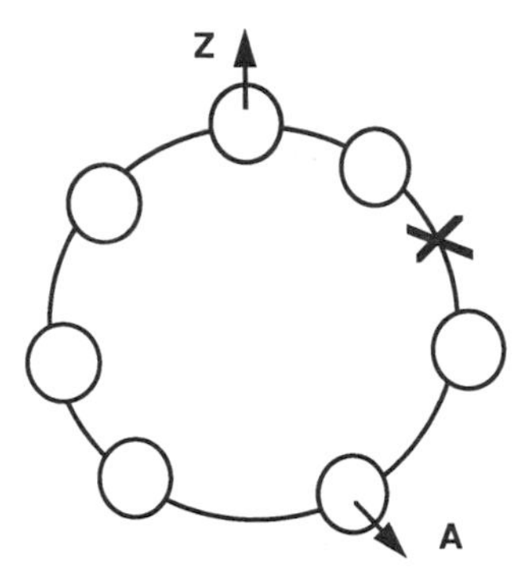

Fig. 9.4 In a path-switched ring, switching is based on the health of each path where it exists on the ring, and switching is done on a per-path basis where the path exits on the ring. In a line-switched ring, switching is based on the health of the line between each pair of nodes, and when a line is faulty, the entire line is switched to a protection loop at the boundaries of the failure.

both directions (termed a *head-end bridge*). The disadvantage of the path-switched ring is that bandwidth utilization of the ring is relatively poor. Because every signal that enters the ring creates two signals on the ring, the available bandwidth of the ring gets used up relatively quickly.

2. Line-Switched Rings

A line-switched ring avoids the bandwidth limitation of path-switched rings by using the protection bandwidth only for those signals that need protection. There are two types of line-switched rings: two fiber and four fiber. The following discussion describes two-fiber line-switched rings. The ring bandwidth between any two ring nodes is divided into two pieces — a service piece and a protection piece. The total bandwidth can be divided so that one-half of it is dedicated to service and one-half to protection. For example, for an optical carrier (OC)-48 two-fiber line-switched ring (see Fig. 9.3), there are two fibers between every ring node. These two fibers have a total bandwidth of 2.488 Gb/s. This bidirectional line can be thought of as two lines — a 1.244-Gb/s service line and a 1.244-Gb/s protection line. How does protection switching work on this type of ring?

In Fig. 9.5 a failure is created at point x. This failure prevents the data from getting between points A and B on the ring. When the failure occurs, the terminals at points A and B both detect a failure condition. In a line-

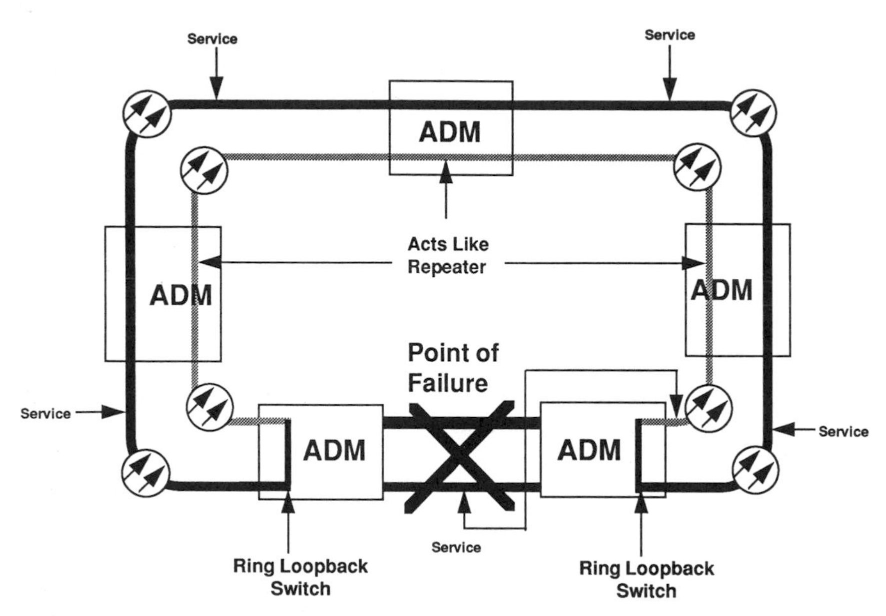

Fig. 9.5 Two-fiber line-protection switching — ring loop-back functions. The protection loop forms a real-time diversion route for the failed span. ADMs at the end of the failed link still act like ADMs, and the other ADMs act like regenerators on the protection loop. The ring loop-back switching protects the failed east (west) with protection loop capacity to the opposite side, loops through circuits onto protection, and gives the ring loop-back switching ADMs access to circuits to be dropped from the other side of the failed link.

switched ring, when this failure is detected the ring terminal takes all the service data that were to be sent in the direction of the failure and instead uses the protection bandwidth to send the data in the opposite direction. This happens at both site *A* and site *B* (this switch is termed a *loop-back switch*). The rings adjacent to the failure that perform the switches are termed *switching nodes.* The data that were meant to travel from *A* to *B* on the ring (through the failure) can still get to the correct destination, but now the path that the data take travels the "long way" around the ring. The switching nodes must recognize that the data that were to be provided on the service bandwidth from the direction of the failure are now available on the protection bandwidth from the opposite direction. Another switch is required to select this data for the local drop data.

This protection system has the advantage over path switching of using the protection bandwidth only when a failure is occurring. During other times, the protection bandwidth is available for "extra" traffic. In addition, the line-switched ring has the ability to continue to protect traffic even in the presence of multiple failures. The Bellcore standard that describes the protection switching protocol has examples with one, two, three, and more simultaneous ring failures.

Four-fiber line-switched rings are an extension of two-fiber line-switched rings. In this configuration, the ring has separate service and protection fiber pairs between every ring node. The main additional attribute of the four-fiber ring is span switching. Span switching allows the protection equipment and fiber to be used between ring nodes to protection switch around failed equipment. Span switching can occur simultaneously with ring switching. Therefore, two types of failures can be protected. This extra protection allows the four-fiber ring to have somewhat greater availability than the two-fiber ring. This can become important in rings that are very large in geographic size.

III. Optical Amplifier System Design

A. *APPLICATIONS*

The optical amplifier provides a method of increasing the optical output power from an end-terminal system by more than 10 dB. In addition, the optical amplifier can be used in a repeater configuration. These two features allow the optical amplifier to be used to increase the distance between end-terminal equipment. As shown in Fig. 9.6, when a system using traditional regenerators with typical regenerator spacing is compared with an amplified system, it is obvious that the amplified system greatly reduces the amount of equipment needed in the network. The cost of the network is, therefore, reduced. The application, then, for optical amplifier systems occurs in those systems where the amplifier can replace many repeater sites. Which systems require many repeaters? These types of systems include

(1) Systems that cover large physical distances. Examples of these are undersea systems and the terrestrial systems that extend across continents (New York to Los Angeles, for example).

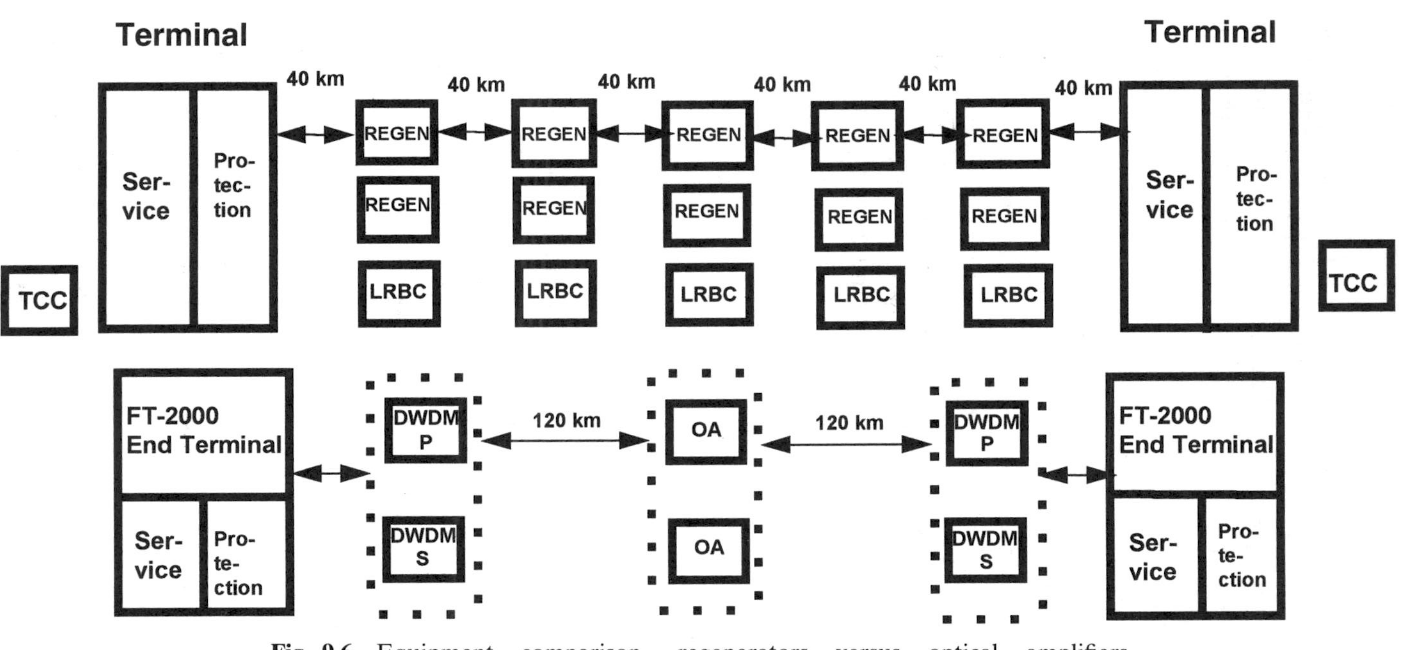

Fig. 9.6 Equipment comparison — regenerators versus optical amplifiers. DWDM, dense wavelength-division multiplex; LRBC, line repeater bay controller; OA, optical amplifier; REGEN, regenerator; TCC, telemetry channel controller.

(2) Systems that use optical fiber that has very high loss. The expense of replacing that fiber is very high. To use the fiber would require regenerators every few miles. An optical amplifier system can extend the distances between repeater sites by two, three, or four times.

(3) Systems that make use of WDM technology. These components typically add 10–20 dB of loss between terminals. The loss in the WDM technology can be overcome with the addition of an optical amplifier.

B. *TERMINAL DESIGN*

A simplified block diagram of typical system connections is shown in Fig. 9.7. Each component of the block diagram is described in a subsequent section.

1. Optical Amplifier

An optical amplifier is characterized by parameters that are familiar to any analog amplifier engineer: gain, noise figure, bandwidth, gain flatness. Even

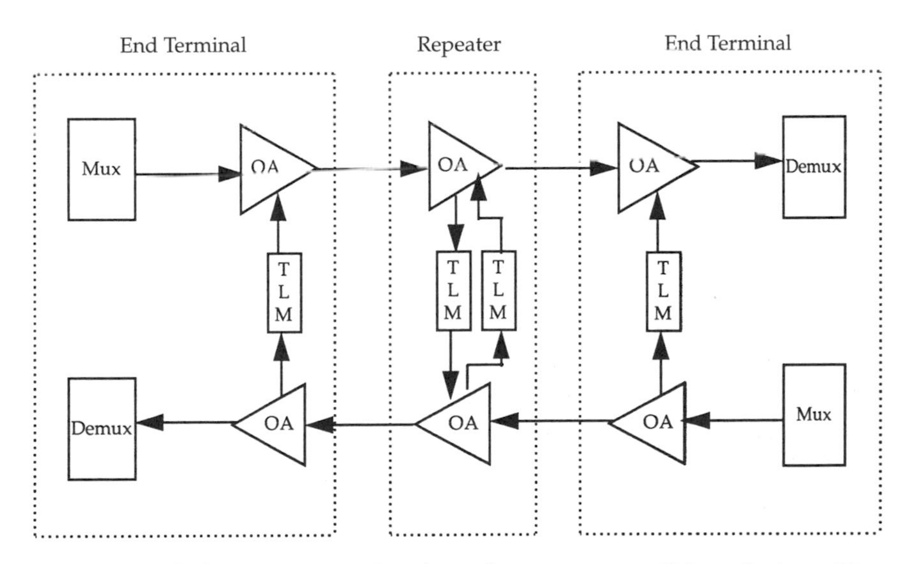

Fig. 9.7 Typical system connections in a telemetry system. OA, optical amplifier; TLM, telemetry.

though the amplifier is operating on optical signals, the noted parameters still control the system performance. The particular choice of amplifier parameters can be changed to optimize the amplifier for a particular application. For example, a system with a long chain of amplifiers (e.g., those that might be used in an intercontinental system) requires minimized noise figure and maximized bandwidth parameters. Noise figure should be minimized because as a signal passes through a chain of amplifiers, the signal is degraded by the addition of noise from each amplifier. This signal-to-noise degradation will eventually increase to the point where error-free transmission cannot be obtained. Minimizing the noise figure minimizes the noise that each amplifier contributes, thus allowing longer chains of amplifiers. The bandwidth should be maximized because passing through a long chain of band-limited amplifiers causes the total end-to-end available bandwidth to be decreased at every amplifier site. That is, each amplifier acts somewhat like a band-pass filter. Those signals on the edge of the passband will get less gain than those signals in the middle of the band. As the signals pass through more and more amplifiers, the signals in the middle of the amplifier band become enhanced relative to those at the edge of the passband. In a long string of amplifiers, the signals at the edge of the band may become unusable because of this effect.

For amplifier applications where the object is to only increase the output power of a laser so as to increase the single span distance available, noise figure is not an important design parameter. In this case, only one amplifier is used. Therefore, the noise contribution of that one amplifier is not so important. Instead, it would be optimum to maximize the gain and total available output power from the amplifier.

2. Power Equalization

When amplifiers are used in a long chain with many optical channels, the effects caused by the limited amplifier bandwidth become important. For example, the Lucent optical line system (OLS) can amplify up to eight optical signals in the 1550-nm optical band. This system uses a signal spacing of approximately 1.6 nm per channel. The total bandwidth required is 12.6 nm. If the amplifier has a 3-dB bandwidth of 14 nm, it can be seen that the signals at the edge of the amplifier bandwidth may become attenuated (relative to the middle channels) as the signals pass through many amplifiers. To solve this problem, it is possible to groom or set the powers of the input signals to inversely compensate for the band-narrowing effect. That is, the power in the first and eighth channels can be accentuated relative to the power in the middle channels. This will allow the signals to pass through several amplifiers and still maintain an approximately equal power level.

3. Mux–Demux

The multiplex and demultiplex are also important components in an optical amplifier system. As was mentioned previously, the optical amplifier is often used in concert with a WDM system. The optical amplifier is used to overcome the inherent 10- to 20-dB loss incurred by using the optical multiplexer–demultiplexer. The optical multiplexer can consist of a simple combiner, or it can be an optical filter. The optical combiner has the advantages of simplicity, relatively low cost, and low loss. The disadvantage is that there is no protection against optical misconnection. That is, an optical combiner does not prevent a user from trying to send two signals of the same wavelength through a combiner. Obviously, when this is done, the signals mix destructively and neither is recoverable.

An optical filter has the advantage that it prevents the aforementioned mixing from occurring. The disadvantages are that they are not as simple as an optical combiner and they are not as low in cost. In addition, when filters are used as both multiplexer and demultiplexer, it becomes important that the center frequencies of the two filter components be stable and well aligned with each other.

4. Telemetry–Supervisory Channel

Systems that span large geographic distances must be able to send maintenance information (telemetry) from one part of the system to another. This is necessary so that the system can be self-monitoring, and so that a communication channel can be provided for intersystem communication (order wire) and alarm–status information. This communication channel allows system information to be obtained from any of the terminal sites in the system. This capability is called *single-ended maintenance.* In traditional regenerated systems, these features are provided by adding extra overhead information to the customer data and sending this overhead data on the transmission line. Access to the overhead data would be gained at each regenerator site. The data could be provided to a local user, processed locally, and/or passed through unchanged. Using the information in the overhead data would thus provide the features of order wire and single-ended maintenance.

How are these features offered in an optical amplifier system? These systems do not have access to the customer data to add overhead bytes. How then is telemetry provided? One solution that has been proposed is to provide an extra optical channel for telemetry. This channel would be positioned in a region in which optical amplifiers provide a secondary gain peak. That is, optical amplifiers have a broad flat-gain spectrum in the region of 1520 ± 15 nm and have a narrow gain peak in the 1530-nm region. This narrow gain peak can be used to send telemetry information from site to site.

5. Maintenance

How are amplifier systems maintained? Amplifier systems present some new problems to the system designer. In any maintenance system, the problem is broken down into three pieces: measurement of performance, signaling, and maintenance actions. As was mentioned in Section II.A, regenerator systems have access to the data stream passing through them for both measurement of signal performance and signaling that performance. But an amplifier system does not have access to the data stream for either function. So how is system performance determined? A method that can be used to address these questions has been proposed by the International Telecommunications Union (ITU). It is termed *toned modulation monitor*.

The transmitters that are used for dense WDM systems have certain requirements. Among the transmission requirements are the following: the signal must have a very clean spectrum, the wavelength must be very well controlled, and, if tone modulation is used, the transmitter must provide a small amount of low-frequency (5- to 50-kHz) amplitude modulation (AM) on the transmitted signal. This signal will allow downstream amplifiers to detect and monitor the signal strength of the transmitted optical signal. This technique is described in a contribution to the ITU [3-1].

The tone modulation method of performance monitoring relies on the fact that the small AM signal that is superimposed on the primary signal will be amplified and attenuated to the same degree as the primary signal. That is, if the optical signal is amplified by 10 dB, the small tone signal will also be amplified by 10 dB. In a WDM system with optical amplifiers, each transmitter driving the WDM system would use a different tone frequency for each of the output channels. The amplifier system would detect and monitor these tones in each amplifier. Tracking changes in the detected power of the tone signals would allow changes in the optical signal levels to be inferred.

For example, a WDM system that uses four wavelengths could provide transmitters with tones of 3, 5, 7, and 9 kHz. The amplifiers in this system would take a small sample of the available optical power and provide that to an optical-to-electrical conversion circuit. Following optical-to-electrical conversion, the resultant signal, which would contain the electrical information from all four WDM channels combined, would be filtered to contain information in only the 1- to 10-kHz region. This signal could then be further processed to detect and monitor the signal strength of the expected tones. The presence or absence of the expected tones could be monitored, as well as the signal strength of the tones that were present.

Such a measurement system would allow the system to provide several maintenance functions. For one, the signal strength of the tones could be monitored over time. This allows the system to track and report changes in signal level. Recall that the basis of this monitor is that changes in the level of the tone signal are directly related to changes in the optical signal level. Therefore, tracking changes in tone levels also would allow changes in the optical signal level to be tracked. This information could be used to determine where in a system optical signal levels might be changing. This could be used for system fault locate functions. A system that is detecting and tracking tone levels could send that information to the end terminals of the system by using the supervisory channel as previously discussed.

C. ENGINEERING RULES

The introduction of the optical amplifier and WDM technologies requires that the traditional approach of designing a transmission line system, based upon a calculation of the power budgets, be supplemented with an additional methodology based upon the optical signal-to-noise ratio (SNR). This new approach reflects the fact that fiber attenuation, being compensated for by the optical amplifiers, is not the dominant factor for the system design. Instead, the system impairment due to optical noise, dispersion, and nonlinearity needs to be taken into account. A two-step approach must be considered. First, it is necessary to determine what SNR is required by the receiver to achieve the required bit error rate (BER) performance. The optical signal coming out of the final amplifier in the chain is degraded by amplified spontaneous emission (ASE) noise. The optical noise is converted to electrical noise in the photodetector. Because the photodetector is a nonlinear device, the optical noise beats with itself and the signal to produce both spontaneous–spontaneous and signal–spontaneous beat noises. Other noise components in the receiver also include the signal shot noise, the ASE shot noise, and the receiver thermal noise. Care must be taken in receiver design to take into account the effect of these noise elements. A receiver that is designed for this high optical power, high optical noise environment can give good BER performance. A receiver that is not designed for this environment will not perform well. For details of receiver design for the optical amplifier environment, refer to Chapter 3 of Volume IIIB.

An additional factor in engineering these systems is the penalty introduced by optical dispersion. The use of amplifiers will allow long system lengths between signal regeneration. Therefore, signal dispersion will become an important factor. Typically, a system will be specified to achieve a certain performance level with an assumption of a certain maximum dispersion. That is, a system allocation of performance degradation measured in decibels will be allocated assuming a certain amount of dispersion penalty. For example, a system may be specified to have a less than 1-dB dispersion penalty given that the maximum eye closure from dispersion is 10% (~40 ps at 2.5 Gb/s). This allocation for dispersion must be added to the SNR previously determined. Therefore, if it was determined that the receiver required a 21-dB SNR to achieve a $1 \times 10E^{-15}$ error rate, the system should be designed to achieve 22 dB when the dispersion penalty is added.

A mathematical model of an EDFA can then be used to calculate the signal and noise through a chain of optical amplifiers. The ASE noise from each amplifier along the chain adds up quadratically, so that the total noise per channel after N stages increases with the square root of N. The net result is that the SNR falls steeply after the first few amplifiers, and less steeply farther down the chain. As a result, there is a range of SNR values from channel to channel at each stage of the optically amplified system. This is a result of the non-flat-gain and saturation characteristics of the amplifiers. In addition, extra system margins need to be added to protect against the potential system fault conditions expected to happen within the lifetime of the system (i.e., pump failures), so that acceptable performance can be maintained.

IV. WDM Issues

In the design of lightwave systems for terrestrial transport, there are different choices of technologies. As explained in Section I, with the maturity of EDFAs and WDM for commercial application after many experiments and feasibility trials in both laboratory and field environments, WDM is becoming a viable alternative to the traditional approach of time-division multiplexing (TDM) in upgrading terrestrial lightwave systems.

A. *WDM VERSUS TDM*

During the transition from the first stage of transport network evolution to the next, the most salient change, besides the increase in the transmission speed from megabit to multigigabit, is the adoption of the international SONET–SDH standards by all systems suppliers and users. The proprietary and region-specific plesiochronous digital hierarchy (PDH) systems are gradually replaced by the standardized synchronous networking systems. This results in a significant improvement in the functional capabilities of transport infrastructure, continuing further on the historical trend of unit cost reduction, yet carrying with it a major overhaul in the network management philosophy, multiplexing scheme, and architecture. There are currently two contending schools of thought regarding the evolution strategy of terrestrial lightwave transport beyond the straightforward single-channel upgrade of the system using OC-48 standard terminals. One concept, recognized now by the service operators, is to apply existing OC-48 terminal technology in multiple wave-channel configurations to increase system capacity to support new service growth. The other approach is to apply TDM to increase the individual link speed from OC-48 to OC-192 to provide a fourfold increase in capacity. Each of these two approaches can support the general objective of capacity expansion for the transmission and each has its individual merits and disadvantages.

For a TDM OC-192 system, there is the issue of choosing between bit interleaving or byte interleaving [4-1]. There is also the question of international standardization on OC-192 systems, which is now being actively discussed by expert groups in the ITU. The argument for TDM OC-192 is based upon the availability of critical high-speed circuits and chips, whether they are integrated over bipolar silicon, GaAs, or bipolar complementary metallic oxide semiconductor (Bi-CMOS). Different circuit designs and high bit-rate operations above 10 Gb/s for equalizing amplifiers, clock-recovery decision circuits, and multiplexer–demultiplexers were all successfully demonstrated, and complete OC-192 lightwave systems are now becoming commercially available [4-2]. It has been claimed that the cost for the first available OC-192 terminals is expected to be approximately 2.5 times the cost of current OC-48 terminals [4-3]. Furthermore, many architectural and management features already implemented for OC-48 can be extended naturally to OC-192, such as add–drop multiplexing (ADM) and self-healing rings.

For a WDM system, the advantage lies in its flexibility for incremental growth and the leverage over technology breakthroughs in EDFA and WDM components. With the EDFA ability to simultaneously amplify multiple wavelengths over a broad optical window and to transmit this multiplexed optical signal for long span distances without electronic regeneration, there are significant savings in regenerator equipment costs and lifetime maintenance and operations of the outside plant that may favor this technology choice. In any case, both technologies, WDM and TDM, open up two dimensions for network capacity growth. To optimize the planning of a lightwave transport network, capacity growth can be pursued by increasing either the number of wavelength channels or the data speed per wavelength channel or both. It is expected that the deciding factors for the selection of specific technology systems for certain fiber routes in the network would be primarily economics and local deployment environment, which will include not only the terminal equipment conditions, but also the fiber medium constraints discussed in Section V. A WDM system allows for a present with multiple OC-48 signals and a future with the addition of OC-192 signals.

B. STABILIZATION OF WAVELENGTHS AND FREQUENCY REFERENCE

To design a WDM system, many parameters must be considered carefully. These include transmitter source characteristics, optical amplifier performance, multiplexer–demultiplexer requirements, and receiver design. First, the source wavelengths must lie within the useful gain region of the amplifiers, and they must be stabilized to maintain channel spacing with acceptable accuracy [4-4]. At the receiver end, the WDM signals need to be demultiplexed. The crucial parameters are center frequency and adjacent channel rejection. The center frequency of the receiving filter and the source frequency must be aligned and must remain aligned over the lifetime of the system. Most WDM systems will require frequency control of the optical demultiplexer. This can be accomplished by using carefully designed and calibrated devices that operate at fixed frequencies and are stable over all operating conditions. Passive filters can then be built with sufficient stability and reproducibility to make WDM operable at standardized frequencies. The optical spectrum can then be allocated similar to the rf spectrum, with sources and filters built for the standard frequencies [4-5]. International standards on standardized optical frequencies have been proposed and discussed in the ITU forum to bring the industry to a consensus on this issue. The most recent draft recommendation presented before the ITU expert groups regarding the frequency allocation for multichannel systems (G.mcs) contains a grid of 25-GHz-spaced frequencies and the center reference frequency at 193,100 GHz [4-6]. Other important standards for both WDM and TDM systems are actively discussed in ITU and other standards bodies [4-7 to 4-9]. The closure and agreement on various technical issues by the expert and study groups in the ITU will definitely have a significant impact on the pending introduction of products in either technology category by the global suppliers of lightwave transport equipment into the competitive marketplace (see Figs. 9.8 through 9.10, which depict ITU recommendations).

C. WDM TECHNOLOGY CHOICES

There are several technology choices for multiplexing multiple wavelength signals into the same fiber. A simple, available way is to combine the signals from various transmitters by using a star coupler followed by a power amplifier, which compensates for the splitting loss of the star coupler and boosts the power of the signal to enable long span transmission. A more elegant way is to apply a frequency-selective optical multiplexer with little loss based upon a planar dispersive waveguide array design. This can be accomplished through a silica integrated-wave-guide device fabricated with silicon-optical-bench (SiOB) technology, where an array of phosphorous-doped SiO_2 waveguides are deposited on a silicon substrates. Another way is to use a power splitter followed by optical filters — for example, of Fabry-Perot type — to select the individual channels. Recently, a new filter design using fiber Bragg grating has been proposed.

Each of these technology alternatives has its advantages and disadvantages. For example, the silica waveguide device is dependent on polariza-

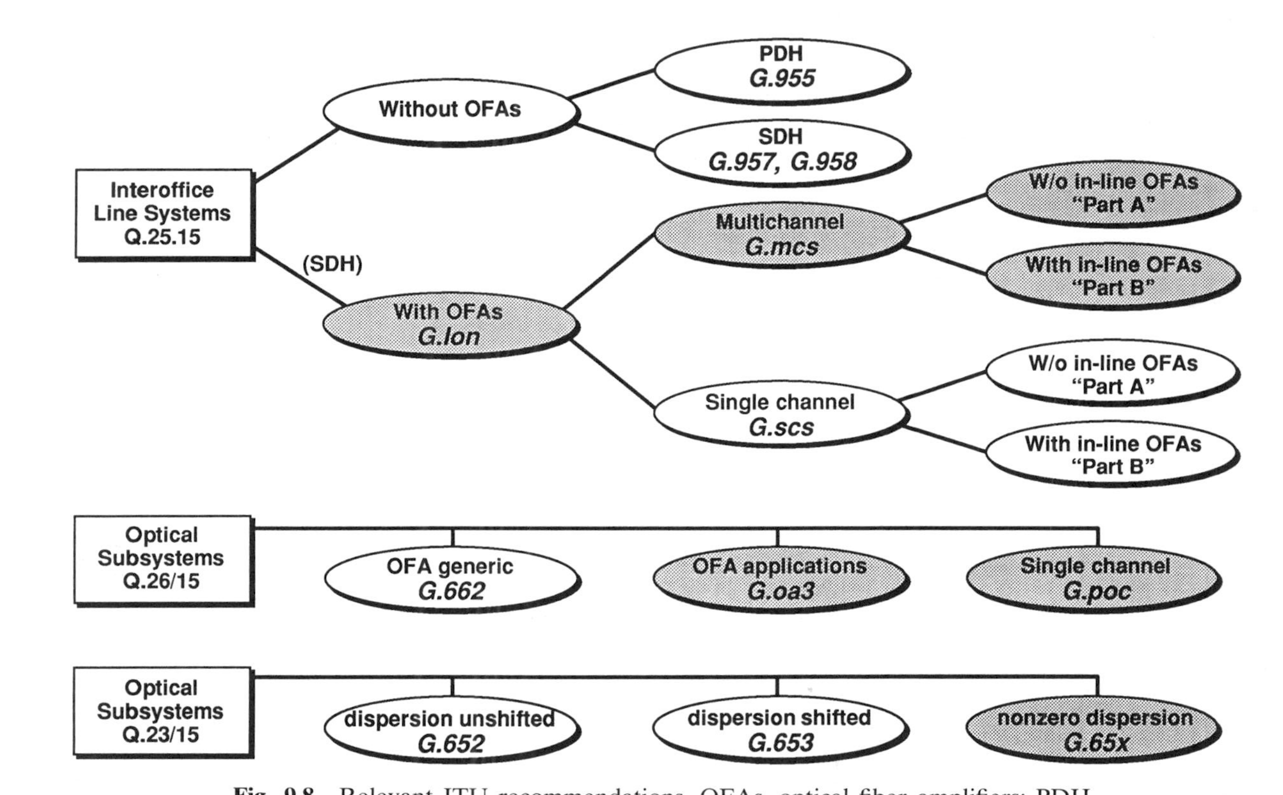

Fig. 9.8 Relevant ITU recommendations. OFAs, optical fiber amplifiers; PDH, plesiochronous digital hierarchy; SDH, synchronous digital hierarchy.

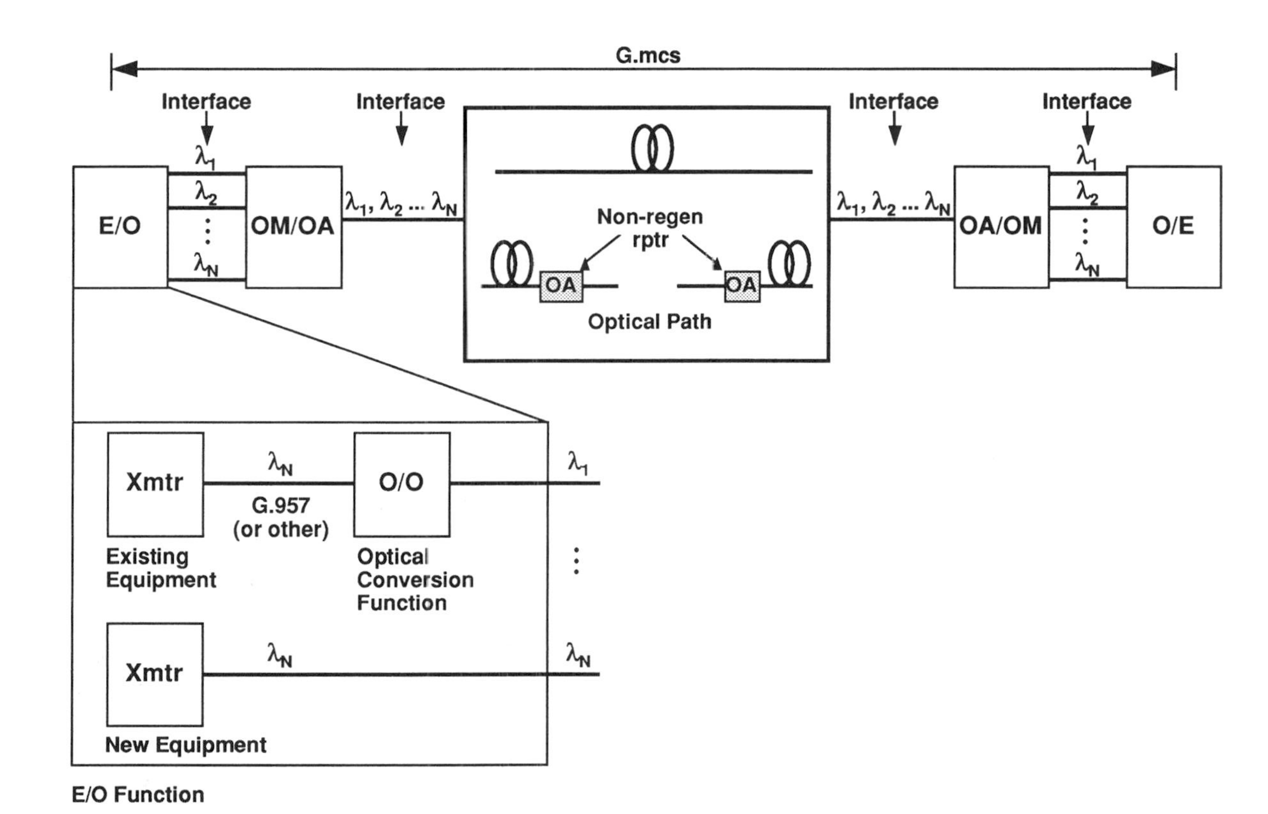

Fig. 9.9 Optical interfaces in ITU draft recommendation G.mcs. E/O, electrical-to-optical conversion; OA, optical amplification; O/E, optical-to-electrical conversion; OM, optical multiplexing; O/O, optical-to-optical conversion.

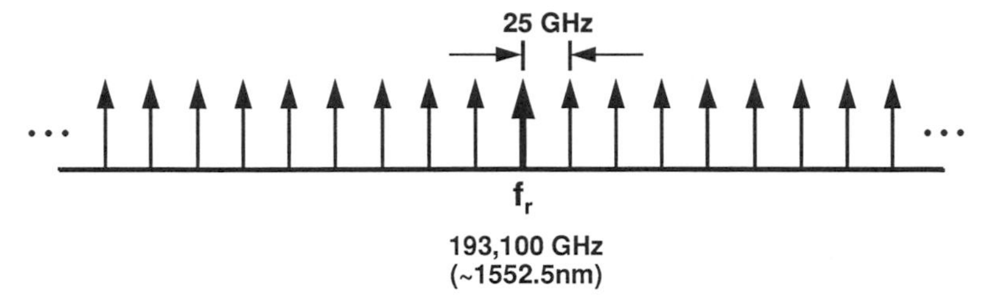

Fig. 9.10 ITU draft recommendation G.mcs: frequency grid and wavelength allocation. Issues to be considered in channel spacing discussions are optical fiber amplifier gain flatness, laser wavelength stability, WDM component technology, and optical fiber nonlinearities. Proposal: An overlay of more widely spaced channels would provide suppliers and operators with some measure of flexibility to select the optical channel most suited to their implementation. The minimum channel separation is 100 GHz, where channels are separated by integer multiples of 100 GHz.

tion and wavelength resolution, whereas the fiber grating is sensitive to temperature variation. These technology issues have an impact upon the reliability and manufacturability of the various proposed device designs, and need to be carefully investigated as a part of lightwave systems development. A more detailed analysis of this subject is presented in Chapters 7 and 8 of Volume IIIB.

D. BENEFITS OF WDM ADD–DROP

With the advance of WDM technology and its successful application to point-to-point optical transmission, the benefits of extending WDM into more complex architectures and functions for terrestrial transport are becoming prominent. For example, a multipoint connection with add–drops can be implemented through WDM using the same passive planar waveguide or grating filter device technologies. Although its fixed hardware design for wavelength channel selection cannot be reconfigured, it can still find application in situations when the bulk traffic pattern for the optical transmission between offices is rather stable. In the normal design of transport networks, service nodes of digital cross-connects are connected by lightwave transmission links. At every office, all the high-capacity bundles of interoffice traffic at a gigabit rate (in units of OC-48) are demultiplexed into lower rate facilities such as DS-3, or synchronous transport module, level 1 (STM-1). These lower rate service facilities may get groomed into even finer tributaries of service traffic and get dropped locally, or cross-connected to the appropriate termination for the next stop before reaching its intended destination. This process involves a series of optical-to-electronic conversions and stages of multiplexing and demultiplexing in addition to the cross-connect function for all the traffic entering and exiting the office. The use of WDM add–drop can therefore provide significant economic savings to the transport network because it precludes the requirements for electronic terminations for the digital cross-connect. For the through traffic that does not need to be processed by the local office, it may pass through, totally bypassing the cross-connect, while only the traffic in certain wavelengths will be dropped locally and get processed by the cross-connect or multiplexing equipment. This potential application of WDM in add–drop configurations to reduce cross-connect terminations, combined with the recent deployment of SONET rings to render the self-healing capability for the fiber network, has, in reality, relieved or even temporarily removed the long-standing concerns for congestion in the central office due to the increasing installation of large electronic cross-connect equipment.

V. Optical Fiber Constraints

The physical characteristics of the optical fiber, as the transmission medium, impose important boundary conditions on the lightwave system design.

A. *LOSS AND CHROMATIC DISPERSION*

From the earlier years of fiber optics research, most investigation and developmental efforts were preoccupied with a constant search for the low-loss fiber material. The recent discovery and introduction of fiber amplifiers into the lightwave system contributed to the removal of fiber loss as the dominant design constraint to optical transmission distance. In the new-generation lightwave system, chromatic and polarization mode dispersion (PMD), and fiber nonlinearities due to the Kerr effect loom more prominent as the limiting factors to system performance, and must be carefully mitigated through ingenious designs.

For fused silica fiber, a 1.5-μm region has the advantage over a 1.3-μm region because of its smaller fiber attenuation, but the fiber chromatic dispersion remains the most serious system impairment. To help overcome this impairment, one can either (1) use a dispersion-shifted single-mode fiber or (2) use a narrow-width single-frequency optical source such as a distributed feedback (DFB) laser or distributed Bragg reflector (DBR) laser. The first option is applicable for only new fiber route construction. To upgrade the system with reuses of the existing embedded standard fiber, only the second option, which involves replacement of the transmitters and receivers at the terminals, is possible. For high-speed transmission beyond 2.5 Gb/s, even with the multiple quantum well (MQW) DFB laser, the chromatic dispersion of the fiber, inducing a line broadening due to the chirp phenomenon associated with direct modulation of the laser at high frequency, will impose a significant dispersion penalty to prevent long span transmission over standard fiber beyond 80 km. The most effective technique to minimize this deleterious effect is to allow the laser to run continuously and modulate the optical carrier externally with a discrete device or a monolithically integrated device such as an electroabsorption-type semiconductor modulator based upon the MQW structure. For instance, assuming that a value of dispersion, D, is 17 ps/nm/km at 15 μm, the externally modulated laser is transform-limited in its width; then, for a 2-dB penalty, the system distance for 2.5 Gb/s can reach 1170 km [5-1]. For 1.3-μm transmission, loss remains an important issue because a fiber amplifier operating in this optical window is still not available commercially. After traveling a span distance of 40 km, the optical signal needs to be regenerated. Because of this, it is considered uneconomical to apply a coarse WDM technique to multiplex 1.3- and 1.5-μm channels together in a single fiber to double the system capacity, because doing so requires separate sets of regenerators and amplifier repeaters in a complicated configuration.

B. POLARIZATION MODE DISPERSION

A single-mode fiber with an ideal circular core has no polarization dispersion. However, in real fibers, there is always some amount of ellipticity, which removes the degeneracy of orthogonal polarized modes and gives rise to different group velocities for the two modes, leading to broadening of the transmitted pulses at the receiver. In Chapter 6 of Volume IIIA, various aspects of the polarization effects are described in great detail. The following section briefly discusses the impact of PMD on terrestrial lightwave system design. For systems using dispersion-shifted fiber (DSF) to alleviate chromatic dispersion at 1.56 μm, PMD is the primary limitation to high bit-rate transmission at or beyond 10 Gb/s. In the design of a 1.5-μm lightwave system for standard fiber, PMD will become a limitation equally restrictive as the chromatic dispersion, if PMD values approach 2 ps/(km)$^{1/2}$.

It is important to recognize that in the conventional standard fiber, PMD is inherently a stochastic process. It is the least well known and least controlled fiber parameter and has to be characterized by well-planned measurements. The system fading caused by PMD is manifest by random fluctuation of the BER, dependent on the environmental conditions of the fiber, such as stress and ambient temperature. In severe cases, this degradation in BER may result in random service outages of unacceptable duration for system operation. The limitation due to PMD is of utmost concern when the designer is evaluating various alternatives to upgrade the existing terrestrial lightwave system operating with conventional fiber. Because the span-distance limit is inversely proportional to the square of the bit rate, it is advantageous to remain at a lower bit rate of 2.5 Gb/s while applying WDM to expand system capacity, rather than increasing to 10 Gb/s with a drastic reduction of span distance by a factor of 16. PMD limitations are different for systems with chirped and chirp-free sources. Because second-order PMD effects become significant, chirped systems require more stringent PMD specifications on the fiber. This is of particular interest with regard to potential applications of DSF to new fiber routes. Compared with standard fiber limited by chromatic dispersion, DSF offers the option of direct modulation at multigigabit rates. However, the more restrictive PMD requirements for a chirped system may force the use of external modulation with DSF and thereby eliminate the cost advantage it may claim.

It is clear that the design and performance of terrestrial lightwaves system for multigigabit transmission are influenced by the characteristics of the fiber medium. The variation in PMD is complicated by the random occurrence of core asymmetries and internal stress introduced during fabrication and cabling, and the unpredictable local environment conditions of the buried and aerial cable. To adequately design a lightwave system that can be operated at 10 Gb/s and higher, a characterization of the PMD for installed cables in the fiber network must be made. Different measurement techniques for PMD need to be assessed and applied, and data collected and fully analyzed, to generate the basis for engineering judgment on how to mitigate the PMD impairment.

C. FIBER NONLINEARITY EFFECTS

There are five optical nonlinear effects that can cause degradation of the transmitted signal. They have an impact on lightwave system performance and design in different ways and to different degrees. The system parameters, such as repeater spacing, signal intensity, source line width, channel spacing and number, bit rate, and dispersion characteristics of the fiber, all influence the severity of the nonlinear behavior experienced in long-distance transmission systems.

Among the fiber nonlinear effects, stimulated Brillouin scattering (SBS) and stimulated Raman scattering (SRS) arise from parametric interaction between light and acoustic or optical photons (due to lattice or molecular vibrations), whereas self-phase modulation (SPM), cross-phase modulation (XPM), and four-wave mixing (FWM) are caused by the Kerr effect, where high-intensity modulation of a signal is converted through refractive index nonlinearity to modulation of phase, leading to excessive pulse broadening.

The SBS process has long been known to limit the maximum power that can be transmitted in low-loss optical fiber. Of all fiber nonlinearities, SBS has potentially the lowest power threshold, and it is a single-channel behavior independent of the number of wavelength channels. Therefore, it is critical that its adverse effects are clearly understood and mitigated during the design of the lightwave system. Although it was observed that the power threshold for SBS can be raised by increasing the source linewidth, it is considered undesirable because it will degrade the dispersion characteristics. A simple and practical method to suppress SBS has been demonstrated, which involves a low-frequency sine wave dither to the laser prebias current. Because the dither frequency is outside the receiver bandwidth, it does not degrade the signal in the presence of dispersion; therefore, SBS will not pose any system limitation — if proper care is taken in the laser transmitter design [5-2].

SRS differs from SBS in that amplification occurs bidirectionally and over a wider bandwidth. It imposes a fundamental limitation on the design of WDM systems by setting an upper bound to the number of channels that can operate with acceptable performance. It also places an upper limit on the wavelength channel spacing and total signal power. These threshold values depend critically on the amplifier repeater and regenerator spacing, and the data bit rate. Therefore, important economic trade-off decisions must be made regarding different alternatives while the designer is specifying the system capacity, performance needs, and growth flexibility for the multichannel system design [5-3 to 5-5].

SPM occurs independent of the interaction between wavelength channels. It may improve transmission performance owing to pulse narrowing, and as the total fiber dispersion builds up, severe degradation is likely for high launch power. XPM will degrade the performance of a multichannel system with conventional dispersion-unshifted fiber. The channel spacing must be designed to exceed a minimum to avoid XPM. Both SPM and XPM can be controlled through implementation of dispersion compensation at appropriate intervals along the length of the system.

FWM is most prominent in DSF. It causes sideband harmonics and beat signals to appear between WDM channels that interfere with the signals. As the signal propagates along the fiber, the sideband waves will grow at the expense of the original waves. It is found that FWM effects depend on channel separation and fiber dispersion. Increasing channel separation or fiber dispersion reduces the mixing efficiency for the sidebands. If a sideband of significant amplitude falls on an operating channel, catastrophic buildup of the FWM mixing products can result if there is no dispersion to destroy the phase match of the interacting waves. For this reason, it is not desirable to use DSF in WDM lightwave systems for multigigabit transmission. To mitigate the debilitating effect of FWM on lightwave system performance, several alternatives can be considered. Neither the option to reduce the repeater spacing and thus reduce the optical power nor the option to increase channel spacing, which leads to a smaller number of channels, is considered desirable because each downgrades the system capability. The third option, to arrange uneven channel spacing, will add complexity to the system without solving the issue of power depletion by the sideband mixing products. Therefore, the most natural solution is to develop and apply a new type of fiber with a small amount of non-zero residual chromatic dispersion to counterbalance the effect of FWM, so that smooth evolution of the multichannel WDM system to multigigabit high-speed transmission can be supported [5-6].

Unlike DSF that is best suited for a single-channel OC-192 system, but would encounter problems such as FWM for multichannel operation, or the conventional fiber that can support WDM systems of multiple OC-48

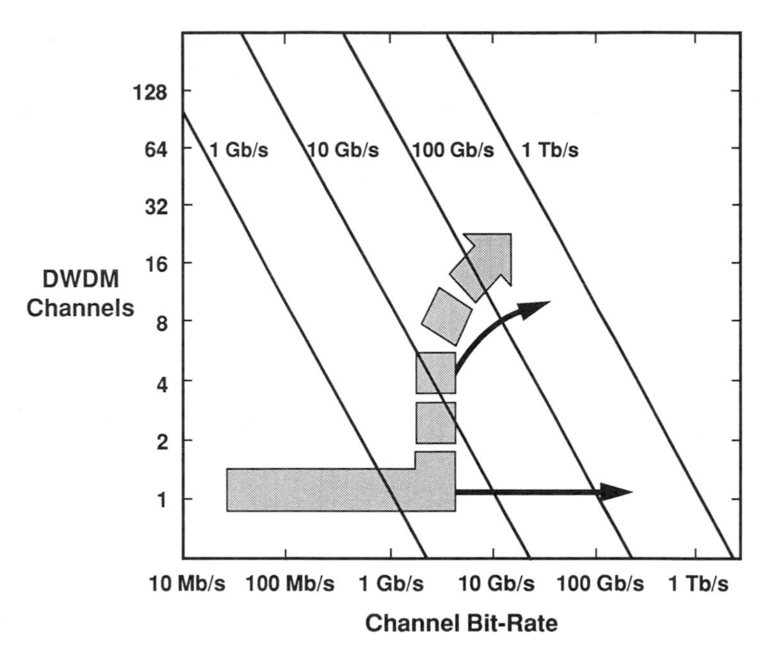

Fig. 9.11 Lightwave capacity upgrade strategies. DWDM, dense wavelength-division multiplex.

channels, while being dispersion limited for upgrading to OC-192, the new design of non-zero-dispersion fiber will be able to support both near-term moderate capacity growth and long-term very high demands requiring multiple OC-192s. The utilization of non-zero-dispersion fiber in the deployment of directly modulated systems yields immediate benefits by allowing a portion of the power budget, otherwise earmarked for compensating for the dispersion penalty, to be used in extending the repeater span distance. It also requires less equalization at OC-192 when compared with DSF. The impact of fiber selection on network upgrade strategies for an optically amplified lightwave system is shown in Fig. 9.11.

VI. Advanced Experiments and Field Trials

There have been many successful laboratory and field experiments that demonstrate the feasibility of simultaneously applying EDFA, WDM, and SONET–SDH terminals to terrestrial lightwave systems. After the initial pioneering work at a 2.5-Gb/s rate performed in the early 1990s [6-1 to 6-3], the attention of recent publications has now been shifted mainly to OC-192 systems. For example, a system of two wavelength channels, each carrying 10 Gb/s operating at 1552 and 1558 nm, was tested at Roaring Creek, Pennsylvania. Optical amplifiers with 120-km spacing were used with dispersion-compensating fiber. The system was error free continuously for 5 days [6-4]. To overcome the limitation due to chromatic dispersion on a 10-Gb/s intensity modulated/direct detection (IM/DD) system, a method of midway phase conjugation was used to allow transmission over standard fiber for up to a distance of 6000 km with line amplifiers placed every 50 km [6-5]. The effects of chirp and fiber nonlinearity in an IM/DD 10-Gb/s system was investigated. It was found that SPM can be equalized by anomalous dispersion of the optical fiber, whereas pulse broadening by chirp can be compensated for by normal dispersion. Optimum compensation using equalizing fiber is always realizable for 10-Gb/s transmission up to 1000 km [6-6].

VII. Future Trends

A. *TECHNOLOGY*

As a result of the research advances and technology innovations in the mid-1980s to mid-1990s, an extra dimension has been added to the evolution path of the terrestrial lightwave network (Fig. 9.12). The telecommunications industry now benefits from the availability of a new generation of lightwave systems, not only with higher transmission speed and longer repeater spacing, but also with an ever-increasing number of wavelength channels for optical signal transport. Although some of the research and development community may still dwell on the debate over the comparative merit of multichannel OC-48 systems versus single-channel OC-192 systems, it is clear that the recent progress in fiber optics, passive photonics, and integrated electronics, as stimulated by the large-scale application of fiber amplifier and WDM development, will all vigorously continue to make a further impact on the future of the terrestrial transport network. The economic and business factors will ultimately answer the question of what mix of TDM and WDM systems to deploy for upgrading the current network infrastructure.

It is expected that along all three dimensions of growth — speed, distance, and wavelength channel — many new network elements will be de-

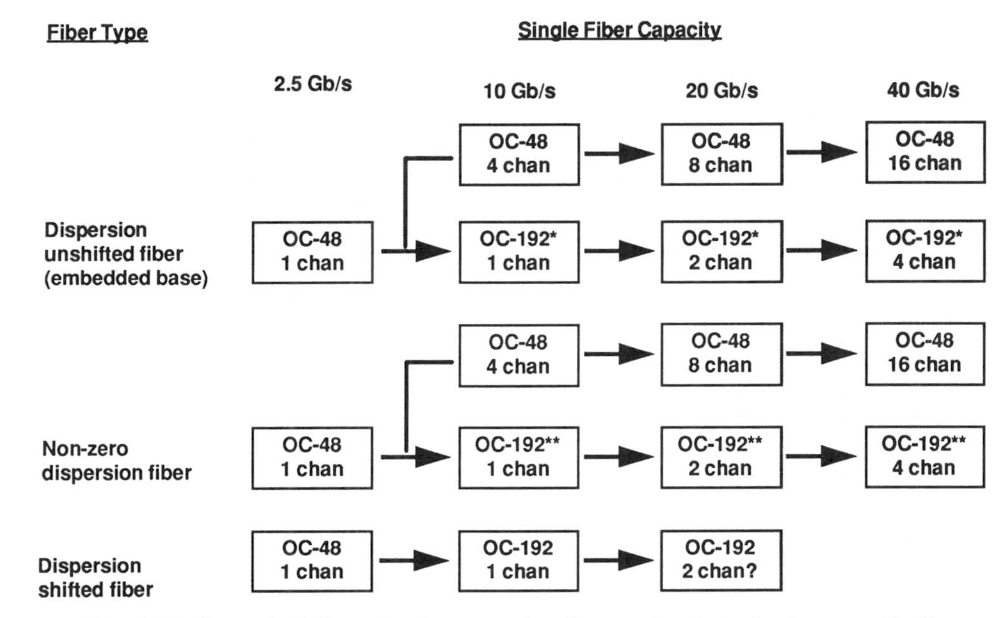

Fig. 9.12 Impact of fiber selection on network upgrade strategies for an optically amplified system. *, chromatic dispersion compensation required for unregenerated system lengths longer than approximately 50 km; **, chromatic dispersion compensation required for unregenerated system lengths longer than approximately 300 km.

signed to collectively offer the advantages of transparency, flexibility, and simplicity of the optical networking technologies. As described in Chapter 15 of Volume IIIA and Chapter 7 of Volume IIIB, this may include such prominent candidates as 1.3-μm fluoride-fiber praseodymium-doped amplifiers to replace existing regenerators [7-1, 7-2], fiber amplifiers of remote pumping [7-3], distributed and bidirectional configurations to increase further repeater spacing and system capacity [7-4], and multiple-wavelength laser sources, as well as passive optic multiplexing elements that are tunable or arrayed to support wider application in WDM architectures [7-5]. Many new dispersion-compensation and management techniques will become viable for commercialization to enable the transmission of ultra-high-gigabit systems over long distances. Novel nonlinear optical devices may emerge to provide the basis for wavelength conversion and optical switching required to support the development of an optical transport network layer.

B. ARCHITECTURE

The current trend of applying SONET self-healing rings to ensure high network survivability will continue with the feasibility investigation and development of WDM rings [7-6, 7-7], which can add the benefit of increasing capacity while sharing fiber in expensive and congested metropolitan areas. A new transport network architecture based upon the use of optical node elements will be established after the successful development of critical technologies, such as wavelength shifters and routers, a monolithic matrix on an InP-based substrate, tunable lasers and filters, and so forth. A new layer, consisting of WDM cross-connects [7-8 to 7-10] and optically amplified paths, will form the foundation of the SONET transport infrastructure, upon which ATM-based switched services may be constructed. To address the simultaneous concerns for network reliability and economics, this WDM transport layer may be used to support a new hybrid network topology that can fully leverage over high-speed and distributed intelligence requirements to implement the sophisticated algorithm for restoration of services in different layers.

C. APPLICATIONS

A further trend, affecting the long-haul system, is the dynamic growth in advanced information-transport services that increasingly demand ubiquitous, flexible, and versatile utilization of bandwidth available to the end users. Even though the industry rush to personal communications service

(PCS) deployment and the public's captivation by the Internet's World Wide Web® may have shifted network operators' attention away from the anticipated buildup of the broadband integrated services digital network, the need to open up the transport bottlenecks at access interconnections with low-cost light pipes seems even more acute and urgent. It is, therefore, apparent that the prospect of mining the unlimited bandwidth resources through a transparent optical network will become so appealing that the historical trend of transmission unit cost reduction will accelerate further to make bandwidth into a virtually free commodity. Although distance may be irrelevant, lightwave transport can still add value to the communications network in the form of many innovative services designed to meet the customer needs of different market segments. Whether it is computer-to-computer interconnect or direct video or graphics delivery to end users, fiber optics-based high-performance and large-bandwidth bit transport still provides a vital alternative and relief to the indstry's relentless and exhaustive exploitation of coding compression as a panacea to complex communications problems.

References

[1-1] Kao, K. C., and G. A. Hockham. 1966. Dielectric surface waveguides for optical communications. *Proc. Inst. Electr. Eng.* 113:1151–1158.

[1-2] Jacobs, I., and J. R. Stauffer. 1978. FT-3 — A metropolitan trunk lightwave system. *Bell Syst. Tech. J.* 57:1857–1879.

[1-3] Cochrane, P., and D. J. T. Heatley. 1991. Future directions in long haul optical-fiber systems. *BT Eng.* 9:268–280.

[1-4] Fan, C. 1994. Planning the next generation lightwave network for capacity upgrade and service growth. In *Conference on Optical Fiber Communication,* 1994 Technical Digest Series, Vol. 1 (Optical Society of America, Washington, DC, 1994).

[1-5] Ballart, P., and Y. C. Ching. 1989. SONET — Now it's the standard optical network. *IEEE Commun. Mag.* 27(3):16–21.

[1-6] Kaplan, M. 1995. Rings take flight. *Telephony*, 7 August, 22–26.

[1-7] Brackett, C. A. 1990. Dense wavelength division multiplexing networks: Principles and applications. *IEEE J. Select. Areas Commun.* 8(6):948–964.

[1-8] Li, T. 1993. The impact of optical amplifiers on long distance lightwave telecommunications. *Proc. IEEE* :1568–1579.

[1-9] Hill, G. R., P. J. Chidgey, F. Kaufhold, T. Lynch, O. Sehlen, M. Gustavsson, M. Janson, B. Lagerström, G. Grasso, F. Meli, S. Johansson, J. Ingers,

L. Fernandez, S. Rotolo, A. Antonielli, S. Tebaldini, E. Vezzoni, R. Caddedu, N. Caponio, R. Testa, A. Scavennec, M. J. O'Mahoney, J. Zhou, A. Yu, W. Schler, U. Rust, and H. Herrman. 1993. A transport network layer based on optical network elements. *J. Lightwave Tech.* 11(5/6):667.

[1-10] Alexander, S., R. S. Bondurant, D. Byrne, V. Chan, S. Finn, R. Gallager, B. Glance, H. Hauss, P. Humblet, R. Jain, I. Kaminow, M. Kasol, R. Kennedy, A. Kirby, H. Le, A. Saleh, J. Shapiro, N. Shankaranarayanan, R. Thomas, R. Williamson, and R. W. Wilson. 1993. A pre-competitive consortium on wide-band all-optical network. *J. Lightwave Tech.* 11(5/6):714.

[1-11] Fan, C., and L. Clark. 1995. Impact of lightwave technologies on terrestrial longhaul network evolution. *Opt. Photon. News* (Feb):26–32.

[2-1] Desurvire, E. 1993. *Erbium-doped fiber amplifiers.* New York: Wiley.

[2-2] Giles, C. R. 1992. System applications of optical amplifiers. In *Tutorial sessions, OFC.* Paper TuF. Washington, DC: Optical Society of America.

[2-3] Bellcore GR-253-CORE. 1994. Synchronous optical network (SONET) transport systems: Common generic criteria, A module of TSGR, FR-NW-000440. (December).

[2-4] ANSI T1.105.01-1994. 1994. *American national standard for telecommunications — SONET — Automatic protection switch.*

[3-1] ITU-T Study Group 15 — Contribution, Question 25/15. 1994. Tone modulation for suppressing stimulated Brillouin scattering and for channel identification on systems using in-line OFAs and WDM. (February).

[4-1] Yamabayashi, Y., *et al.* 1993. Line terminating multiplexers for SDH optical networks. *J. Lightwave Tech.* 11(5/6):875.

[4-2] Ichino, H., M. Togashi, M. Ohhata, Y. Imai, N. Ishihara, and E. Sano. 1994. Over-10-Gb/s IC's for future lightwave communications. *J. Lightwave Tech.* 12(2):308–317.

[4-3] Radick, T. 1995. More power to you. *Telephony*, 7 August, 27.

[4-4] Chung, Y. C. 1992. Frequency-stabilized lasers and their applications. In *Technical Conference 1837 SPIE, 16–18 November.*

[4-5] Knight, D. J. E., P. S. Hansell, H. C. Leeson, G. Duxbury, J. Meldau, and M. Lawrence. 1992. Review of user requirements for, and practical possibilities for, frequency standards for the optical fiber communication bands (paper presented at SPIE technical conference 1837, Frequency-Stabilized Lasers and Their Applications, 16–18 November).

[4-6] ITU-T SG-15 WP4 Q.25 Draft Recommendations: G.mcs. *Optical interfaces for multichannel systems with optical amplifiers.*

[4-7] ITU-T SG-15 WP4 Q.25 Draft Recommendations: G.Ion. *Functional characteristics of interoffice and longhaul systems using optical amplifiers, including optical multiplexing.*

[4-8] ETSI-STC TM1 Draft Technical Report (ETR). *Technical possibilities of SDH long-haul systems with 10 Gb/s capacity.*

[4-9] Fee, J., M. Betts, K. Mahon, R. Cubbage, and J. Morgan. 1993. *Proposal for OC-192 rate and format standards.* Contribution to T1 Standards Project, April 30, 1993.

[5-1] Fishman, D. 1993. Design and performance of externally modulated 1.5 μm laser transmitter in the presence of chromatic dispersion. *J. Lightwave Tech.* 11(4):624.

5-2] Fishman, D., and J. Nagel. 1993. Degradations due to stimulated Brillouin scattering in multigigabit intensity-modulated fiber optic systems. *J. Lightwave Tech.* 11:1721–1728.

[5-3] Chraplyvy, A. R. 1990. Limitation on lightwave communications imposed by optical-fiber nonlinearities. *J. Lightwave Tech.* 8(10):1548–1557.

[5-4] Chraplyvy, A. R., R. W. Tkach. 1993. What is the actual capacity of single-mode fibers in amplified lightwave systems? In *Summer topical meeting on impact of fiber nonlinearities on lightwave systems, 1993 OSA Summer Topical Meeting Digest.* Paper T1.5.

[5-5] Tkach, R. W. 1994. Strategies for coping with fiber nonlinearities in lightwave systems. In *Tutorial sessions, OFC.* Washington, DC: Optical Society of America.

[5-6] Clark, L., D. W. Peckham, and A. A. Klein. 1994. Impact of fiber selection and nonlinear behavior on network upgrade strategies for optically amplified long interoffice routes (paper presented at National Fiber Optics Engineering Conference, June 12–16).

[6-1] Taga, H., Y. Yoshida, N. Edgawa, S. Yamamoto, and H. Wakabayashi. 1990. 459 km, 2.4 Gbits/s four wavelength multiplexing optical fiber transmission experiment using six Er-doped fiber amplifiers. *Electron. Lett.* 26:500–501.

[6-2] Forrester, D. S., A. M. Hill, R. A. Lobbett, R. Wyatt, and S. F. Carter. 1991. 39.81 Gbits/s, 43.8 million-way WDM broadcast network with 527 km range. *Electron. Lett.* 27:2051–2053.

[6-3] Nagel, J. A., S. M. Bahsoun, D. A. Fishman, D. R. Zimmerman, J. J. Thomas, and J. F. Gallagher. 1992. Optical amplifier system design and field trial. In *Optical amplifiers and their applications.* Paper ThA1. Technical Digest. Washington, DC: Optical Society of America.

[6-4] Park, Y. K., P. D. Yeates, J-M. P. Delavaux, O. Mizuhara, T. V. Nguyen, L. D. Tzeng, R. E. Tench, B. W. Hakki, C. D. Chen, R. J. Nuyts, and K. Ogawa. 1995. A field demonstration of 20 Gb/s capacity transmission over 360 km of installed standard (non-DSF) fiber. *IEEE Photon. Tech. Lett.* 7(7):816–818.

[6-5] Zhang, X., F. Ebskamp, and B. F. Jorgensen. 1995. Long-distance transmission over standard fiber by use of mid-way phase conjugation. *IEEE Photon. Tech. Lett.* 7(7):819–821.

[6-6] Suzuki, N., and T. Ozeki. 1993. Simultaneous compensation of laser chirp, Kerr effect and dispersion in 10 Gb/s long-haul transmission systems. *J. Lightwave Tech.* 11:1486–1494.

[7-1] Yamada, M., *et al.* 1995. Low-noise and high-power Pr^{3+}-doped fluoride fiber amplifier. *IEEE Photon. Tech. Lett.* 7(8):869–871.

[7-2] Whitley, T., R. Wyatt, D. Szebesta, and S. Davey. 1993. High outpower from an efficient praseodymium-doped fluoride fiber amplifier. *IEEE Photon. Tech. Lett.* 4(4).

[7-3] da Silva, V. L., D. L. Wilson, G. Nykolak, J. R. Simpson, P. F. Wysocki, P. B. Hansen, D. J. DiGiovanni, P. C. Becker, and S. G. Kosinski. 1995. Remotely pumped erbium-doped fiber amplifiers for repeaterless submarine system. *IEEE Photon. Tech. Lett.* 7(9):1081–1083.

[7-4] van Devanter, M. O., and O. J. Koning. 1995. Unimpaired transmission through a bidirectional erbium-doped fiber amplifier near lasing threshold. *IEEE Photon. Tech.* 7(9):1078–1080.

[7-5] Kano, F., and Y. Yoshikuni. 1995. Broadly tunable semiconductor laser. *NTT Rev.* 7(2):96–101.

[7-6] Gay, E., M. J. Chawki, D. H. B. Hoa, and V. Tholey. 1995. Theoretical simulation and experimental investigation on a WDM survivable unidirectional open ring network using tunable channel selecting receivers. *J. Lightwave Tech.* 13(8):1636–1647.

[7-7] Irshid, M., and M. Kavehrad. 1992. A fully transparent fiber optic ring architecture for WDM networks. *J. Lightwave Tech.* 10(1):101.

[7-8] Tabiani, M., and M. Kavehrad. 1993. A novel integrated-optic WDM cross-connect for wide area all optical networks. *J. Lightwave Tech.* 11(3):512.

[7-9] Johansson, S., M. Lindblom, P. Granestrand, B. Lagerström, and L. Thylen. 1993. Optical crossconnect system in broadband network: System concept and demonstrator's description. *J. Lightwave Tech.* 11(5/6):688.

[7-10] Brackett, C., A. Acampora, J. Sweitzer, G. Tangonan, M. Smith, W. Lennon, K. Wang, and R. Hobbs. 1993. A scalable multiwavelength multihop optical network: A proposal for research on all-optical networks. *J. Lightwave Tech.* 11(5/6):736.

Chapter 10 | Undersea Amplified Lightwave Systems Design

Neal S. Bergano

AT&T Laboratories, Holmdel, New Jersey

10.1 Introduction

The erbium-doped fiber amplifier (EDFA) has had a profound impact on the design, operation, and performance of transoceanic cable transmission systems and is central to the expected proliferation of undersea cable networks. The first long-haul amplifier systems were installed in 1995 with a single 5-Gb/s optical channel, twice the capacity of the most advanced digital regenerator-based fiber system in use. Laboratory experiments have demonstrated in excess of 100 Gb/s over transoceanic distances using wavelength-division multiplexing (WDM) techniques. This large transmission capacity will make possible the deployment of new systems concepts such as optical undersea networks. Systems based on the new optical amplifier technology promise to satisfy the international transmission needs into the 21st century.

Undersea cable systems have a long history of providing international connectivity of terrestrial communications networks. The first successful transoceanic cable system was deployed in 1866 in the North Atlantic and provided telegraph service from North America to Europe (Thiennot, Pirtio, and Thomine 1993). It took about 90 years from the time of the first successful telegraph cable until the first transatlantic telephone cable was installed. The first transatlantic telephone cable was commissioned in 1956 and provided 48 voice channels using coaxial cable and repeaters with vacuum tube amplifiers (Ehrbar 1986). Since the first telephone cable systems were installed, the capacity of transatlantic cable's circuits

OPTICAL FIBER TELECOMMUNICATIONS,
VOLUME IIIA

ISBN: 0-12-395170-4

has increased at an annual rate of 20–27% (Fig. 10.1). New innovations, rather than a single technology, have allowed capacity to keep pace with demand. Coaxial systems technology progressed through the 1960s and 1970s to include better cables and transistor amplifiers, eventually providing 10,000 voice circuits in systems deployed in the late 1970s and early 1980s.

The first fiber optic systems were installed in the Atlantic and Pacific oceans in 1988 to 1989 and operated at 280 Mb/s per fiber pair. These were actually hybrid optical systems in the sense that the repeaters converted the incoming signals from optical to electrical, regenerated the data with high-speed integrated circuits, and retransmitted the data with a local semiconductor laser. The transmission capacity of the regenerated fiber cables eventually increased to 2.5 Gb/s, and repeater spacing increased with the switch from 1.3-μm multifrequency lasers to 1.55-μm single-frequency laser diodes. The first fiber systems greatly improved the quality of international telephony; however, the ability of the regenerator systems to exploit the large fiber bandwidth were limited by the capacity bottleneck in the high-speed electronics of undersea

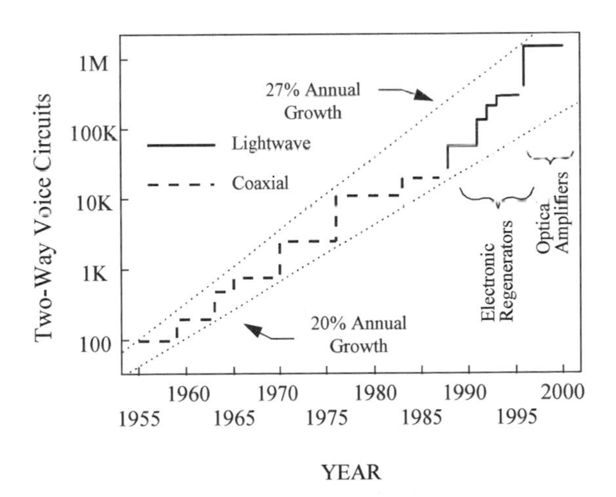

Fig. 10.1 This figure shows the cumulative number of *voice* circuits installed across the Atlantic by AT&T and its partners during the past four decades. For the coaxial cables, a circuit multiplication factor of two is assumed for each 3-kHz channel using Time Assignment Speech Interpolation (TASI). For the digital cables, a circuit multiplication factor of five is assumed for each 64-kB/s channel using digital speech compression.

repeaters. Undersea lightwave systems with EDFA repeaters remove the electronic bottleneck and provide the first clear optical channel connectivity between the world's continents.

This chapter reviews important concepts for the design of long-haul transmission systems based on optical amplifier repeaters and the non-return-to-zero (NRZ) modulation format. When possible, the reader is directed to other chapters of this book for an expanded explanation of concepts, such as amplifier design. The first sections give a comparison of the different modulation formats considered useful to long-haul communications systems. The next few sections lead the reader through a description of amplifier chains, dispersion management, measures of system performance, important polarization effects, and long-haul transmission measurements. Finally, a description of a typical long-haul optical amplifier system is given.

10.2 Transmission Formats

Although lightwave systems are known to be at the cutting edge of technology, the basic signaling format for most systems is simple. Most lightwave systems use direct detection of amplitude shift keyed light pulses to send binary information. Binary ones and zeros are sent by the presence or the absence of light pulses, much as one could imagine using a flashlight with an on–off switch to convey a string of ones and zeros. A transmission system based on this simple pulse scheme is referred to as a *unipolar pulse system* (Bell Telephone Laboratories 1982). Most often, the light pulse that is used is a rectangular pulse that occupies the entire bit period, which is referred to as an *NRZ format* (Fig. 10.2). The term *NRZ* attempts to describe the waveform's constant value characteristic when consecutive binary ones are sent (Fig. 10.3). Alternatively, a string of binary data with optical pulses that do not occupy the entire bit period are described generically as *RZ,* or *return-to-zero.* Two common examples of RZ signaling pulses are the rectangular pulse that occupies one-half of the bit period, and a hyperbolic secant squared pulse (or soliton) with a pulse width of about one-fifth the time slot.

The two interesting signaling formats for use in long-haul transmission systems are NRZ and solitons (see Chapter 12 in Volume IIIA for a complete description of solitons). Table 10.1 lists some of the strengths and weaknesses of both formats. Most long-haul lightwave systems in use today

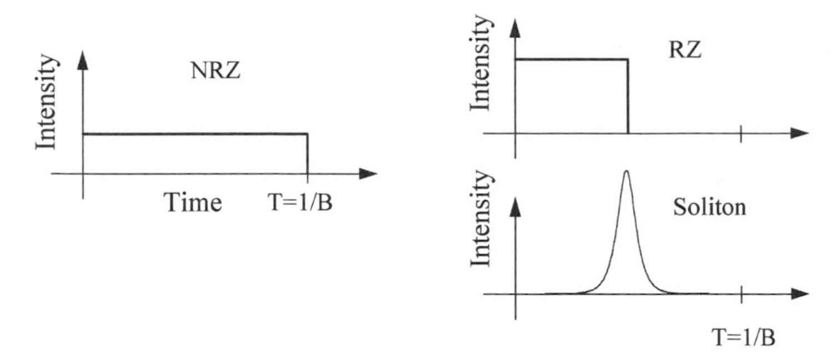

Fig. 10.2 Different unipolar binary code signaling pulses. NRZ, non-return to zero; RZ, return to zero.

use the NRZ signaling format because it is easy to generate, detect, and process. The NRZ transmitter is simpler to construct than the soliton transmitter. Actually, the electronics for the two are similar; however, the soliton transmitter requires an optical pulse source. The quality of the optical pulse source depends on the soliton control technique being used in the transmission line. Soliton systems will probably require some form of jitter control (Mecozzi *et al.* 1991; Kodama and Hasegawa 1992; Mollenauer, Gordon, and Evangelides 1992), which complicates the design of the amplifier chain. Line monitoring (Chapter 2 in Volume IIIB) is an issue for soliton systems, whereas it is well understood for NRZ. The single-channel

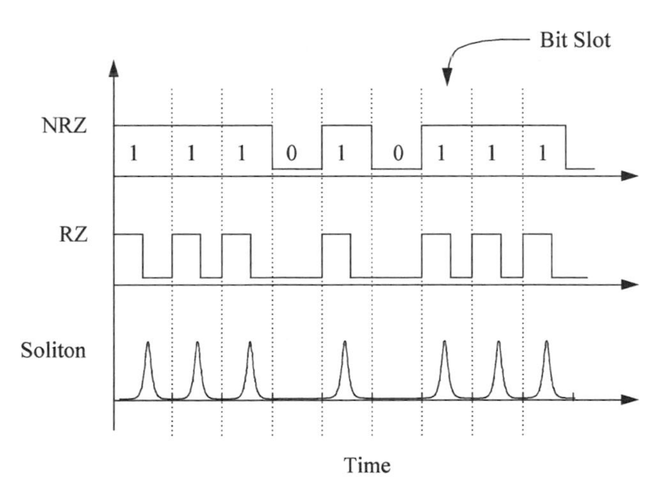

Fig. 10.3 Data waveforms for three pulse formats.

Table 10.1 **Comparison between the NRZ and Soliton Transmission Formats**

NRZ Format	*Soliton Format*
Simpler transmitter	More complex transmitter
Simpler amplifier chain	More complex amplifier chain
More tolerant to power fluctuations	Less tolerant to power fluctuations
Line monitoring understood	Line monitoring an issue
Single-channel capacity of ~10 Gb/s	Single-channel capacity of ~20 Gb/s
WDM $\Delta\lambda > 0.5$ nm	WDM $\Delta\lambda > 0.3$ nm
Amplifier gain spectrum more important for WDM systems	Amplifier gain spectrum less important for WDM systems
Wavelength stability less important	Wavelength stability more important
Synergy with existing systems	

capacity of soliton systems is probably larger than that of NRZ systems; however, large-capacity growth will come through WDM techniques. The question of which transmission format will be more useful for future systems is still open and depends on the system's requirements. For example, a system that needs to achieve the maximum total transmission capacity while utilizing the full EDFA optical bandwidth might benefit from the NRZ format. Alternatively, a system having fewer channels with the maximum bit rate per channel might make better use of the soliton format. Ultimately, the transmission format of choice will be the one that can best use the available EDFA bandwidth in a simple and reliable design.

It is interesting to ask "What ever happened to coherent communications for long-haul systems?" The introduction of the EDFA into the long-haul market greatly reduced research and development of coherent communications because many of the potential benefits of using such communications could be achieved more economically by using EDFAs. For example, very high receiver sensitivity is easily achieved using standard transmission equipment with a simple EDFA preamplified receiver. Also, coherent transmission does not work well with optically amplified lines. In optical amplifier systems, optical phase is not well maintained because of the nonlinear mixing of signal and noise (Gordon and Mollenauer 1990). Thus, direct detection of amplitude shift keying has been the clear winner in long-haul systems to date.

10.3 Amplifier Chains

The EDFA is a nearly ideal building block for providing optical gain in a lightwave communications system (Li 1993). EDFAs can be made with a variety of gains in the low-loss 1550-nm wavelength window of telecommunications fibers, with nearly ideal noise performance (see Chapter 2 Volume IIIB). EDFAs amplify high-speed signals without distortion or cross talk between wavelengths, even when the amplifier is operated deep in gain compression (Giles, Desurvire, and Simpson 1989). Because the EDFA is a fiber device, it can be easily connected to telecommunications fiber with low loss and low polarization dependence. Most important, EDFAs can be manufactured with the 25-year reliability that is required for use in undersea systems.

The undersea portion of the transmission system is essentially a chain of concatenated amplifiers and cable sections. The EDFA provides optical gain to overcome attenuation in the lightwave cable, similar to the way that electrical amplifiers were used to overcome attenuation in coaxial cable in the older style of analog undersea cables. The design of the amplifier chain must provide control of the optical power level, must address control of noise accumulation, must provide an adequate optical bandwidth for the data channels, and must minimize pulse distortion caused by chromatic dispersion and nonlinear effects.

The optical power level launched into the transmission fiber can be controlled by the natural gain saturation of the EDFAs (Fig. 10.4). Self-

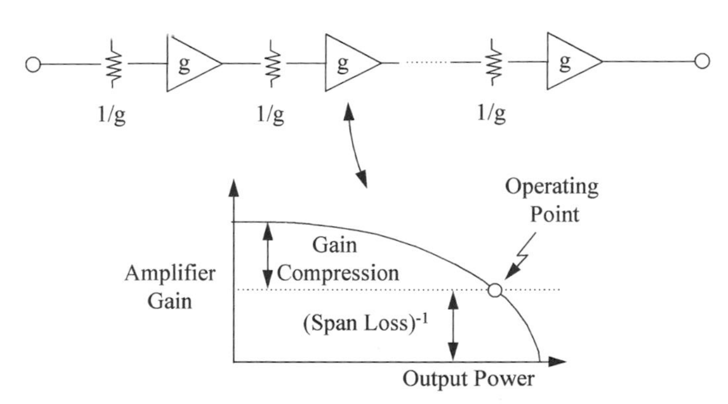

Fig. 10.4 The amplifier's output power is controlled by operating each erbium-doped fiber amplifier (EDFA) with 3–5 dB of gain compression. Steady-state operation occurs where gain equals inverse attenuation in the spans.

regulation is achieved by making the EDFA's small signal gain several decibels larger than is needed to compensate for the span's attenuation. This creates a stable operating point where amplifier gain equals span loss. If the input power to the amplifier decreases, its gain increases, and vice versa, thus the power level remains stable (Fig. 10.5).

The natural automatic gain control forces a balance of amplifier's gain and span's attenuation for signals near the system's gain peak, which occurs at the wavelength of maximum gain for the amplifier–span combination. When a 1480-nm pumped EDFA is operated with several decibels of gain compression, the gain peak occurs at the long-wavelength edge of the pass-band near 1558 nm. Figure 10.6 shows the gain and amplified spontaneous emission (ASE) noise of an EDFA operated with 5 dB of gain compression. The wavelength of the gain peak depends most strongly on the average inversion in the erbium-doped fiber (Chapter 3 in Volume IIIB), and to a lesser extent on the gain shape of the components that make up the remainder of the amplifier chain.

The usable bandwidth of a single EDFA is generally accepted to span a wavelength range of about 35 nm (1530–1565 nm). However, for a long chain of 1480-nm pumped amplifiers operated in gain compression, the usable bandwidth is substantially less. A transatlantic length system of 6300 km would require 140 amplifier spans of 45 km. The 10-dB bandwidth of a system using 140 1480-nm pumped EDFAs with 9.5-dB net gain is about 3.5 nm as shown in Figure 10.7. Unless some form of gain equalization is used, only about 10% of the EDFA's intrinsic bandwidth is available for the transmission of data. This bandwidth is large enough for a single-

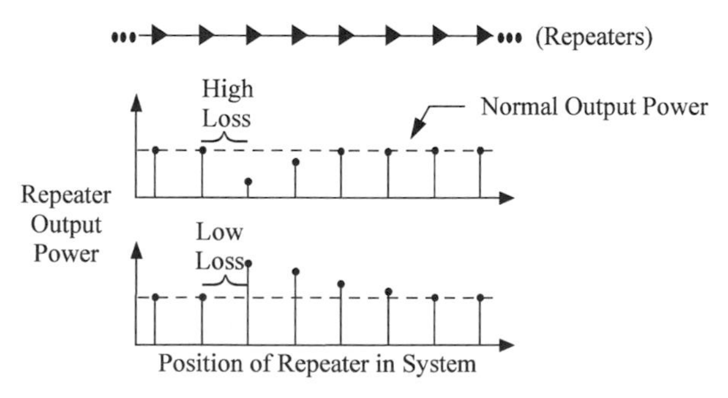

Fig. 10.5 Repeater output power versus transmission distance. The repeater's output power is controlled by gain compression in the amplifiers.

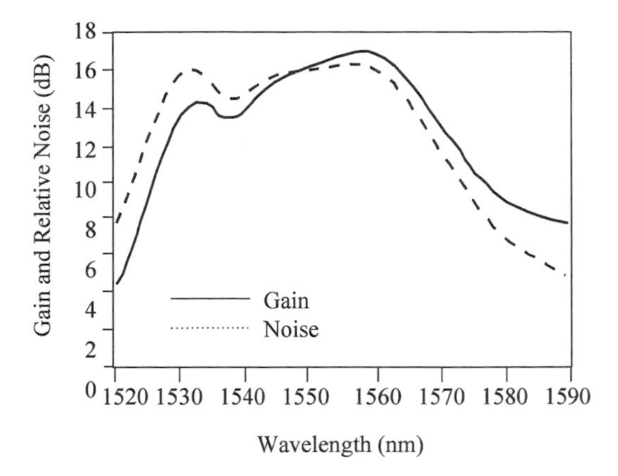

Fig. 10.6 Gain and relative output noise from a single EDFA in gain compression. Note that the gain peak occurs near 1558 nm.

channel system, but is inadequate for many WDM channels. Thus, some form of gain equalization is needed.

The usable bandwidth of a long amplifier chain can be significantly increased by using passive gain equalizing filters. The idea of gain equalization is conceptually simple; gain equalizing filters are designed to approximate the inverse characteristic of the combination of EDFA and fiber span.

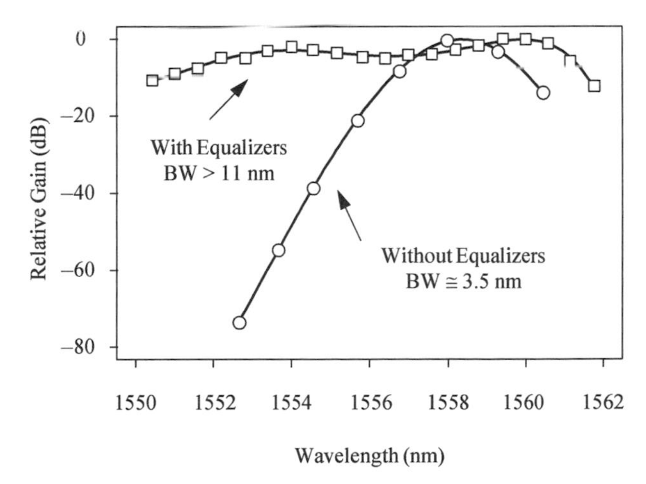

Fig. 10.7 Relative gain versus wavelength for a 6300-km amplifier chain, with and without passive gain equalization. Each symbol shows a gain measurement.

In practice, it is often more simple to design gain equalizers that correct for several amplifier spans. This has the advantage of using fewer devices, at the expense of a larger insertion loss for each equalizer. One promising technique of making gain equalizing filters uses long-period fiber grating filters (Vengsarkar *et al.* 1995), which can be made temperature insensitive (Judkins *et al.* 1996) with low back reflection, low polarization dependent loss, and low polarization mode dispersion. For example, in the WDM experiments outlined in Section 10.7, long-period fiber grating equalizers were designed to correct for the gain-shape of 8 to 10 amplifiers, with a goal of expanding the usable bandwidth by two or three times. As a practical matter, the midband loss of the filters were made similar to a typical fiber span (9.5 dB). Thus, the excess loss in the equalizing filter could be compensated using an additional amplifier of the same design as the span amplifier. Figure 10.7 shows the gain profile of the amplifier chain with and without the gain equalizers. At 6300 km, the 10-dB spectral width of the system was over 11 nm with the filters, compared to 3.5 nm without the filters. Thus, for this arrangement more than a three-fold improvement in the unstable optical bandwidth was obtained.

In a long transmission line, ASE noise generated in the EDFAs can accumulate to power levels similar to the data carrying signal. The accumulated noise can influence the system's performance by reducing the level of the signal and the signal-to-noise ratio (SNR). The noise power out of an optical amplifier is proportional to the amplifier's gain and is given by

$$P_n = 2n_{sp}h\nu(g - 1)B_o, \tag{10.1}$$

where $h\nu$ is photon energy, g is gain in linear units, n_{sp} is the excess noise factor related to the amplifier's noise figure, and B_o is optical bandwidth. As a consequence, the spectral density of the accumulated noise at the end of the system depends on the repeater gain and fiber loss. Consider the system with a fiber loss of 0.2 dB/km. A 150-km repeater spacing would require 30-dB amplifiers, whereas a 50-km spacing would need three times as many 10-dB amplifiers. A 30-dB amplifier (1000x gain) generates about 100 times more noise per unit bandwidth than that of a 10-dB gain amplifier. Thus, the 30-dB gain system would have 33 times more noise than that of the 10-dB gain system. The relationship between accumulated noise and amplifier gain imposes an interesting tradeoff; longer systems require shorter repeater spacing to keep the same output SNR.

The excess noise generated as a consequence of the amplifier's gain was described by Gordon and Mollenauer (1991). The excess noise is the factor by which the amplifier's *output power* must increase to maintain a constant received SNR. Lichtman (1993a) embellished this description to include excess loss in the amplifiers. The excess noise is given by

$$\text{Excess noise} = \frac{g\beta - 1}{\beta \ln g}, \tag{10.2}$$

where β is postamplifier loss (Fig. 10.8). The β term is added because in a real amplifier there is always some loss that follows the erbium-doped fiber, such as an isolator or a wavelength selective coupler. The most important result of adding the postamplifier loss is that the optimum repeater spacing is not zero (pure distributed gain); rather, it occurs between 10 and 20 km.

The total accumulated noise power depends on the amount of noise generated at each amplifier (Eq. [10.1]) times the number of amplifiers in the system. As stated previously, the automatic control obtained through the amplifier's gain saturation fixes the total output power. This total power contains both signal and accumulated noise. Because the total power is fixed and noise accumulates, the signal's power must decrease as the signal propagates down the amplifier chain (Giles and Desurvire 1991). Figure 10.9 shows a measurement of signal power and accumulated noise power versus transmission distance for a single 5-Gb/s channel propagating over 10,000 km.

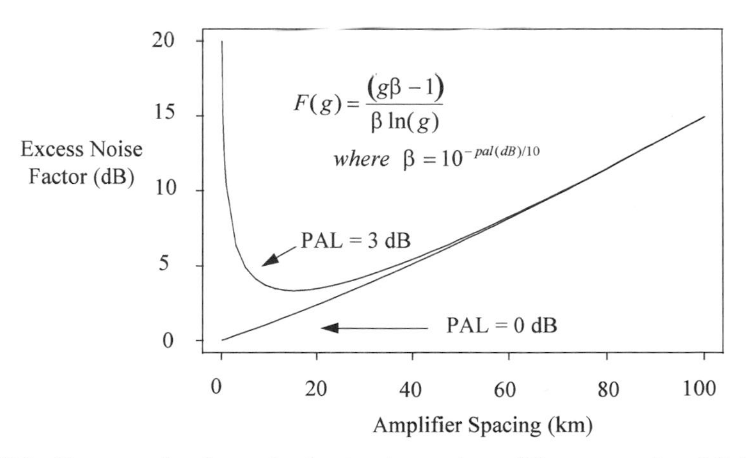

Fig. 10.8 Excess noise factor for having lumped amplifiers, assuming 0.2-dB/km fiber loss. The top curve is calculated for a postamplifier loss (PAL) of 3 dB, and the bottom curve for 0 dB.

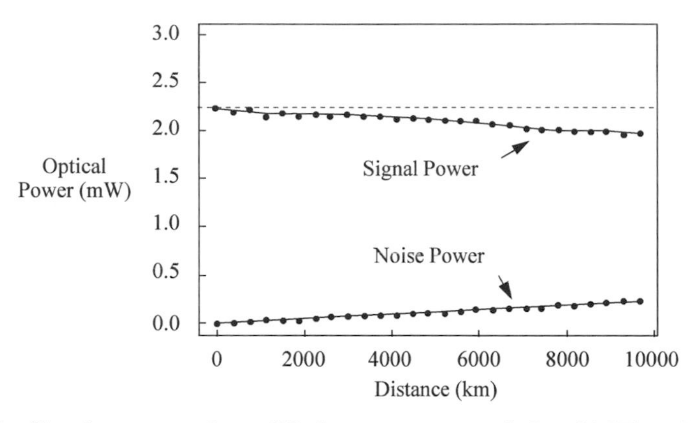

Fig. 10.9 Signal power and amplified spontaneous emission (ASE) noise power versus transmission distance for a single 5-Gb/s data signal. The *dashed horizontal line* indicates the initial launch signal power.

A simple but useful estimate of the SNR at the output of a chain of N amplifiers is

$$SNR_o = \frac{P_L}{NF(gh\nu B_o N)}, \tag{10.3}$$

where P_L is the average optical power launched into the transmission spans, NF is noise figure, g is the amplifier's gain, and B_o is optical bandwidth in hertz. Equation (10.3) assumes that all the amplifiers are identical and that there is no signal decay from noise accumulation.

10.4 Dispersion and Nonlinearity Management

Several phenomena limit the transmission performance of long-haul optical transmission systems, including noise, dispersion, and nonlinearities. For long systems, the nonlinear refractive index can couple different signal channels and can couple the signal with noise. Single-mode fibers are slightly nonlinear as a result of the dependence of the fiber's index on the intensity of the light propagating through it. This behavior is expressed as

$$n = n_0 + \frac{N_2 P}{A_{eff}}, \tag{10.4}$$

where n_0 is the linear part of the refractive index, N_2 is the nonlinear coefficient ($\sim 2.6 \times 10^{-16}$ cm^2/W), P is the light power in the fiber, and A_{eff} is the effective area over which the power is distributed (Marcuse, Chraplyvy, and Tkach 1991). This nonlinear behavior causes only minute changes in the group velocity of the fiber. For example, the presence or absence of a 1-mW signal changes the index by 35 trillionths of a percent. The equivalent length change of 3.5 μm over the 10,000-km system is of no direct consequence to system performance; however, the time shifts are significant when one is considering optical phase. The nonlinear index leads to important phenomena such as self-phase modulation, cross-phase modulation, and four-wave mixing (Chapter 8 in Volume IIIA).

Until recently, the conventional thinking was that the maximum system bandwidth occurred for operation around the zero-dispersion wavelength in the fiber. However, this rule of thumb is not complete when one is considering fiber nonlinearities. When the system is operated at the fiber's zero-dispersion wavelength, the data signals and the amplifier noise (with wavelengths similar to the signal) travel at similar velocities. Under these conditions, the signal and noise waves have long interaction lengths and can mix. Chromatic dispersion causes different wavelengths to travel at different group velocities in single-mode transmission fiber (see, for example, Agrawal 1989a or Kaiser and Keck 1988). Chromatic dispersion can reduce phase matching, or the propagation distance over which closely spaced wavelengths overlap, and can reduce the amount of interaction through the nonlinear index in the fiber. Thus, in a long undersea system, the nonlinear behavior can be managed by tailoring the dispersion accumulation so that the phase-matching lengths are short and the end-to-end dispersion is small. This technique is known as *dispersion mapping* (see Chapter 8 in Volume IIIA). An example is shown in Fig. 10.10. This figure shows the accumulated dispersion versus transmission distance for an eight-channel WDM transmission experiment. About 900 km of single-mode dispersion-shifted fiber with a λ_0 = 1585 nm is used for every 100 km of conventional single-mode fiber with λ_0 = 1310 nm. The signal wavelengths are in the range of 1556–1560 nm. Thus, the nonlinear mixing is minimized by reducing the interaction lengths, and the distortion of the data is minimized by ensuring that the total dispersion returns to zero at the end of the system.

An important observation for WDM systems is that the accumulated dispersion returns to zero for only one wavelength near the average zero-dispersion wavelength for the transmission line. This is seen in Fig. 10.10

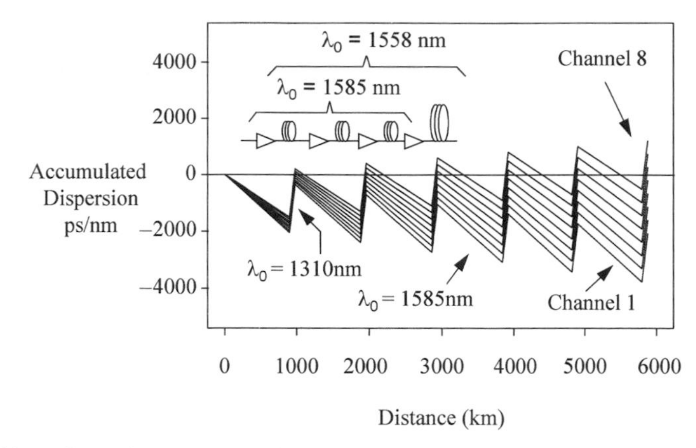

Fig. 10.10 Dispersion map showing the accumulated chromatic dispersion versus transmission distance for eight channels of a wavelength-division multiplexing (WDM) transmission experiment. Most amplifier spans use negative-dispersion fiber with $\lambda_0 = 1585$ nm and $D \cong -2$ ps/km-nm. The dispersion is compensated for every 1000 km using conventional single-mode fiber (i.e., $\lambda_0 = 1310$ nm).

by the diverging lines for channels 1 through 8. Obviously, this is of no consequence for single-channel systems where the end-to-end zero-dispersion wavelength is simply made to coincide with the operating wavelength. However, it is important for WDM systems. This differing accumulated dispersion for the WDM channels results from the nonzero slope of the dispersion curve. The linear approximation for dispersion versus wavelength (near λ_0) is

$$D(\lambda) = SL(\lambda - \lambda_0) \,, \tag{10.5}$$

where D is dispersion, S is dispersion slope (~0.07 ps/km-nm^2), L is fiber length, λ_0 is zero-dispersion wavelength, and λ is wavelength. To minimize signal distortion, the accumulated dispersion for the channels away from λ_0 can be compensated for with the opposite dispersion at the receiver.

10.5 Measures of System Margin

The most important feature of a digital transmission system is the ability to operate with a small bit error rate (BER). In fact, most digital transmission systems operate with BERs that are too small to be practically measured. This is especially true at the beginning of life for an undersea system where extra margin has been added to the design to allow for the system to age. When a system is installed, it is necessary that it performs with a low error rate, but this is not a sufficient characteristic to ensure that the system has an adequate margin against SNR fluctuations and aging. What is required is an accurate measure of system margin. System margin is defined as the difference (measured in decibels) in the received SNR and the SNR required to maintain a given BER. In long-haul lightwave systems, the BER is set by a combination of the electrical SNR of the data signal at the decision circuit, and any distortions in the data's waveform. The BER is degraded by optical noise, fiber chromatic dispersion, polarization mode dispersion, fiber nonlinearities, and changes in the receiver. Also, the BER can fluctuate with time as a result of polarization effects in the transmission fiber and the amplifier's components.

A measure of the system's performance must properly include both the random noise and the pattern-dependent effects that can degrade the BER. Accumulated noise adds random fluctuations to the received data and has the effect of closing the received eye diagram with an unbounded noise process. Noise accumulation alone does not change the underlying shape of the transmitted waveform. The primary source of the noise is the accumulated ASE from the EDFAs. Pattern dependence, or intersymbol interference (ISI), limits transmission performance by changing the shape of the received waveform. Many effects contribute to ISI, including nonideal transmitters and receivers, chromatic dispersion, polarization dispersion, and fiber nonlinearity. The effects of added noise and waveform distortions are depicted in Fig. 10.11, where data waveforms were recorded using an analog oscilloscope.

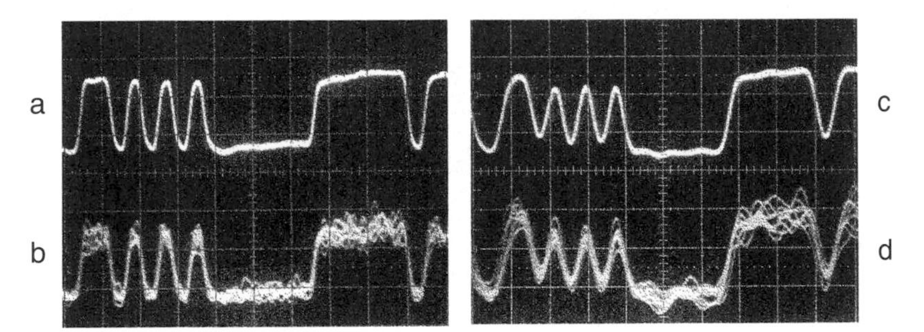

Fig. 10.11 Waveforms showing noise and intersymbol interference (ISI) effects. Waveform with (a) low noise and low distortion, (b) large noise and low distortion, (c) low noise and large distortion, and (d) large noise and large distortion.

The most accurate method of measuring margin in long-haul lightwave systems is the decision circuit method of measuring the Q factor (Bergano, Kerfoot, and Davidson 1993). The Q factor (adapted from Personick's work in 1973 on calculating the performance of receivers in lightwave links) is the argument to the normal error function for the purpose of calculating the BER. This is shown schematically in Figure 10.12. The Q factor is estimated using the regenerator's decision circuit to probe the rails of the data eye, and it includes the ISI present in the regenerator's linear channel as well as that generated in the system from dispersion and fiber nonlinearity. Figure 10.13 shows a Q-factor measurement of 7.2:1 (linear ratio), or 17.2 dB, for a 9000-km transmission experiment using 33-km amplifier spacing. The Q factor is estimated by measuring the BER at different threshold settings in the decision circuit; then the data are fitted with an ideal curve, assuming Gaussian noise statistics:

$$BER(V) = \frac{1}{2}\left[erfc\left(\frac{|\mu_1 - V|}{\sigma_1}\right) + erfc\left(\frac{|V - \mu_0|}{\sigma_0}\right)\right], \tag{10.6}$$

where $erfc(x) = (1/\sqrt{2\pi}) \int_x^{\infty} e^{-\alpha^2/2}\, d\alpha$. The curve fit gives the equivalent values for the means ($\mu_{1,0}$) and standard deviations ($\sigma_{1,0}$) of the voltages on the marks and spaces in the data eye, and the Q factor is formed as

$$Q \equiv \frac{|\mu_1 - \mu_0|}{\sigma_1 + \sigma_0}. \tag{10.7}$$

The Q factor given in Eq. (10.7) is unitless quantity expressed as a linear

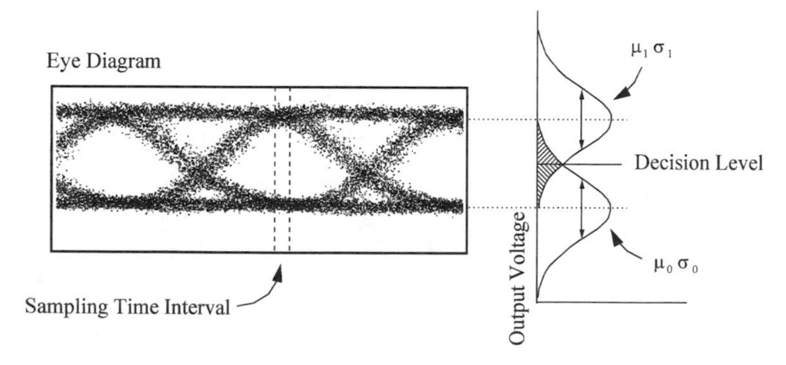

Fig. 10.12 Typical received eye diagram for an undersea lightwave system operating at 5 Gb/s. A voltage histogram is schematically shown to indicate the parameters that are included in the definition of the Q factor.

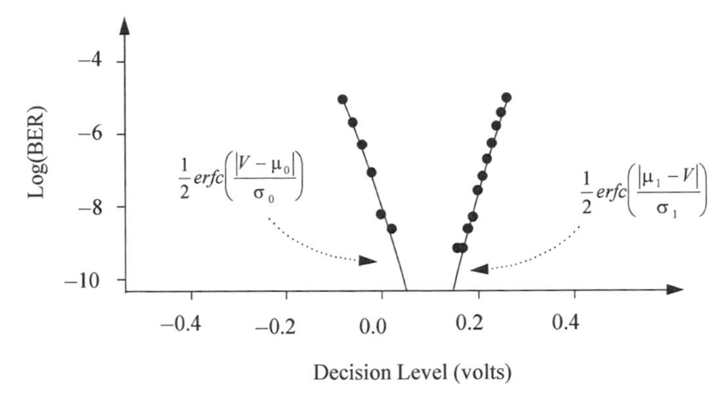

Fig. 10.13 Typical Q-factor measurement for 5-Gb/s, 9000-km operation. The data show the bit error rate (BER) versus the decision threshold at 9000 km. The *solid lines* show the fit of Eq. (10.6) to the data.

ratio, or it can be expressed in decibels as 20 log(Q). The factor of 20 (or 10 log[Q^2]) is used to maintain consistency with the linear noise accumulation model. For example, a 3-dB increase in the average launch power in all the spans results in a 3-dB increase in the Q factor (ignoring signal decay and fiber nonlinearity).

It is well known that the electrical noise at the decision circuit is not exactly Gaussian (Marcuse 1990); however, the Gaussian approximation can lead to close BER estimates (Humblet and Azizoglu 1991). Figure 10.14 shows the measured voltage histogram of a detected optical signal emerging from a long lightwave system operating at 5 Gb/s. For this measurement, 1-million voltage samples were recorded for a zero bit and a one bit in a 2^7-1 data pattern. The non-Gaussian probability density function is apparent when the actual density is compared with a best fit Gaussian. The measurement of the Q factor as described previously measures only a subset of the distributions located near "inside" rails of the received eye, or the voltages that are close to the decision circuit. Thus, the inside edges of the eye are fitted with an equivalent Gaussian function, and the underlying SNR is extrapolated from the fit.

Mazurczyk and Duff (1995) have identified the inability of the decision circuit Q-factor measurement to measure large margin using long data patterns. Pattern-dependent effects cause the Q-factor measurement to underestimate the actual Q factor for long pseudo-random data patterns with large margin. The cause of the effect is shown schematically in Fig.

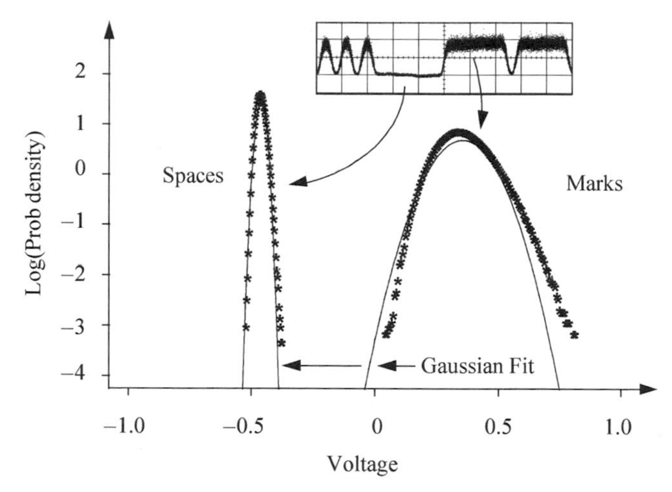

Fig. 10.14 Typical voltage histogram of a 5-Gb/s NRZ data signal for the ones and zeros rails.

10.15. In the figure, the upper part of the received eye diagram is expanded to show the ISI. In this diagram, the pattern dependence of the data causes different bits to have different mean voltages at the decision circuit's timing

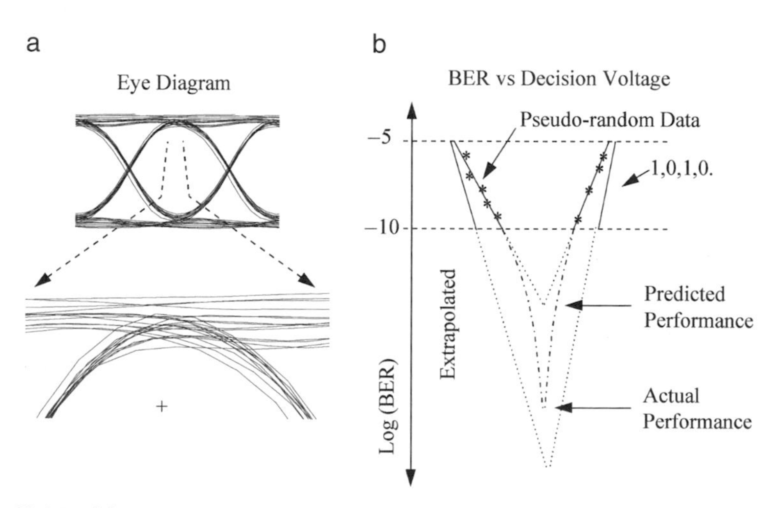

Fig. 10.15 (a) Expanded view of the upper rail of an eye diagram showing ISI. (b) The resulting BER versus decision level curve can have a slope change causing the Q-factor measurement to underestimate the actual value.

point. The resulting BER versus decision voltage curve does not follow a simple Gaussian characteristic; rather, it follows the rules of total probability given each bit's probability density function. For large margins, the resulting curve can exhibit a slope change at BERs less than what is practical to measure, and the extrapolated BER and Q factor are then underestimated. In practice, this is not a serious limitation to characterizing a working system, because typical values of beginning-of-life optical margins are less than 5 or 6 dB. A practical engineering fix to this problem is to measure the Q factor for a series of word lengths.

The decision circuit method of measuring the Q factor is the most accurate measurement of system margin because the measurement is based on the BER, which considers every bit in the data sequence. The BER is routinely and accurately measured to 1 part in 10^{10}. Alternatives to this technique include other electrical SNR measurements such as voltage histograms and optical SNR measurements using optical spectrum analyzers. A voltage histogram down the center of the eye can be measured with a digital sampling oscilloscope to estimate the Q factor. However, this technique fails to give a good correlation between the measurement of the Q factor and the BER, because the variation seen around each rail represents a mix of pattern effects and noise. Such effects in turn artifically broaden the estimates of $\sigma_{1,0}$, thus giving erroneous results. In addition, this method operates on a limited set of bits (i.e., the data arrive at 5 Gb/s, while the oscilloscope's analog-to-digital converter samples at 10–100 kHz). Alternatively, the voltage histograms can be made at a specific point in the pattern as opposed to the data eye. This eliminates the pattern effects from the measurements of $\sigma_{1,0}$ but yields a potentially inaccurate measure of $\mu_{1,0}$. Also, this approach has the drawback of recording even fewer bits than the measurement in the eye, and it is not practical in a real transmission system, where the data bits are random.

It is often useful to know the ideal Q factor as a starting place for systems calculations. Humblet and Azizoglu (1991) described the ideal Q factor, considering only accumulated noise impairments in terms of the optical SNR_o (Eq. [10.3]) as

$$Q(dB) = 20 \log \left[\frac{2SNR_o\sqrt{B_o/B_e}}{1 + \sqrt{1 + 4SNR_o}} \right] \tag{10.8}$$

where B_o and B_e are the optical and electrical bandwidths in the receiver.

10.6 Polarization Effects

Several polarization effects in lightwave systems can combine to degrade the performance of long-haul lightwave systems. These effects can both reduce the mean received SNR (Lichtman 1993b; Bruyere and Audouin 1994) and cause the SNR to fluctuate with time (Yamamoto *et al.* 1993). Standard telecommunications optical fibers do not maintain the state of polarization (SOP) of the transmitted signal. Random perturbations along the fiber's length can couple the transmitted signal between the two polarization modes and give rise to the time-varying SOP and polarization mode dispersion (PMD). The unstable polarization can interact with polarization-dependent loss (PDL) in the EDFA's components and polarization hole burning (PHB) in the erbium fiber and give rise to the SNR fluctuations. (See Chapter 6 in Volume IIIA for a more complete treatment of polarization effects in lightwave systems.)

SNR fluctuations in long-haul amplifier-based systems are unavoidable because all practical systems use nonpolarization-maintaining transmission fibers. SNR fluctuations must be accounted for in the design of the system by reducing PDL and PMD where possible, and by building in additional margin at the beginning of life of the system. Figure 10.16 shows a typical plot of the received Q factor versus time for a transatlantic-length transmission at 5 Gb/s. Peak-to-peak fluctuations of 1.5 dB are observed without polarization scrambling at the transmitter (Taylor and Penticost 1994).

PHB can cause the ASE noise to accumulate in the polarization orthogonal to the signal faster than along the parallel axis (Fig. 10.17). Noise accumulates faster than would be predicted by the simple noise accumulation theory, and, as a result, the signal decays at the expense of the noise. PHB results from an anisotropic saturation created when a polarized saturating signal is launched into the erbium-doped fiber. The PHB effect was first observed by Taylor (1993) as an excess noise accumulation in a chain of saturated EDFAs, and it was later isolated in a single amplifier and identified by Mazurczyk and Zyskind (1993). The gain difference caused by PHB is small in a single amplifier, with a typical value of about 0.07 dB for an amplifier with 3 dB of gain compression. Although PHB is a very small effect in a single EDFA, its effect on the overall performance of an optical amplifier transmission line can be several decibels in the received Q factor. This illuminates the basic difference between optical amplifier and regenerative transmission systems: subtle effects in the amplifiers and/or fibers can *accumulate* to cause significant impairments of the system's performance.

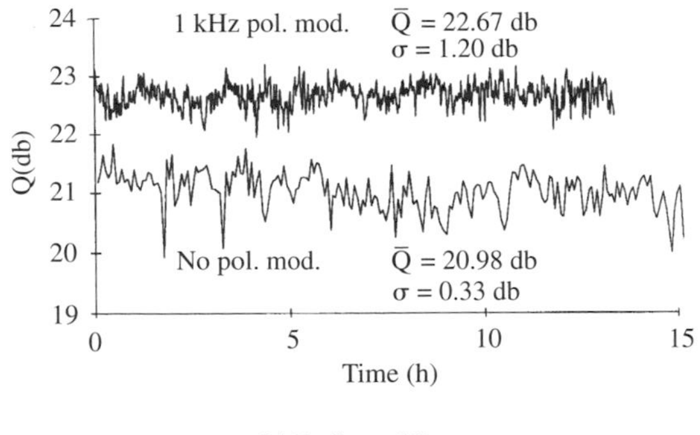

1 kHz Scrambling

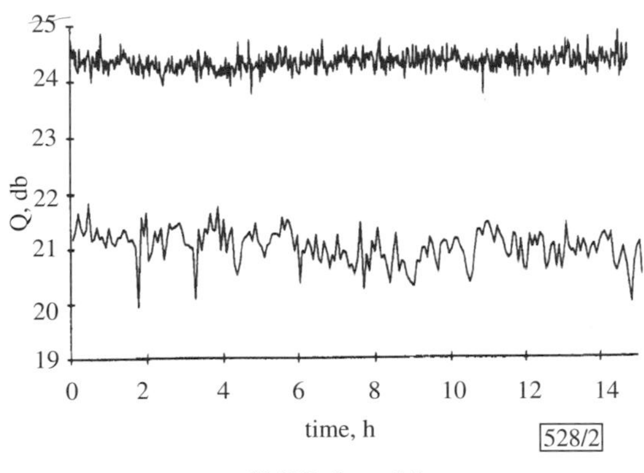

10 GHz Scrambling

Fig. 10.16 Q-factor fluctuations using different polarization scrambling techniques. [*Top,* Adapted with permission from Taylor, M. G. 1994. Improvement in Q with low frequency polarization modulation on transoceanic EDFA link. *IEEE Photon. Tech. Lett.* 6(7):860, © 1994 IEEE; *bottom,* adapted with permission from Taylor, M. G., and S. J. Penticost, 1994. Improvement in performance of long haul EDFA link using high frequency polarization modulation. *Electron. Lett.* 30(10):805.]

Fortunately, the deleterious effects of PHB can be avoided by polarization modulation (or scrambling) of the signal at a rate faster than that to which the EDFA can respond. When the signal's polarization is scrambled faster than the amplifier can respond, there is no preferred polarization axis for the gain to be depleted, and the transmission performance returns to the expected value. The characteristic time constant associated with PHB is similar to the time constants that govern the large signal response of the EDFA, or about 130–200 μs. To reduce the negative effects of PHB on transmission systems, the SOP of the transmitted optical data signal should be scrambled at a rate that is high compared with the amplifier's response

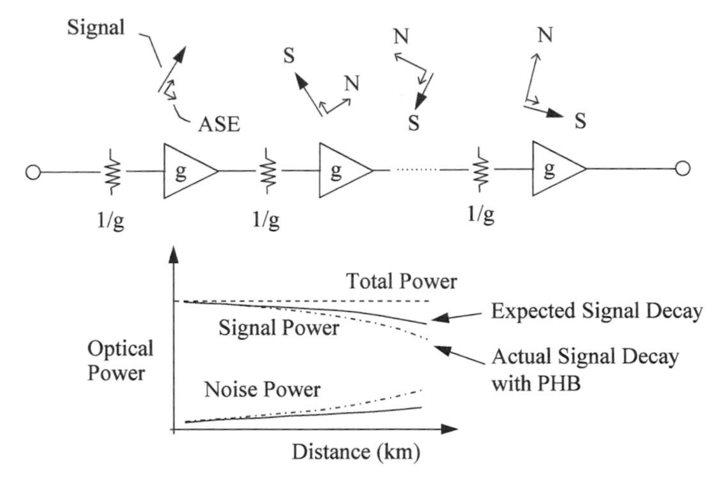

Fig. 10.17 Polarization hole burning (PHB) causes the noise in the orthogonal polarization to have an excess gain of ΔG per amplifier stage.

time. Therefore, polarization scrambling should interchange the optical signal between orthogonal polarizations at a frequency higher than (1/130 μs, or about 8 kHz. Thus, the frequency of polarization scrambling (on a great circle route around the Poincaré sphere; Kliger, Lewis, and Randall 1990) should be greater than 4 kHz to reduce the negative effects of PHB.

Although scrambling in the kilohertz range improves the PHB, it can also cause other effects that can negate any potential improvement. These include amplitude modulation (AM) of the data signal resulting from polarization modulation to AM conversion through the PDL elements in the system, and jitter caused from the interaction of the polarization scrambling and PMD (Bergano 1995b). Considering these phenomena, it is more desirable to scramble at rates equal to or faster than the data's bit rate so that the AM modulation occurs outside the receiver's bandwidth.

The first report of polarization scrambling to improve the performance of long EDFA transmission systems used a two-wavelength transmitter (Bergano, Davidson, and Li 1993). Since that time, other reports have confirmed the results using high-speed lithium niobate scramblers (Taylor and Penticost 1994; Fukada, Imai, and Mamoru 1994) and low-speed scramblers (Bergano *et al.* 1994). The optimal choice of the scrambling frequency is the clock frequency that defines the bit rate of the transmitter (Bergano *et al.* 1995). This technique is particularly important for the efficient use of optical bandwidth in WDM systems. Bit-synchronous polarization scrambling is the optimal trade-off between the two regimes of low-speed and high-speed scrambling. Scrambling the polarization at speeds lower than the bit rate reduces the effects of PHB but introduces unwanted AM through PDL in the optical elements of the system (Lichtman 1995). Scrambling at frequencies higher than the bit rate reduces the PDL problem (Taylor and Penticost 1994) but causes an increase in the transmitted bandwidth, which can limit the number of channels packed into a fixed optical bandwidth. In addition, synchronous polarization scrambling with superimposed phase modulation (PM) can dramatically increase the eye opening of the received data pattern. The increase in eye opening results from the conversion of PM into bit-synchronous AM through chromatic dispersion and nonlinear effects in the fiber.

10.7 Transmission Experiments

Most long-haul transmission experiments using optical amplifiers fall into one of three categories: circulating loops, test beds, and special measurements performed on installed systems. Circulating loop experiments were first performed in 1991 to demonstrate the feasibility of optical amplifier transmission systems. Then, in 1992 to 1993, long amplifier chains were constructed for laboratory use as a test bed for establishing design parameters and feasibility of monitoring systems concepts. Finally, special measurements were performed on the first installed amplifier systems in 1994 to 1995 to determine the feasibility of upgrading the system after its installation. This upgrade potential is unique to amplified systems because there are no bit-rate-limiting elements in the undersea amplified line.

Optical loop experiments were performed as early as 1977 to study pulse propagation in multimode fiber (Tanifuji and Ikeda 1977), jitter accumulation in digital fiber systems (Trischitta, Sannuti, and Chamzas 1988), optical soliton pulse propagation (Mollenauer and Smith 1988), and pulse propagation in single-mode fiber (Malyon *et al.* 1991). Loop transmission experiments became useful for optical amplifier feasibility demonstrations after techniques were developed to measure the BER of long pseudo-random data patterns (Bergano *et al.* 1991a, 1991b). A loop experiment attempts to simulate the transmission performance of a long system by reusing or recirculating an optical data signal through a modest-length amplifier chain ranging from tens to hundreds of kilometers. In the loop experiment (Fig. 10.18), optical switching is added to allow data to flow into the loop (the *load* state) or to allow data to circulate (the *loop* state). The data circulates

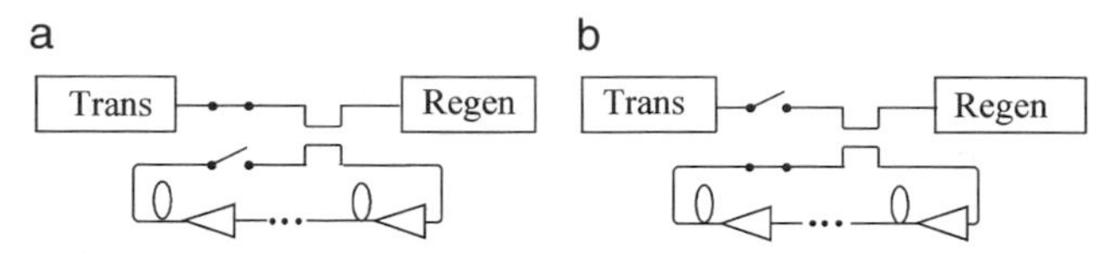

Fig. 10.18 Simplified block diagram of a loop transmission experiment, showing (a) the load state and (b) the loop state.

for a specified time, after which the state of the experiment toggles, and the load–loop cycle is repeated.

Figure 10.19 shows the BER versus transmission distance for a single-channel loop transmission experiment using the 264-km amplifier chain. For the 5 Gb/s transmission, a 10^{-9} BER was achieved at a transmission distance of 20,000 km, or 76 circulations through the amplifier chain. At this distance, the error counting duty cycle was 136:1; thus, an error-free interval of 68 s (real time) was required to demonstrate a 10^{-9} BER with 90% confidence. At 10 Gb/s, the 10^{-9} BER intercept distance was 10,400 km.

Circulating loop techniques can also be used to study WDM transmission. Figure 10.20 shows the block diagram of a 100-Gb/s transmission experiment where 20 5-Gb/s NRZ data channels were transmitted over 6300 km in 11.4 nm of optical bandwidth using a gain-flattened EDFA chain (Bergano *et al.* 1995). The transmission of many WDM channels over transoceanic distances can be limited by the finite bandwidth of the EDFA repeaters

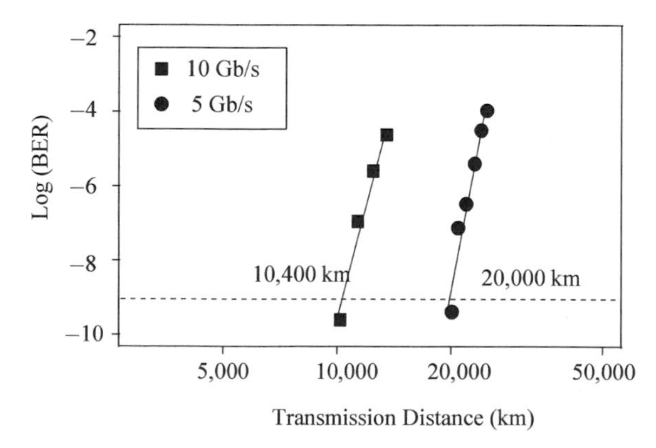

Fig. 10.19 BER versus transmission distance for the 264-km amplifier chain, at 5 and 10 Gb/s.

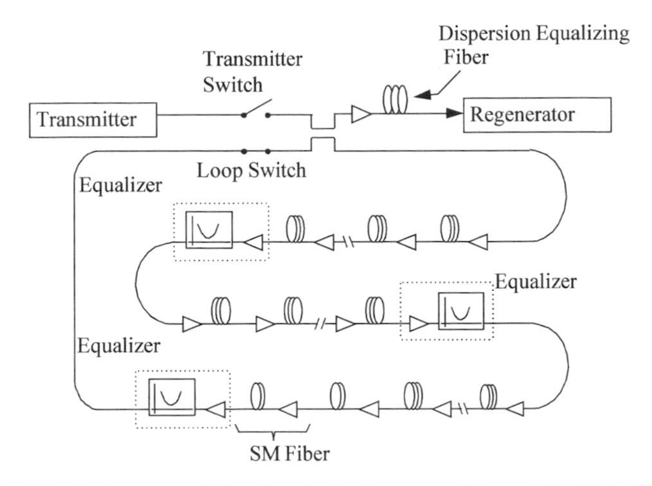

Fig. 10.20 Block diagram of the circulating loop used for a 100-Gb/s WDM transmission experiment.

and the nonlinear interactions between channels. In this experiment, the usable EDFA bandwidth was increased by a factor of 3 using long-period fiber grating filters (Vengsarkar, Lemaire, Jacobvitz, *et al.* 1995; Vengsarkar, Lemaire, Judkins, *et al.* 1995) as gain equalizers. The nonlinear interactions between channels were suppressed by using a transmission fiber with −2 ps/km-nm of dispersion (Bergano *et al.* 1995). Figure 10.21 shows the optical spectrum of the 20 WDM channels before and after propagation through 6300 km (i.e., five passes through the 1260-km amplifier chain).

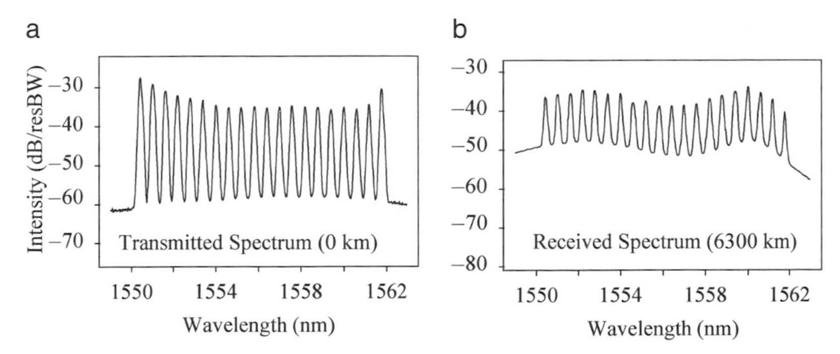

Fig. 10.21 Optical spectrum of 20 5-Gb/s data channels (a) before and (b) after transmission through 6300 km of fiber.

All 20 channels had a time-averaged BER better than 6×10^{-10}. The average Q factors ranged from 15.9 to 17.7 dB.

The performance of the 100-Gb/s WDM experiment was eventually extended to 9300 km using a low noise 980-nm pumped amplifier chain and a variety of the techniques described in this chapter. The transmission format used was NRZ with a combination of synchronous polarization modulation and amplitude modulation (Bergano *et al.* 1996). The gain equalized amplifier chain used 45-km amplifier spacing with a dispersion map similar to that shown in Fig. 10.10. The 980-nm EDFAs had an average noise figure of 4 dB and an output power of +8 dBm. Figure 10.22 shows measured Q-factors and corresponding bit error ratios for 20 5-Gb/s channels after 9300 km. At least 10 individual measurements were performed per channel over a time period long enough to have different polarization states. This procedure ensured an accurate measure of the average BER, considering the fluctuations of the received SNR from polarization effects. The *left-hand arrow* indicates the average Q-factor, and the *right-hand arrow* indicates the average BER over the entire measurement period. All 20 channels had a time-averaged BER lower than 10^{-11}, and the average Q-factor for all 20 channels was 17.4 dB.

Since the time of the initial feasibility studies using circulating loop experiments, several long test-bed experiments have been reported, where

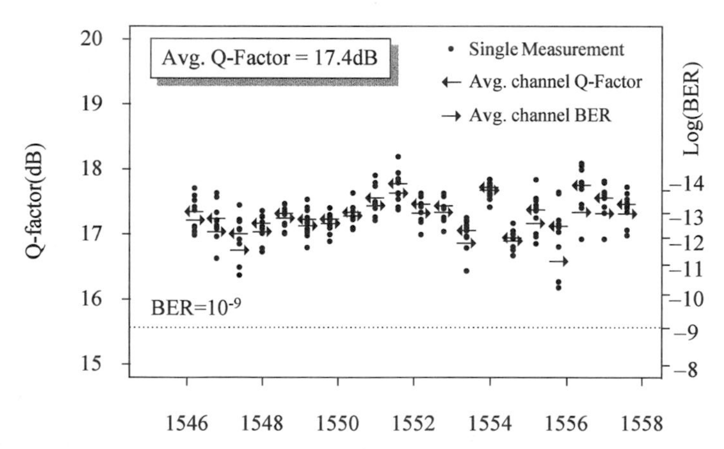

Fig. 10.22 The measured Q factor and corresponding BERs versus channel wavelength for the 20 5-Gb/s WDM NRZ channels at 9300 km. The average Q factors are indicated by the *left-hand arrows*, and the average BER are indicated by the *right-hand arrows*. The Q factor averaged over all 20 channels was 17.4 dB.

hundreds of amplifiers were concatenated to form amplifier chains up to 10,000 km long (Bergano *et al.* 1992; Imai *et al.* 1992). These experiments provided invaluable information on many of the design parameters in a long-haul system, including many of the entries in the impairment budget. The test-bed experiments provided the most accurate representation of the actual system performance; however, given the amount of equipment involved, test-bed experiments tend to be expensive. By contrast, transmission experiments performed using loop techniques require far fewer components, cost much less, and provide the flexibility of making measurements that are impractical in test-bed experiments. For example, the side-by-side comparison of two transmission fiber types or amplifier designs is more easily and economically made in a loop measurement than in a test bed, given the amount of equipment involved.

AT&T installed its first optical amplifier undersea systems in 1994 in the Caribbean. Two cables, each about 2100 km long, were installed as part of the Americas and Columbus systems. The Americas-1 North cable joined Vero Beach, Florida, to Magens Bay in St. Thomas, and the Columbus IIB cable joined West Palm Beach, Florida, to Magens Bay, St. Thomas. The Columbus-IIB segment contains two fiber pairs and 26 repeaters with an average repeater spacing of 80 km. Each pair operates with a single 2.5-Gb/s NRZ optical channel at 1558.5 nm. The amplifiers have an output power of about +5 dBm at 18 dB of gain and 5.7 dB of noise figure. Measurements performed with a higher bit-rate terminal and using WDM techniques demonstrated that this system could operate at up to 15 Gb/s, or six times its designed capacity (Jensen *et al.* 1995).

Similar measurements were also performed on a 4200-km installed cable section of the Trans-Pacific Cable (TPC-5) system (Feggeler *et al.* 1996). Segment G of TPC-5 is installed between San Luis Obispo, California, and Keawaula, Hawaii. Tests performed at transmission rates up to 10 Gb/s and at distances up to 16,800 km (by looping the received signal back at each end) demonstrated that the cable segment is fully capable of operating at twice the design line rate of 5 Gb/s.

10.8 Transmission Systems

Thus far, we reviewed several aspects of optical amplifier transmission technology used in undersea cable systems. This section attempts to put the pieces together by reviewing the design of a typical transoceanic 5-Gb/s

NRZ transmission system. To this end, we use the TAT-12/13 cables as a model. The TAT-12/13 cables form a ring network in the north Atlantic with two transatlantic cables of 5900 km and 6300 km, and two interconnection cables, each a few hundred kilometers in length (Trischitta *et al.* 1996) (Fig. 10.23). The ring architecture allows for mutual restoration capability *within the network* with the use of network protection elements at each of the four landing sights. In the unlikely event of a cable failure, the customer's traffic is automatically rerouted in the opposite direction around the ring to bypass a fault in any of the undersea segments. The name TAT-12/13 signifies that they are 12th and 13th transatlantic telephone systems installed in the Atlantic by AT&T and its partners. The TAT-12 cable is one of the first major applications of EDFA technology in the undersea market. The TAT-12/13 cable system is owned and operated by a consortium of 44 partners, and was designed and manufactured by AT&T Submarine Systems, Alcatel Submarine Networks, and Toshiba. The network was placed into service in September 1996.

Undersea cable systems are designed with a 25-year life expectancy. The goal of the system design is to have an adequate beginning-of-life margin to allow for Q-factor fluctuations, component aging, and system repairs. Key design parameters are the repeater spacing, the launch power, and the dispersion management of the repeatered line. The trade-offs in the selection of the repeater spacing are cost and performance. Increasing the repeater spacing reduces cost and improves reliability because fewer repeaters

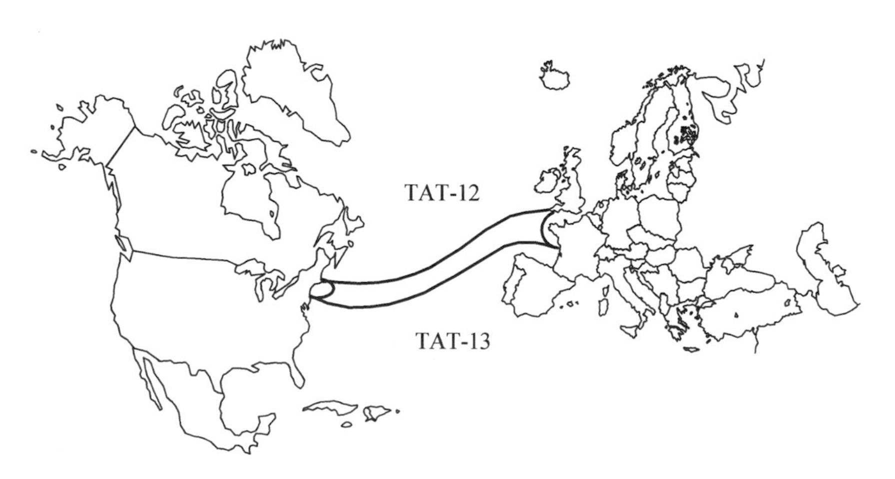

Fig. 10.23 TAT-12/13 cable system.

are needed; however, fewer repeaters mean longer repeater spacing, which translates into more accumulated noise. For a particular system, the repeater spacing is chosen as the largest length possible, satisfying the end-of-life requirement, given in the impairment budget. Performance is measured against the Q factor of 16.1 dB or the value relating to a BER of 1×10^{-10}.

The impairment budget is a design tool used to aid in the selection of span length by accounting for all the expected impairments over the system's lifetime. An impairment budget is given in Table 10.2. The starting point is the ideal Q factor, which is calculated considering only optical noise using Eqs. (10.3) and (10.8). In our example, this value is 30.5 dB, assuming a 6300-km system with 45-km repeater spacing, +3-dBm output power, a 5.4-dB noise figure, and 0.2-dB/km fiber attenuation. From this starting point the values of all expected degradations are subtracted. For example, the 8.7-dB value in line 2 includes effects arising from the fiber's nonlinear index, added noise from optical reflections, ISI, and Q-factor fluctuations caused by polarization effects. This value is obtained using data from laboratory experiments and computer modeling. The values in the table are recalculated for different span lengths until the 1-dB end-of-life figure is

Table 10.2 **Performance Budget for a 5-Gb/s Transmission System**[a]

Line	*Parameter*	*Value or Penalty (dB)*
1	Ideal Q factor (dB)	30.5
2	Interactive and fluctuation impairments	−8.7
3	Worst case nonimpaired Q factor (1 + 2)	21.8
4	Beginning-of-life (BOL) impairments: Line design margin, transmitter–receiver design margin, loop-back path impairments, etc.	−3.5
5	Unallocated margin	−0.5
6	Minimum BOL performance (3 + 4 + 5)	17.8
7	Aging and repairs	−0.7
8	Minimum end-of-life (EOL) performance (6 + 7)	17.1
9	Required Q factor for bit error rate (BER) $<1 \times 10^{-10}$	16.1
10	EOL margin (8 − 9)	1.0

[a] Adapted from Schesser *et al.* 1995.

reached. If forward error-correction coding is used in the terminals, the value on line 9 could be lowered to 11.4 dB, corresponding to an uncorrected BER of 10^{-4}.

Long undersea cable systems include a monitoring capability to ascertain the performance of the cable and repeaters and to locate faults in the event of a cable break or component failure (Chapters 2 and 3 in Volume IIIB). In the AT&T system design, high-loss loop-back paths joining the amplifiers in a pair allow a small amount of signal (−45 dB) to leak back in the opposite direction with a unique round-trip delay (Jensen *et al.* 1994). The amplitude of the out-going 5-Gb/s data signal is modulated with a low-frequency pseudo-random bit pattern located on a 2-MHz subcarrier. When digital signal processing techniques are used at the terminal, the small amount of in-coming signal is correlated with the transmitted signal, and the gain of each loop-back path is measured. When the system is in service, the depth of modulation is only a few percent. For out-of-service operation, the depth of modulation is increased to 100% to speed up the measurement time.

Each transatlantic cable has four chains of amplifiers, grouped as two bidirectional pairs. Each of the four fibers carries a single 5-Gb/s NRZ data signal. The amplifiers are spaced at 45-km intervals along the cable (Table 10.3) and are located in repeater pressure vessels that can withstand up to 800 atmospheres of pressure found at the ocean bottom. Repeaters are powered by passing a 0.9-A DC current through a power conductor within the cable. The amplifiers have a nominal gain of 9.5 dB and a noise figure of 5 dB, at an output power of about +3 dBm, and at 5 dB of gain compression. The transmission fiber is single-mode dispersion-shifted fiber, selected to have a dispersion value of about −0.3 ps/km-nm on average.

Table 10.3 **Repeater Spacing for Existing Optical Amplifier Undersea Transmission Systems**

System	*Bit Rate (Gb/s)*	*System Length (km)*	*Repeater Spacing (km)*
TPC-5 segment J	5.0	8620	33
TAT-12/13	5.0	5900/6300	45
TPC-5 segment G	5.0	4200	68
TPC-5 segment T1	5.0	1300	82

This small amount of negative dispersion reduces the nonlinear mixing between the signal and the optical noise. The accumulated negative dispersion is compensated for every 500 km using normal single-mode fiber with $\lambda_0 = 1310$ nm.

Each fiber pair carries a single NRZ channel operating at a bit rate of 4.97664 Gb/s, comprised of two STM-16 (2.48832 Gb/s) bit interleaved. Thus, each cable with its two fiber pairs carries a total of 4 STM-16 channels. Since the TAT-12/13 network is designed to have mutual restoration capability within the network, each cable is configured with one pair designated for normal traffic and one pair for restoration. Each STM-16 channel consists of 30,240 64-kB/s digital circuits (calculated by multiplying 30 64-kB/s circuits per E1, 63 E1s in an STM-1, and 16 STM-1s in an STM-16). The network has 4 STM-16 primary channels for normal traffic, which is 120,960 64-kB/s circuits. These circuits configured to carry voice traffic only would support about 600,000 simultaneous conversations (using a compression ratio of 5:1), with an additional 600,000 voice circuits available for standby traffic on the protection lines.

10.9 Summary

The first generation of optical amplifier undersea lightwave systems has now been deployed in both the Atlantic and the Pacific oceans. These first systems operate at 5 Gb/s, which is about 300,000 equivalent voice circuits per fiber pair. Technical innovations such as WDM transmission will allow the transmission capacity of undersea cable systems to increase 20-fold in the next few years. As the demand for newer advanced digital services expands in the national markets, so too will the demand grow for these services in the international market. Lightwave transmission systems based on optical amplifier repeaters will provide the international conduits for these new services.

The reader who is interested in a more detailed understanding of optical amplifier undersea cable systems is directed to review an issue of *IEEE Communications Magazine* that includes information on Global Undersea Communication Networks (Trischitta and Marr 1996), a special issue of the *AT&T Technical Journal* (AT&T 1995), and Thiennot, Pirtio, and Thomine (1993). For a good review of amplifier systems in general, see Li (1993).

References

Agrawal, G. P. 1989a. Group-velocity dispersion. In *Nonlinear fiber optics,* Chapter 3. Boston: Academic Press, 51–73.

Agrawal, G. P. 1989b. *Nonlinear fiber optics.* Boston: Academic Press.

AT&T. 1995. Undersea communications systems. *AT&T Tech. J.* 74(1).

Bell Telephone Laboratories. 1982. *Transmission systems for communications.* 5th ed. Holmdel, NJ: Bell Telephone Laboratories, Inc., 741.

Bergano, N. S., J. Aspell, C. R. Davidson, P. R. Trischitta, B. M. Nyman, and F. W. Kerfoot. 1991a. Bit error rate measurements of a 14,000 km 5 Gb/s fiber-amplifier transmission system using a circulating loop. *Electron. Lett.* 27(21):1889.

Bergano, N. S., J. Aspell, C. R. Davidson, P. R. Trischitta, B. M. Nyman, and F. W. Kerfoot. 1991b. A 9000 km 5 Gb/s and 21,000 km 2.4 Gb/s feasibility demonstration of transoceanic EDFA systems using a circulating loop. In *OFC '91, San Diego, CA.* Postdeadline paper PD13, PD13-1.

Bergano, N. S., and C. R. Davidson. 1995. Polarization scrambling induced timing jitter in optical amplifier systems. In *OFC '95, San Diego, CA.* Paper WG3, 122.

Bergano, N. S., C. R. Davidson, and F. Heismann. 1996. Bit-synchronous polarization and phase modulation scheme for improving the transmission performance of optical amplifier transmission system. *Electron. Lett.* 32(1).

Bergano, N. S., C. R. Davidson, G. M. Homsey, D. J. Kalmus, P. R. Trischitta, J. Aspell, D. A. Gray, R. L. Maybach, S. Yamamoto, H. Taga, N. Edagawa, Y. Yoshida, Y. Horiuchi, T. Kawazawa, Y. Namihira, and S. Akiba. 1992. 9000 km, 5 Gb/s NRZ transmission experiment using 274 erbium-doped fiber-amplifiers. In *Optical amplifiers and their applications topical meeting, Santa Fe, NM.* Postdeadline paper PD11.

Bergano, N. S., C. R. Davidson, and T. Li. 1993. A two-wavelength depolarized transmitter for improved transmission performance in long-haul EDFA systems. In *LEOS '93 annual meeting, San Jose, CA, November 1993.* Postdeadline paper PD2.2, 23.

Bergano, N. S., C. R. Davidson, *et al.* 1995. 40 Gb/s WDM transmission of eight 5 Gb/s data channels over transoceanic distances using the conventional NRZ modulation format. In *OFC '95, San Diego, CA.* Paper PD19.

Bergano, N. S., C. R. Davidson, A. M. Vengsarkar, B. M. Nyman, S. G. Evangelides, J. M. Darcie, M. Ma, J. D. Evankow, P. C. Corbett, M. A. Mills, G. A. Ferguson, J. R. Pedrazzani, J. A. Nagel, J. L. Zyskind, J. W. Sulhoff, and A. J. Lucero. 1995. 100 Gb/s WDM transmission of twenty 5 Gb/s NRZ data channels over transoceanic distances using a gain flattened amplifier chain. In *European Conference on Optical Communication (ECOC '95), Brussels, Belgium,* Paper Th.A. 3.1, 967.

Bergano, N. S., F. W. Kerfoot, and C. R. Davidson. 1993. Margin measurements in optical amplifier systems. *IEEE Photon. Tech. Lett.* 5(3):304.

Bergano, N. S., V. J. Mazurczyk, and C. R. Davidson. 1994. Polarization scrambling improves SNR performance in a chain of EDFAs. In *OFC '94, San Jose, CA.*

Bruyere, F., and O. Audouin. 1994. Penalties in long-haul optical amplifier systems due to polarization dependent loss and gain. *IEEE Photon. Tech. Lett.* 6(5):654.

Ehrbar, R. D. 1986. Undersea cables for telephony. In *Undersea lightwave communications,* chapter 1. New York: IEEE Press.

Feggeler, J. C., D. G. Duff, N. S. Bergano, C. Chen, Y. Chen, C. R. Davidson, D. G. Ehrenberg, S. J. Evangelides, G. A. Ferguson, F. L. Heismann, G. M. Homsey, H. D. Kidorf, T. M. Kissell, A. E. Meixner, R. Menges, J. L. Miller, O. Mizuhara, T. V. Nguyen, B. M. Nyman, Y. K. Park, W. W. Patterson, and G. F. Valvo. 1996. 10 Gb/s WDM transmission measurements on an installed optical amplifier undersea cable system. In *OFC '96, San Jose, CA.* Paper TUN3, 72.

Fukada, Y., T. Imai, and A. Mamoru. 1994. BER fluctuation suppression in optical in-line amplifier systems using polarization scrambling technique. *Electron. Lett.* 30(5):432.

Giles, C. R., and E. Desurvire. 1991. Propagation of signal and noise in concatenated erbium-doped fiber amplifiers. *J. Lightwave Tech.* 9(2):147.

Giles, C. R., E. Desurvire, and J. Simpson. 1989. Transient gain and cross talk in erbium-doped fiber amplifier. *Opt. Lett.* 14(16):880.

Gordon, J. P., and L. F. Mollenauer. 1990. Phase noise in photonic communications systems using linear amplifiers. *Opt. Lett.* 15(23):1351.

Gordon, J. P., and L. F. Mollenauer. 1991. Effects on fiber nonlinearities and amplifier spacing on ultra-long distance transmission. *J. Lightwave Commun.* 9(2):170.

Humblet, P. A., and M. Azizoglu. 1991. On the bit error rate of lightwave systems with optical amplifiers. *J. Lightwave Tech.* 9:1576.

Imai, T., M. Murakami, Y. Fukada, M. Aiki, and T. Ito. 1992. Over 10,000 km straight line transmission system experiment at 2.5 Gb/s using in-line optical amplifiers. In *Optical amplifiers and their applications topical meeting, Santa Fe, NM.* Postdeadline paper PD12, PD-12.

Jensen, R. A., C. R. Davidson, D. L. Wilson, and J. K. Lyons. 1994. Novel technique for monitoring long-haul undersea optical-amplifier systems. In *Optical Fiber Communications Conference, San Jose, CA.* Paper ThR3.

Jensen, R. A., D. G. Duff, J. J. Risko, and C. R. Davidson. 1995. Possibility for upgrade of the first installed optical amplifier system. In *Topical meeting on optical amplifiers and their applications, Davos, Switzerland.* Paper ThB2, 9.

Kaiser, P., and D. B. Keck. 1988. Fiber types and their status. In *Optical fiber telecommunications II,* ed. S. E. Miller and I. P. Kaminow, 29–51. Boston: Academic Press.

Kawai, S., K. Iwatsuki, K. Suzuki, S. Nishi, M. Saruwatari, K. Sato, and K. Wakita. 1994. 10 Gbit/s optical soliton transmission over 7200 km by using a monolithically integrated MQW-DFB-LD/MQW-EA modulator light source. *Electron. Lett.* 30(3):251.

Kliger, D. S., J. W. Lewis, and C. E. Randall. 1990. *Polarized light in optics and spectroscopy.* New York: Academic Press.

Kodama, Y., and A. Hasegawa. 1992. Generation of asymptotically stable optical solitons and suppression of the Gordon-Haus effect. *Opt. Lett.* 17:31.

Li, T. 1993. The impact of optical amplifiers on long-distance lightwave telecommunications. *Proc. IEEE* 18(11):1568.

Lichtman, E. 1993a. Optimal amplifier spacing in ultra-long lightwave systems. *Electron. Lett.* 29:2058.

Lichtman, E. 1993b. Performance degradation due to polarization dependent gain and loss in lightwave systems with optical amplifiers. *Electron. Lett.* 29(22):1971.

Lichtman, E. 1995. Limitations imposed by polarization-dependent gain and loss on all-optical ultra-long communication systems. *J. Lightwave Tech.* 13(5).

Malyon, D., T. Widdowson, E. G. Bryant, S. F. Carter, J. V. Wright, and W. A. Stallard. 1991. Demonstration of optical pulse propagation over 10,000 km of fiber using recirculating loop. *Electron. Lett.* 27(2):120.

Marcuse, D. 1990. Derivation of analytical expressions for the bit-error probability in lightwave systems with optical amplifiers. *J. Lightwave Tech.* 8:1816.

Marcuse, D., A. R. Chraplyvy, and R. W. Tkach. 1991. Effects of fiber nonlinearity on long-distance transmission. *J. Lightwave Tech.* 9(1):121.

Mazurczyk, V. J., and D. G. Duff. 1995. Effect of intersymbol interference on signal-to-noise measurements. In *OFC '95, San Diego, CA.* Paper WQ1, 188.

Mazurczyk, V. J., and J. L. Zyskind. 1993. Polarization hole burning in erbium doped fiber amplifiers. *CLEO '93, Baltimore.* Postdeadline paper CPD26.

Mecozzi, A., J. D. Moores, H. A. Haus, and Y. Lai. 1991. Soliton transmission control. *Opt. Lett.* 16:1841.

Mollenauer, L. F., J. P. Gordon, and S. G. Evangelides. 1992. The sliding-frequency guiding filter: An improved form of soliton jitter control. *Opt. Lett.* 17:1575–1577.

Mollenauer, L. F., M. J. Neubelt, M. Haner, E. Lichtman, S. G. Evangelides, and B. M. Nyman. 1991. Demonstration of error-free soliton transmission at 2.5 Gb/s over more than 14,000 km. *Electron. Lett.* 27(22):2055.

Mollenauer, L. F., and K. Smith. 1988. Demonstration of soliton transmission over more than 4000 km in fiber with loss periodically compensated by Raman gain. *Opt. Lett.* 13(8):675.

Personick, S. D. 1973. Receiver design for digital fiber optic communications systems. *Bell Syst. Tech. J.* 52(6).

Runge, P. K., and P. R. Trischitta. 1986. The SL undersea lightwave system. In *Undersea lightwave communications,* New York: IEEE Press.

Schesser, J., S. M. Abbott, R. L. Easton, and M. S. Stix. 1995. Design requirements for the current generation of undersea cable systems. *AT&T Tech. J.* 74(1):16.

Taga, H., M. Suzuki, Y. Yoshida, H. Tanaka, S. Yamamoto, and H. Wakabayashi. 1991. 2.5 Gb/s optical transmission using electroabsorption modulator over 11,000 km EDFA systems. In *Conference on Lasers and Electro-Optics.* Paper CPDP38. Technical Digest. Washington, DC: Optical Society of America.

Tanifuji, T., and M. Ikeda. 1977. Pulse circulation measurement of transmission characteristics in long optical fibers. *Appl. Opt.* 16(8):2175.

Taylor, M. G. 1993. Observation of new polarization dependence effect in long haul optically amplified system. In *OFC '93, San Jose, CA.* Post deadline paper PD5.

Taylor, M. G. 1994. Improvement in Q with low frequency polarization modulation on transoceanic EDFA link. *IEEE Photon. Tech. Lett.* 6(7):860.

Taylor, M. G., and S. J. Penticost. 1994. Improvement in performance of long haul EDFA link using high frequency polarization modulation. *Electron. Lett.* 30(10):805.

Thiennot, J., F. Pirtio, and J. B. Thomine. 1993. Optical undersea cable systems trends. *Proc. IEEE* 81(11):1610.

Trishitta, P. R., M. Colas, M. Green, G. Wuzniak, and J. Arena. 1996. The TAT-12/13 cable network. *IEEE Communications Magazine.* 34(2):24.

Trischitta, P. R., and W. C. Marra. 1996. Global undersea communications networks. *IEEE Communications Magazine.* 34(2).

Trischitta, P. R., P. Sannuti, and C. Chamzas. 1988. A circulating loop experimental technique to simulate the jitter accumulation of a chain of fiber optic regenerators. *IEEE Trans. Commun.* COM-36:2.

Vengsarkar, A. M., P. J. Lemaire, G. Jacobovitz, J. J. Veselka, V. Bhatia, and J. B. Judkins. 1995. Long-period fiber gratings as gain-flattening and laser stabilizing devices. In *Proceedings of the IOOC '95,* vol. 5, 3–4. Hong Kong.

Vengsarkar, A. M., P. J. Lemaire, J. B. Judkins, J. E. Sipe, and T. Erdogan. 1995. Long-period fiber gratings as band-rejection filters. In *OFC '95, San Diego, CA.* Paper PD4.

Widdowson, T., and D. J. Malyon. 1991. Error ratio measurements over transoceanic distances using recirculating loop. *Electron. Lett.* 27(24):2201.

Yamamoto, S., N. Edagawa, H. Taga, Y. Yoshida, and H. Wakabayashi. 1993. Observation of BER degradation due to fading in long-distance optical amplifier system. *Electron. Lett.* 29(2):209.

Chapter 11 | Advances in High Bit-Rate Transmission Systems

Kinichiro Ogawa
Liang D. Tzeng
Yong Kwan Park

Lucent Technologies, Bell Laboratories, Breinigsville, Pennsylvania

Eiichi Sano

NTT LSI Laboratories, Atsugi, Japan

11.1 Introduction

The bit rate of optical systems has increased almost six times, from 1.7 Gb/s in the 1980s to 10 Gb/s in the 1990s. This capacity increase is a result of technical progress in high-speed electronics as well as optical devices, and was stimulated by the postdivestiture (of the Bell System) competitive communications market in the United States. There is one major obstacle in the U.S. market, however. Many communications providers would like to make a system upgrade based on the already-installed single-mode fiber cable network. These "embedded" fibers limit the traditional transmission distance to a range of about 40 km because of fiber loss at a transmission wavelength of 1.3 μm or chromatic dispersion at a wavelength of 1.5 μm.

With the emerging new technologies such as the erbium-doped fiber amplifier (EDFA) (see Chapter 4 in Volume IIIA and Chapter 2 in Volume IIIB) and the dispersion-compensation techniques (Chapter 7, Volume IIIA), the problem of chromatic dispersion at 1.55 μm is at least manageable. Consequently, for economical purposes, the capability of future upgrades, or differentiation against competition, line providers want to include these new technologies to push for transmission through a more than 300-km distance at 1.55 μm for a 10-Gb/s system design. These subjects are discussed further in Section 11.3, using examples from 10-Gb/s field experiments.

OPTICAL FIBER TELECOMMUNICATIONS,
VOLUME IIIA

ISBN: 0-12-395170-4

Besides the "straight-through" 10-Gb/s time-division multiplexing (TDM) system configuration, an alternative approach for high-capacity transmission will be the wavelength-division multiplexing (WDM) system. At present, eight-channel WDM of 2.5-Gb/s is being actively pursued in the United States. Compared with the TDM system, the WDM system has many advantages, such as that it provides an easier solution to avoid the limitation imposed by chromatic dispersion and that it can share the same electronic technology platform with the low bit-rate (2.5-Gb/s) system. However, in general, the TDM system has an economical advantage as long as high-speed electronics are available. In this chapter, we concentrate on system-related technologies and design considerations for 10-Gb/s applications. The subject of high-speed electronics is discussed in Section 11.4.

One important difference between the 1.7-Gb/s system and the 10-Gb/s system is that direct modulation on the laser diode at a speed of 10-Gb/s is very difficult for implementation, except for a short-distance link. External modulation technologies such as the Ti:$LiNbO_3$ modulators and integrated electroabsorption modulators are currently indispensable components for 10-Gb/s applications. Discussions on these components and their applications to high-speed transmitters are included in Chapters 4 and 9 of Volume IIIB and therefore are not repeated in this chapter.

Another critical issue in high-speed systems is the bandwidth management for the overall system performance. This subject is covered in Section 11.2, where we discuss the effect of the optical receiver bandwidth on system eye-margin performance.

11.2 High-Speed Receiver Design

11.2.1 RECEIVER SYSTEM SETUP

Shown in Fig. 11.1 is a typical arrangement for an optical receiver system. The incident lightwave signal is converted to an electrical signal by the *optical receiver front end*, which contains a photodetector and a preamplifier circuit. Usually, to enhance the receiver sensitivity, some means of increasing the average number of photoelectrons generated by the photodetector per incident photon is also included in the receiver setup. Schematically, this process is represented by the gain block, G, as illustrated in Fig. 11.1. This preamplification process can be accomplished in several ways. For a direct detection system, the most commonly adopted method is to use an

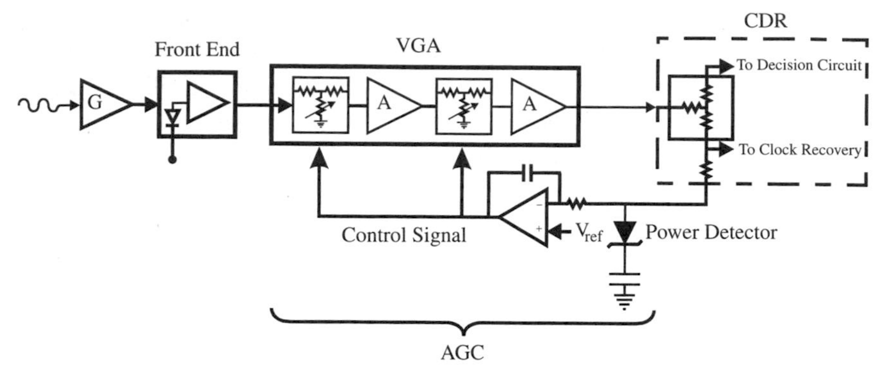

Fig. 11.1 Typical optical receiver system setup. AGC, automatic gain controlled; CDR, clock–data recovery; VGA, variable gain amplifier.

avalanche photodiode (APD) as the photodetector. Statistically, for every incident photon, the average number of photoelectrons generated by the APD is ηM, where η is the quantum efficiency and M is the avalanche gain of the APD. The typical M value is about 8–12 for most of the receivers used in telecommunications systems.

Another way of implementing the preamplification process is to use an optical amplifier, either a semiconductor optical amplifier[1,2] or a fiber amplifier, such as the EDFA. Because of the excellent achievable gain (>30 dB), low noise characteristic, and low insertion loss, the EDFA has been the prevailing choice for implementing the *optical preamplifier receiver* in long-haul system experiments, particularly at high bit rates (e.g., 5 or 10 Gb/s). Currently, the major disadvantages of using the EDFA as the optical preamplifier are (1) high cost, (2) high power consumption, and (3) relatively bulky size compared with that of a simple APD detector. For moderate link length (<50-km, for example) system applications, the APD receiver[3] will still be a very attractive option.

In general, the preamplifier in the optical front end is an analog circuit. With a fixed gain preamplifier, the front-end output signal level will follow the variation of the input optical power. This kind of signal level variation will impair the performance of the clock-recovery and decision circuit subsystem, indicated as the CDR (clock–data recovery) block in Fig. 11.1. Also, at the low input optical power range, the output signal level from the front end is usually not high enough to be processed by the decision circuit, which typically requires an input signal peak-to-peak value of at least several hundred millivolts. Therefore, a postamplifier is needed after the optical front end to minimize any extra degradation in the performance of the CDR section. The main functions of this postamplifier are to provide an adequate signal amplification and to maintain (quantize) a stable output signal level.

There are two commonly used methods of implementing this kind of quantization amplifier. One is to use a variable gain amplifier (VGA) as the postamplifier and to adjust its gain according to the input signal power. This is the so-called automatic gain controlled (AGC) setup. As illustrated in Fig. 11.1, an example for the VGA is constructed by cascading a chain of variable attenuators and fixed-gain amplifiers. A downstream power detector (or peak detector) is used to monitor the output signal level from the VGA. This power detector output is then compared with a predetermined reference voltage to generate the control signal, which will then adjust the amount of attenuation in the VGA. Therefore, under the closed-loop condition, the VGA automatically adjusts its overall gain to maintain a constant output signal level.

Another form of quantization amplifier is a limiting amplifier. The simplest form of the limiting amplifier can be pictured as a high-gain amplifier followed by a digital flip-flop. An ideal AGC amplifier is, by definition, an analog circuit. The signal spectrum of an AGC output should be a scaled replica of that of the input signal. The limiting amplifier, on the other hand, is inherently a device with digital output(s). There are subtle differences in the bit error rate (BER) determination and system eye-margin characterization processes between a receiver system with an AGC amplifier and a receiver system with a limiting amplifier.

For a receiver system with an AGC amplifier, the BER determination process occurs at the decision circuit. On the other hand, in a receiver system with a limiting amplifier, the "error-making" process should include both the limiting amplifier and the decision circuit.[4] For a practical system application, besides receiver sensitivity, system *eye margin* and *phase margin* are very important parameters. Referring to Fig. 11.2, let P_1 and P_0 represent the probability distribution functions for the mark and space (or one and zero) of the detected signal, respectively, at an optimum decision timing position. d_1 and d_0 are the two corresponding decision levels for a predetermined value (e.g., 1×10^{-15}) of the BER. Assuming that the average signal peak-to-peak level is S, the percentage eye margin is defined as $100 \times [(d_1 - d_0)/S]\%$. Similarly, for a clock signal applied to the decision circuit at the optimum decision threshold, the phase margin can be defined (in units of degrees) as the free traveling time interval within which the BER is below this predetermined value. Experimentally, if an AGC postamplifier

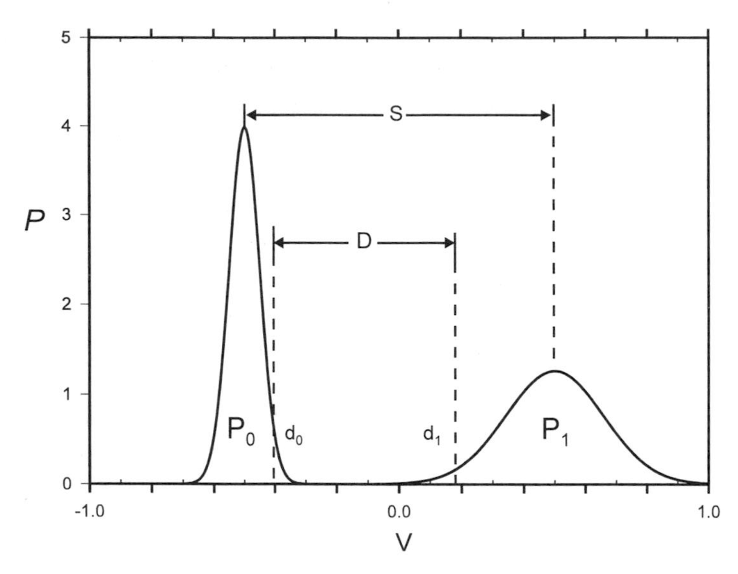

Fig. 11.2 Eye-margin definition. P_1 and P_0 are the probability density distributions for the one and zero, respectively, of the received non-return-to-zero (NRZ) signal. S is the average signal peak-to-peak value. d_1 and d_0 are the two decision threshold boundaries within which the bit error rate (BER) falls below a predefined value (see text).

is used, the eye margin can be measured by observing the changes in the BER while varying the decision threshold. In a receiver system employing a limiting amplifier, this type of eye-margin determination is not possible. Indirect measurement procedures, such as the S-over-X^4 or noise loading technique,* would have to be used instead.

11.2.2 RECEIVER BANDWIDTH AND INTERSYMBOL INTERFERENCE PENALTY

For a low bit-rate system (e.g., optical carrier (OC)-12 and lower), it is relatively easy to obtain high-quality (such as small pulse distortion, high extinction ratio, and wide bandwidth) transmitters. The receiver bandwidth is usually designed to be lower than the system bit rate in order to minimize the receiver circuit noise. For higher bit-rate systems, the situation changes

* By introducing a controlled amount of noise power and mixing with the input signal, from the measured sensitivity penalty, one can estimate the original system eye margin. However, like the S-over-X measurement, one would have to be careful in interpreting the measurement results because of the effects of asymmetrical noise distribution.

noticeably. At present, the equivalent small signal bandwidth of a typical transmitter for a 10-Gb/s application is on the order of (or even lower than) 10 GHz. Because the transmitter bandwidth is limited, a smaller receiver bandwidth will inevitably introduce an ISI (intersymbol interference) penalty. To qualitatively illustrate the relationship between the ISI and system bandwidth, a computer simulation study for the eye-margin estimations at 10 Gb/s is summarized next.

In this study, we characterize the transmitter bandwidth with a simple parameter t_{rf}, the rise and fall time (10–90%) of the transmitter output pulse. We also restrict ourselves to the non-return-to-zero (NRZ) random data format and assume that no pulse distortion effects such as overshoot or undershoot are associated with the transmitter output pulse shape. At the receiver end, we assume that the received optical signal power (-10 dBm for this example) is adequately higher than the receiver sensitivity. The eye margin is calculated for a BER of 1×10^{-15}. The receiver model used in this simulation program is based on a measured receiver small signal frequency response (to be discussed in the next section). We then hypothetically scale the receiver bandwidth to see its effects on the eye-margin calculations. In this bandwidth scaling process, we intentionally treat the receiver amplitude bandwidth and phase bandwidth independently. Although this is not a truly attainable process, it does provide a better guide for setting up criteria in receiver design.

It should be pointed out that the phase bandwidth is defined as the frequency where the linear phase frequency response (or equivalently the group delay) has deviated from its DC value by 20°. This parameter can be easily obtained with a sweep frequency measurement using a network analyzer. For a high-speed receiver, this type of phase linearity deviation is most commonly due to the physical connection, such as the wire bond, between the photodiode and the preamplifier. The inductance associated with this interconnection and the total input capacitance of the preamplifier will tend to provide some level of inductance peaking to the amplitude frequency response of the front end. This inductance peaking will also tend to create a nonuniform group delay (or equivalently a phase deviation) at the corresponding resonant frequency region.

The two curves of Fig. 11.3a represent the eye-margin estimations* as functions of the receiver amplitude bandwidth, with the assumption that the receiver phase response is perfectly linear (the phase deviation is zero).

* Here we have made the assumption of using an ideal decision circuit. In reality, the ambiguity level and nonlinear effects of the decision circuit will also degrade the eye-margin measurement.

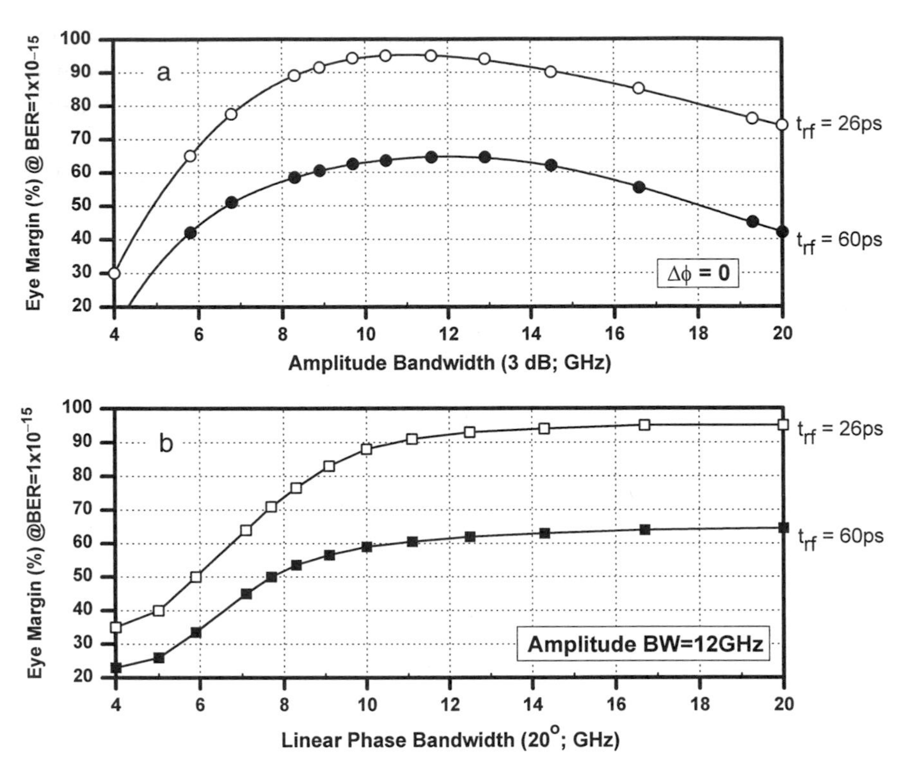

Fig. 11.3 (a) Eye margin versus the receiver amplitude bandwidth with the assumption that the phase bandwidth is infinite. (b) Eye margin versus the receiver phase bandwidth with the assumption that the amplitude bandwidth equals 12 GHz. t_{rf} represents the rise and fall time of the transmitter output pulse shape.

The open and solid circles are for a transmitter with $t_{rf} = 26$ ps and $t_{rf} = 60$ ps, respectively. When the receiver bandwidth is lower than the system bit rate (10 Gb/s), the eye closure is mainly due to the ISI. When the receiver bandwidth is excessively larger than the system bit rate, the ISI effect is minimized, yet the receiver circuit noise will cause the reduction of sensitivity and eye margin. Now, let us fix the receiver amplitude bandwidth to 12 GHz and vary the phase bandwidth. The results are shown in Fig. 11.3b. The important point is that any phase deviation within the amplitude bandwidth will cause additional eye closure.

Figures 11.3a and 11.3b, together with the traditional receiver noise calculations,[6,7] provide the guidelines for designing an optical receiver for 10-Gb/s applications. In choosing the receiver bandwidth, one would have to make a compromise between optimizing the receiver sensitivity and pushing for the highest attainable eye margin.

11.2.3 RECEIVER CIRCUIT AND PERFORMANCE

The schematic of Fig. 11.4 shows a front-end circuit evolved from the original work of Ogawa and Chinnock.[8] The FET drain voltage (V_{DS1}) of Q1 is set to a V_{BE} (~0.75 V) above the base voltage of the PNP transistor, Q2. Under normal operating conditions, the emitter current of Q2 is arranged to be smaller than the drain current of Q1. Therefore, the drain current of Q1 can be approximately determined from the voltage drop across the resistor R2. These DC bias conditions set the gate voltage (V_{GS1}) of Q1 accordingly. Under the assumption that the gate leakage current of Q1 is negligible, the source voltage of Q3 is approximately the same as V_{GS1} when the input photocurrent is small. The value of R7 and the negative supply voltage set the drain current of Q3. With a known drain current of Q3 and the resistor value of R6, the value of V_{DS3} and V_{GS3} can then be determined. Again, this value of V_{GS3} and resistor R5 can be used to calculate the collector current of Q2 and complete the DC feedback loop.

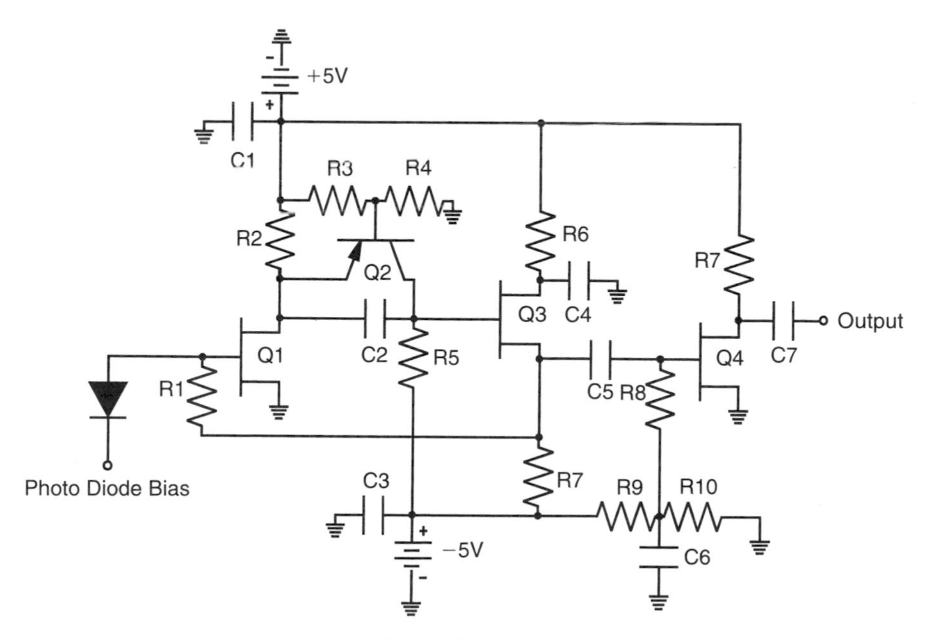

Fig. 11.4 Schematic of a 10-Gb/s receiver front-end circuit.

The last stage FET, Q4, acts as a simple buffer amplifier that provides a 50-Ω output impedance match. In operation, the input photocurrent is mainly shunt through the feedback resistor R1. For a large amount of photocurrent, the voltage drop across resistor R1 will lower the drain current of Q3 and eventually cause this stage to run into saturation. Equivalently, the front-end circuit is overloaded, or at its dynamic range limit.

Ogawa's circuit was originally designed for a relatively lower bit-rate application. In its original form, an NPN transistor (emitter follower), instead of an FET (source follower), was chosen for Q3, and the capacitor C2 was not included. The PNP transistor Q2 not only provides a bridge in the DC bias scheme but also is part of the input cascode amplification stage (Q1 and Q2). This type of circuit has been used successfully in commercial applications for data speed up to the OC-48 rate (2.488 Gb/s). However, the major difficulty of extending this circuit to a higher bit rate is the lack of availability of high-speed PNP transistors. For 10-Gb/s applications, we have incorporated a shunt capacitor (C2) in parallel with Q2. The DC bias condition is still established through Q2 while the RF signal is transferred through capacitor C2. As for Q3, the FET or HEMT (high electron mobility transistor) is preferred over the HBT (heterojunction bipolar transistor) because of the higher input impedance value, particularly at the low-frequency range. Therefore, we do not need to use a large value capacitor for C2 in order to achieve an acceptable low-frequency cutoff.

High-quality pseudomorphic HEMTs are used for the FETs indicated in Fig. 11.4 to construct an optical receiver front end for 10-Gb/s applications. In selecting the photodetector, one would have to consider not only the electrical parameters such as the junction capacitance and dark current, but also the practical parameters such as the reliability, quantum efficiency, coupling efficiency, and ease of assembling. In this example, we chose the PIN diode with a 50-μm-diameter active area.[9] The diode capacitance (including the stray capacitance of the photodiode package) is about 0.12 pF, and the combined value of the coupling efficiency and quantum efficiency is about 93%. Figure 11.5 shows the measured small signal frequency responses (both amplitude and phase) and the equivalent input noise current spectral density. This receiver front end has an effective transimpedance gain of about 650 Ω, an amplitude bandwidth (3 dB) of about 12.5 GHz, a linear phase bandwidth (20°) of about 10 GHz, and a maximum optical signal overloading capability of about +2 dBm.

To match the operation of this receiver front end, the required gain controllable range of the postamplifier is about 40 dB. If we refer to Fig.

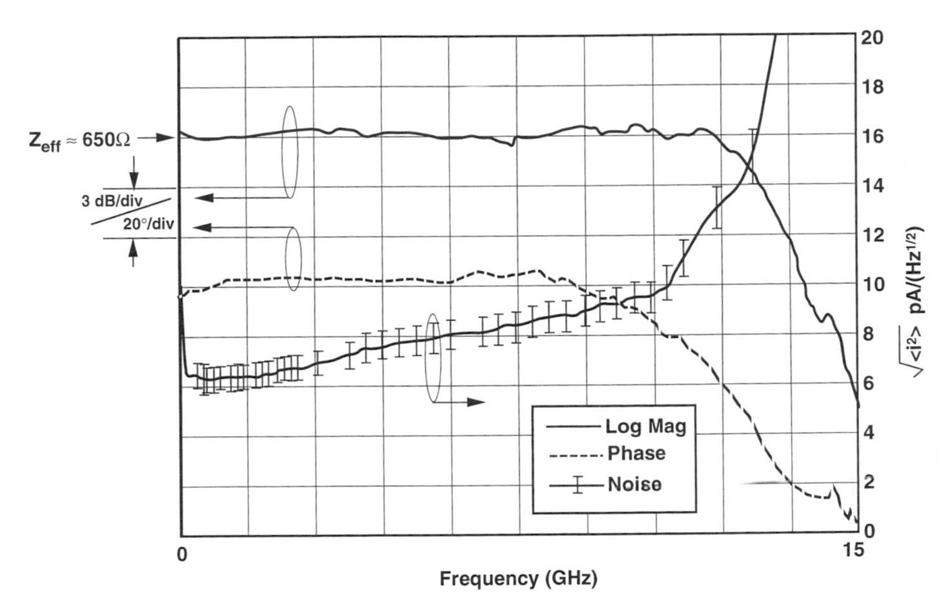

Fig. 11.5 Measured receiver frequency response and noise spectral density. The *solid curve* is for the amplitude frequency response (3 dB/div.), and the *dashed curve* is for the phase frequency response (20°/div.) The *curve with the error bars* is the measured input current noise spectral density with the units shown by the abscissa at the right.

11.1, we see that this can be accomplished by using two fixed gain blocks with a power gain of 20 dB each and two variable attenuators, also with a variable attenuation of about 20 dB each. Figures 11.6a and 11.6b show the amplitude and phase frequency responses of the VGA with several different overall gain settings. A complete model includes the receiver front end, and the VGA is shown in Fig. 11.7.

11.2.4 RECEIVER NOISE AND SENSITIVITY ESTIMATION

The measured total input noise current power for this receiver linear channel setup is about 9×10^{-13} (A^2). Without including any possible ISI penalty, the *expected* PIN receiver sensitivity at a BER of 1×10^{-9} should be about -23 dBm for a signal wavelength of 1.55 μm. Figure 11.8 shows the measured receiver sensitivity with (solid squares) and without (solid circles) including an optical preamplifier. For the case of the PIN receiver, the measured sensitivity is about 1.2 dB away from the expectation. This finding

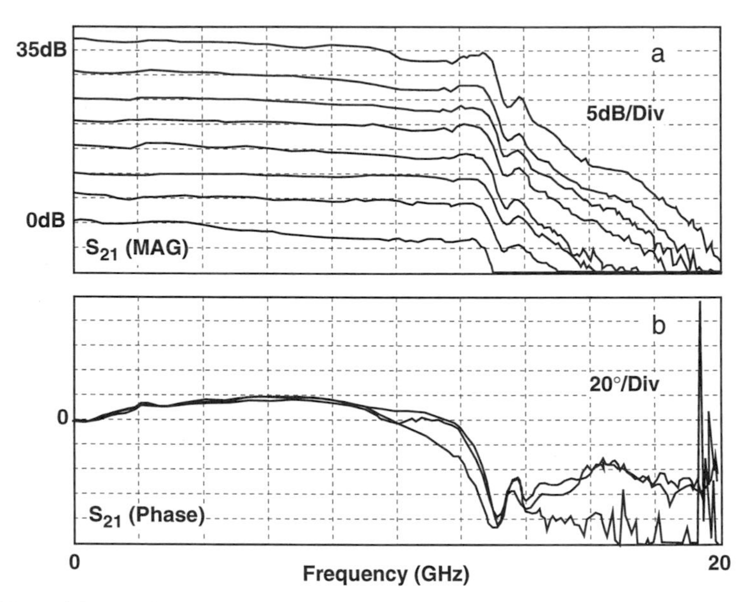

Fig. 11.6 (a) Amplitude frequency response and (b) phase frequency response of a variable gain amplifier (VGA) with different overall gain values.

is consistent with the estimation of about 1 dB of ISI due to a finite linear phase bandwidth and a 40-ps transmitter t_{rf}.

To estimate the receiver sensitivity with an EDFA optical preamplifier, we first calculate the various noise terms, as functions of input optical power, according to the work of Olsson.[10] It should be mentioned that in Olsson's original formulas, the optical filter and electrical filter (i.e., the receiver frequency response) are both treated as having a square shape. In reality, to account for the effect of practically achievable filter band shape, one can incorporate filter factors, in the form of Personick integrals[6] for the various frequency-dependent terms. For now, we assume that the effective bandwidth for the optical filter is about 42 GHz and that the bandwidth for the electrical filter is about 9.5 GHz (convoluted bandwidth including the front end and the postamplifier). The other parameters used for the optical preamplifier receiver sensitivity calculation are:[11] (1) the EDFA amplifier gain is about 37 dB, (2) the noise figure of the EDFA is about 3.5 dB, (3) the insertion loss for the optical amplifier is assumed to be zero, (4) the effective quantum efficiency (including the coupling efficiency) of the photodetector is about 93%, (5) the total circuit noise of

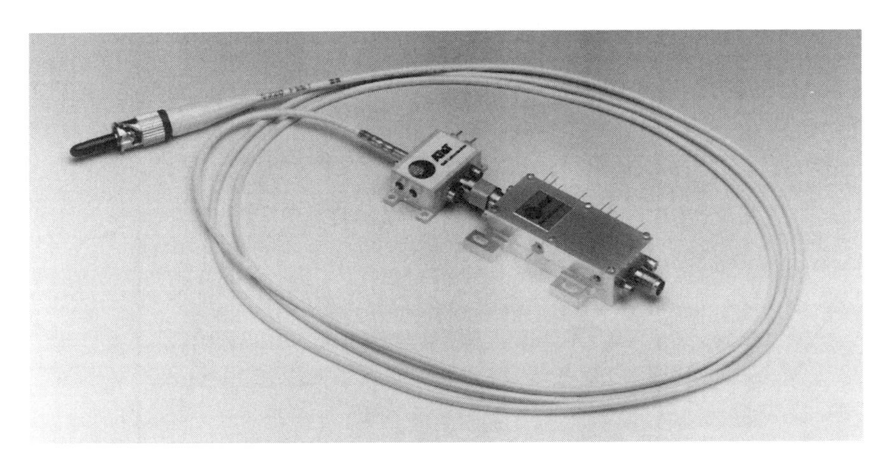

Fig. 11.7 Model of an optical front end and VGA assembly for 10-Gb/s system applications.

the receiver is 9×10^{-13} A^2, and (6) the transmitter (with a $Ti{:}LiNbO_3$ external modulator) extinction ratio is about 15 dB.

According to these parameters, the various noise terms are plotted in Fig. 11.9 for an input optical power range from −40 to −20 dBm, where $N_{S\text{-}SP}(1)$ and $N_{S\text{-}SP}(0)$ represent the total noise power of the signal–spontaneous beat noise terms for mark and space (one and zero), respectively. Similarly, $N_{SHOT}(1)$ and $N_{SHOT}(0)$ are for the corresponding shot noise terms. N_{SPSP} represents the total noise power for the spontaneous–spontaneous beat noise and N_{RX} represents the total receiver circuit noise power. It is clearly shown that the most dominant noise terms are $N_{S\text{-}SP}(1)$ and $N_{S\text{-}SP}(0)$. From these noise terms, the *expected* receiver sensitivity with such an optical preamplifier is about −38.2 dBm. The best measured result, as indicated in Fig. 11.8, shows a sensitivity of about −37 dBm.

11.3 System Performance

11.3.1 SYSTEM PERFORMANCE MEASURE

In many digital transmission systems, receiver sensitivity is frequently quoted as a primary measure of system performance. However, from a practical point of view, opening (or closure) of the eye diagram of the received signal[12] is equally, if not more, important because it represents a

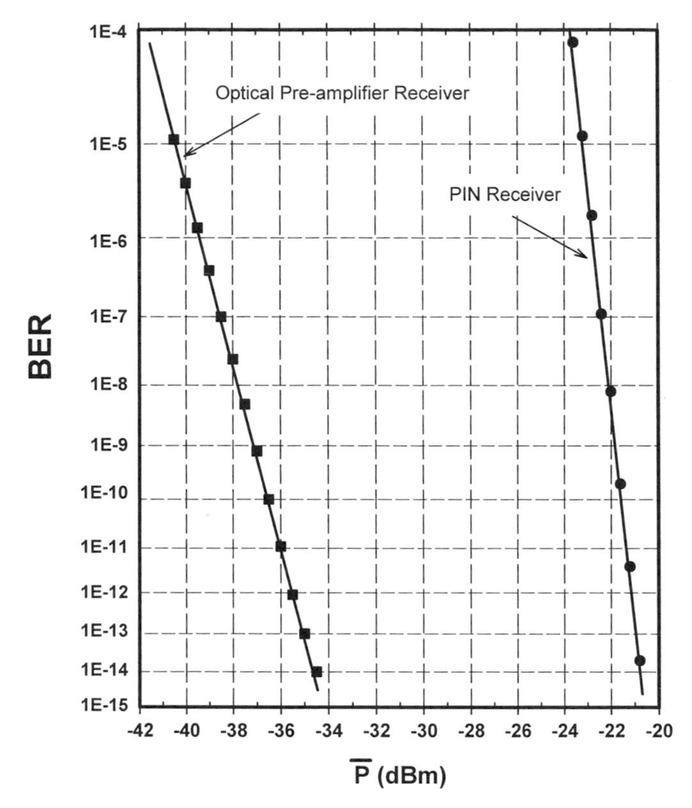

Fig. 11.8 PIN receiver sensitivity measurements with (*solid squares*) and without (*solid circles*) including an optical preamplifier.

direct measure of the system's performance degradation caused by transmission impairment, environmental change, and aging. An eye diagram measured by directly coupling the output signal of a 10-Gb/s transmitter into a receiver (referring to back-to-back measurement) and an eye diagram of the same signal after transmission through some length of dispersive fiber and optical amplifiers are shown in Figs. 11.10a and 11.10b, respectively. It is clearly seen that the fully opened eye diagram (Fig. 11.10a) closes significantly (Fig. 11.10b) as a result of the amplifier spontaneous noise (ASE) of the optical amplifiers and the ISI caused by the dispersion effect.

A more quantitative expression of the eye opening is the use of the eye margin and the phase margin (see Section 11.2.1). For the regenerators employing an AGC postamplifier, the eye margin can be derived from the plot of the BER as a function of threshold voltage (V_{th}) of the integrated decision circuit at an optimum decision timing position. Similarly, the phase

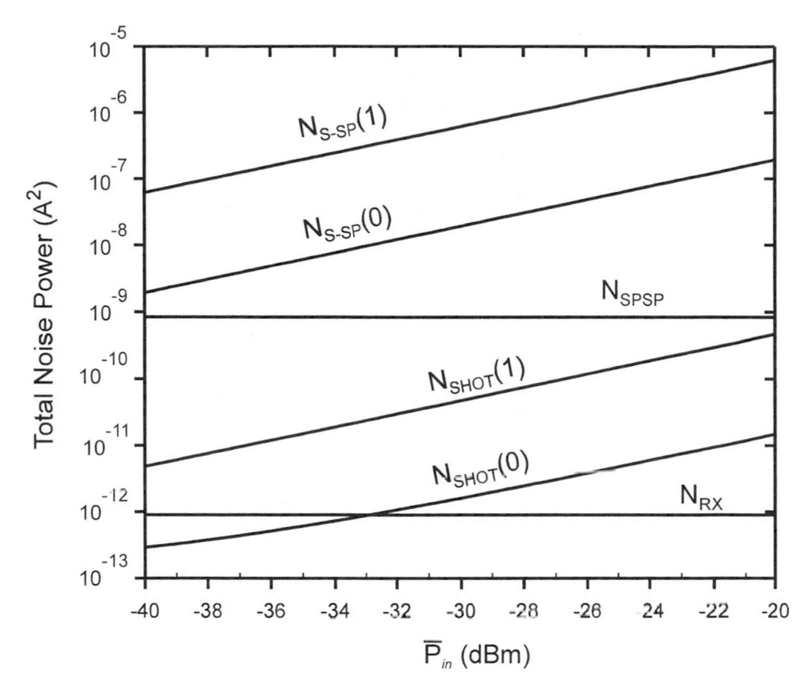

Fig. 11.9 Noise terms for the optical preamplifier receiver versus input optical power. The system parameters and the nomenclature are explained in the text.

margin is derived from the plot of the BER versus the phase of the clock signal applied to the decision circuit at the optimum V_{th}. For example, the actual measurement of the eye margin is accomplished by vertically moving the rectangular box (which represents the decision circuit with finite ambiguity in threshold voltage and phase) inside the eye diagrams in Fig. 11.10 at a constant received optical power. Figure 11.11 shows the measured eye margin of the eye diagrams in Fig. 11.10 at a -8 dBm received optical power. The back-to-back measurement (*a*s) shows an eye opening of 440 mV for a 1×10^{-15} BER, which corresponds to a 55% eye margin (the rail-to-rail voltage of the signal at the decision circuit is 800 mV). The main sources of this eye closure from the ideal 100% eye margin are a finite extinction ratio of the transmitter output signal, receiver circuit noise, decision circuit ambiguity, and ISI penalties caused by nonideal frequency response characteristics of the transmitter and the receiver–regenerator. As is discussed later, it is important to design the transmitter and the receiver–regenerator hardware to have the largest possible back-to-back

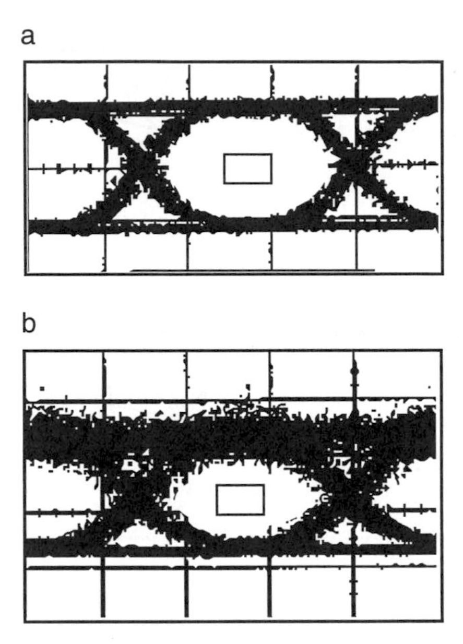

Fig. 11.10 Eye diagram of a 10-Gb/s optical data signal measured for (a) the back-to-back operation of the transmitter and receiver and (b) after the signal is transmitted through the fiber and optical amplifiers.

eye margin. The transmitted signal (*b*s) shows an eye margin of 30%. The 25% reduction from the back-to-back measurement is due to transmission impairment such as amplifier ASE noise and the ISI penalties caused by the fiber dispersion and nonlinear self-phase modulation (SPM).

11.3.2 OPTIMUM DECISION THRESHOLD SETTING

In Fig. 11.11 we notice that the eye closure of the transmitted signal is more severe at the one level than at the zero level. This is because, in most optical amplifier transmission systems, the main noise term is the signal–spontaneous beat noise, which is a signal-dependent noise. Therefore, the optimum V_{th} (defined as the midpoint of the eye opening) must be shifted toward the zero level in order to ensure "error-free" operation (defined as a less than 1×10^{-15} BER) with the maximum system tolerance — that is, a shift from V_{th}(a) to V_{th}(b) in Fig. 11.11.[13] This asymmetrical behavior of the eye closure is very common, even in the transmission systems where ISI penalties are larger than the ASE noise penalties. This

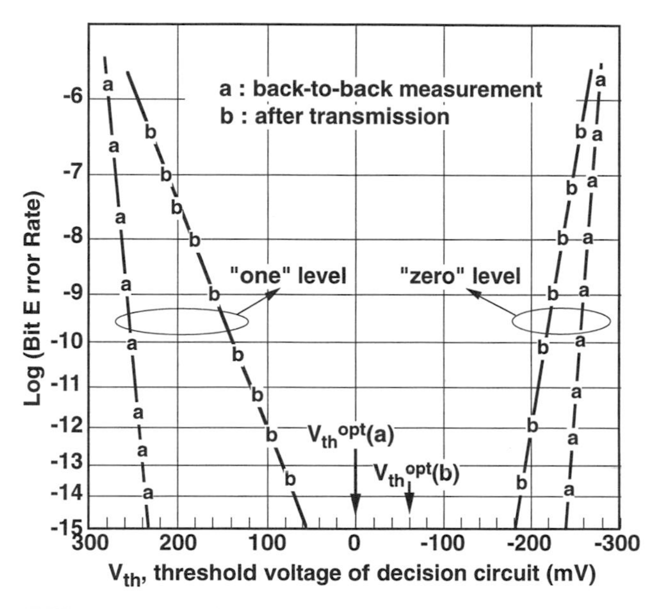

Fig. 11.11 BER as a function of the decision threshold voltage (V_{th}) at the optimum phase position. The eye margin and the optimum V_{th} are derived from these plots.

implies that the optimum V_{th} must be reset to a new value whenever the transmission distance of the system is changed.

Figure 11.12 shows an example of the effect of the optimum V_{th} setting on the receiver sensitivity for a transmission system with the dispersion effect only. In this experiment, an integrated electroabsorption (EA) modulator–laser transmitter operating at 10 Gb/s and a PIN receiver–regenerator are used with conventional non-dispersion-shifted fibers (non-DSFs). The receiver sensitivities measured with the V_{th} optimized for each fiber length are plotted with circles. Naturally, this case offers the best sensitivity at each fiber length. The receiver sensitivities measured for the V_{th} optimized at 65 km are 1 dB (or less) worse than the best case in the measured range from 0 to 65 km. On the other hand, the receiver sensitivities with the V_{th} optimized for the back-to-back operation yield huge sensitivity penalties, particularly for the long transmission distance. From a practical standpoint, it is very inconvenient to constantly reoptimize V_{th} whenever the system's transmission distance is changed. It is more practical to maintain a predetermined V_{th} throughout the entire range of the transmission distance. The result shown in Fig. 11.12, then, clearly indicates that the initial setting of

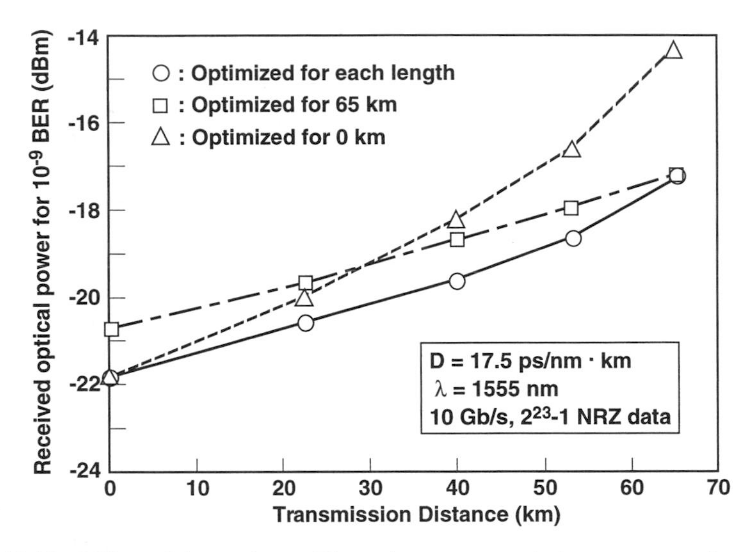

Fig. 11.12 Effect of the optimum V_{th} setting on transmission system performance.

V_{th} should be done for the longest possible transmission distance, not for the 0-km fiber length.

11.3.3 SYSTEM PERFORMANCE RESULTS

The progress of high bit-rate transmission systems (10 Gb/s and beyond) has been advancing so rapidly that many experimental results obtained from the laboratories become obsolete within a couple of months after they are reported. Therefore, instead of describing those laboratory results, it makes more sense to discuss the results obtained from the field because the field experiments offer important milestones toward the actual deployment of a transmission system. In many cases, the successful result of the field experiments is considered a better measure of technology advancement than that of the laboratory experiments because it serves as solid evidence that the system has sufficient system margin and that the prototype hardware under testing is sufficiently stable and robust to meet the performance requirements in the adverse field environment. So far, the highest bit rate that has been attempted in the field experiments is a 10-Gb/s line rate. In the following sections, some of the 10-Gb/s field experiments and their performance results are described, with emphasis on the eye margin and the optimum V_{th} setting.

11.3.4 TERRESTRIAL 10-Gb/s FIELD EXPERIMENTS

Early field experiments of 10-Gb/s terrestrial transmission have been performed on newly installed DSFs, which have negligible dispersion in the 1550-nm wavelength region.[14] Unfortunately, most of the installed fiber lines underground are non-DSFs, which limits the 10-Gb/s line rate transmission to less than 50 km because of high chromatic dispersion (see Chapter 7 in Volume IIIA). This problem can be alleviated significantly by using a new optical repeater concept that simultaneously compensates for chromatic dispersion and loss of the fiber. Equally important is designing the transmitter–receiver hardware to have the maximum possible back-to-back eye margin because the large back-to-back eye margin can withstand more eye closure.

A configuration of the field transmission experiments conducted on the embedded fiber cable connecting between AT&T's Roaring Creek and Dalmatia stations in Pennsylvania[15] is shown in Fig. 11.13. In addition, Fig. 11.14 shows one of the shelves under testing. It contains a 10-Gb/s transmitter, a receiver–regenerator, and an optical amplifier. The transmitter is based on a Ti:$LiNbO_3$ modulator and driven by 10-Gb/s, 2^{23}-1 NRZ

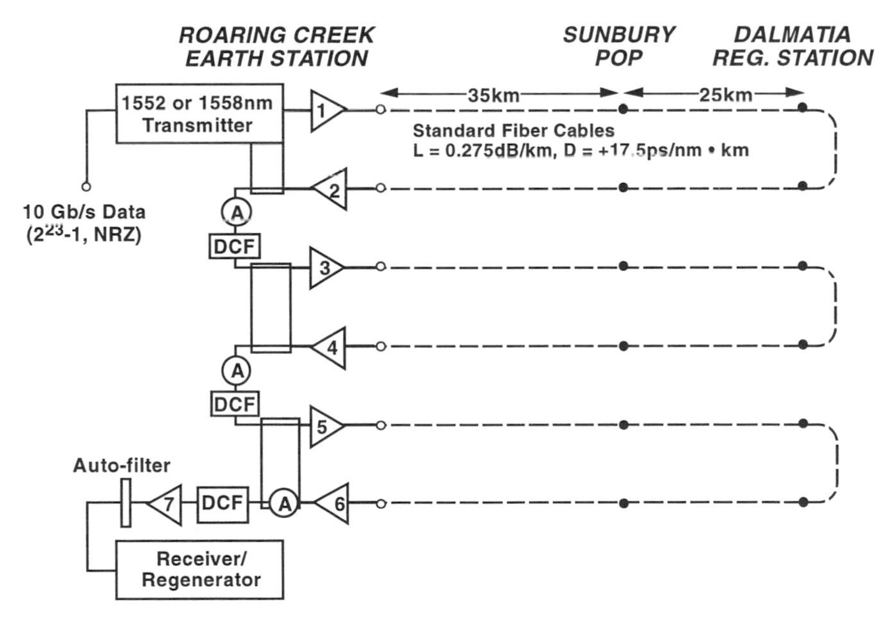

Fig. 11.13 Configuration of a 10-Gb/s terrestrial field experiment using non-dispersion-shifted fibers (non-DSFs). DCF, dispersion-compensating fiber.

Fig. 11.14 Prototype 10-Gb/s transmitter, receiver–regenerator, and optical amplifier circuit packs tested in the field experiment.

PRBS (non-return to zone pseudo random bit sequence) data. Each optical repeater consists of a compact dispersion-compensating fiber (DCF) circuit pack sandwiched between two EDFA modules. The repeater spacing is 120 km (three times the typical electronic repeater spacing), with typical fiber dispersion and loss of +2100 ps/nm and 33 dB at 1558 nm, respectively. Therefore, the total transmission distance, dispersion, and loss of the embedded non-DCF line in the system are 360 km, +6300 ps/nm, and 99 dB, respectively.

Figure 11.15 shows eye diagrams of the data signal at various locations along the 360-km path. The nearly full restoration of the eye opening after each repeater clearly demonstrates the effectiveness of the chromatic dispersion compensation by the DCFs. The eye margin and the long-term BER performance of the system measured after 360-km transmission are shown, and compared with the back-to-back operation, in Figs. 11.16a and 11.16b. The eye margin (measured for a 1×10^{-15} BER) is decreased from 55 to 20% after 360 km, mainly because of the degradation of the optical signal-to-noise ratio (SNR) as evidenced from the severe eye closure at the one level. The 360-km BER performance clearly shows "error-free" operation for 11 consecutive days (corresponding to a less than 1×10^{-16} BER) even though it experiences a slight degradation from the baseline

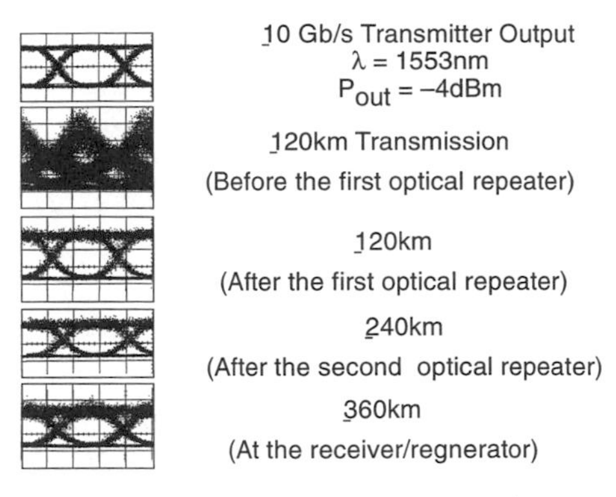

Fig. 11.15 Eye diagrams of the 10-Gb/s signal at each repeater along the 360-km span.

back-to-back performance. Note that the BER data for 0 and 360 km in Fig. 11.16b were taken at a fixed V_{th} (i.e., at $V_{th} = -120$ mV), which was optimized for the 360-km transmission. If both BER data were taken with the V_{th} optimized for 0 km (i.e., at $V_{th} = 0$ mV), the 360-km performance would suffer a BER floor at 1×10^{-13} instead.

A plot of the optical SNR at each optical amplifier along the 360-km path is shown in Fig. 11.17. The SNR values measured for 0.1-nm optical bandwidth (B_o) agree well with a theoretical expression of SNR:

$$\frac{1}{[(\mathrm{SNR})_{\text{after } N_{th} \text{ amplifier}}]} = \sum_{j=1}^{n} \frac{2h\nu B_o N_{sp}(j)}{P_j^{in}}, \tag{11.1}$$

where $N_{sp}(j)$ and P_j^{in} are the spontaneous noise factor and the input power of the jth amplifier, respectively. Although this is a very simple expression, it is very effective to predict the degradation of the optical SNR along the transmission path in optical amplifier systems.

The large eye margin of the 10-Gb/s system also permits two 10-Gb/s WDM transmissions through the same embedded 360-km-long fiber cable. In this field experiment,[16] two pairs of prototype transmitters and receiver–regenerators are used with the same optical repeaters and the repeater spacing as the ones in the single-channel experiment. Figure 11.18 shows typical spectra (1552 and 1558 nm) and eye diagrams of those two channels after 360-km transmission. The BER performance of both channels is shown in Fig. 11.19. The "error-free" operation of both channels for 5 consecutive days clearly indicates the system's high stability and robustness. The optical

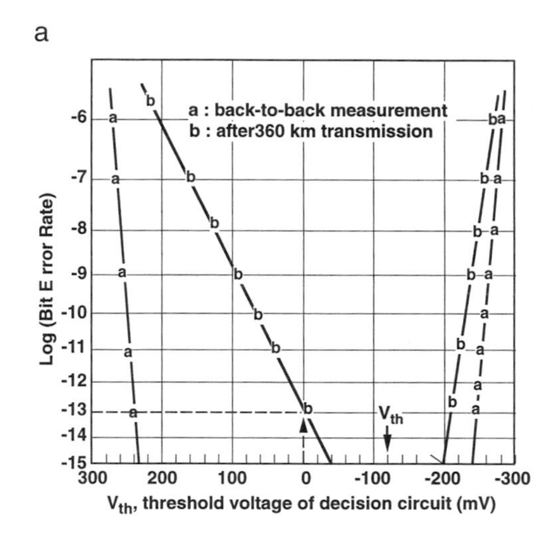

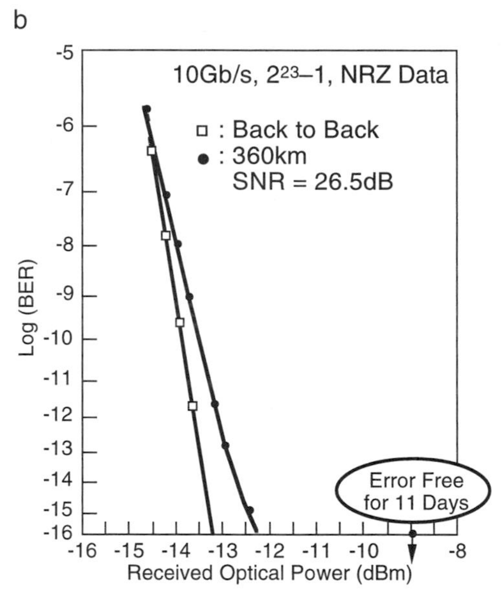

Fig. 11.16 (a) BER versus decision threshold voltage and (b) long-term BER performance for the 360-km operation. The results of the back-to-back operation are compared. SNR, signal-to-noise ratio.

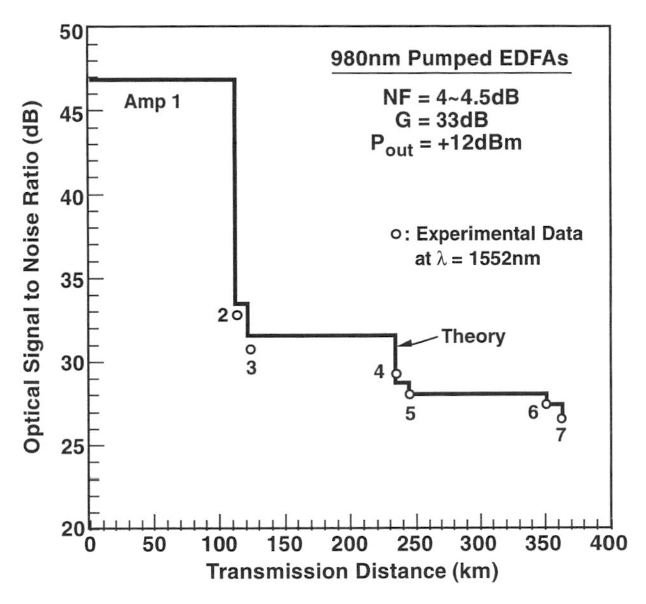

Fig. 11.17 Optical SNR after each amplifier along the 360-km path. NF, noise figure.

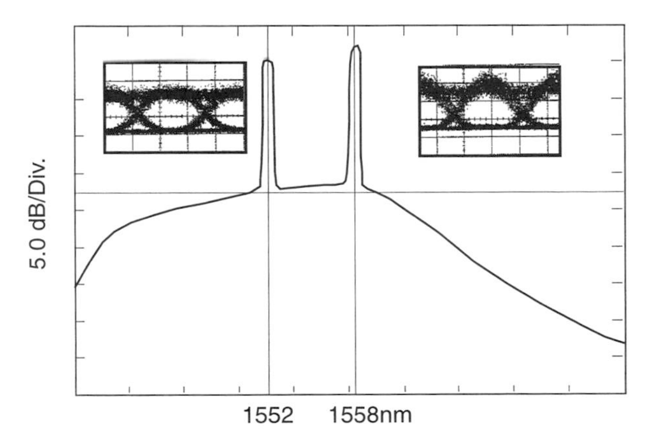

Fig. 11.18 Typical spectra and eye diagrams for two 10-Gb/s wavelength-division multiplexed (WDM) channels after 360-km transmission.

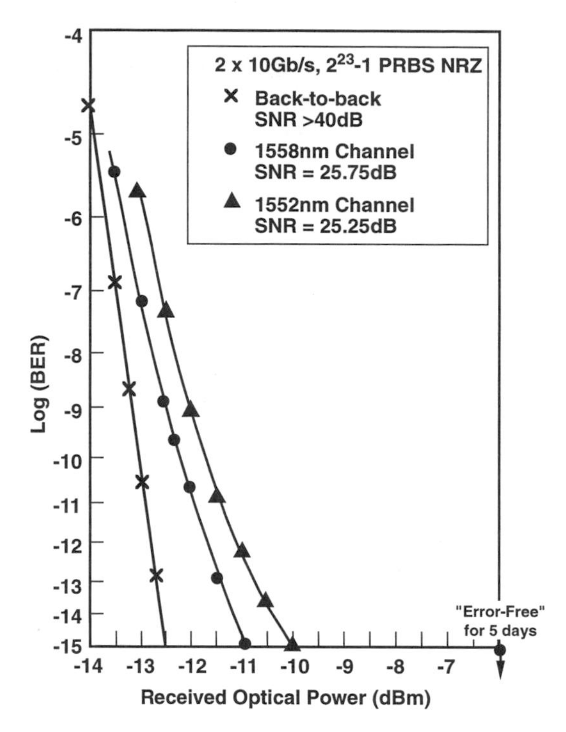

Fig. 11.19 BER performance of two 10-Gb/s WDM transmissions through the 360-km embedded non-DSF.

SNRs for the 1552- and 1558-nm channels are 25.25 and 25.75 dB, respectively, which are again consistent with the prediction from the previously mentioned optical SNR expression. It should also be pointed out that the effects of wavelength cross talk and unequal dispersion compensation between the two channels are negligible for the channel separation ranged from 1 to 6 nm.

11.3.5 UNDERSEA 10-Gb/s FIELD EXPERIMENTS

Figure 11.20 depicts a configuration of the undersea field experiments conducted in 1995 on the newly installed Trans-Pacific Cable (TPC-5) G section connecting San Luis Obispo, California, with Keawaula, Hawaii. The TPC-5G section is 4230 km long with optical amplifier repeaters spaced 68 km apart.[17] The prototype 10-Gb/s transmitters and receiver–regenerators are similar to the ones used in the terrestrial field experi-

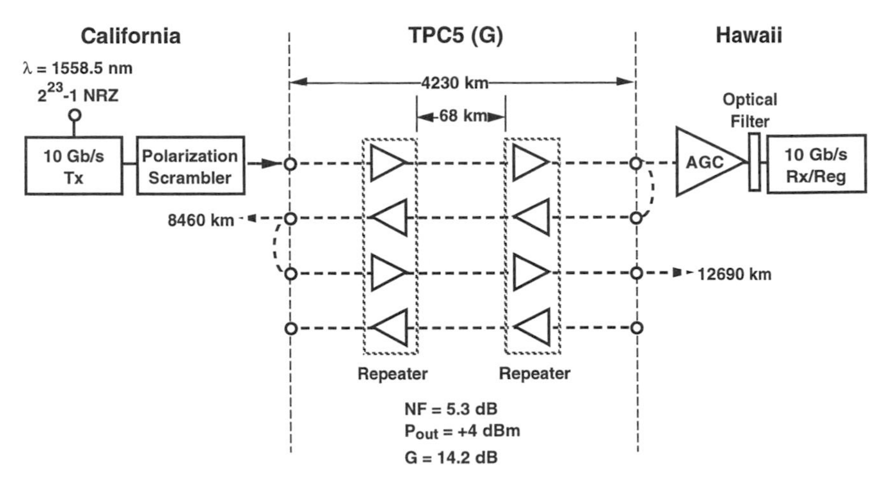

Fig. 11.20 Configuration of the undersea 10-Gb/s field experiment on the newly installed 4230-km-long TPC-5G section fiber cable.

ments. The transmitter is operated at a 1558.5-nm wavelength, where the dispersion of the installed fibers is near minimum and the amplifier chain has its gain peak. In addition, a high-speed synchronous polarization scrambler is used to reduce the polarization hole burning (PHB) effect[18] (see Chapter 6 in Volume IIIA).

Figure 11.21 shows the BER versus received optical power at the preamplifier receiver and an eye diagram after 4230 km. Reducing the input power to the preamplifier results in a decrease of the optical SNR at the p-i-n receiver. Long-term "error-free" operation is obtained at $P^{in} = -24.5$ dBm. The measured optical SNR, 19.3 dB, at the receiver agrees well with the prediction from the SNR expression:

$$\frac{1}{[(\mathrm{SNR})_{\text{after 62 amplifiers}}} = 124hB_oN_{sp}/P^{in}. \quad (11.2)$$

Figure 11.22 shows the BER versus the V_{th}, indicating that the eye margins for the back-to-back operation and the 4230-km transmission are 48 and 31%, respectively, for a 1×10^{-15} BER. Again, there is more eye closure at the one level than at the zero level, and naturally the optimum V_{th} is set at -30 mV for the long-term operation. The data for the 8460- and 12,690-km transmissions measured using loop-back configurations show complete eye closure (i.e., a BER floor) at a 1×10^{-11} and a 1×10^{-5} BER, respectively. In terms of the Q value, which is often used as a measure

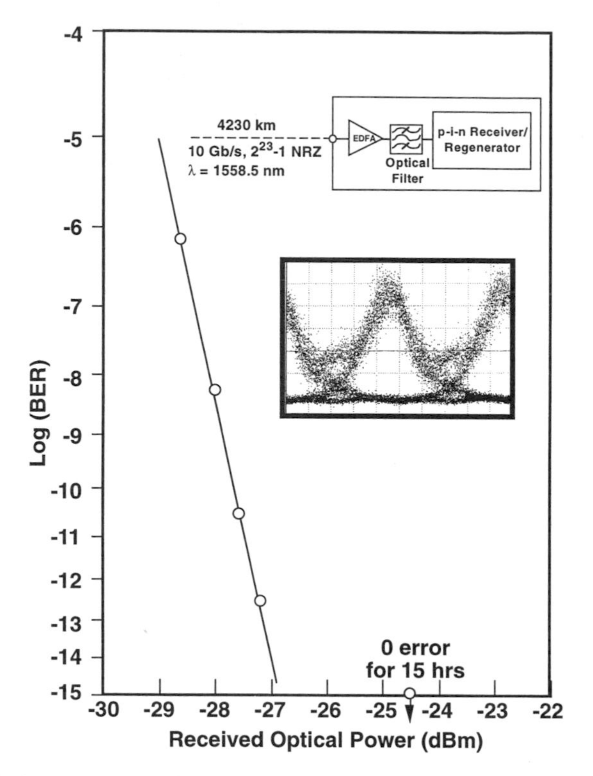

Fig. 11.21 BER versus received optical power at the preamplifier receiver and an eye diagram after 4230-km transmission at 10 Gb/s.

of undersea system performance (see Chapter 10 in Volume IIIA), it is 20, 16.5, and 12.5 dB for 4230-, 8460-, and 12,390-km transmission, respectively.

11.3.6 SOURCES OF SYSTEM PERFORMANCE DEGRADATION

In Section 11.3.1, we pointed out that the limited eye margin of the back-to-back operation is attributed solely to the characteristics of the transmitter and receiver–regenerator pair and that transmission impairment further reduces the eye margin. The main sources of this eye margin reduction in typical optical amplifier transmission systems are the accumulated ASE noise of the amplifiers and the ISI penalty caused by the fiber dispersion and SPM. In order to identify the sources of the performance degradation of the transmission system, we need to separate the ISI contribution from the ASE noise contribution to the performance degradation. A straightforward method is shown in Fig. 11.23. In this method, we plot the BER floor (i.e., lowest attain-

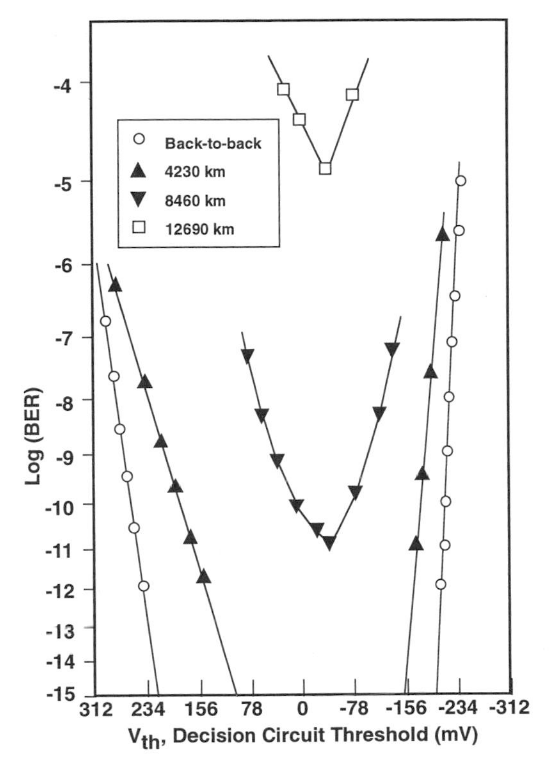

Fig. 11.22 BER versus decision threshold voltage for 0-, 4230-, 8460-, and 12,690-km transmission of 10-Gb/s data.

able BER) as a function of the optical SNR at the receiver, with and without the 360-km transmission fiber in the terrestrial 10-Gb/s field experiment. The BER floors are determined from the BER versus V_{th} plots for each optical SNR value. It shows that at a given optical SNR (e.g., at 24 dB) the floor level for the 360-km transmission rises from a1×10^{-15} (for the back-to-back operation) to a 1×10^{-13} BER. Another way to look at it is that the minimum required optical SNR must be increased from 24 to 25 dB to ensure a less than 1×10^{-15} BER for the 360-km transmission. This degradation is attributed to the ISI caused by the SPM effect and imperfect dispersion compensation for the 360-km-long non-DSF.

Similar plots obtained from the undersea 10-Gb/s field experiment are shown in Fig. 11.24. This time, contrary to the results shown in Fig. 11.23, the minimum required optical SNR to obtain a BER floor less than 1×10^{-15} for the 4230-km transmission is decreased by approximately 4 dB from that for the back-to-back operation. That is, for a given optical SNR,

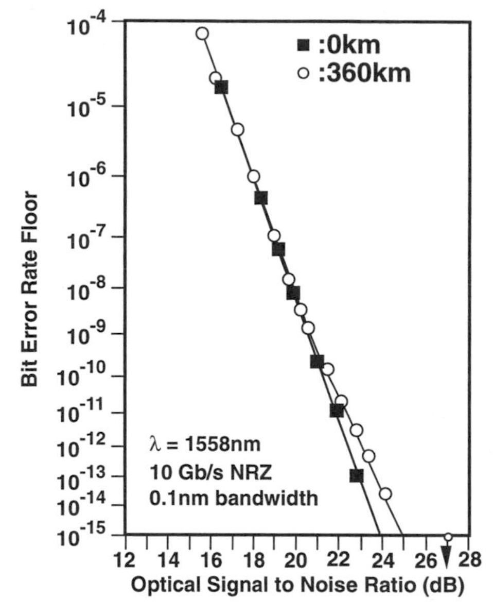

Fig. 11.23 BER floor versus the optical SNR for a terrestrial 10-Gb/s field experiment over 360-km embedded non-DSF.

the eye margin after 4230-km transmission is larger than that of the back-to-back operation (called a *negative ISI penalty*). This is because the SPM in the fiber compresses the NRZ pulse and, hence, opens the eye wider. This point is clearly seen from the eye diagrams (insets), both of which were taken at the optical SNR of 19.3 dB.

11.4 High-Speed Electronics

Very high-speed optical fiber transmission experiments have been extensively conducted, and speeds have reached 100 Gb/s. Such extremely high-speed transmission resulted from the introduction of the EDFA and all-optical multiplexer–demultiplexer technologies. Nevertheless, as mentioned in Section 11.1, electronic devices continue to have some advantages over optical devices, namely high functionality, small size, low cost, and high reliability. Thus, our main interest here is to investigate the use of electronic devices for the highest possible bit rates. We have fabricated a family of 10-Gb/s integrated circuits using self-aligned GaAs MESFETs

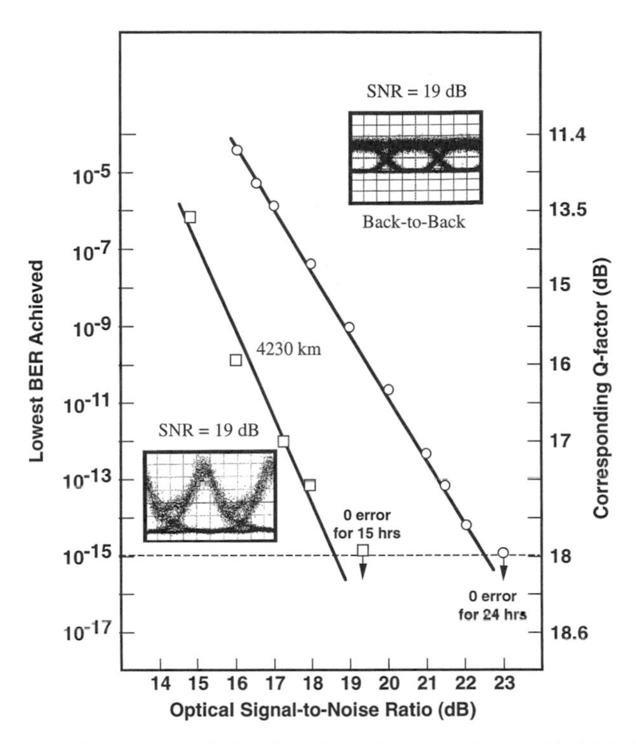

Fig. 11.24 BER floor versus the optical SNR for an undersea 10-Gb/s field experiment over 4230 km of installed fiber.

with a 0.2-μm gate length, and we successfully used this in a field transmission experiment.[14] Silicon bipolar transistors (BJTs) can also be applied to 10-Gb/s lightwave communications integrated circuits.[19] A lot of work has been conducted on more than 20-Gb/s lightwave communications ntegrated circuits using BJTs, MESFETs, HBTs, and high electron mobility transistors (HEMTs). Figure 11.25 shows the highest operating speeds reported to date for lightwave communications integrated circuits, which have reached 20–50 Gb/s.[20] In this section, we discuss the limitations of electronic devices and provide a perspective of integrated circuit performance.

11.4.1 LIMITATIONS OF DEVICES

The performance of FETs and HEMTs has improved mainly as a result of reducing the gate lengths of devices. Figure 11.26 shows the dependence of current gain cutoff frequency f_T on gate length reported for GaAs MES-

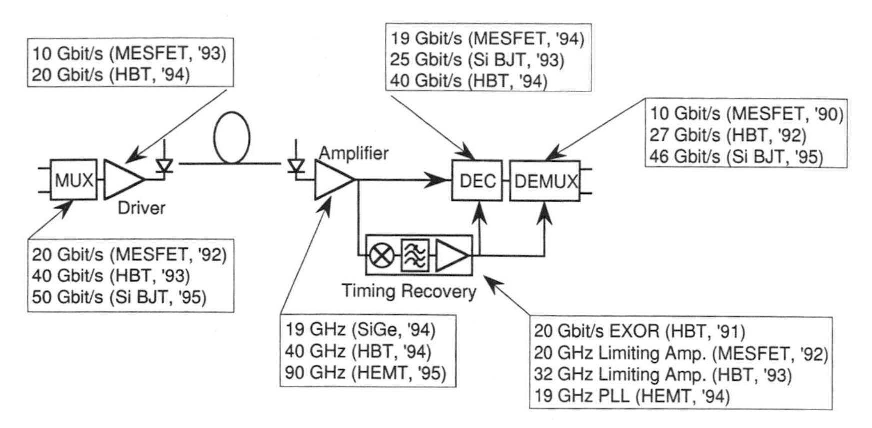

Fig. 11.25 State-of-the-art lightwave communications integrated circuit performance. BJT, bipolar transistor; HBT, heterojunction bipolar transistor; HEMT, high electron mobility transistor.

FETs, GaAs-based HEMTs, and InP-based HEMTs. InP-based HEMTs give the highest f_T of all the devices because of their larger electron velocity and smaller short-channel effect (reductions in threshold voltage and transconductance with decreasing gate length). A high f_T value of 343 GHz has been reported for an InP pseudomorphic HEMT with a gate length of 0.05 μm.[21] From Fig. 11.26, we can expect an f_T of 400 GHz by scaling

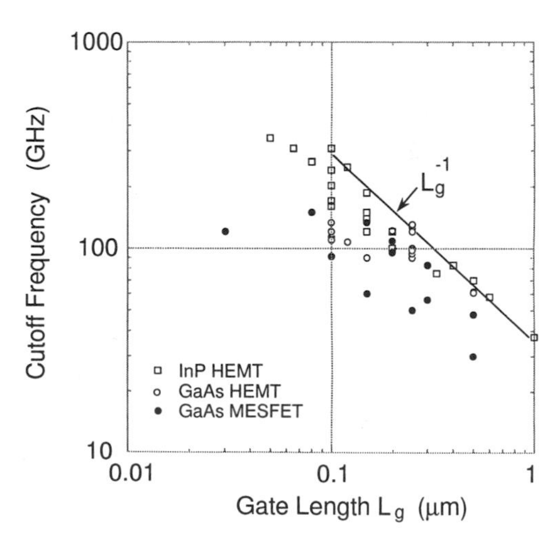

Fig. 11.26 Dependence of current gain cutoff frequency f_T on gate length.

down the gate length to 0.03 μm. InP-based HBTs also have high potential and the f_T has reached 228 GHz.[22] Both f_T and f_{max} (maximum oscillation frequency) over 300 GHz should be obtainable by optimizing the structure of the epitaxial layer and reducing the width of the emitter.[23] Not only RF performance such as f_T but also device breakdown voltage is important in producing integrated circuits. Figure 11.27 depicts the relationship between RF performance and breakdown voltage. A breakdown voltage of more than 3 V is needed for a digital integrated circuit with a logic swing of 1 V. An optical modulator driver demands a higher breakdown voltage, as is discussed later. Enhancing the breakdown voltages for InP-based HEMTs while maintaining RF performance is indispensable if ultra-high-speed integrated circuits are to be produced.

11.4.2 INTEGRATED CIRCUIT DESIGN

Emitter-coupled logic (ECL) and source-coupled FET logic (SCFL) are commonly employed in high-speed digital integrated circuits. The delay time for an ECL inverter is approximated as

$$t_{pd} \approx \alpha \left[\frac{1}{f_T} + \sqrt{2A}\, \frac{1}{f_{max}} + (2 + A)\, \frac{f_T}{4f_{max}^2} \right], \tag{11.3}$$

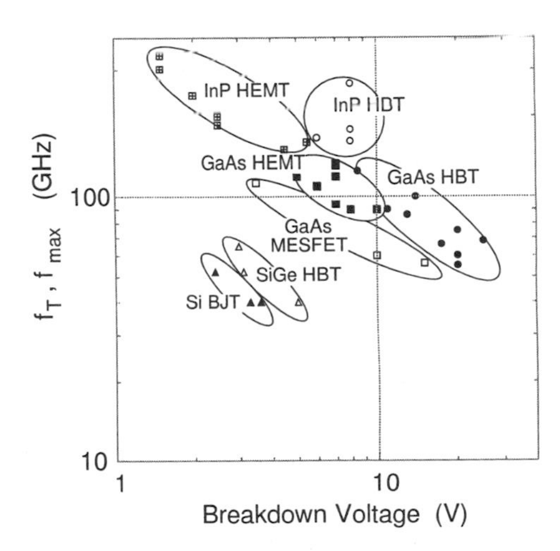

Fig. 11.27 Relationship between RF performance and breakdown voltage. The terms f_T and f_{max} are measured for FETs and bipolar transistors, respectively.

where A is the voltage gain of the current switch and α is a constant.[24] If the gate resistance and the drain conductance can be neglected, the delay time for an SCFL inverter is given by

$$t_{pd} \approx \frac{C_{gs}}{g_m} + (2 + 3A)\frac{C_{gd}}{g_m}, \tag{11.4}$$

where g_m is transconductance and C_{gs} and C_{gd} are the gate-to-source capacitance and gate-to-drain capacitance, respectively. For simplicity, if we take $2(f_T^{-1} + f_{max}^{-1})$ for bipolar transistors and f_T for FETs as device figures of merit, this leads to the relationship shown in Fig. 11.28 between device figures of merit and the maximum operating speeds of the D-type flip-flop (D-F/F) integrated circuits reported in previous studies. As pointed out by Ichino *et al.*,[25] the differences between Si BJT and GaAs HBT circuits might be due to the difference in circuit optimization levels. In fact, a double-feedback emitter-follower configuration is used, and circuit parameters are optimized to increase the operating speed of Si BJT D-F/F integrated circuits. From Fig. 11.28, we can expect that SiGe HBTs and InP HBTs will be applied to 40- and 100-Gb/s D-F/Fs, respectively, when they reach maturity. Not only high-performance devices but also new circuit configurations such as the HLO (high-speed latching operation) D-F/F

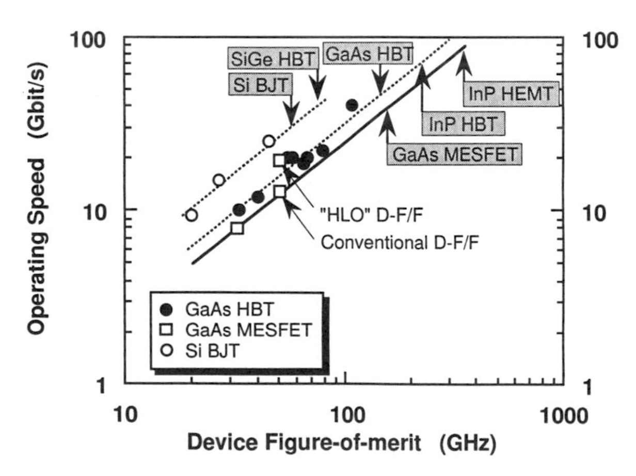

Fig. 11.28 Relationship between maximum operating speeds of D-type flip-flop (D-F/F) integrated circuits and device figures of merit. The terms $2(f_T^{-1} + f_{max}^{-1})^{-1}$ and f_T are measured for bipolar transistors and FETs, respectively. HLO, high-speed latching operation.

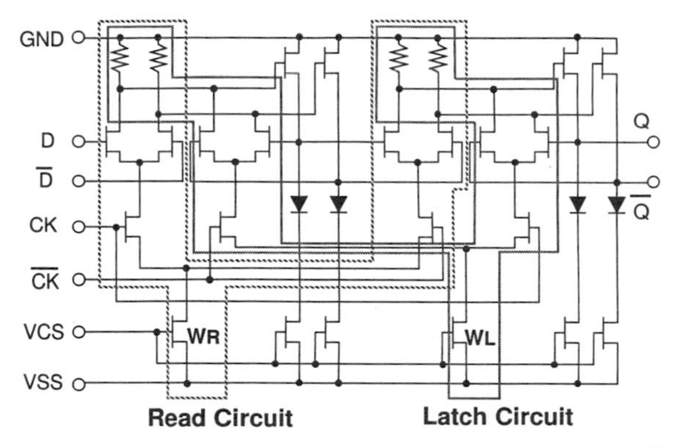

Fig. 11.29 Configuration of an HLO D-F/F circuit.[26]

shown in Fig. 11.29 are indispensable in achieving high-speed integrated circuits. The key feature of HLO D-F/F is that the transistors in the latch circuit have a smaller gate width than those in the read circuit. This reduces the logic swing and load capacitance, which in turn yields a higher speed than is possible with a conventional F/F. Experiments on 0.2-μm-gate GaAs MESFETs have demonstrated that this new circuit configuration is 1.5 times faster.[26] An operation of 100 Gb/s can be expected by using HLO D-F/F with InP-based HEMTs.

Base-band amplifiers are another important component in lightwave communication systems. Figure 11.30 shows the relationship between the device figures of merit and the bandwidth of base-band amplifiers having a gain greater than 15 dB. The figures of merit are again f_T for FETs but f_{max} for bipolar transistors. Although a 40-GHz bandwidth has been obtained for a GaAs HBT feedback amplifier,[27] a 100-GHz bandwidth cannot be achieved using any conventional amplifier. Innovative circuit technology is needed to reach the 100-GHz bandwidth. As a means to enhance the bandwidth, Kimura *et al.*[28] proposed the distributed base-band amplifier shown in Fig. 11.31. Distributed amplifiers are common in microwave and millimeter-wave applications. This is because they inherently have a wide bandwidth, which is determined by transistor capacitances and transmission line inductances. However, conventional distributed amplifiers are band limited and cannot be used as base-band amplifiers. One reason is that the gain in a conventional distributed amplifier is poor in the low-frequency region because of the drain conductance of FETs. In this special distributed base-band amplifier, frequency-dependent termination is introduced to

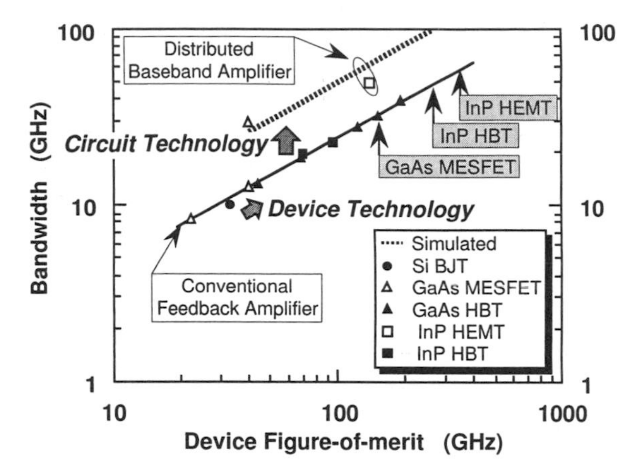

Fig. 11.30 Relationship between bandwidths of base-band amplifiers and device figures of merit. The terms f_{max} and f_T are measured for bipolar transistors and FETs, respectively.

compensate for the degradation and flattening of the gain-versus-frequency characteristics. In addition, a drain peaking line and a loss compensation circuit help to extend the bandwidth. These techniques provide flat-gain characteristics. They enable us to increase the number of sections in a distributed amplifier to improve gain while maintaining bandwidth. By using these techniques, Kimura *et al.*[29] demonstrated a DC to 90-GHz bandwidth with 0.1-μm-gate InP HEMTs. A bandwidth of more than 100 GHz can be obtained by using sub-0.1-μm-gate InP HEMTs.

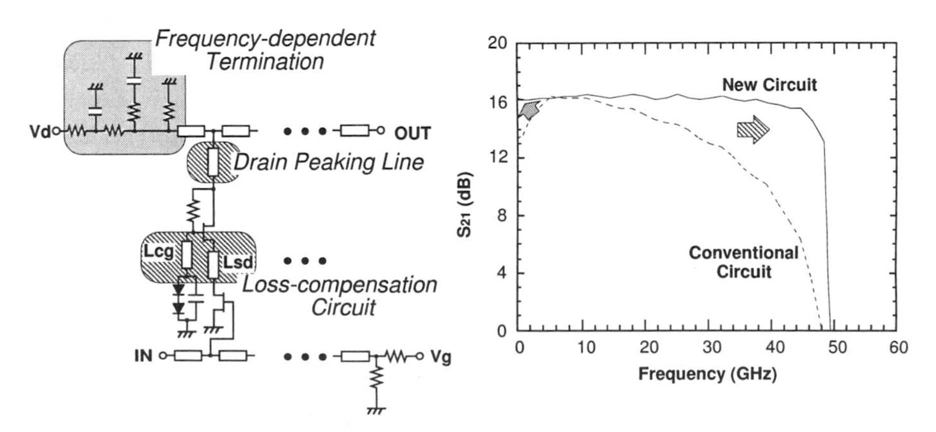

Fig. 11.31 Circuit configuration of a distributed base-band amplifier.[28]

Progress in optical modulator drivers having a large output voltage swing lags behind that of other lightwave communications integrated circuits. This is because of the trade-off shown in Fig. 11.27 between breakdown voltage and RF performance. Figure 11.32 plots modulator bandwidth f_{mod} and driver speed f_{drv} versus the driving voltage V_{drv}. Modulator and driver figures of merit are expressed as f_{mod}/V_{drv} and $f_{drv}V_{drv}$, respectively. The two lines in Fig. 11.32 indicate the highest figures of merit for modulators and drivers reported to date.[30,31] Their intersection shows the highest obtainable bit rate for optical modulation. The most reasonable way to increase the bit rate would seem to be to develop devices using a driving voltage of around 2 V. In terms of driving voltage, InP DHBTs are very attractive to obtain the higher bit rates, as indicated from Fig. 11.27.

As we discussed, it is necessary to choose device technologies from the point of view of reproducibility, reliability, and cost, as well as performance. Although some problems remain for epitaxial-based devices, intensive study should solve these concerns. Another important issue is the interconnection problem, at both the interchip and intrachip levels for the 100-GHz bandwidth applications. Proposals such as the chip-size cavity package[32] and an inverted microstrip line integrated circuit structure[33] are candidates for solving the interconnection problem. By combining the device and circuit technologies as we described, we anticipate 100-Gb/s lightwave communications integrated circuits to be developed early in the 21st century.

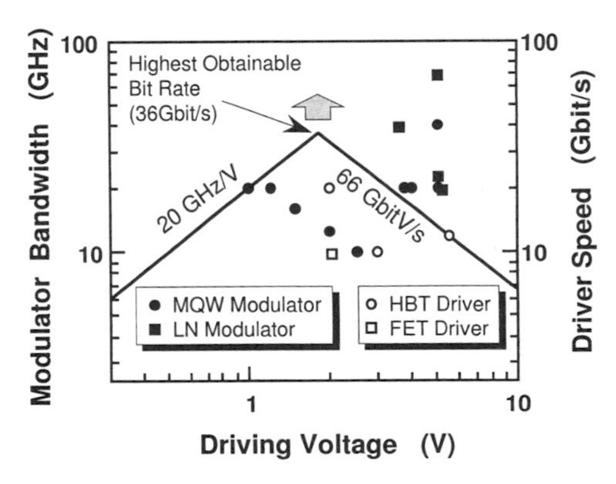

Fig. 11.32 Modulator bandwidths and driver speeds versus driving voltage characteristics. MQW, multiple quantum well.

References

1. Olsson, N. A., and P. Garbinski. 1986. High-sensitivity direct-detection receiver with a 1.5 μm optical preamplifier. *Electron. Lett.* 22(21):1114–1116.
2. O'Mohony, M. J., I. W. Marshall, H. J. Westlake, and W. G. Stallard. 1986. Wideband 1.5 μm optical receiver using traveling-wave laser amplifier. *Electron. Lett.* 22(23):1238–1240.
3. Ishikawa, H., I. Watanabe, T. Suzaki, M. Tsuji, S. Sugou, K. Makita, and K. Taguchi, 1993. High sensitivity 10 Gbit/s optical receiver with superlattice APD. *Electron. Lett.* 29(21):1874–1875.
4. Tzeng, L. D. 1994. Design and analysis of a high-sensitivity optical receiver for SONET OC-12 systems. *J. Lightwave Tech.* 12(8):1462–1470.
5. Nuyts, R., L. D. Tzeng, and O. Mizuhara. 1995. Internal report. Breinigsville, PA: AT&T Bell Laboratories.
6. Smith, R. G., and S. D. Personick. 1982. Receiver design for optical fiber communication systems. In *Topics in applied physics*, vol. 39, ed. H. Kressel, 89–160. New York: Springer-Verlag.
7. Kasper, B. L. 1988. Receiver design. In *Optical fiber telecommunications II*, ed. S. E. Miller and I. P. Kaminow, 689–722. Boston: Academic Press.
8. Ogawa, K., and E. L. Chinnock. 1979. GaAs FET transimpedance front-end design for a wideband optical receiver. *Electron. Lett.* 15(20):650–652.
9. Frahm, R., D. R. Zolnowski, and O. G. Lorimor. 1995. Private communication. Murray Hill, NJ: AT&T Bell Laboratories.
10. Olsson, N. A. 1989. Lightwave systems with optical amplifiers. *J. Lightwave Tech.* 7(7):1071–1082.
11. Park, Y. K., O. Mizuhara, and L. D. Tzeng. 1994. Multigigabit repeaterless optical fiber transmission systems. *ECOC'94, Florence, Italy*. Paper WE.B.1.1.
12. Henry, P. H., R. A. Linke, and A. H. Gnauck. 1988. Introduction of lightwave systems. In *Optical fiber telecommunications II*, ed. S. E. Miller and I. Kaminow, 791. Boston: Academic Press.
13. Park, Y. K., and S. W. Granlund. 1994. Optical preamplifier receivers: Application to long-haul digital transmission. In *Optical fiber technology*, vol. 1, 59–71. Boston: Academic Press.
14. Nakagawa, K., S. Hagimoto, S. Nishia, and K. Aoyama. 1991. Bit rate flexible transmission field trial over 300 km installed cables employing optical fiber amplifiers. In *Topical meetings on optical amplifiers and their applications*. Paper PDP 11.
15. Chen, C. D., J. P. Delavaux, B. W. Hakki, O. Mizuhara, T. V. Nuygen, K. Ogawa, Y. K. Park, R. E. Tench, L. D. Tzeng, and P. D. Yeates. 1994. Field experiment of 10 Gb/s, 360 km transmission through embedded standard fiber cables. *Electron. Lett.* 30:1159–1160.

16. Park, Y. K., P. D. Yeates, J. P. Delavaux, O. Mizuhara, T. V. Nguyen, L. D. Tzeng, R. E. Tench, B. W. Hakki, C. D. Chen, R. J. Nuyts, and K. Ogawa. 1995. A field demonstration of 20 Gb/s capacity transmission over 360 km of installed standard (non-DSF) fiber. *IEEE Photon. Tech. Lett.* 7:816–818.
17. Park, Y. K., T. V. Nguyen, O. Mizuhara, C. D. Chen, L. D. Tzeng, P. D. Yeates, F. Heismann, Y. C. Chen, D. G. Ehrenberg, and J. C. Feggeler. 1996. Field demonstration of 10-Gb/s line-rate transmission on an installed transoceanic submarine lightwave cable. *IEEE Photon. Tech. Lett.* 8:425–427.
18. Taylor, M. G. 1993. Observation of new polarization dependence effect in long-haul optically amplified systems. *IEEE Photon. Tech. Lett.* 5:1244–1246.
19. Suzaki, T., M. Soda, T. Morikawa, H. Tezuka, C. Ogawa, S. Fujita, H. Takemura, and T. Tashiro. 1992. Si bipolar chip set for 10-Gb/s optical receiver. *IEEE J. Solid-State Circ.* 27:1781–1786.
20. Sano, E., Y. Imai, and H. Ichino. 1995. Lightwave-communication ICs for 10 Gbit/s and beyond. *OFC'95 Tech. Dig.* 36–37.
21. Nguyen, L. D., A. S. Brown, M. A. Thompson, and L. M. Jelloian. 1992. 50-nm-self-aligned-gate pseudomorphic AllnAs/GaInAs high electron mobility transistors. *IEEE Trans. Electron. Devices* 39:2007–2014.
22. Yamahata, S., K. Kurishima, H. Ito, and Y. Matsuoka. 1995. Over-220-GHz-f_T- and f_{max} InP/InGaAs double-heterojunction bipolar transistors with a new hexagonal-shaped emitter. In *1995 IEEE GaAs IC Symposium.*
23. Ishibashi, T. 1995. Personal communication.
24. Sano, E., Y. Matsuoka, and T. Ishibashi. 1995. Device figure-of-merits for high-speed digital ICs and baseband amplifiers. *IEICE Trans. Electron.* E78-C:1182–1188.
25. Ichino, H., M. Togashi, M. Ohhata, Y. Imai, N. Ishihara, and E. Sano. 1994. Over-10 Gb/s IC's for future lightwave communications. *J. Lightwave Tech.* 12:308–319.
26. Murata, K., T. Otsuji, M. Ohhata, M. Togashi, E. Sano, and M. Suzuki. 1994. A novel high-speed latching operation flip-flop (HLO-FF) circuit and its application to a 19 Gb/s decision circuit using 0.2 μm GaAs MESFET. In *Technical Digest 1994 GaAs IC Symposium, October*, 193–196.
27. Sano, E., S. Yamahata, and Y. Matsuoka, 1994. 40 Ghz bandwidth amplifier IC using AlGaAs/GaAs ballistic collection transistors with carbon-doped bases. *Electron. Lett.* 30:635–636.
28. Kimura, S., Y. Imai, Y. Umeda, and T. Enoki. 1994. A 16-dB DC-to-50-GHz InAlAs/InGaAs HEMT distributed baseband amplifier using a new loss compensation technique. In *Technical Digest 1994 GaAs IC Symposium, October*, 96–99.
29. Kimura, S., Y. Imai, Y. Umeda, and T. Enoki. 1995. 0–90GHz InAlAs/InGaAs/InP HEMT distributed baseband amplifier IC. *Electron. Lett.* 31:1430–1431.

30. Devaux, F., S. Chelles, A. Ougazzaden, A. Mircea, M. Carre, F. Huet, A. Carenco, Y. Sorel, J. F. Kerdiles, and M. Henry. 1994. Full polarization insensitivity of a 20 Gb/s strained-MQW electroabsorption modulator. *IEEE Photon. Tech. Lett.* 6:1203–1206.
31. Yamauchi, Y., K. Nagata, T. Makimura, O. Nakajima, H. Ito, and T. Ishibashi. 1994. 10 Gb/s monolithic optical modulator driver with high output voltage of 5 V using InGaP/GaAs HBTs. In *Technical Digest 1994 GaAs IC Symposium, October*, 207–210.
32. Shibata, T., S. Kimura, H. Kimura, Y. Imai, Y. Umeda, and Y. Akazawa. 1994. A design technique for a 60 GHz-bandwidth distributed baseband amplifer IC module. *IEEE J. Solid-State Circ.* 29:1537–1544.
33. Yamaguchi, S., Y. Imai, T. Shibata, T. Otsuji, M. Hirano, and E. Sano. 1995. An inverted microstrip line IC structure for ultra-high-speed applications. *IEEE MTT-S Dig.* (May):1643–1646.

Chapter 12 | Solitons in High Bit-Rate, Long-Distance Transmission

L. F. Mollenauer
J. P. Gordon
P. V. Mamyshev
Lucent Technologies, Bell Laboratories, Holmdel, New Jersey

1. Introduction

In the development of the long-distance transmission technologies made possible by the invention of the erbium fiber amplifier, there have been two very different philosophies on how best to deal with the effects of fiber nonlinearity. The first one attempts, by various means, such as the use of special dispersion maps, and by holding signal intensities to the lowest possible level, to make nonlinear penalties acceptably small. The result is the non-return-to-zero (NRZ) transmission mode reported in Chapters 8, 9, and 10 of Volume IIIA. The second philosophy, by contrast, embraces the fiber's nonlinearity and attempts to extract the maximum possible benefits from it. It is this second approach, based on the uniquely stable and nondispersive pulse known as the *soliton,* that is the proper subject of this chapter.

Soliton transmission makes positive use of the fiber's nonlinearity in a number of ways: First, as is well known, the soliton owes its existence to the fiber nonlinearity. That is, as we shall detail shortly, for the soliton, the effects of the nonlinearity more or less continuously cancel the usual pulse-broadening effect of chromatic dispersion. Second, it makes use of the fact that solitons can regenerate themselves, from the nonlinear effect, while traversing a transmission line containing narrow band optical filters, even when the peaks of such guiding filters gradually shift frequency with distance along the line. That amenability to passive regeneration, again unique to solitons, enables a great reduction of the error-producing effects of noise

OPTICAL FIBER TELECOMMUNICATIONS,
VOLUME IIIA

ISBN: 0-12-395170-4

and a further stabilization of the transmission. Third, it makes use of the nearly perfect transparency of solitons to one another in the collisions that occur between pulses of different channels in wavelength-division multiplexing (WDM). Fourth, the interaction between the guiding filters and the fiber nonlinearity engenders a powerful, automatic regulation of the relative signal strengths among the various channels in WDM. Fifth, with solitons, the fiber nonlinearity effectively counteracts the dispersive effects, or polarization mode dispersion, of the residual birefringence of the fiber.

Thus, it should come as no surprise that solitons are the undisputed long-distance champions for both single-channel (Fig. 12.1) and WDM transmission (Fig. 12.2), or that certain modes, such as single-channel rates greater than 10 Gb/s or massive WDM at a per-channel rate of 10 Gb/s, are their exclusive domains.

Beyond this leadership in sheer performance, however, soliton transmission has certain other properties that make it highly attractive. For example, in contrast to NRZ, which tends to require lumped dispersion compensation specific to each distance and each wavelength, the continuous dispersion compensation of soliton transmission renders the data immediately readable and/or injectable at any node of a network. The extreme return-to-zero (RZ) format of the soliton transmission further enhances this compatibility with networking, because it enables all-optical manipulation of the data, with attendant high speed, simplicity, convenience, and low cost.

Finally, as we shall attempt to show, the basic physics of solition transmission is straightforward, highly predictable, and easy to understand. One

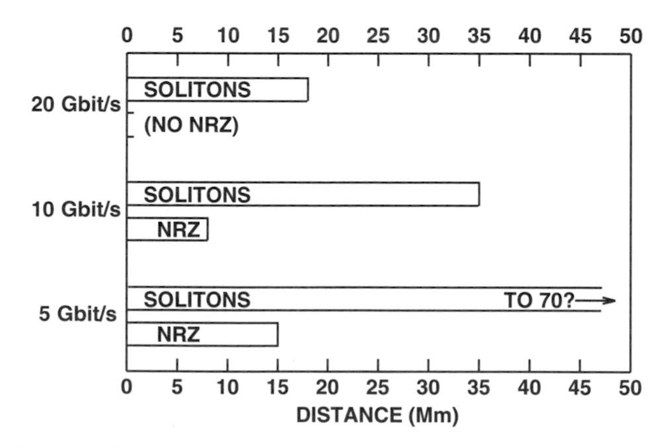

Fig. 12.1 Achieved error-free distances in a single channel, for soliton transmission using sliding frequency guiding filters and for the non-return-to-zero (NRZ) mode.

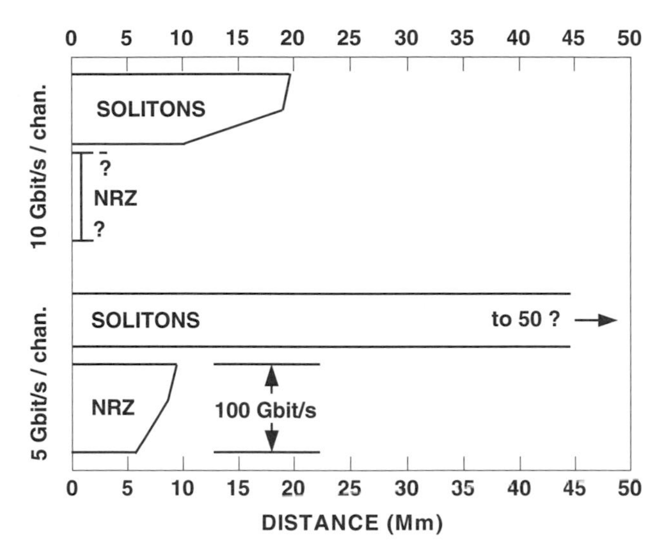

Fig. 12.2 Achieved error-free distances in massive wavelength-division multiplexing (WDM), for solitons and for NRZ. The vertical width of each band is directly proportional to the total achieved capacity in gigabits per second.

consequence of this understanding and ease of analysis is that optimum system designs can be quickly established, and their performance reliably predicted, with a minimum of time-consuming numerical simulation. In this age of rapidly shifting expectations and demands, that is no small advantage.

2. Pulse Propagation and Solitons in Optical Fibers: A Tutorial

2.1 APOLOGIA

This section constitutes a brief tutorial on the theory of pulse propagation in optical fibers. In it, we discuss the fundamental dispersive and nonlinear properties of fiber, and from these we derive the fundamental propagation equation (the nonlinear Schrödinger equation). We then discuss not only the origin and fundamental properties of the soliton, but also other closely related issues such as soliton units and path-average solitons. Because the material in this section is fundamental to all that follows, a careful reading is urged, unless one is already thoroughly familiar with the concepts.

2.2 DISPERSION RELATIONS AND RELATED VELOCITIES

If weak monochromatic light at some angular frequency ω enters a fiber, the wavelength λ_{fiber} of the resulting lightwave in the fiber is determined by the refractive index of the fiber and to a lesser extent by its guiding properties. The phase ϕ of the lightwave has the form $\phi(z, t) = kz - \omega t$, where the wave number k is equal to $2\pi/\lambda_{fiber}$, and z is the distance along the fiber. Central to the problem of lightwave propagation in the fiber is the dependence of the wave number on the frequency. This is the dispersion relation $k = k(\omega)$. For plane waves in vacuum, it is simply $k = \omega/c$. For plane waves in an isotropic transparent medium, it is $k = n\omega/c$, where n is the refractive index at frequency ω. For a single-mode transmission line consisting of a fiber core and cladding, we can also use the same form, $k = n\omega/c$, with the caveat that n now has an effective value intermediate between the values for core and cladding, depending on the transverse mode shape.

An observer moving with velocity $v = dz/dt$ will observe the phase ϕ to change with time according to

$$\frac{d\phi(z, t)}{dt} = kv - \omega. \tag{2.1}$$

If we require the phase to be constant in Eq. (2.1), the needed velocity is $v = \omega/k$. This is the velocity with which any point of constant phase on the wave travels down the fiber. It is called the phase velocity v_p, and its value is just c/n. If we add a second wave at a slightly different frequency, the combined wave will be modulated, with greater amplitude where the two frequency components are in phase and add, and lesser where they are out of phase and subtract. To follow the modulation envelope, our observer must travel at a velocity such that the rate of change of the phase difference between the two waves is zero. From Eq. (2.1) we thus require $k_1 v - \omega_1 = k_2 v - \omega_2$, where the indices 1 and 2 refer to the two waves. The required velocity is $v = (\omega_1 - \omega_2)/(k_1 - k_2)$. The group velocity $v_g(\omega)$ is this velocity in the limit of a small-frequency difference. It satisfies

$$v_g^{-1}(\omega) = \frac{dk}{d\omega}. \tag{2.2}$$

The phase and group velocities usually differ, hence one will generally observe the phase moving with respect to the envelope of a modulated wave. Finally, consider a pulse having a continuous band of frequency components. Such a pulse is most highly peaked if all of its frequency components have the same phase at some common time and place. The pulse envelope will tend to move with its average group velocity, but if the group velocity varies with frequency — i.e., there is some group velocity dispersion — then the pulse will spread as it propagates, becoming chirped as its various frequency components separate in time. We shall see this behavior in more detail later.

In dealing with lightwave propagation in fibers, it is natural to observe the wave as a function of time at various locations along the fiber. To record the progress of a pulse, we therefore plot power versus t for a succession of values of z. In order to keep the pulse in sight, one's time window must be moved as z is varied. However, we can easily manage this movement. A pulse traveling with the group velocity v_g will appear to be stationary in a retarded time frame t' such that $t' = t - v_g^{-1}z$. This is a standard trick used to simplify the analysis.

2.3 *FIBER NONLINEARITY*

The induced polarization in a nonlinear dielectric takes the form

$$\mathbf{P} = \varepsilon_0[\chi^{(1)} \cdot \mathbf{E} + \chi^{(2)} : \mathbf{EE} + \chi^{(3)} \cdot \mathbf{EEE} + \cdots],$$

where $\mathbf{P}$ and $\mathbf{E}$ are the polarization and electric field vectors, respectively, and the susceptibilities $\chi^{(n)}$ are nth rank tensors. Because the glass of optical fibers is isotropic, one has simply $\chi^{(1)} = n^2 - 1$, where n is the index of refraction, while $\chi^{(2)} = 0$. The effects of $\chi^{(3)}$ of interest here are nonlinear refraction and four-wave mixing. Raman scattering becomes important for shorter pulses than we consider here; third harmonic generation is negligibly small.

In silica glass fibers, because of their isotropy, and because of the relatively small value of $\chi^{(3)}$, the index can be written with great accuracy as

$$n(\omega, |\mathbf{E}|^2) = n(\omega) + n_2|\mathbf{E}|^2, \tag{2.3}$$

where n_2 is related to $\chi^{(3)}$ by

$$n_2 = \frac{3}{8n} \chi^{(3)}_{xxxx}, \tag{2.4}$$

where $\chi^{(3)}_{xxxx}$ is a scalar component of $\chi^{(3)}$, appropriate to whatever polarization state the light may have at the moment.

Even the highest quality transmission fibers are mildly birefringent, however, so that the polarization states of the light tend to change significantly on a scale of no more than a few meters. In contrast, the nonlinear effects of interest in long-distance transmission tend to require many kilometers of path for their development. Thus, in general, with optical fibers we are usually interested only in n_2 as suitably averaged over all possible polarization states. In silica glass fibers, if we write the nonlinear index as $n_2 I$, where I is the intensity in W/cm^2, then the lastest measurements of n_2 yield such a polarization-averaged value of about 2.6×10^{-16} cm^2/W.

2.4 *FUNDAMENTAL PROPAGATION EQUATION*

2.4.1 Derivation

We now consider lightwave propagation in a fiber that has both group velocity dispersion and index nonlinearity. Let the lightwave in the line be represented by a scalar function $U(z, t)$ proportional to the complex field amplitude, such that the power P in the line is given by

$$P = P_c |U|^2. \tag{2.5}$$

The proportionality constant P_c can be considered as a power unit. For frequencies near some central frequency ω_0, the generic dispersion relation $k = n\omega/c$ can be expanded to the approximate form

$$k = k_0 + k'(\omega - \omega_0) + \tfrac{1}{2}k''(\omega - \omega_0)^2 + k_2 P, \tag{2.6}$$

where we have used Eq. (2.3). This equation has the form of a Taylor series expansion of $k(\omega, P)$ in the neighborhood of $(\omega_0, 0)$. It adequately describes the propagation of monochromatic waves $U = u_0 \exp(ikz - i\omega t)$ in the line so long as the frequency does not stray too far from ω_0. It leads directly to the nonlinear Schrödinger (NLS) equation. For solitons in fibers, the last two terms of Eq. (2.6) are of comparable importance; k_2 is positive, k'' is negative, and succeeding higher order terms (e.g., $k'''[\omega - \omega_0]^3$, $k_2'[\omega - \omega_0]P$, etc.) can be neglected or adequately treated as perturbations. For the moment, we assume that the line has no loss or gain — i.e., that the constants k_0, k', k'', and k_2 are all real.

The expression for the reciprocal group velocity, namely,

$$v_g^{-1} = \frac{\partial k}{\partial \omega} = k' + k''(\omega - \omega_0), \tag{2.7}$$

identifies k' as the reciprocal group velocity at frequency ω_0 and k'' as its frequency dispersion constant. The dispersion parameter D used widely to describe fibers is the wavelength derivative of v_g^{-1}, and so is related to k'' or the refractive index by

$$D = \frac{d}{d\lambda}(v_g^{-1}) = -\frac{2\pi c}{\lambda^2} k'' = -\frac{\lambda}{c}\frac{d^2 n}{d\lambda^2}. \tag{2.7a}$$

The term $k_2 P$ in Eq. (2.6) represents the primary nonlinear effect, self-phase modulation, resulting from the intensity dependence of the refractive index of the fiber. Note from Eq. (2.7) that the dependence of the group velocity on power is among the higher order terms not included in Eq. (2.6).

If we now remove the central frequency and wave number from U by defining

$$u(z, t) = Ue^{i(\omega_0 t - k_0 z)} \tag{2.8}$$

so that when U is written out explicitly, $u(z, t)$ becomes

$$u(z, t) = u_0 e^{i[(k-k_0)z-(\omega-\omega_0)t]}, \tag{2.8a}$$

then the wave equation for u that is necessary and sufficient to reproduce exactly our initial dispersion relation (Eq. [2.6]) is

$$-i\frac{\partial u}{\partial z} = ik'\frac{\partial u}{\partial t} - \frac{1}{2}k''\frac{\partial^2 u}{\partial t^2} + k_2 P_c |u|^2 u. \tag{2.9}$$

This can be shown by inserting Eq. (2.8a) in Eq. (2.9).

The standard form of the propagation equation is generated from Eq. (2.9) by transforming to the retarded time frame (this eliminates the k' term) and by choosing unit values of time and distance such that $k'' = -1$ and $k_2 = 1$ when measured in those units, and the power unit already mentioned. The appropriate new variables are

$$\begin{aligned} t' &= (t - k'z)/t_c \\ z' &= z/z_c, \end{aligned} \tag{2.10}$$

where the unit values t_c, z_c, and P_c satisfy the relations

$$\begin{aligned} t_c^2/z_c &= -k'' = \lambda^2 D/(2\pi c) \\ z_c P_c &= 1/k_2. \end{aligned} \tag{2.11}$$

The resulting propagation equation (after we drop the primes on z and t) is the NLS equation

$$-i\frac{\partial u}{\partial z} = \frac{1}{2}\frac{\partial^2 u}{\partial t^2} + |u|^2 u. \tag{2.12}$$

Clearly, the first term on the right-hand side of Eq. (2.12) is the dispersive term, whereas the second term is the nonlinear one.

As its name suggests, Eq. (2.12) has a form similar to the well-known Schrödinger equation of quantum mechanics. Here, of course, it is based instead on Maxwell's classical field equations.

There is an important arbitrariness left in the definitions of the three unit values z_c, t_c, and P_c, because there are only two relations (Eq. [2.11]) that they must satisfy. One unit value may be chosen freely, and thus different real-world fields can be represented by the same solution of Eq. (2.12), and vice versa. In particular, if one solution of Eq. (2.12) is $u(t, z)$, then different scalings of the same real-world field give other solutions of the form $Au(At, A^2z)$, where A is the ratio of the values of t_c. This scaling transformation of the solutions of Eq. (2.12) can be verified by direct substitution.

Broadband gain and/or loss can be accommodated by adding a third term, $-i(\alpha/2)u$, to the right side of Eq. (2.12), where α is the coefficient of energy gain per z_c (negative values of α represent loss). In that case, Eq. (2.12) becomes

$$-i\,\frac{\partial u}{\partial z} = \frac{1}{2}\frac{\partial^2 u}{\partial t^2} + |u|^2u - i(\alpha/2)u. \tag{2.12a}$$

2.4.2 Soliton Units

Equation (2.12) is often referred to as a *dimensionless* form of the NLS equation. It is more useful to think of it as having specific dimensions, with z, for example, being a distance always measured in units of z_c rather than in meters or kilometers or any other standard unit. Thus, $z = 2$ means a distance of 2 times z_c. Similarly, t_c and P_c become the units of time and power, respectively. Because here we are primarily interested in solitons, it is convenient to tie these three units (which so far are very general in meaning) to the specific requirements of solitons. The canonical single-soliton solution of the NLS equation, Eq. (2.12), is

$$u(z, t) = \mathrm{sech}(t)\exp(iz/2). \tag{2.13}$$

This soliton has a full width at half maximum (FWHM) power of $\Delta t = 2\cosh^{-1}\sqrt{2} = 1.7627\ldots$. In order for this form to represent some soliton whose FWHM power is τ (in picoseconds, for example), we need simply to take

$$t_c = \frac{\tau}{1.7627\ldots}. \tag{2.14}$$

The unit distance, z_c, is a characteristic length for effects of the dispersive term and is given by

$$z_c = \frac{1}{(1.7627\ldots)^2} \frac{2\pi c}{\lambda^2} \frac{\tau^2}{D}, \tag{2.15}$$

where c and λ are the light velocity and wavelength in vacuum, respectively, and where D is the dispersion constant, as already described (in Section 2.4). (D is often expressed as picoseconds of change in transit time, per nanometer change in wavelength, per kilometer of fiber length. Also, note that $D > 0$ corresponds to anomalous dispersion.) When D is expressed in those units, τ in picoseconds, and for $\lambda = 1557$ nm (corresponding to the longer wavelength erbium amplifier gain peak), Eq. (2.15) becomes

$$z_c \approx 0.25\ \tau^2/D, \tag{2.15a}$$

where z_c is in kilometers. Note that for the pulse widths ($\tau \sim 15$–50 ps) and dispersion parameters ($D \sim 0.3$–1 ps/nm-km) most desirable for long-distance soliton transmission, z_c is hundreds of kilometers. Finally, the unit of power, P_c, is just the soliton peak power, and is given by the formula

$$P_c = \frac{A_{eff}}{2\pi n_2} \frac{\lambda}{z_c} = \left(\frac{1.7627\ldots}{2\pi}\right)^2 \frac{A_{eff}\lambda^3}{n_2 c} \frac{D}{\tau^2}, \tag{2.16}$$

where A_{eff} is the effective area of the fiber core, and where n_2, the nonlinear coefficient, has the polarization-averaged value already cited (see Section 2.3). Thus, for $A_{eff} \sim 50\ \mu\text{m}^2$ and $\lambda = 1557$ nm, one has $P_c \approx 0.476/z_c$, where P_c is in watts, and z_c is in kilometers. Note that for a z_c of hundreds of kilometers, the peak soliton power is just a few milliwatts.

2.4.3 Pulse Motion in the Retarded Time Frame

Another important transformation of the solutions of Eq. (2.12) is that produced by a carrier frequency shift. Because the inverse group velocity dispersion constant has the value -1 in the soliton unit system, a frequency shift produces an inverse group velocity shift of equal magnitude. Thus, for the same solution $u(t, z)$ as mentioned previously, one finds yet other solutions, frequency shifted by Ω (in units of t_c^{-1}), of the form

$$u(t + \Omega z, z)e^{-i(\Omega t + \Omega^2 z/2)}. \tag{2.17}$$

This transformation also applies to *any* solution of Eq. (2.12).

2.4.4 A Useful Property of Fourier Transforms

Shortly, we shall have need of the following simple relation between the Fourier transforms of $u(t)$ and those of its time derivatives. Let $u(t)$ and $\tilde{u}(\omega)$ be Fourier transforms of each other, i.e.,

$$u(t) = \frac{1}{\sqrt{2\pi}} \int_{-\infty}^{+\infty} \tilde{u}(\omega) e^{-i\omega t} d\omega. \tag{2.18}$$

We are using the tilde symbol to imply a function of frequency. Successive time differentiations of Eq. (2.18) show that $\partial u(t)/\partial t$ and $-i\omega\tilde{u}(\omega)$ are also Fourier transforms of each other, as are $\partial^2 u(t)/\partial t^2$ and $-\omega^2\tilde{u}(\omega)$, and so on.

2.4.5 Action of the Dispersive Term in the NLS Equation

To obtain the action of the dispersive term alone, we temporarily turn off the nonlinear term, so that Eq. (2.12) becomes

$$\frac{\partial u}{\partial z} = \frac{i}{2} \frac{\partial^2 u}{\partial t^2}. \tag{2.19}$$

The problem is most easily solved in the frequency domain. The Fourier transform of this last equation yields

$$\frac{\partial \tilde{u}}{\partial z} = -\frac{i}{2} \omega^2 \tilde{u}, \tag{2.20}$$

and its solution is

$$\tilde{u}(z, \omega) = \tilde{u}(0, \omega) e^{-i\omega^2 z/2}. \tag{2.21}$$

From the form of this general solution, it should be clear that the dispersive term merely rearranges the phase relations among existing frequency components; it adds no new ones. To find how the dispersion affects a pulse, we must transform back to the time domain. An example that has an instructive analytic solution is the Gaussian pulse. Taking $u(0, t) = e^{-t^2/2}$, we have $\tilde{u}(0, \omega) = e^{-\omega^2/2}$, and upon turning the crank we get

$$u(z, t) = \frac{1}{\sqrt{1 + iz}} \exp\left(\frac{-t^2}{2(1 + z^2)} (1 - iz)\right). \tag{2.22}$$

In the near field ($z \ll 1$), the field gets some chirp but the pulse shape does not change (see Fig. 12.3). In the far field ($z^2 \gg 1$), the field approaches what can be shown to be a general relation for an initially narrow pulse

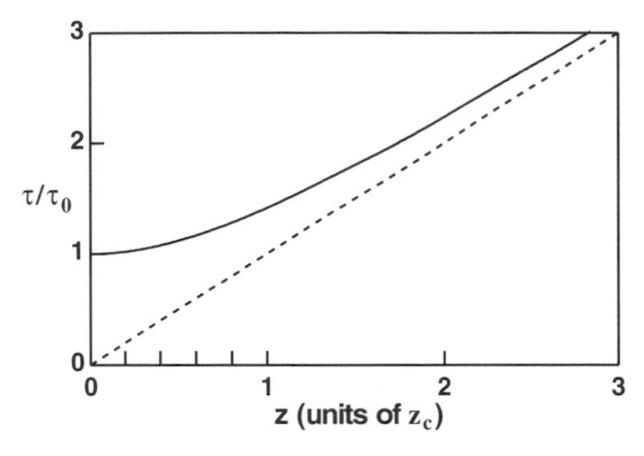

Fig. 12.3 Dispersive broadening of a Gaussian pulse with distance.

$$u(z \gg 1, t) \approx \frac{1}{\sqrt{iz}}\tilde{u}(0, \omega = -t/z)\exp\left(i\frac{t^2}{2z}\right), \tag{2.22a}$$

which shows how dispersion fans the field out into a spectrum of its various frequency components. As depicted in Fig. 12.3, the intensity envelope $|u|^2 \propto \exp(-t^2/(1 + z^2))$. Thus the pulse width grows as

$$\tau = \tau_0\sqrt{1 + z^2}, \tag{2.23}$$

where τ_0 is the initial, minimum pulse width. We see that the change in τ is only to second order in z at the origin. This may also be seen directly from the differential equation, where we note that if $u(t)$ has a constant phase, then $\partial u/\partial z$ is everywhere in quadrature with u.

2.4.6 Action of the Nonlinear Term in the NLS Equation

To observe the action of the nonlinear term of Eq. (2.12) alone, we turn off the dispersive term, so the equation becomes simply

$$\frac{\partial u}{\partial z} = i|u|^2 u. \tag{2.24}$$

The problem is most naturally solved in the time domain, where the general solution is

$$u(z, t) = u(0, t)e^{i|u(0,t)|^2 z}. \tag{2.25}$$

The nonlinear term modifies $\phi(t)$, but not the intensity envelope. Thus, it adds only new frequency components. To get the spectral spreading, we must transform back to the frequency domain. Once again, for example, let $u(0, t) = e^{-t^2/2}$. In that case, one has

$$\tilde{u}(z, \omega) = \frac{1}{\sqrt{2\pi}} \int_{-\infty}^{\infty} u(0, t) e^{i|u|^2 z} e^{i\omega t}\, dt$$

$$= \frac{1}{\sqrt{2\pi}} \int_{-\infty}^{\infty} e^{-t^2/2} e^{ize^{-t^2}} e^{i\omega t}\, d\omega. \quad (2.26)$$

For $z \gg 1$, this integral produces a multipeaked spectrum, where the number of peaks and the overall spectral width increase directly with z (see Fig. 12.4). However, for $z \ll 1$, the integral is approximately

$$\frac{1}{\sqrt{2\pi}} \int_{-\infty}^{\infty} e^{-t^2/2}(1 + ize^{-t^2}) e^{i\omega t}\, dt = \tilde{u}(0, \omega) + \frac{iz}{\sqrt{3}} \tilde{u}(0, \omega)^3. \quad (2.27)$$

Note that once again the new component is in quadrature with the original pulse, so the increase in net spectral width scales only as z^2. Thus, the initial increase, here in bandwidth, is also only to second order in z. *This behavior is equally important to the creation of the soliton as was the increase in pulse width from the dispersive term.*

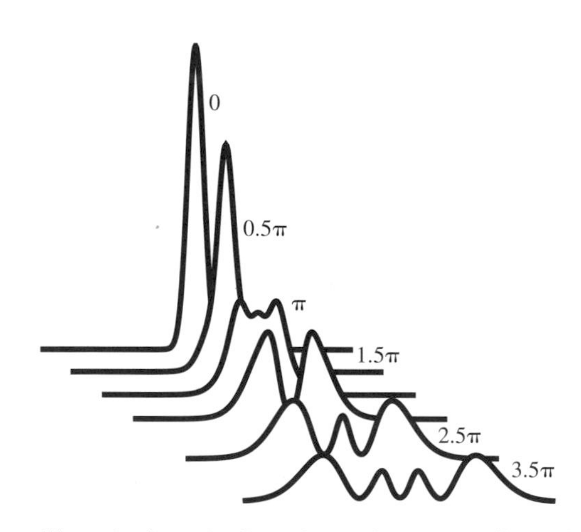

Fig. 12.4 Spectral broadening of a Gaussian pulse at zero dispersion. The numbers at each spectrum indicate the peak nonlinear phase shift.

2.5 *THE SOLITON*

2.5.1 Origin of the Soliton

We are now finally in a position to discuss the origin of the soliton. In Section 1, we stated that the soliton is that pulse for which the nonlinear and dispersive terms of the NLS equation cancel each other's effects. At first, it may seem mysterious that the tendencies to spectral and temporal broadening can cancel each other. As we have just taken pains to show, however, whenever one starts from a transform-limited pulse, such as sech(t), there is no broadening of either kind to first order in z (to be thought of as dz). Instead, the first order effects of both terms are just complementary phase shifts $d\phi(t)$. We have already seen how the nonlinear term generates $d\phi(t) = |u(t)|^2 dz$. For the dispersive effect, first we recognize that if $f(z, t)$ is real, then the general equation

$$\frac{\partial u}{\partial z} = if(z, t)u \tag{2.28}$$

simply generates the phase change $d\phi(t) = f(0, t)\, dz$ in the distance dz. We then write the reduced NLS equation in the form

$$\frac{\partial u}{\partial z} = \left(\frac{i}{2u}\frac{\partial^2 u}{\partial t^2}\right) u. \tag{2.29}$$

Thus, the dispersive term generates

$$d\phi = \left(\frac{1}{2u}\frac{\partial^2 u}{\partial t^2}\right) dz. \tag{2.30}$$

For $u = \text{sech}(t)$, these terms are, respectively,

$$d\phi_{NL} = \text{sech}^2(t)\, dz \qquad \text{and} \qquad d\phi_{disp.} = [\tfrac{1}{2} - \text{sech}^2(t)]\, dz. \tag{2.31}$$

Note that these differentials sum to a constant (see Fig. 12.5), which, when integrated, simply yields a phase shift of $z/2$ common to the entire pulse. In this way, we arrive at the simplest form for the soliton, already displayed in Eq. (2.13).

It should also be noted that a common phase shift does nothing to change the temporal or spectral shapes of a pulse. Thus, as already advertised, the soliton remains completely nondispersive in both the temporal and frequency domains. Nevertheless, the associated wave-number shift of $(2z_c)^{-1}$, or simply $\frac{1}{2}$ in soliton units, is important in understanding the interaction of the solition with perturbing nonsoliton field components.

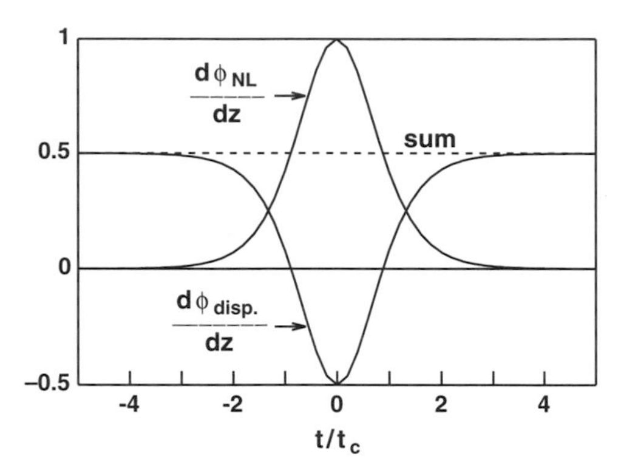

Fig. 12.5 Dispersive and nonlinear phase shifts of a soliton pulse, and their sum.

2.5.2 Path-Average Solitons

For reasons of economy, the loss-canceling optical fiber amplifiers of a long fiber transmission line are usually spaced apart by a distance, which we shall call the *amplifier span,* or L_{amp}, of several tens of kilometers. This spacing results in a large periodic variation in the signal intensity, as illustrated in Fig. 12.6. In addition, the dispersion parameter D may vary significantly within each amplifier span (again, see Fig. 12.6). Clearly, in that case, the differential phase shifts of the dispersive and nonlinear terms (see Eq.

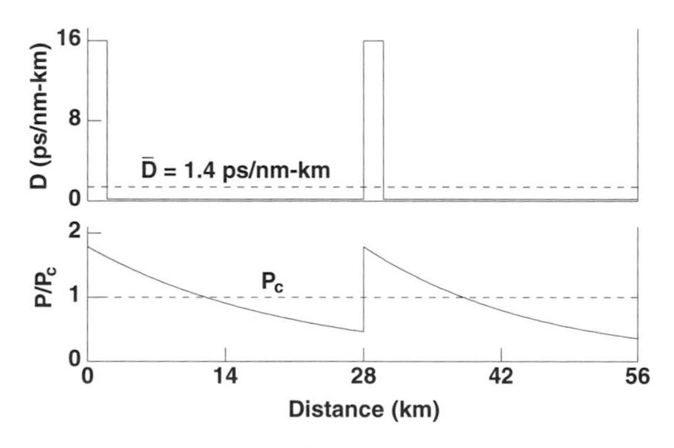

Fig. 12.6 Sample of a transmission line used for numerical testing of the path-average soliton concept. As in certain real-world experiments, the desired $\overline{D}$ is obtained by combining short lengths of high D fiber with dispersion-shifted fiber (for which $D \sim 0$), so there are large variations in D, periodic with the amplifier spacing, as well as in the pulse power.

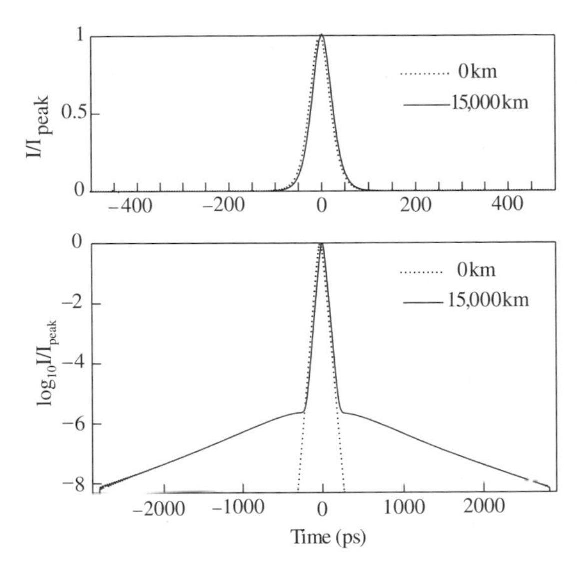

Fig. 12.7 Solitons, for which $z_c \sim 440$ km, at input and after traversing 15,000 km of the transmission line of Fig. 12.6. Note that the defects (dispersive tails on the pulse) are extremely small. The defects have also been computed analytically.

[2.31]) do not cancel in every element dz of the fiber. Nevertheless, if the condition

$$z_c \gg L_{amp} \tag{2.32}$$

is satisfied, and if, furthermore, the path-average values, $\bar{I}$ and $\bar{D}$ of the intensity and dispersion, respectively, are the same for every amplifier span, then one can still have, at least in a practical sense, perfectly good solitons. The reason is that when the inequality of Eq. (2.32) is well satisfied, as already shown, neither the temporal nor the spectral shapes of the soliton are significantly affected within each span. Thus, all that matters is that, over each amplifier span, the path-average dispersive and nonlinear phase shifts cancel (sum to a standard constant). Figure 12.7 illustrates how very well this concept of *path-average solitons** can work [1–5]. Through numerical simulation, it shows $\tau = 50$ ps solitions before and after traversal of 15,000 km of transmission line whose spans are those of Fig. 12.6. For this

* Although the name tends to be obscure, the "guiding center solitons" of Ref. 5 are essentially path-average solitons. The latter name is preferred, however, because it is more accurately descriptive, is better known, and avoids confusion with the more apt use of the word *guiding* in connection with jitter-reducing filters (see Section 4).

case, $z_c/L_{amp} \approx 16$. The difference between the output and input solitions, which can be seen only on the logarithmic plot, appears in the form of very low-intensity tails on the pulse. This defect, usually known as *dispersive wave radiation,* represents the nonsolition component of the pulse. When it is as small as shown here, it is usually of no practical import. Even when z_c/L_{amp} is as small as 3 or 4, the path-average solition concept still works fairly well.

There is another complementary, insightful way of understanding the behavior of path-average solitons, wherein the periodic fluctuations of the amplifier spans are seen as a perturbation to provide phase matching between the solitons on the one hand, and the linear, or dispersive, waves on the other. As we have just seen in the foregoing, the dispersion relation for solitons is just $k_{sol.} = \frac{1}{2}$ (in soliton units), whereas that for the linear waves is $k_{lin.} = -\frac{1}{2}\omega^2$. Clearly, the amplifier spans provide $k_{pert.} = z_c/L_{amp}$. The phase matching condition is

$$k_{pert.} = k_{sol.} - k_{lin.}. \tag{2.33}$$

If, as illustrated in Fig. 12.8, $k_{pert.}$ is so large that the phase matching occurs only where the spectral density of the soliton is small, then the path-average

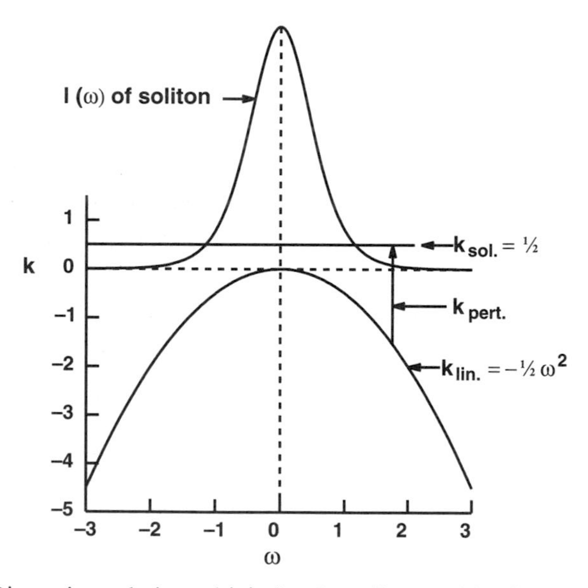

Fig. 12.8 Dispersion relations, $k(\omega)$, for the soliton and for linear waves, and the spectral density of the soliton. A perturbation of wave vector $k_{pert.}$ phase matches linear waves of frequency ω to the corresponding region of the soliton's spectrum.

solitons work well. On the other hand, if $k_{pert.}$ is small enough, then the phase matching will be to a region of high spectral density, where a large fraction of the soliton's energy will drain away into dispersive (linear) waves, and the path-average soliton will not work well. It is interesting that, historically, the concept of path-average solitons and the associated resonance condition for disaster ($k_{pert.} = \frac{1}{2}$) were first encountered in these terms of phase matching [1].

2.5.3 Soliton Transmission in Dispersion-Tapered Fiber

With an ever-increasing bit rate, eventually the soliton pulse width and hence z_c become so short that it is no longer possible to satisfy the inequality of Eq. (2.21) in a satisfactory way. Nevertheless, in principle at least, there is still a way to have perfect soliton transmission with lumped amplifiers, and that is to taper the fiber's dispersion parameter $D(z)$ to the same exponential decay curve as that of the intensity itself. That is, $D(z)$ should be given by

$$D(z) = \frac{\alpha L_{amp} \overline{D}}{1 - \exp(-\alpha L_{amp})} e^{-\alpha z} = D_0 e^{-\alpha z}, \tag{2.34}$$

so that Eq. (2.12a) becomes

$$-i\frac{\partial u}{\partial z} = \frac{D_0}{2\overline{D}} e^{-\alpha z} \frac{\partial^2 u}{\partial t^2} + |u|^2 u - i(\alpha/2)u. \tag{2.35}$$

Clearly, the phase shifts generated by the dispersive and nonlinear terms of Eq. (2.12b) will cancel in each and every segment dz of the amplifier span, so the solitons will be without perturbation. It is also easily understood that an N-step approximation to the ideal dispersion taper of Eq. (2.34) can be tremendously helpful, even when N is as small as two or three. Later, in Section 5, we shall show that tapered dispersion is perhaps of even greater importance to avoid excessive growth of four-wave mixing components.

2.5.4 More General Forms for the Soliton

Any single real-world soliton can be expressed in the simple form of Eq. (2.13) by the appropriate choice of scale and frequency. Alternatively, application to Eq. (2.13) of the scale and frequency transformations discussed in Sections 2.4.1 and 2.4.3, respectively, yields the more general form

$$u = A\,\mathrm{sech}(A(t + \Omega z))e^{i(\phi - \Omega t)}, \tag{2.36}$$

where

$$\phi = (A^2 - \Omega^2)z/2.$$

The form given by Eq. (2.36) is necessary to the consideration of perturbation or multisoliton problems, such as soliton–soliton collisions in WDM, for example.

2.6 *NUMERICAL SOLUTION OF THE NLS EQUATION: THE SPLIT-STEP FOURIER METHOD*

The NLS equation is generally difficult to solve analytically. Numerical solution, however, can be remarkably efficient, when it is based on the split-step Fourier method shown in Fig. 12.9. The method is based on the fact that the effects of the dispersive term are most naturally dealt with in the frequency domain, whereas those of the nonlinear term are best handled in the time domain. Thus, each increment h in z is treated in two consecutive steps, as follows:

$$\text{Step 1:}\quad u(z, t) \longrightarrow \tilde{u}(z, \omega); \qquad \tilde{u}(z, \omega)e^{-i(\omega^2/2)h} = \tilde{u}(z + h, \omega)$$

and

$$\text{Step 2:}\quad \tilde{u}(z + h, \omega) \longrightarrow u_{new}(z, t); \qquad u_{new}(z, t)e^{i|u|^2 h} = u(z + h, t).$$

That is, in Step 1, $u(z, t)$ is Fourier transformed to $\tilde{u}(z, \omega)$, and then, to reflect the dispersive effects of the element h, $\tilde{u}(z + h, \omega)$ is computed from $\tilde{u}(z, \omega)$ according to the *analytic* solution of Eq. (2.20). In Step 2, $\tilde{u}(z + h, \omega)$ is first Fourier transformed back to make a "new" version of $u(z, t)$. Then, from that new $u(z, t)$, $u(z + h, t)$ is computed according to the *analytic* solution of Eq. (2.24), so that it now reflects the nonlinear effect of the element h as well. On the basis of the ideas just discussed with respect to path-average solitons, one can easily see that reasonable accuracy can often be obtained with relatively large step sizes. Finally, note that fiber loss, amplifier gain, filter response functions, and other linear

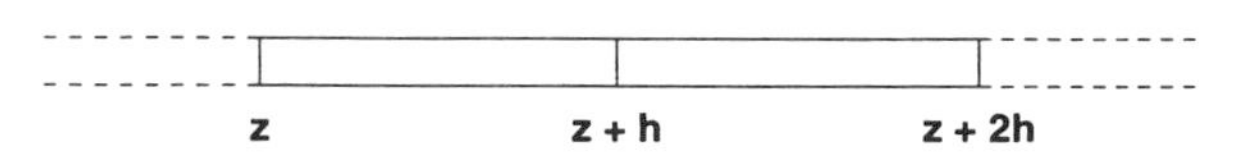

Fig. 12.9 Scheme of the split-step Fourier method.

frequency-dependent factors are most easily applied in the frequency domain.

3. Spontaneous Emission and Other Noise Effects

3.1 GROWTH OF SPONTANEOUS EMISSION NOISE IN A BROADBAND TRANSMISSION LINE

In long-distance fiber optic transmission systems, limits to the bit rate and distance for error-free transmission are set mainly by the various effects of noise fields in the line [6, 7]. Although we shall later show how the proper use of narrow band filters can greatly reduce those noise fields and their effects, it is nevertheless instructive to first consider a broadband system in which the response of the transmission line is essentially flat over the entire bandwidth of the soliton. The prototypical system (see Fig. 12.10) consists of single-mode fiber segments of length L_{amp} and loss coefficient α, so that the power loss per segment is $\exp(-\alpha L_{amp})$, connected by amplifiers whose gain $G = \exp(\alpha L_{amp})$ offsets this loss. The path-average signal is thus maintained at the same high level, from the transmitter right through to the receiver. Detector noise is overwhelmed, and the most important noise is the accumulated, or "amplified," spontaneous emission (ASE) created by the amplifiers. In some cases, dispersive-wave radiation created by path-average solitons (Section 2.5.2), and residual four-wave mixing fields from soliton–soliton collisions in WDM, can also make a significant contribution to the noise. Here, however, we shall concentrate on the inevitable ASE noise.

In the system of Fig. 12.10, each amplifier contributes to its output an additive Gaussian ASE noise field whose mean power per unit bandwidth is

$$P_{amp}(\nu) = (G - 1)n_{sp}h\nu, \tag{3.1}$$

where $n_{sp} \geq 1$ is the excess spontaneous emission factor (close to unity if the amplifier populations are highly inverted), and $h\nu$ is the photon energy.

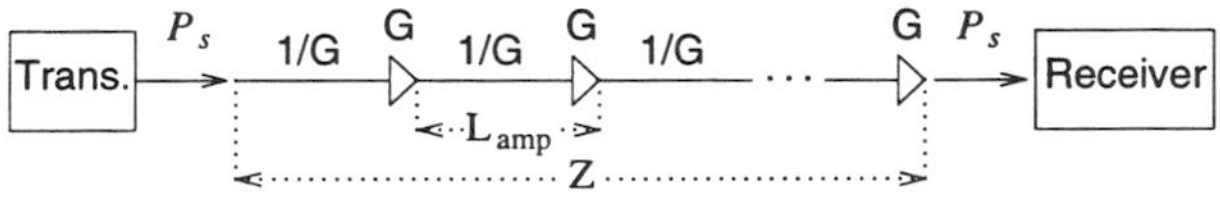

Fig. 12.10 Prototypical all-optical transmission line containing amplifiers of power gain G interleaved with fiber spans of loss $1/G$.

One may note that $P_{amp}(\nu)$ has units of energy. In general, if $P(\nu)$ is the power per unit bandwidth of an effectively white-noise field, then it is also the noise energy emitted (or received) in any time T (e.g., the bit period) and in a corresponding bandwidth $\delta\nu = 1/T$. It is independent of one's choice of time unit and is known as the *equipartition energy*. (Note that we are invoking only classical transmission line theory — i.e., no mention of discrete photons, etc. This is always adequate when the equipartition noise energy is much greater than the photon energy, as it is in systems that use in-line coherent amplifiers. A more sophisticated semiclassic theory involving zero-point fields is conceptually better, but in the present context it is unnecessary.)

Because there is unity gain from the output of each amplifier in Fig. 12.10 to the receiver, the accumulated ASE noise at the receiver is just the value given by Eq. (3.1) multiplied by N, the total number of amplifiers in the chain. To compare the noise with the soliton signal, we must use the path-average value for the noise, just as we do for the solitons. The path-average power is equal to the power at the output of an amplifier multiplied by the average from 0 to L_{amp} of $\exp(-\alpha z)$, which can be expressed as $(G - 1)/(G \ln G)$. If we also write N as $\alpha Z/\ln G$, then the path-average noise at system output takes the form

$$\overline{P}_N(\nu) = \alpha Z h\nu n_{sp} F(G), \tag{3.2}$$

where the overbar on $\overline{P}$ symbolizes the path average, Z is the system length, and the function

$$F(G) = \frac{1}{G}\left[\frac{(G - 1)}{\ln G}\right]^2 \tag{3.3}$$

(see Fig. 12.11) represents an important noise penalty incurred simply by using long spans and high gain amplifiers [6]. Systems designers, accustomed to the economics of regenerated systems, would like to place amplifiers no closer together than about 100 km. For systems of transoceanic length, however, the more than 7-dB ASE noise penalty one must pay for such large spacing is excessive, and smaller spacings are usually needed.

3.2 ENERGY ERRORS

There are two main sources of error that affect the soliton system; fluctuations of the pulse energies and of their arrival times. At each amplifier, the addition of the ASE noise changes the energy, central frequency, mean

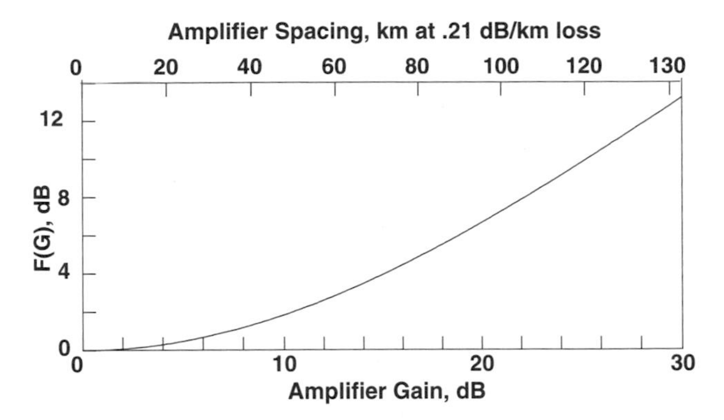

Fig. 12.11 Penalty function, $F(G)$, as a function of amplifier gain.

time, and phase of the solitons in statistically random ways. The changes in mean time and phase are of little importance in the present context. The other two changes can be analyzed separately. We shall focus on the energy fluctuations in this section and on the frequency changes and the resultant jitter in arrival times in the next.

The energy fluctuations are similar to those that occur in a linear system. The argument is as follows: The system is effectively linear over short distances, so there is no difference in the way the noise field is injected into the system. The only difference in the soliton system is that the energy changes of first order in the noise field (the so-called signal–spontaneous noise) are captured by the solitons, which then reshape themselves as they propagate. This reshaping is done with no significant change in energy. Thus, the energy fluctuations at the receiver at like those that would occur if the system were linear and dispersion free.

To evaluate the errors incurred by the energy fluctuations, some model detector must be chosen. For simplicity we shall assume that the detector consists first of an optical filter of bandwidth B_0, followed by a photodetector, followed by an integrator, so that in effect the total energy that passes the optical filter in each time slot T is measured. The detectors actually used in most systems do not work this way, but the good ones give similar results. Let $m = KB_0T$, where $K = 1$ or 2 is the number of polarization states to which the receiver is sensitive. The sampling theorem says that the detected optical field has approximately $2m$ independent degrees of freedom (DOFs), and without loss of generality the soliton may be considered to occupy just one of these. The mean ASE noise energy per DOF is one-half the equipartition energy, or $\overline{P}_N(\nu_0)/2$. Thus, if we let S be the ratio of the total energy in a bit period to the equipartition energy $\overline{P}_N(\nu_0)$, then S becomes the sum of the squares of $2m$ independent Gaussian random field variables, each with variance equal to one-half. All but one of these have zero mean values. The exception has a mean value that is the square root of the normalized unperturbed soliton energy. One can think of S as the square of the radius to a point in a $2m$-dimensional Euclidean space whose coordinates are the real amplitudes of the normalized DOF field components. On the basis of this picture, we get the following results. The mean and variance of the distribution of S for a zero (soliton absent) are both equal to m, and for a one (soliton present) are equal, respectively, to $S_1 + m$ and $2S_1 + m$, where S_1 is the ratio of the unperturbed soliton energy to the noise equipartition energy. In the variances, the term m represents what is often called the *spontaneous–spontaneous beat noise*, whereas the term $2S_1$ represents the *signal–spontaneous beat noise.*

It is standard practice to characterize error rates using the quantity Q, which is related to the bit error rate (BER) through the complementary error function

$$\text{BER} = (\tfrac{1}{2}\,\text{erfc}\,[Q/\sqrt{(2)}] \approx [2\pi(Q^2 + 2)]^{-1/2}\exp(-Q^2/2). \tag{3.4}$$

The approximate form can be used when $Q \geq 3$. If the energy distributions are assumed Gaussian and the optimum decision level is chosen, then the value of Q is related to the means, μ, and variances, σ^2, of the distributions of ones and zeros by $Q = (\mu_1 - \mu_0)/(\sigma_1 + \sigma_0)$. On this heuristic assumption of Gaussian energy distributions, one gets error rates as a function of the signal to equipartition noise ratio S, as shown by the dashed line in Fig. 12.12.

In more detail, the true probability distributions for S are given by [6]

$$Prob(S) = \frac{S^{m-1}}{(m-1)!}\exp(-S) \tag{3.5}$$

for a zero, and

$$\begin{aligned} Prob(s) &= 2s(s/s_1)^{m-1}\exp[-(s^2 + s_1^2)]I_{m-1}(2ss_1)^* \\ &\approx \pi^{-1/2}(s/s_1)^{m-1/2}\exp[-(s - s_1)^2] \end{aligned} \tag{3.6}$$

for a one, where s and s_1 are, respectively, the square roots of S and S_1, whereas I_{m-1} signifies a modified Bessel function of the first kind. The approximate asymptotic form is valid in the tail of the distribution needed

* We thank Curtis Menyuk for pointing out the exact form.

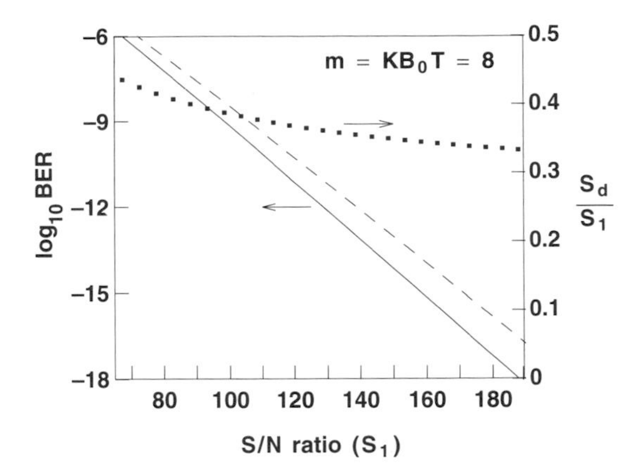

Fig. 12.12 Bit error rate (BER) for amplitude or energy errors, and the optimum decision energy level, as functions of the signal-to-noise ratio, for $m = 8$. (S_d is the decision energy level.) Also shown is the error rate estimate using the Gaussian approximation (*dashed line*).

for the calculation of errors. Using these probability distributions, one determines error rates by choosing a decision level S_d that equates the probability that $S > S_d$ for a zero with the probability that $s < \sqrt{S_d}$ for a one. The results of this computation are also plotted in Fig. 12.12. Note that the more accurate computation of error rates gives fewer errors than the Gaussian approximation does. For a larger value of m (larger optical bandwidth), the difference would be smaller.

3.3 GORDON–HAUS EFFECT

The ASE noise also acts to produce random variations of the solitons' central frequencies. The fiber's chromatic dispersion then converts these variations in frequency to a jitter in pulse arrival times, known as the *Gordon–Haus effect* [8]. Such timing jitter can move some pulses out of their proper time slots. Thus, the Gordon–Haus effect is a fundamental and potentially serious cause of errors in soliton transmission.

The calculation of the jitter can be summarized as follows: Recall that each DOF of the noise field produced by an amplifier has a mean path-average energy of $(\frac{1}{2})\overline{P}_{amp}(\nu)$. The field of one such DOF shifts the frequency of the soliton. From perturbation theory, one can deduce that the effective noise-field component has the form $\delta u = iau_{sol}\tanh(t)$ (see Fig. 12.13), and that it shifts the soliton's frequency by an amount $\delta\Omega = 2a/3$. Here, a is a

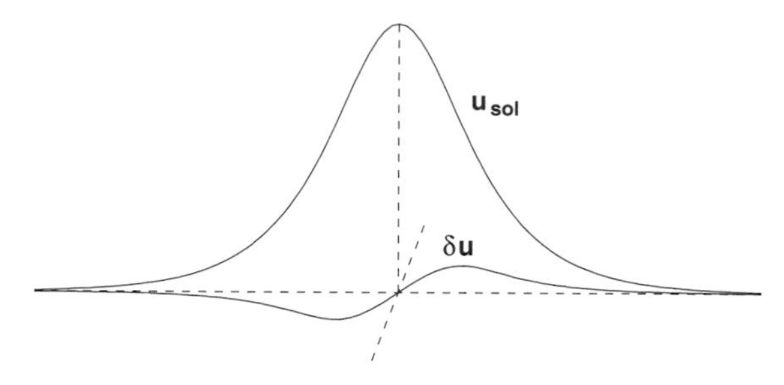

Fig. 12.13 Noise component that modifies the frequency of a soliton, in relation to the field envelope of the soliton. The two field components are in quadrature.

real random Gaussian variable whose variance $(\frac{3}{4})\overline{P}_{amp}(\nu)$ is determined by the DOF's mean energy requirement. The soliton's random frequency shift therefore has a variance of

$$\langle \delta\Omega^2 \rangle_{amp} = \tfrac{1}{3}\,\overline{P}_{amp}(\nu). \tag{3.7}$$

Because in soliton units (see Section 2.4.3) the inverse velocity shift is numerically just -1 times the frequency shift, the net time shift of a given pulse is

$$\delta t = -\sum_{amps} \delta\Omega_n z_n, \tag{3.8}$$

where z_n is the distance from nth amplifier to the end. On the right side of Eq. (3.8) we have the sum of N independent variables, each of which has a Gaussian distribution. In such a case, the sum also has a Gaussian distribution whose variance is the sum of the variances of the individual terms. Thus, the variance of δt is

$$\langle \delta t^2 \rangle = \langle \delta\Omega^2 \rangle_{amp} \sum_{amps} z_n^2 = \frac{\overline{P}_{amp}(\nu)}{3}\,\frac{Z^3}{3L_{amp}}, \tag{3.9}$$

where, for the second step in Eq. (3.9), Z represents the total system length, and we have approximated the discrete sum over the (many) amplifiers by an integral. Now substituting ln G/α for L_{amp} and using the path average form of Eq. (3.1), we obtain

$$\langle \delta t^2 \rangle = \tfrac{1}{9}\,\alpha n_{sp}\, F(G) h\nu Z^3. \tag{3.10}$$

(From now on in this chapter, we shall write the variance in arrival times, $\langle \delta t^2 \rangle$, as σ^2, so σ then becomes the corresponding standard deviation.)

Translated from soliton units into practical units (see Eqs. [2.14] and [2.15], Eq. (3.10) becomes

$$\sigma_{gh}^2 = 3600\, n_{sp} F(G) \frac{\alpha}{A_{eff}} \frac{D}{\tau} Z^3, \qquad (3.11)$$

where σ_{gh} is in picoseconds, n_{sp} and $F(G)$ are as defined previously, the fiber loss factor α is in inverse kilometers, the effective fiber core area A_{eff} is in square microns, the group delay dispersion D is in picoseconds per nanometer per kilometer, τ is the soliton FWHM intensity in picoseconds, and Z is the total system length in megameters (1 Mm = 1000 km). (The numerical constant in Eq. (3.11) is not dimensionless.) We can deduce from Eq. (3.11) that σ_{gh}^2 is proportionial to the energy of the solitons because the latter is also proportional to D/τ.

To get a feeling for the size of the effect, consider the example $Z = 9$ Mm (trans-Pacific distance), $\tau = 20$ ps, $D = 0.5$ ps/nm-km, $A_{eff} = 50\ \mu m^2$, $\alpha = 0.048\ km^{-1}$, $n_{sp} = 1.4$, and $F = 1.19$ (~30-km amplifier spacing). Equation (3.11) then yields $\sigma_{gh} = 11$ ps.

The BER from the Gordon–Haus effect is the probability that a pulse will arrive outside the acceptance window of the detection system. If the window width is $2w$ and we assume that these errors affect only the ones, then the BER has a Q value of w/σ (see Eq. [3.4]. For example, this implies that for an error rate no greater than 1×10^{-9}, $2w \geq 12\sigma_{gh}$. Now, the upper bound on $2w$ is just the bit period, although practical considerations may make the effective value of $2w$ somewhat smaller. Note, therefore, that for the previous example, where $\sigma_{gh} = 11$ ps, the quantity $12\sigma_{gh}$ corresponds to a maximum allowable bit rate of about 7.5 Gb/s.

3.4 THE ACOUSTIC EFFECT

Traditionally, the Gordon–Haus effect is considered to be the dominant source of timing jitter. There is, however, another contribution, one arising from an acoustic interaction among the pulses. Unlike the bit-rate-independent Gordon–Haus jitter, the acoustic jitter increases with bit rate, and as we shall soon see, it also increases as a higher power of the distance. Thus, the acoustic jitter tends to become important for the combination of great distance and high bit rate. In this section, we briefly review what is known about the acoustic effect.

The acoustic effect appeared in the earliest long-distance soliton transmission experiments [9] as an unpredicted "long-range" interaction: one that enabled pairs of solitons separated by at least several nanoseconds (and which were thus far beyond the reach of direct nonlinear interaction) to significantly alter each other's optical frequencies, and hence to displace each other in time. Shortly thereafter, Dianov *et al.* [10] correctly identified the source of the interaction as an acoustic wave, generated through electrostriction as the soliton propagates down the fiber (see Fig. 12.14). Other pulses, following in the wake of the soliton, experience effects of the index change induced by the acoustic wave. In particular, they suffer a steady acceleration, or rate of change of inverse group velocity with distance, dv_g^{-1}/dz, proportional to the local slope of the induced index change (again, see Fig. 12.14). In a broadband transmission line, when this steady acceleration is integrated over z, it yields $\delta v_g^{-1} \propto z$, and a second such integration yields a time displacement $\delta t \propto z^2$. It can be shown that the standard deviation of the acoustic effect for a fiber with $A_{\mathrm{eff}} = 50\ \mu\mathrm{m}^2$ is approximately [11, 12]

$$\sigma_a \approx 8.6 \frac{D^2}{\tau} \sqrt{R - 0.99} \frac{Z^2}{2}, \tag{3.12}$$

where σ_a is in picoseconds, D is in picoseconds per nanometer per kilometer, τ is in picoseconds, R is in gigabits per second, and Z is in megameters. Comparing Eq. (3.12) with the square root of Eq. (3.11), note the different

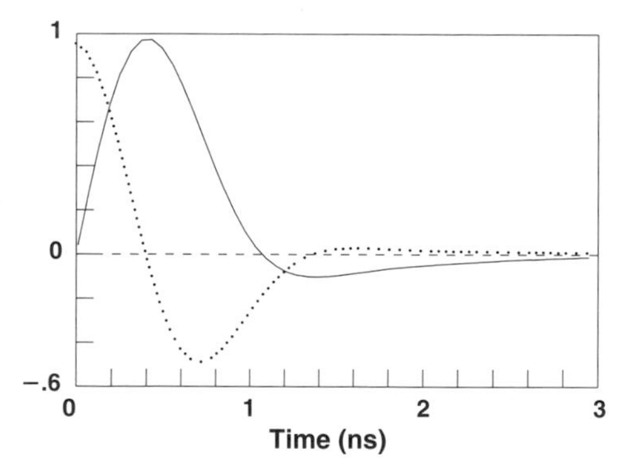

Fig. 12.14 *Solid curve:* Relative index change due to the acoustic effect following passage of a soliton at $t = 0$. *Dotted curve:* Relative force acting on the following soliton at t; this curve is proportional to the time derivative of the relative index curve. For these curves, the interacting solitons are assumed to have a common state of polarization. The effect is only weakly dependent on their relative polarizations, however.

power dependencies on D, τ, R, and Z, most of which have already been discussed for both the Gordon–Haus and the acoustic effects. For the acoustic jitter, the scaling of σ_a as D^2/τ is easily understood, because it is clearly in direct proportion to the soliton energy, or D/τ, and the extra factor in D is required for the conversion of frequency shifts into timing shifts.

3.5 OPTIMIZATION OF THE SOLITON ENERGY FOR BEST BER PERFORMANCE

It should be clear from the discussion in Sections 3.2 and 3.3 that energy errors decrease, whereas errors from the Gordon–Haus jitter increase, with increasing soliton pulse energy W_{sol}. Thus, there will be an optimum value of W_{sol} for which the combined error rates are a minimum. Because $W_{sol} \propto D/\tau$, one can hope to attain that optimum value of W_{sol} by adjusting D and τ. There are, of course, certain limitations on the practical ranges for both parameters. For example, lack of perfect uniformity of fiber preforms and other factors tend to limit the smallest values of D that can be produced reliably. To avoid significant interaction between nearest-neighbor soliton pulses, τ can be no more than about 20–25% of the bit period. Nevertheless, D/τ can usually be adjusted over a considerable range.

The optimum value of W_{sol} can be most efficiently found from a diagram [6] like that shown in Fig. 12.15, where, for a fixed value of the transmission distance, the rates for both energy and timing errors are plotted as a function of the parameter τ/D. Proceeding from the far right, where W_{sol} is smallest, note that at first, only energy errors are significant, but as W_{sol} increases, those errors fall off exponentially. Eventually, timing errors become significant and then dominate. Also note that although the energy errors are bit-rate independent, the timing errors are not, because the allowable size of the acceptance window in time is determined by the bit period. Note that for transmission at 5 Gb/s, the optimum value of $\tau/D \approx 70$ nm-km. If we choose $D = 0.5$ ps/nm-km, a value large enough to be reproducible, then we have $\tau = 35$ ps, a value short enough relative to the (200-ps) bit period to allow for negligible pulse interactions.

The curves of Fig. 12.15 would seem to imply a maximum allowable bit rate not much greater than 5 Gb/s for trans-Pacific soliton transmission over a broadband transmission line (at least not for the specific choice of parameters reflected there). As already noted in Section 1, and as will soon be thoroughly explored in the following section, the technique of passive regeneration known as *guiding filters* has enabled that limit to be surpassed

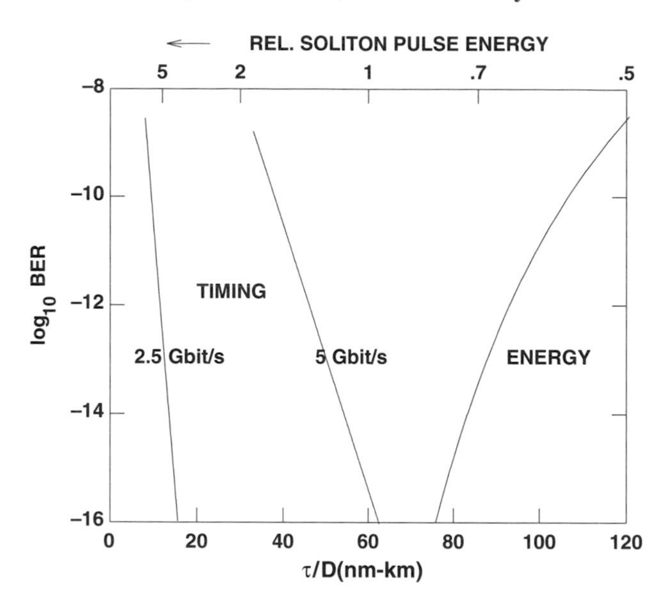

Fig. 12.15 BER, for energy and timing errors, as a function of the parameter τ/D, for a transmission distance of 9 Mm (the trans-Pacific distance). The other assumed parameters are as follows: fiber loss rate, 0.21 dB/km; L_{amp} = 30 km; n_{sp} = 1.5; A_{eff} = 50 μm^2; m = 8 (see Section 3.2).

by a large factor. Thus, the theory in this section, and its predictions, are largely of interest as background for the understanding of transmission using filters. Nevertheless, for the record, we close this section by citing the results of an experimental test of single-channel transmission at 5 Gb/s that was made a number of years ago [13]. Although made with guiding filters, the filters were only of the weak, fixed tuned type, so the results that would have been obtained without filters may reasonably be projected from them. See Fig. 12.16.

4. Frequency Guiding Filters

4.1 INTRODUCTION

In mid-1991, two groups independently suggested the idea that the Gordon–Haus jitter and other noise effects could be significantly suppressed in soliton transmission systems simply through a narrowing of the amplifier

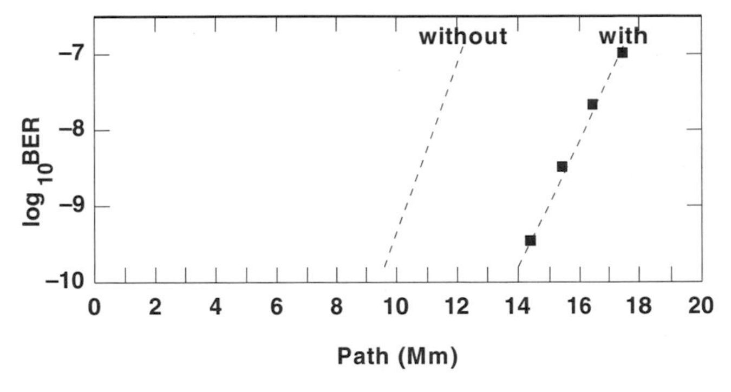

Fig. 12.16 Experimentally measured BER for a single-channel rate of 5 Gb/s, as a function of transmission distance, and with a 2^{13}-bit-long random sequence. *Curve labeled "with"*: Weak, fixed-frequency guiding filters used. *Curve labeled "without"*: Scaled back projection with no filters. The other parameters are as follows: fiber loss rate, 0.21 dB/km; L_{amp} = 28 km; D = 0.7 ps/nm-km; n_{sp} = 1.6; τ = 40 ps; A_{eff} = 35 μm^2. At the receiver, the effective window width was about 170 ps.

gain bandwidth [14, 15]. In practice, this means the use of narrow band filters, typically one per amplifier. Figure 12.17 shows appropriate filter response curves in comparison with the spectrum of a 20-ps-wide soliton. The fundamental idea is that any soliton whose central frequency has

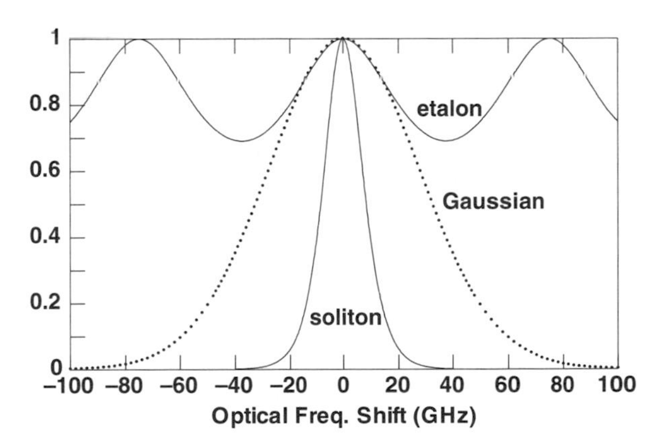

Fig. 12.17 Intensity response curves of a practical etalon guiding filter and a Gaussian filter with the same peak curvature, compared with the spectrum of a 20-ps soliton. The etalon mirrors have R = 9%, and their 2.0-mm spacing creates the 75-GHz free spectral range.

strayed from the filter peak will be returned to the peak, in a characteristic damping length delta (Δ), by virtue of the differential loss that the filters induce across its spectrum. The resultant damping of the frequency jitter leads in turn to a corresponding damping of the jitter in pulse arrival times. For example, in Eq. (3.9) for the variance of the Gordon–Haus jitter, when guiding filters are used, the quantities z_n^2 in the sum are all replaced with the common factor Δ^2. Thus, the factor $Z^3/3$ in the final expression is replaced by the (potentially much smaller) factor $Z \times \Delta^2$.

The filters also cause a reduction in amplitude jitter. Consider, for example, a pulse with greater than normal power; that pulse will be narrower in time, have a greater bandwidth, and hence experience greater loss from passage through the filter than the normal pulse. The opposite will occur for a pulse of less than standard power. Thus, amplitude jitter also tends to be dampened out, as is detailed later, in essentially the same characteristic length Δ as is the frequency jitter.

Because the major benefit comes from the filter response in the neighborhood of its peak, the etalon filter whose shallow response is shown in Fig. 12.17 provides almost as much benefit as the Gaussian filter of the simplest theory [14, 15]. But the etalon, with its multiple peaks, has the great fundamental advantage that it is compatible with extensive WDM. The etalons also have the practical advantage that they are simple, are low in cost, and can be easily made in a rugged and highly stable form.

It should be understood that linear pulses cannot traverse a long chain of such filters: after a sufficient distance, their spectra will be greatly narrowed, and the pulses correspondingly spread out in time. The solitons survive because they can regenerate the lost frequency components, more or less continuously, from the nonlinear term of the NLS equation. On the other hand, the amplifiers must supply a certain excess gain to compensate for the net loss that the solitons suffer from passage through the filters. As a result, noise components at or near the filter peak grow exponentially with distance. To keep the noise growth under control, the filters can be made only so strong, so the maximum possible benefit from them is limited. For example, Fig. 12.18 shows the standard deviation of timing jitter, as a function of distance, for systems with the optimum strength filters (those experimentally observed [13] to produce the best BER performance), and for those with no filters. Note that at the trans-Pacific distance of 9 Mm, the filters reduce the standard deviation of the jitter by a factor somewhat less than two times.

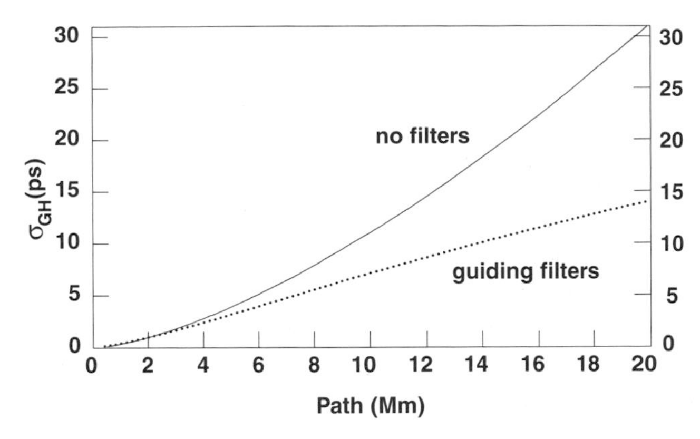

Fig. 12.18 Computed standard deviation, σ_{GH}, of pure Gordon–Haus jitter as a function of total path, for a broadband transmission line ("no filters") and for one having optimum strength, fixed-frequency guiding filters. The optimum filter strength, determined experimentally, corresponds to one uncoated, 1.5-mm-thick solid quartz etalon filter every 78 km. The other pertinent transmission line and soliton parameters in the strength-determining experiment were as follows: D = 0.7 ps/nm-km, n_{sp} = 1.4, and τ = 40 ps. (The BER data of a Fig. 12.16 were obtained in the same expriments [13].)

4.2 SLIDING-FREQUENCY GUIDING FILTERS

There is a simple and elegant way [16] to overcome the noise growth, and hence the limited performance, of a system of fixed-frequency filters. The trick is to "slide," i.e., translate, the peak frequency of the filters with distance along the transmission line (see Fig. 12.19). As long as the sliding is gradual enough, the solitons will follow, in accord with the same "guiding" principle that dampens the jitter. On the other hand, the noise, being

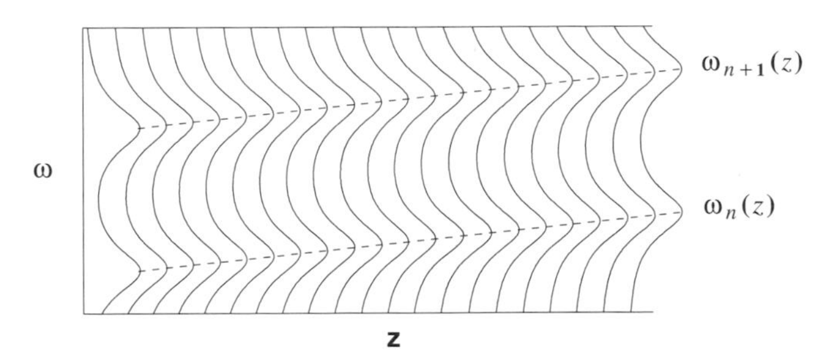

Fig. 12.19 Transmission of sliding-frequency guiding filters versus z.

essentially linear, can follow only the horizontal path in Fig. 12.19. Thus, the sliding creates a transmission line that is opaque to noise for all but a small, final fraction of its length, yet remains transparent to solitons. In consequence, the filters can be made many times stronger, and the jitter reduced by a corresponding factor, with the final result that the maximum bit rate can be increased at least several-fold over that possible without sliding.

The sliding-frequency filters provide many other important benefits beyond the simple suppression of timing and amplitude jitter. Note, for example, that they suppress all noiselike fields, whatever the source, such as dispersive-wave radiation from imperfect input pulses or other perturbations. They provide tight regulation of all the fundamental soliton properties, such as energy, pulse width, and optical frequency. As is detailed later, in WDM the filters suppress timing shifts and other defects from soliton–soliton collisions, and they provide a powerful regulation of the relative signal strengths among the channels in the face of wavelength-dependent amplifier gain. Thus, in short, the sliding-frequency guiding filters can be regarded as an effective form of passive, all-optical regeneration, and one that is uniquely compatible with WDM.

Finally, it should be noted that the required set of hundreds of sliding-frequency filters can be supplied more easily and at less cost than the corresponding set of fixed-frequency filters. That is, unlike the fixed-frequency mode, in which all filters must be carefully tuned to a common standard, the sliding-frequency mode does not require any tuning. Rather, only a statistically uniform distribution of frequencies is needed. Provided the distribution is over at least one or more free spectral ranges, simple ordering of the filters should be able to provide any reasonable desired sliding rate.

4.3 ANALYTIC THEORY OF GUIDING FILTERS

In the mathematical representation of a transmission line with filters, in general, only numerical solutions are possible when the exact response functions of real filters are used. Nevertheless, analytic solutions are possible when the filter response is approximated by a truncated series expansion [16]. When expanded in a Taylor series, the logarithm of the filter response function F takes the general form

$$\ln F(\omega - \omega_f) = i\zeta_1(\omega - \omega_f) - \zeta_2(\omega - \omega_f)^2 - i\zeta_e(\omega - \omega_f)^3 + \cdots, \quad (4.1)$$

where ω_f is the filter peak frequency and the constants ζ are all real and positive. The first-order term can be ignored because the linear phase shift that it provides serves only to translate the pulses in time. Although higher order filter terms can have important effects, the most fundamental features are revealed by analytic solutions to the simplified propagation equation employing only the second-order term:

$$\frac{\partial u}{\partial z} = i\left[\frac{1}{2}\frac{\partial^2 u}{\partial t^2} + u^* u^2\right] + \frac{1}{2}\left[\alpha - \eta\left(i\frac{\partial}{\partial t} - \omega_f\right)^2\right] u, \tag{4.2}$$

where α is the gain required to overcome the loss imposed on the solitons by the filters and $\eta = 2\zeta_2$. (Both continuously distributed quantities α and η are easily converted into lumped, periodic equivalents.) Without filter sliding ($d\omega_f/dz \equiv \omega_f' = 0$), and where, for convenience, we set $\omega_f = 0$, the exact stationary solution is

$$u = \sqrt{P}\,\mathrm{sech}(t)\exp(i\phi)$$
$$\text{where} \quad \phi = Kz - \nu \ln\cosh(t), \tag{4.3}$$

and the parameters (α, η, P, ν, K) must satisfy

$$\nu = \frac{3}{2\eta}\left[\left(1 + \frac{8\eta^2}{9}\right)^{1/2} - 1\right] = \frac{2}{3}\eta - \frac{4}{27}\eta^3 + \cdots \tag{4.4a}$$

$$\alpha = (\eta/3)(1 + \nu^2) \tag{4.4b}$$

$$P = (1 + \eta^2)(1 - \nu^2/2) \tag{4.4c}$$

$$K = (\tfrac{1}{2})(1 - \nu^2) + (\nu^2/3)(2 - \nu^2). \tag{4.4d}$$

Note from Eq. (4.3) that the pulse's frequency is chirped — i.e., $\partial\phi/\partial t = -\nu\tanh(t)$. Because of this chirp, the root mean square (rms) bandwidth is increased by the factor $(1 + \nu^2)^{1/2}$.

Numerical simulation involving real filters shows that the principal features (the chirp, extra bandwidth, and increased peak power) of the previous solution are approximately preserved. The major differences lie in slight asymmetries induced in $u(t)$ and in its spectrum by the third-order filter term, and in the fact that, through a complex chain of events, those asymmetries cause the soliton mean frequency to come to rest somewhat above the filter peak.

Numerical simulation has also shown that sliding (within certain limits given subsequently) does not significantly alter this solution. Sliding does, however, have the potential to alter the damping of amplitude and frequency fluctuations. To get some notion of the effects of sliding on damping, we introduce the general unperturbed form for the soliton (Eq. [2.36]) into Eq. (4.2). We then obtain the following pair of coupled, first-order perturbation equations:

$$\frac{1}{A}\frac{dA}{dz} = \alpha - \eta[(\Omega - \omega_f)^2 + \tfrac{1}{3}A^2] \tag{4.5a}$$

$$\frac{d\Omega}{dz} = -\frac{2}{3}\eta(\Omega - \omega_f)A^2. \tag{4.5b}$$

According to Eq. (4.5b), equilibrium at $A = 1$ and at constant ω_f' requires that $d\Omega/dz = \omega_f'$, and hence that the lag $\Delta\Omega \equiv (\Omega - \omega_f)$ of the soliton mean frequency behind the filter frequencies is

$$\Delta\Omega = -\frac{3}{2\eta}\omega_f'. \tag{4.6}$$

Equation (4.6), as written, correctly predicts the difference in lag frequencies for up- versus down-sliding, $\Delta\Omega_u - \Delta\Omega_d$. To acount for the offset in $\Delta\Omega$ produced by the third-order filter term (mentioned previously for the case of no sliding), one must add a positive constant (as determined empirically from numerical simulation) to the right-hand side of Eq. (4.6). Equation (4.6) then correctly predicts $\Delta\Omega$ for all sliding rates. For etalon filters, the offset in $\Delta\Omega$ has been estimated from the third-order filter term [17]. That is, it has been shown from perturbation theory that

$$\Delta\Omega_{offset} \approx \tfrac{6}{5}\zeta_3, \tag{4.7}$$

where ζ_3 is computed as

$$\zeta_3 = \frac{1.762\ldots}{6}\frac{(1+R)}{(1-R)}\frac{\eta}{\tau F} \tag{4.7a}$$

where F is the free spectral range of the etalon filters. Combining Eqs. (4.6), (4.7), and (4.7a), one obtains

$$\Delta\Omega = \left(\frac{1.762\ldots}{5}\frac{(1+R)}{(1-R)}\frac{\eta}{\tau F}\right) - \frac{3}{2\eta}\omega_f'. \tag{4.6a}$$

Note that the two terms in Eq. (4.6a) tend to cancel for up-sliding ($\omega_f' > 0$), whereas they add for down-sliding. Because the damping of the filters is best for the smallest $|\Delta\Omega|$ (see Eq. [4.8]), up-sliding is definitely preferable to down-sliding.*

* *Note:* In the original article on sliding-frequency guiding filters (Ref. 16), because of an improper combination of sign conventions, the numerical simulations reported on there incorrectly yielded the offset $\delta\omega < 0$. As a further result of that error, the article incorrectly recommended down-sliding.

Equations (4.5a) and (4.5b), when linearized in small soliton frequency and amplitude displacements δ and a, respectively, yield two eigenvalues (damping constants):

$$\gamma_1 = \tfrac{2}{3}\,\eta(1 + \sqrt{6}\,\Delta\Omega) \qquad \text{and} \qquad \gamma_2 = \tfrac{2}{3}\,\eta(1 - \sqrt{6}\,\Delta\Omega), \tag{4.8}$$

with corresponding normal modes $x_1 = \delta + \sqrt{\tfrac{2}{3}}a$ and $x_2 = \delta - \sqrt{\tfrac{2}{3}}a$. This implies a monotonic decrease of damping for both frequency and amplitude fluctuations with increasing $|\Delta\Omega|$, and, through Eq. (4.6), the existence of maximum allowable sliding rates for stability. The numerical simulations we have done to date with real filters are at least qualitatively consistent with these predictions.

In principle, on the basis of the damping constants of Eq. (4.8), one can go on to write expressions for the variances in soliton energy and arrival time. It is not at all clear, however, how accurate such expressions would be in predicting the effects of strong, real filters. Nevertheless, because we are primarily interested in the behavior for $\gamma z \gg 1$, where the energy fluctuations have come to equilibrium with the noise, the energy variance can be written as

$$\frac{\langle \delta E_{sol}^2 \rangle}{E_{sol}^2} \approx \frac{N}{\gamma_E E_{sol}} = \frac{N\Delta_E}{E_{sol}}, \tag{4.9}$$

where N is the spontaneous emission noise spectral density generated per unit length of the transmission line, E_{sol} is the soliton pulse energy, and the effective damping length, $\Delta_E = 1/\gamma_E$, is expected to increase monotonically with increasing $|\Delta\Omega|$. Note that Eq. (4.9) implies that as far as the noise growth of ones is concerned, the system is never effectively longer than Δ_E. Because the characteristic damping lengths with sliding-frequency filters are typically about 600 km or less, this means a very large reduction of amplitude jitter in transoceanic systems.

As far as the variance in timing jitter is concerned, we have already seen that, for $\gamma z \gg 1$, the factor $Z^3/3$ is replaced by $Z\Delta_t^2$. In other words, the variance in timing jitter is subject to a reduction factor

$$f(\gamma_t, z) \approx \frac{3}{(\gamma_t Z)^2} = 3\left(\frac{\Delta_t}{Z}\right)^2, \tag{4.10}$$

where Δ_t is also expected to increase monotonically with increasing $|\Delta\Omega|$. Although Δ_E and Δ_t are in general different, nevertheless, for Gaussian filters, both are expected to be approximately equal to $3/(2\eta)$ in the neighborhood of $\Delta\Omega = 0$.

For a transmission line using Fabry–Perot etalon filters with mirror spacing d and reflectivity R, the parameters η, ω_f', and α, in soliton units, are computed from the corresponding real-world quantities as follows [18]:

$$\eta = \frac{8\pi R}{(1 - R)^2}\left(\frac{d}{\lambda}\right)^2 \frac{1}{cDL_f} \tag{4.11a}$$

$$\omega_f' = 4\pi^2 f' c t_c^3/(\lambda^2 D) \tag{4.11b}$$

$$\alpha = \alpha_R t_c^2 2\pi c/(\lambda^2 D). \tag{4.11c}$$

Here, f' and α_R are just $\omega_f'/2\pi$ and α, respectively, but as expressed in "real" units (such as GHz/Mm, for example); $t_c \equiv \tau/1.763$ (Eq. [2.14]); and L_f is the filter spacing.

We can now illustrate the power of sliding-frequency filters through a specific numerical example. Anticipating a bit from the next section, where we discuss the optimum choice of filter parameters, we choose the following soliton and fiber parameters: $D = 0.5$ ps/nm-km, $\tau = 16$ ps, so $z_c = 128$ km. The sliding rate will be 13 GHz/Mm (note that this means that the total sliding will be just about 1 nm in the trans-Pacific distance), so by Eq. (4.11b), $\omega_f' = 0.095$. For the filters, we choose $R = 8\%$, 2-mm air-gap etalons, with $L_f = 50$ km. By Eq. (4.11a), $\eta = 0.52$. Thus, we have $\alpha = 0.185$, and by Eq. (4.11c), $\alpha_R = 1.4$/Mm. For the damping constants of shallow etalons, however, η has a certain functional dependence on τ and must be degraded to about $\eta_{eff} = 0.4$ for the 16-ps pulses to be used in this case.

The relative noise growth with sliding-frequency filters is easily simulated. Figure 12.20 shows the results of such a simulation for the conditions of our example. Note that the sliding keeps the peak spectral density clamped to a value less than that which would be obtained at 10 Mm without filtering, whereas without the sliding, the noise would potentially grow by e^{14}, or about 1.2 million times, in the same distance! (Long before that could happen, however, the amplifiers would saturate.) Also note the spectral narrowness of the noise.

In Fig. 12.21, normalized standard deviations of the soliton energy (ones) and of the noise energy in empty bit periods (zeros) are shown as functions of distance [16]. These curves are obtained from Eq. (4.9), and the estimate $\Delta_E \approx 600$ km derived from $\eta_{eff} \approx 0.4$ and Eq. (4.8), the data of Fig. 12.20, and the analysis of Section 3.2. Note that with the filtering, both standard deviations soon become clamped to small, indefinitely maintained values, corresponding to immeasurably small BERs.

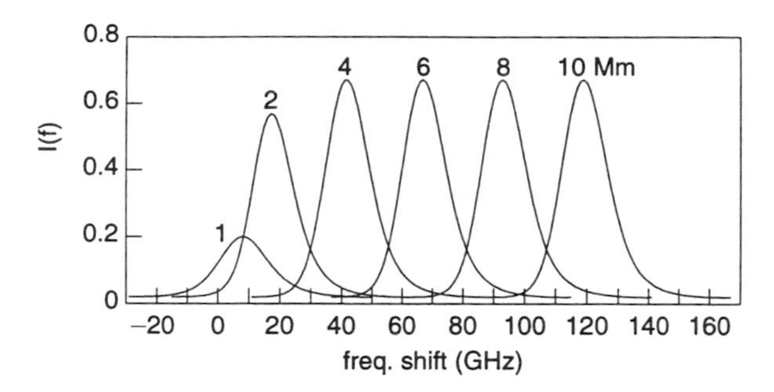

Fig. 12.20 Noise spectral density, normalized to value at 10 Mm with no filtering, as a function of frequency and distance, for the conditions of our example (one R = 8%, 75-GHz free spectral range (FSR) etalon filter per 50 km; sliding rate = 13 GHz/Mm; α_R = 1.4/Mm).

Finally, in Fig. 12.22, the standard deviation of the Gordon–Haus jitter is plotted versus distance, both for when the sliding filters are used and for when there are no filters. Note the nearly 10× reduction in σ_{GH} at 10 Mm, and compare it with the same factor from Fig. 12.18.

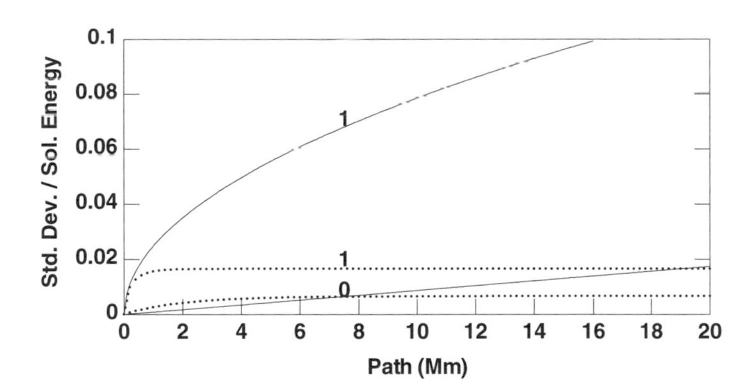

Fig. 12.21 Standard deviations of the soliton energy (ones) and of the noise in empty bit periods (zeros) versus distance, both normalized to the soliton energy itself, for the sliding-filter scheme of Fig. 12.20 (*dotted curves*), and with no filtering save for a single filter at z, passing only eight noise modes (*solid curves*). The assumed fiber loss rate and effective core area are 0.21 dB/km and 50 μm^2, respectively; the amplifier spacing and the excess spontaneous emission factor are ~30 km and 1.4, respectively.

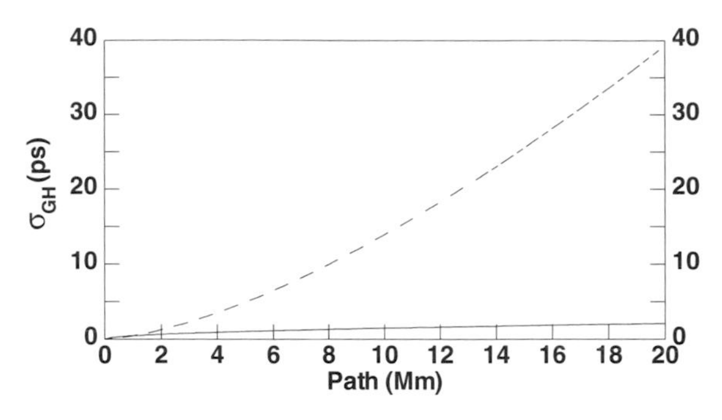

Fig. 12.22 Standard deviation of the Gordon–Haus jitter, σ_{GH}, as a function of total path. *Solid curve:* With strong, sliding-frequency guiding filters. *Dashed curve:* No filters. Conditions: $\tau = 16$ ps; $D = 0.5$ ps/nm-km; $n_{sp} = 1.4$; $F(G) = 1.1$; filter strength parameter $\eta = 0.5$; damping length $\Delta \approx 600$ km.

4.4 *EXPERIMENTAL CONFIRMATION*

4.4.1 Measurement of Noise and Amplitude Jitter

Figures 12.23 and 12.24 refer to experimental transmission with sliding-frequency guiding filters, where the parameters are at least similar, if not identical, to those in the example cited in Section 4.2. Figure 12.23 shows the signal and noise levels during a 15-Mm-long transmission, where the signal train was purposely made not quite long enough to fill the recirculating loop. Thus, once each round-trip, for a period too brief (a few microseconds) for the amplifier populations to change significantly, one sees only

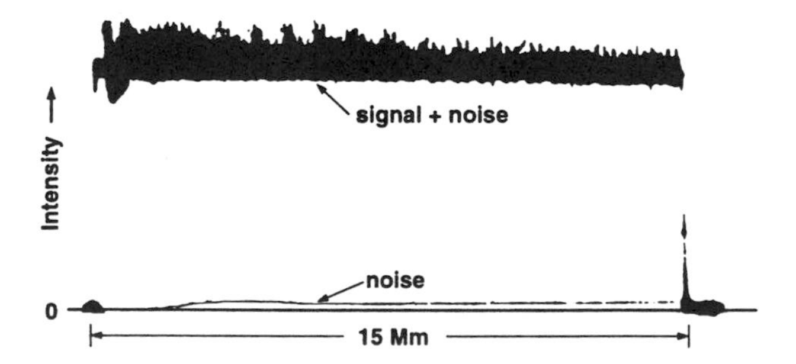

Fig. 12.23 Noise and signal levels during a transmission using strong, sliding-frequency filters. The signal level is represented by the *thick, upper line,* whereas the noise level is represented by the *fine line* immediately above the zero signal level.

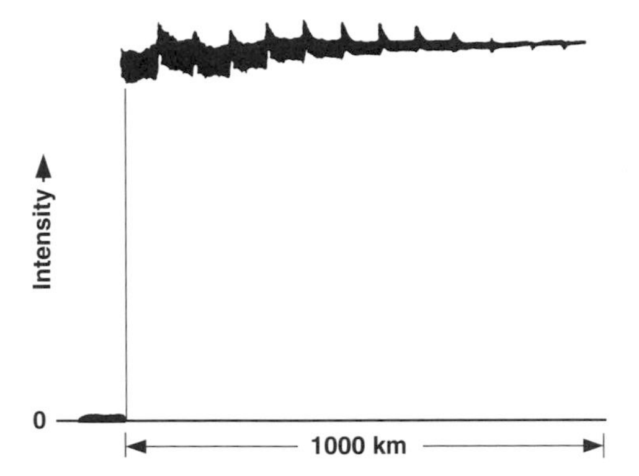

Fig. 12.24 Observed amplitude jitter reduction with successive round-trips in a transmission using sliding-frequency filters.

the noise. Note that, just as in the theoretical model (Fig. 12.20), the noise grows for only a few megameters and then saturates at a steady, low value.

In another early experiment with sliding-frequency filters, the pulse source was a mode-locked, erbium fiber ring laser, which had been purposely maladjusted to produce a substantial amplitude jitter at a few tens of kilohertz. Figure 12.24 shows the very rapid reduction in that amplitude jitter with successive round-trips in the recirculating loop. The data shown in this figure imply a damping length of about 400 km, which is consistent with the filter strength parameter of $\eta \approx 0.6$ and the known dispersion length $z_c = 160$ km (see Eq. [4.8]).

4.4.2 Measurement of Timing Jitter

The timing jitter in a transmission using sliding-frequency filters has been measured accurately by observing the dependence of the BER on the position, with respect to the expected pulse arrival times, of a nearly square acceptance window in time [19]. The scheme, which involved time-division demultiplexing, is shown in Fig. 12.25. The fundamental measurement is of the time span (inferred from the precision phase shifter in Fig. 12.25), for which the BER is 10^{-10} or less for each distance. The resultant spans, or time-phase margins, are plotted in Fig. 12.26, as a function of distance, for three cases: (1) a 2.5-Gb/s data stream (which, because it also passes through the loop mirror, can be thought of as a 10-Gb/s data stream for which only every fourth 2.5-Gb/s subchannel is occupied); (2) a true

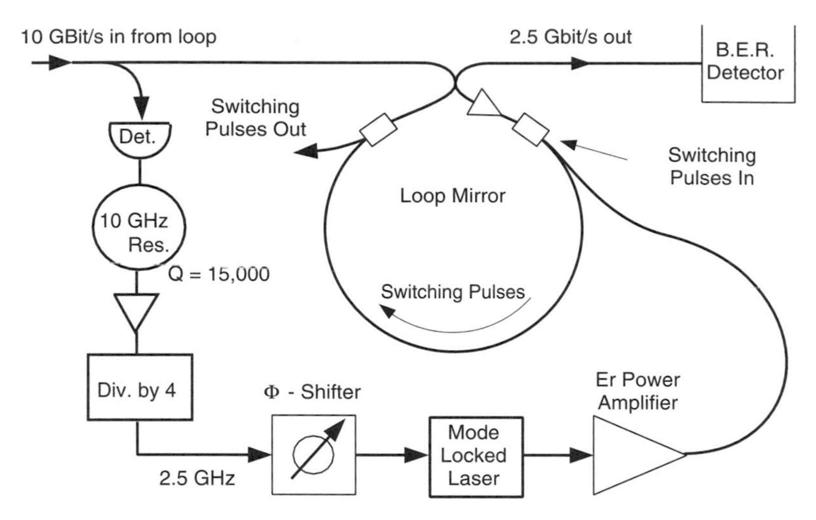

Fig. 12.25 Scheme for the pulse timing measurements. The main elements of the clock recovery are the detector; the high-Q, 10-GHz resonator; and the divide-by-4 chip. The wavelength-dependent couplers in the loop mirror (*small rectangular boxes*) each contain an interference filter that transmits at the signal wavelength (~1557 nm) and reflects the $\lambda = 1534$ nm switching pulses.

10-Gb/s data stream with adjacent pulses orthogonally polarized; (3) a 10-Gb/s data stream with all pulses copolarized. Note that the error-free distances (for which the phase margin first becomes zero) are 48, 35, and 24 Mm, respectively.

From the known properties of the error function, the difference between the effective width (here 82 ps) of the acceptance window and the measured

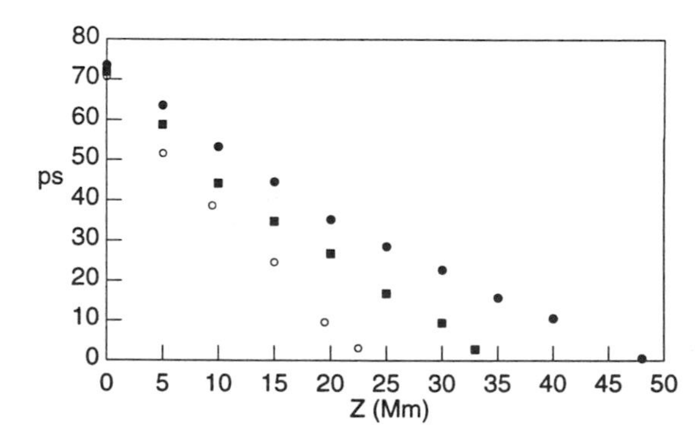

Fig. 12.26 Time-phase margin versus distance. *Bullets:* 2.5 Gb/s; *squares:* 10 Gb/s, adjacent pulses orthogonally polarized; *open circles:* 10 Gb/s, all pulses copolarized.

time-phase margin at a 10^{-10} BER should be about 13 σ of the Gaussian distribution in pulse arrival times. From this fact, one can then obtain the plots of σ shown in Fig. 12.27. Note that in all three cases shown there, the data makes a good fit to a curve of the form

$$\sigma = \sqrt{\sigma_0^2 + \sigma_{GH}^2 + \sigma_{lin}^2}, \tag{4.12}$$

where σ_0, σ_{GH}, and σ_{lin} represent the standard deviations of the source jitter (a constant), filter-damped Grodon–Haus jitter (varies as $z^{1/2}$), and jitter whose σ varies linearly with z, respectively.

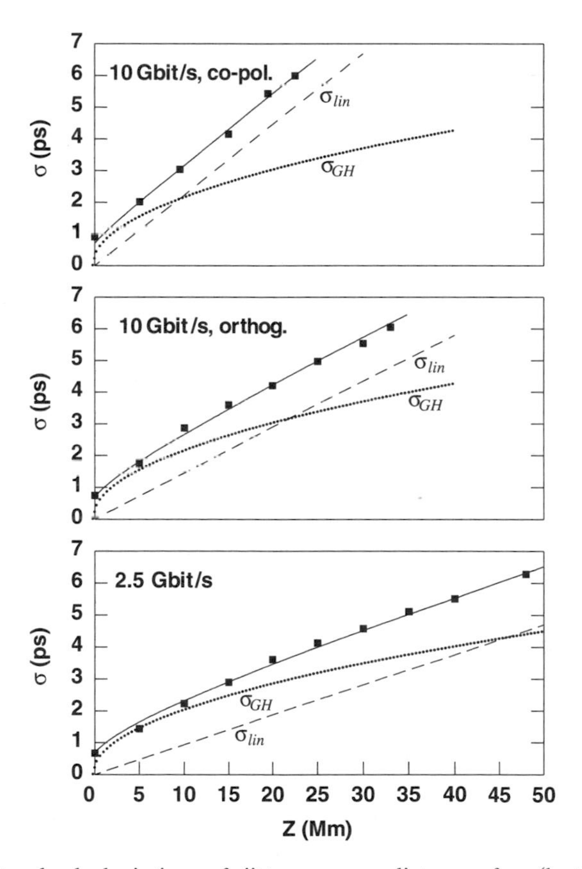

Fig. 12.27 Standard deviation of jitter versus distance for (bottom to top) 2.5 Gb/s, 10 Gb/s with adjacent pulses orthogonally polarized, and 10 Gb/s with adjacent pulses copolarized. *Squares:* Experimental points, as extrapolated from the data of Fig. 12.25; *solid curve:* best fit to theoretical curve of form $\sigma = \sqrt{\sigma_0^2 + \sigma_{GH}^2 + \sigma_{lin}^2}$; *dotted curve:* σ_{GH}; *dashed line:* σ_{lin}.

The best-fit Gordon–Haus term (σ_{GH}) is always about two times greater than expected for the known parameters of the experiment, and for the damping length of approximately 400 km as calculated, and as confirmed by the independent measurement of the damping of amplitude jitter (Fig. 12.24). The frequency offset of the pulse spectra from the filter peaks is very small and can thus at best account for only a small part of the discrepancy, despite speculation to the contrary [20]. Thus, the only likely explanation here is in terms of the noiselike fields of dispersive-wave radiation. Among the perturbations that may be responsible for significant amounts of such noise are the fiber's birefringence (see Section 6) and the periodic intensity associated with the use of lumped amplifiers (Section 2.5.2).

There are two contributions to the linear term; polarization jitter and the acoustic effect (Section 3.4). (The polarization jitter arises from a noise-induced spread in the polarization states of the solitons and the conversion of that spread by the fiber's birefringence into a timing jitter. This effect discussed further in Section 6.) Because both contributions have essentially Gaussian distributions, the effects add as $\sigma_{lin} = \sqrt{\sigma_{pol}^2 + \sigma_a^2}$. For a filtered transmission line, the factor $Z^2/2$ in Eq. (3.12) is replaced with $Z \times \Delta$. Thus modified, Eq. (3.12) becomes:

$$\sigma_a \approx 8.6 \frac{D^2}{\tau} Z\Delta \sqrt{R - 0.99}. \tag{3.12a}$$

Using the bit-rate dependence of σ_a and the slopes of σ_{lin} from the two lower plots of Fig. 12.27, one can easily extract values for σ_{pol} and σ_a. At $Z = 10$ Mm, those values are as follows: $\sigma_{pol} = 0.80$ ps, $\sigma_{a,2.5} = 0.50$ ps, and $\sigma_{a,10} = 1.21$ ps. The experimental values for σ_a are just 7% less than predicted by Eq. (3.12a), a remarkable degree of agreement.

BER measurements have been made at 12.5 and 15 Gb/s, as well as at the 10 Gb/s already cited [19]. Figure 12.28 summarizes those results. Finally, it should be noted that by using sliding-frequency guiding filters, LeGuen *et al.* [21] achieved error-free transmission at 20 Gb/s over more than 14 Mm.

4.5 STABILITY RANGE

The range of soliton pulse energies for which the transmission with sliding-frequency filters is stable and error free will henceforth be simply referred to as the *stability range*. It is important for the stability range to be large enough (at least several decibels) to allow for the aging of amplifier pump lasers, and other factors that may tend to degrade the signal strength with

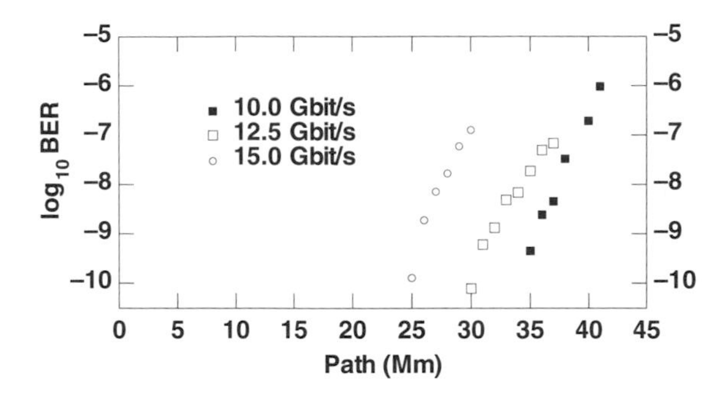

Fig. 12.28 Measured BER as a function of distance, at 10, 12.5, and 15 Gb/s. In all cases, adjacent pulses were orthogonally polarized, and the data stream was a repeated, 2^{14}-bit random word.

time, in real systems. Perhaps not surprisingly, the stability range is a function of both the filter strength parameter (η) and the sliding rate (ω_f'). The following is a brief summary of an experimental determination of that dependence, and of the optimum values for those parameters [18].

The experiment was carried out in a small recirculating loop with piezo-driven etalon sliding-frequency filters having fixed reflectivity ($R = 9\%$) and fixed mirror spacing ($d = 1.5$ mm), and where $L_f = 39$ km. Because of the fact that η is inversely proportional to D (see Eq. [4.11a]), the fiber's third-order dispersion ($\partial D/\partial\lambda = 0.7$ ps/km^2-nm) enabled η to be varied simply through change of the signal wavelength itself. At the same time, the signal power at equilibrium, hence the soliton pulse energies, could be controlled by means of the pump power supplied to the loop amplifiers. Thus, the experiment consisted simply of measuring, for each signal wavelength, and for a fixed sliding rate, the maximum and minimum signal power levels for which a transmission over 10 Mm was stable and error free. The results are shown in Fig. 12.29. Note that although error-free propagation ceases for $\eta \geq 0.8$, the stability range reaches a maximum of nearly two to one for $\eta \approx 0.4$. Figure 12.30 shows the complementary data — i.e., the measured stability range as a function of sliding rate — for fixed $\eta = 0.4$. Note that in this case, too, there is an optimum rate of about 13 GHz/Mm. Essentially the same results as in Figs. 12.29 and 12.30 were obtained for two other values of L_f (26 and 50 km, respectively) and for etalons having an FSR of 75 GHz (as opposed to 100 GHz).

The existence of $E_{\min}$ is easily predicted from the analysis of Section 4.3. That is, for stability, neither of the damping constants can be negative,

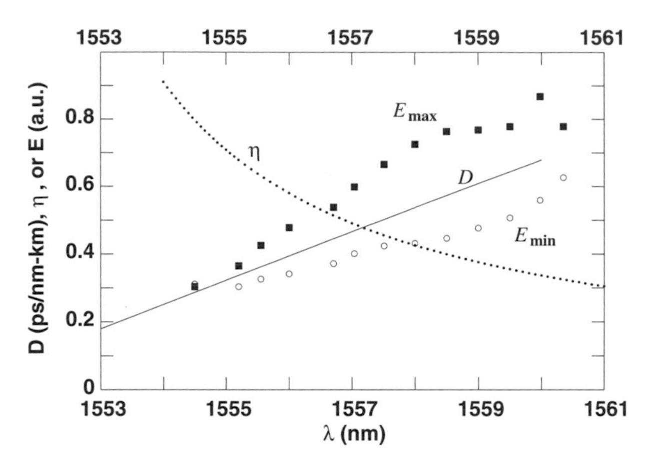

Fig. 12.29 Fiber dispersion D (*solid line*), filter strength parameter η (*dotted line*), and experimentally determined allowable soliton pulse energy limits $E_{\max}$ and $E_{\min}$ (*filled squares* and *open circles*, respectively), as functions of the signal wavelength λ, for a fixed sliding rate of 13 GHz/Mm.

so from Eq. (4.8), one has $|\Delta\Omega| \leq 1/\sqrt{6}$. From Eq. (4.6) or (4.6a), one then gets a maximum allowed sliding rate, ω_f', in soliton units. Finally, from Eq. (4.11b), one sees that, for fixed real sliding rate f' and for fixed D, ω_f' increases as the third power of the pulse width, and hence inversely as the third power of the pulse energy. Perhaps less abstractly, one can easily see that as the pulse energy is lowered, the rate at which the nonlinear term can alter the soliton's frequency will eventually become so low that it can no longer keep up with the filter sliding.

The existence of the upper limit, $E_{\max}$, may be somewhat less obvious, but it has to do with the fact that eventually, as its energy is raised, the soliton's bandwidth, hence its loss from the filters, becomes too great. In this example, numerical simulation was helpful in elucidating the precise failure mechanism. Figure 12.31 shows the simulated pulse intensity evolution at $\eta = 0.4$ for different values of α_R. Note that for α_R below some critical value (here, $\approx$1.45/Mm), there is no stable solution, and the pulse disappears after some distance of propagation; this corresponds to the lower energy limit already discussed. Above this lower limit, there is a range of allowable values of α_R (between 1.5/Mm and 3.5/Mm in Fig. 12.31). Nevertheless, one can see nondecaying oscillations in the pulse intensity

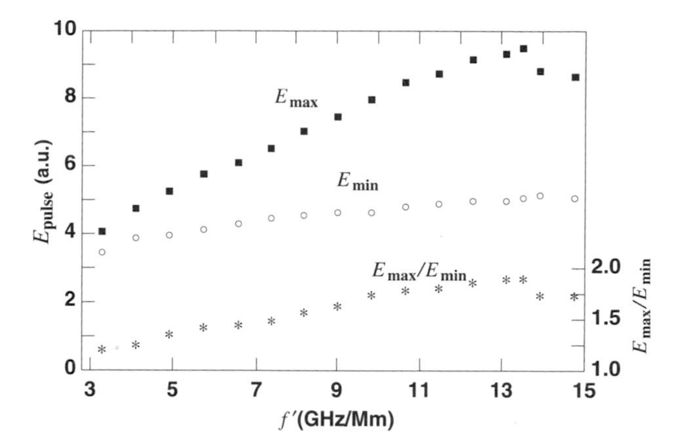

Fig. 12.30 Experimentally determined allowable soliton pulse energy limits E_{max} and E_{min} (*filled squares* and *open circles*, respectively), and their ratio (*asterisks*), as functions of the frequency-sliding rate f', for fixed, optimum filter strength parameter $\eta = 0.4$.

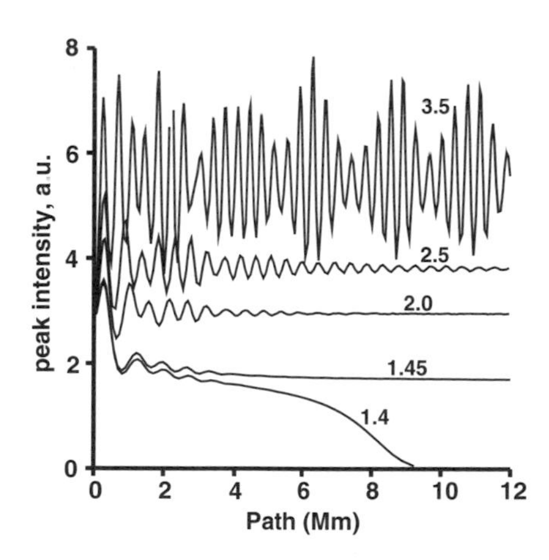

Fig. 12.31 Soliton peak intensities as a function of distance, as determined by numerical simulation, for filter strength $\eta = 0.4$ and for various of excess gain. The number next to each curve represents the excess gain parameter, α_R, in units of Mm^{-1}.

evolution for the higher values of α_R. These oscillations are due to a nonsoliton component, not completely removed by the sliding, and generated by the perturbing effects of the filtering and sliding themselves. If the excess gain is further increased, this nonsoliton component evolves into a second soliton. Clearly, this process determines the upper limit of the excess gain and of the soliton energy.

4.6 FILTERING IN TIME

Another form of optical regeneration for solitons involves the use of intensity modulators periodically placed along the transmission line and timed to open only during the middle of each bit period. The mean position in time of a pulse that is either early or late is thus guided back to the center of its bit period in a manner that is analogous to the guiding in the frequency domain provided by the etalon filters. Note, for example, that like frequency filtering, this "filtering in time" requires excess gain to overcome the loss imposed by the modulators, even to those arriving exactly on time. Unlike frequency filtering, however, filtering in time is not stable by itself; rather, it must always be accompanied by a proportional amount of frequency filtering. The principal advantage of filtering in time is that it corrects timing jitter directly, rather than indirectly as the frequency filtering does. When combined with filtering in the frequency domain, it can offer error-free transmission over an indefinitely long distance, at least in principle [22, 23].

Unfortunately, however, filtering in time shares several of the most fundamental disadvantages of electronic regeneration. First, it is incompatible with WDM. (To do WDM, at each regenerator the N channels must be demultiplexed, separately regenerated, and then remultiplexed, in a process that is at once extremely expensive and an engineering nightmare.) Second, each regenerator requires a quantity of active, failure-prone hardware, including clock recovery, adjustable delay lines, and modulator drive, in addition to the modulator itself. (Compare this with the extremely simple, inexpensive, stable, and strictly passive etalon filters of the pure frequency filtering.) At present, there are serious technical difficulties as well, such as the fact that nonchirping intensity modulators, whose insertion loss is polarization independent, simply do not exist. Combine all that with the fact that through massive WDM the sliding-frequency filters have already enabled a net capacity of nearly 100 Gb/s, whereas single-channel transmission at anything like that rate is at best formidably difficult, and filtering in time no longer looks economically or technically competitive for transmission. On the other hand, filtering in time might be useful in enabling long-term, fast-access data storage in large recirculating fiber loops.

4.7 LUMPED DISPERSION COMPENSATION AND FILTERING

From the discussion in Section 3.3 on the Gordon–Haus effect, we have seen how the fiber's dispersion converts the frequency displacements of the solitons into corresponding timing displacements (see Eq. [3.9]). Thus, to the extent that the accompanying pulse broadening can be tolerated, one might be tempted to consider using dispersion-compensating elements (either fiber having $D < 0$ or some equivalent lumped device) to reduce the net timing jitter. In principle, at least, postransmission dispersion compensation of $-\frac{1}{2}$ the total dispersion of the transmission line itself can reduce the standard deviation of timing jitter by a factor of 2 [24]. As the system length grows beyond a few times z_c, however, to prevent pulse broadening from exceeding the bit period, the compensating element must become weaker, and the jitter reduction correspondingly small. Because in most cases of interest the system is many times z_c long, such dispersion compensation does not buy much improvement.

Very recently, however, Suzuki *et al.* [25] have shown how to make much better use of dispersion compensation. In their experiments, periodically along the transmission line, these researchers insert elements that compensate for 90–100% of the dispersion. In a system N amplifier spans long, and where there is complete compensation after each set of n amplifiers, one can easily show (again, see Eq. [3.9]) that the standard deviation of Gordon–Haus jitter should be reduced by a factor of n/N. In the experiments of Suzuki *et al.*, where $N \approx 300$ and $n = 10$, the Gordon–Haus jitter is thus greatly reduced. To the extent that timing jitter is the dominant source of errors, it is thus not surprising that transmission at 20 Gb/s was error free over as much as 14 Mm. What is surprising is that the pulses were apparently able to tolerate the huge periodic perturbation caused by the dispersion compensation. One would expect that perturbation to generate copious amounts of nonsoliton components. Indeed, apparently it is necessary to use filters as well [25], in order to suppress the growth of such nonsoliton fields. Because the optimum filter strength in this case is much weaker than with sliding-frequency filters, it is not surprising that the decrease in the Q factor with distance is primarily from amplitude jitter [25].

Because of third-order dispersion, any technique involving dispersion compensation tends to work only over a rather narrow wavelength band.

In addition, the discontinuity in D at each point of compensation would tend to wreak havoc with soliton–soliton collisions centered there. Thus, Suzuki *et al.*'s dispersion-compensation technique is likely not suitable for WDM. For single-channel transmission, however, it represents an intriguing concept, and one that may be of practical interest.

5. Wavelength-Division Multiplexing

5.1 SOLITON–SOLITON COLLISIONS IN WDM

In WDM, solitons of different channels gradually overtake and pass through each other. Because the solitons interact with each other then, the time of overlap is known as a *collision* (Fig. 12.32). An important parameter here is the collision length, L_{coll}, or the distance that the solitons must travel down the fiber together in the act of passing through each other. If L_{coll} is defined to begin and end with overlap at the half-power points, then transparently

$$L_{coll} = \frac{2\tau}{D\Delta\lambda}, \tag{5.1}$$

where $\Delta\lambda = \lambda_1 - \lambda_2$. For example, for $\tau = 20$ ps, $D = 0.5$ ps/nm-km, and $\Delta\lambda = 0.6$ nm, $L_{coll} = 133$ km.

The interaction stems, of course, from the nonlinear susceptibility, $\chi^{(3)}$ (see Section 2.3), or equivalently, from the nonlinear term in the NLS equation (Eq. [2.12a]). In a single-channel transmission, the only significant effect of that term is the *self-phase modulation* resulting from the self-induced index change at each pulse. During collisions, however, each pulse experiences an additional nonlinear index change as induced by the other pulse or pulses; as is detailed shortly, the resultant *cross-phase modulation* tends to produce shifts in the mean frequency, or group velocity, of the

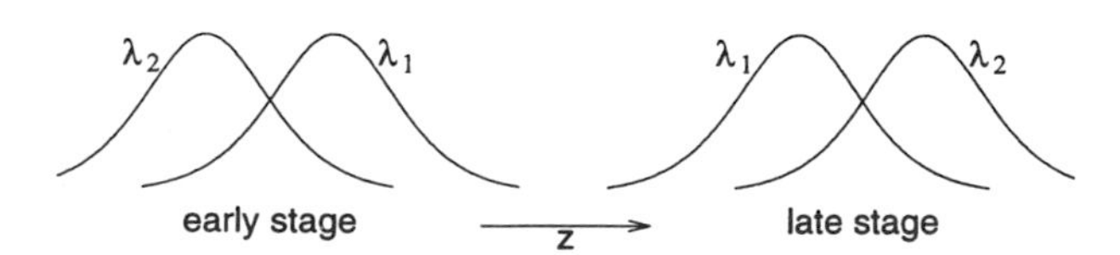

Fig. 12.32 Two stages of a soliton–soliton collision. Because of the anomalous group velocity dispersion, the shorter wavelength soliton (λ_2) gradually overtakes and passes through the longer wavelength one (λ_1).

affected pulse. Finally, the nonlinear term enables the colliding pulses to produce fields at the frequencies $\omega_S = 2\omega_1 - \omega_2$ and $\omega_A = 2\omega_2 - \omega_1$ (the Stokes and anti-Stokes frequencies, respectively) in the process known as *four-wave mixing*. In general, these effects have the potential to bring about a significant exchange of energy and momentum between the pulses and hence to create serious timing and amplitude jitter. Under the right conditions, however, with solitons the nonlinear effects are only transient — i.e., the solitons emerge from a collision with pulse shapes, widths, energies, and momenta completely unchanged. It is this potential for nearly perfect transparency to each other that makes solitons so well adapted for WDM.

5.2 COLLISIONS IN LOSSLESS AND CONSTANT-DISPERSION FIBER

It is useful first to consider the ideal case of lossless and constant-dispersion fiber. It is the simplest example of perfect transparency, and it is the easiest to analyze. It is of more than academic interest, however, because it can also serve as the paradigm for a number of practically realizable situations with real fibers and lumped amplifiers. In particular, it is the exact mathematical equivalent of the (at least approximately) realizable and practically important case of dispersion-tapered fiber spans (see Section 2.5.3).

The colliding solitions are written in the general form of Eq. (2.36), where, for simplicity, we set $A = 1$:

$$u_1 = \mathrm{sech}(t + \Omega_1 z)e^{-i\Omega_1 t}e^{i(1-\Omega_1^2)z/2} \tag{5.2a}$$

$$u_2 = \mathrm{sech}(t + \Omega_2 z)e^{-i\Omega_2 t}e^{i(1-\Omega_2^2)z/2}. \tag{5.2b}$$

Although the solitons' frequencies will change during the actual collision, and the difference between them is determined by their relative velocities, we are perfectly free to choose the zero of frequency. For convenience then, let us set $\Omega_2(-\infty) = -\Omega_1(-\infty) = \Omega \gg 1$; note that this makes the solitons move with equal but opposite velocities in the retarded time frame. Also note that this makes $L_{coll} = 1/\Omega$ ($= 1.762 \ldots \times z_c/\Omega$ in ordinary units).

Now let us insert $u = u_1 + u_2$ into the NLS equation for lossless fiber, expand, and group the terms according to their frequency dependencies. As long as $\Omega \gg 1$, the various frequency terms are independent. Thus, we get four equations: one for each soliton and one each for the terms in $\pm 3\Omega$. The latter correspond to the aforementioned four-wave mixing components. Because for the special case under consideration these components are weak and disappear completely after the collision, they are neglected for now. Assuming, for the moment, that the pulses are copolarized, the equation for u_1 is

$$\frac{\partial u_1}{\partial z} = i\frac{1}{2}\frac{\partial^2 u_1}{\partial t^2} + i|u_1|^2 u_1 + 2i|u_2|^2 u_1. \tag{5.3}$$

The first two terms on the right in Eq. (5.3) correspond to the NLS equation for the isolated pulse, u_1. The last term in Eq. (5.3) corresponds to the cross-phase modulation and is zero except when the pulses overlap. It produces a transient phase shift in u_1 at the rate

$$\frac{d\phi_1(z, t)}{dz} = 2|u_2(z, t)|^2.$$

The corresponding frequency shift $\omega_1(z, t) = \partial\phi_1/\partial t$ in u_1 is thus induced at the rate

$$\frac{\partial \omega_1}{\partial z} = \frac{\partial}{\partial z}\frac{\partial \phi_1}{\partial t} = \frac{\partial}{\partial t}\frac{\partial \phi_1}{\partial z} = \frac{\partial}{\partial t}(2|u_2|^2). \tag{5.4}$$

Note that this induced frequency shift is not uniform across the pulse. Nevertheless, the soliton retains its shape, and we want the shift in its inverse group velocity. Now it can be shown rigorously that the inverse group velocity is given by (-1 times) the mean frequency of the pulse. Thus, we really want $\delta\Omega_1 = \langle\omega_1\rangle$, the time-averaged frequency of u_1.

Using the weighting factor $|u_a(t)|^2$, and using the fact that $\int_{-\infty}^{\infty} \mathrm{sech}^2(x)\, dx = 2$, one obtains

$$\begin{aligned}\frac{\partial \Omega_1}{\partial z} &= 2\int_{-\infty}^{\infty} |u_1|^2 \frac{\partial}{\partial t}|u_2|^2\, dt \\ &= \int_{-\infty}^{\infty} \mathrm{sech}^2(t + \Omega_1 z)\frac{\partial}{\partial t}\mathrm{sech}^2(t + \Omega_2 z)\, dt.\end{aligned} \tag{5.5}$$

(Interchanging subscripts in Eq. (5.5) yields a similar expression for $\partial\Omega_2/\partial z$.)

Equation (5.5) (multiplied by -1) represents the ''acceleration'' — i.e., the rate of change of inverse group velocity with distance into the collision. All that remains now is to evaluate Eq. (5.5) and the corresponding inverse velocity shift. Because for $\Omega \gg 1$, $\delta\Omega/\Omega \ll 1$, in the integrals, we can replace Ω_1 and Ω_2 by their initial values. Equation (5.5) can then be rewritten as

$$\frac{\partial \Omega_{(1,2)}}{\partial z} = (-, +)\frac{1}{2\Omega}\frac{d}{dz}\int_{-\infty}^{\infty} \mathrm{sech}^2(t - \Omega z)\,\mathrm{sech}^2(t + \Omega z)\, dt. \tag{5.5a}$$

Equation (5.5a) is now easily integrated to yield the inverse velocity shift:

$$\delta\Omega_{(1,2)} = (-, +)\frac{1}{2\Omega}\int_{-\infty}^{\infty} \operatorname{sech}^2(t - \Omega z)\operatorname{sech}^2(t + \Omega z)\, dt \qquad (5.6)$$

$$= (-, +)\frac{2}{\Omega}\frac{[2\Omega z\cosh(2\Omega z) - \sinh(2\Omega z)]}{\sinh^3(2\Omega z)}. \qquad (5.6a)$$

Finally, Eq. (5.6) can be integrated over z to yield the net time displacements:

$$\delta t_{(1,2)} = (+, -)\,\Omega^{-2}. \qquad (5.7)$$

The preceding expressions for the acceleration and the inverse velocity shift (Eqs. [5.5] and [5.6]) may not be particularly transparent. When numerically evaluated and graphed, however, they are seen to be simply behaved (see Fig. 12.33). Note, either from the graph in Fig. 12.33 or from the pertinent equations, that the pulses attract each other, whereas their frequencies repel each other. Also note that, as purported, the completed collision leaves the soliton intact, with the same velocity and other properties that it had before the collision. Thus, the only change is the time shift, δt. As is shown later, however, the guiding filters tend to remove even that defect.

The discussion in this section has thus far been almost entirely in terms of soliton units. For convenient future reference, however, we now list formulas for the principal quantities in practical units. First, in terms of the full-channel separation Δf and the pulse width τ, the half-channel separation in soliton units is

$$\Omega = 1.783\ \tau\Delta f. \qquad (5.8a)$$

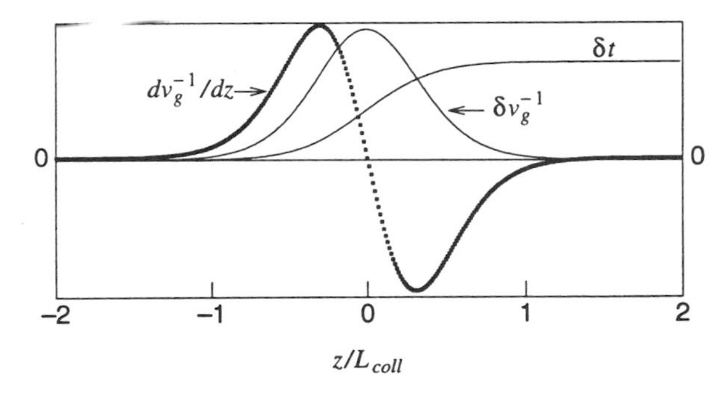

Fig. 12.33 Acceleration (dv_g^{-1}/dz), velocity shift ($\delta v_g^{-1} = -\delta\Omega$), and time shift ($\delta t$) of the slower pulse, during a soliton–soliton collision in a lossless fiber. (For the faster [higher frequency] pulse, turn this graph upside down.) The maximum inverse group velocity shift $(\delta v_g^{-1})_{\max} = 2/(3\Omega)$, and $\delta t_{\max} = \Omega^{-2}$.

The maximum frequency shift during the collision, δf, and the net time displacement, δt, when expressed in practical units, are, respectively,

$$\delta f = \pm \frac{0.105}{\tau^2 \Delta f} \tag{5.8b}$$

and

$$\delta t = \pm \frac{0.1786}{\tau \Delta f^2}. \tag{5.8c}$$

(In this case, the numerical coefficients are all dimensionless and represent various combinations of $\tau/t_c = 1.762\ldots$ and π.) For example, consider a collision between 20-ps solitons in channels separated by 75 GHz (0.6 nm). Then Eqs. (5.8a) through (5.8c) yield, respectively, $\Omega = 2.67$ (more than large enough for effective separation of the soliton spectra), $\delta f = \pm 3.5$ GHz, and $\delta t = \pm 1.59$ ps.

5.3 *EFFECTS OF PERIODIC LOSS AND VARIABLE DISPERSION*

The periodic intensity fluctuations in a system with real fiber and lumped amplifiers can serve to destroy the perfect asymmetry of the acceleration curve of Fig. 12.33, and hence result in a net residual velocity shift and associated timing displacement [26]. For the purposes of illustration, Fig. 12.34 shows an extreme case, where the collision length is short relative to the amplifier spacing, and where the collision is centered at an amplifier. Note that just prior to the amplifier, where the intensity is low, the acceleration curve is correspondingly attenuated, whereas just the opposite happens in the space immediately following the amplifier. Thus, most of the integral of the acceleration curve comes from the right half of the graph, and, as a result, there is a large residual velocity (frequency) shift. The residual frequency shift in this example (~4 GHz) is large. Note that when its wavelength equivalent is multiplied by $D = 0.8$ ps/nm-km and by a characteristic filter damping length of say, 600 km, the resultant time shift is approximately 15 ps. When further magnified by the typical spread of nearly zero to at least several tens of collisions in a transoceanic length, that time shift would result in a completely disastrous timing jitter.

On the other extreme, where the collision length is large relative to the amplifier spacing, one might reasonably expect the velocity curve to look much like that of Fig. 12.33. For example, Fig. 12.35 shows what happens

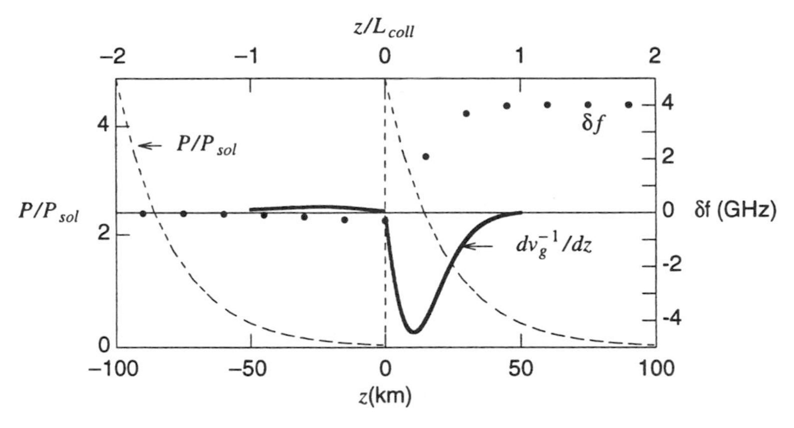

Fig. 12.34 Acceleration (*heavy curve*) and frequency shift ($\propto -v_g^{-1}$) (*dots*) for a collision centered at an amplifier, and where $L_{coll} \ll L_{amp}$ (i.e., 50 vs. 100 km). The *dashed lines* show the relative intensities in each span. The other pertinent parameters here are: $\tau = 16$ ps, $D = 0.8$ ps/nm-km, and $\Delta\lambda = 0.8$ nm ($\Delta f = 100$ GHz). Note the severe asymmetry of the acceleration curve and the resultant large residual velocity shift.

when L_{coll} is just $2.5L_{amp}$. Although the acceleration curve in Fig. 12.35 contains large discontinuities at each amplifier, its integral looks remarkably close to the ideal velocity curve (Fig. 12.33). Most important, the velocity returns almost exactly to zero following the collision. Extensive numerical

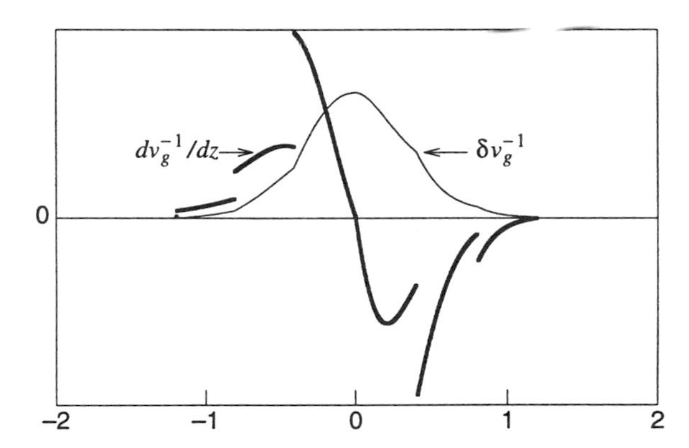

Fig. 12.35 Acceleration and velocity shift for a collision centered at an amplifier, in constant D fiber, and where $L_{coll} = 2.5L_{amp}$. Note how the velocity curve here closely approximates the ideal of lossless fiber (Fig. 12.33).

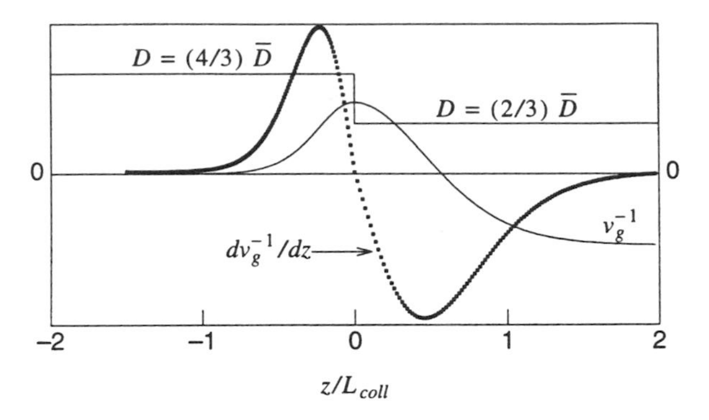

Fig. 12.36 Acceleration and velocity curves for a collision centered at an discontinuity in D, and where, for simplicity, the fiber is lossless.

simulation and related analysis [26] both show that the residual velocity is essentially zero as long as the condition

$$L_{coll} \geq 2L_{amp} \tag{5.9}$$

is satisfied. Note that this condition, combined with Eq. (5.1), puts an upper bound on the maximum allowable channel spacing:

$$\Delta\lambda_{\max} = \frac{\tau}{DL_{amp}}. \tag{5.10}$$

For example, let $\tau = 20$ ps, $D = 0.5$ ps/nm-km, and $L_{amp} = 33$ km. Equation (5.10) then yields $\Delta\lambda_{\max} \approx 1.2$ nm; with a nearest channel wavelength spacing of 0.6 nm, the maximum allowable number of channels is just three. As is shown later, however, this limit can be circumvented.

Variation in the fiber's chromatic dispersion can also upset the symmetry of the collision. For example, Fig. 12.36 illustrates what happens when the collision is centered at a discontinuity in D. To keep the example pure, the fiber is lossless. Note the different length scales on either side of the discontinuity in D. These occur, of course, because the length scale for the collision, or for any part of it, is L_{coll}, which is in turn inversely proportional to D (see Eq. [5.1]).

5.4 DISPERSION-TAPERED FIBER SPANS

From the discussion in Section 5.3, note that a decrease in the acceleration, from decreasing intensity, can be compensated for by a corresponding increase in L_{coll}, hence by a decrease in D. Thus, if the dispersion-tapered fiber spans discussed in Section 2.5.3 were used, it should be clear that the resultant curves of acceleration and velocity (as plotted in soliton units) will look exactly like the symmetrical ideal of Fig. 12.33. Formally, the NLS equation for real fiber with dispersion tapering can be transformed into that for lossless fiber with constant dispersion (see Ref. 26, Appendix). Thus, with that one transformation, it can be seen that the use of dispersion-tapered fiber spans removes *all* perturbing effects stemming from the use of lumped amplifiers. This idea is extremely important for the achievement of massive WDM with solitons.

Fiber spans having the ideal exponential taper in D are not yet available commercially. Thus, at present it is usually necessary to use a stepwise approximation. Figure 12.37 illustrates the optimum three-step approximation to the ideal taper. The length of each step is inversely proportional to the D value of the step. Note that this makes the steps all have equal lengths, as measured in soliton units. Also note that using such an N-step approximation to the ideal dispersion taper increases the limit on maximum channel spacing imposed by Eq. (5.10) by a factor of N. That is, one now has

$$\Delta\lambda_{\max} = \frac{N\tau}{DL_{amp}}. \tag{5.10a}$$

In practice, once N is large enough, the intensity variation over each step becomes so small that the perturbations become acceptably small in any event. In that case, the "limit" of Eq. (5.10a) is no longer significant.

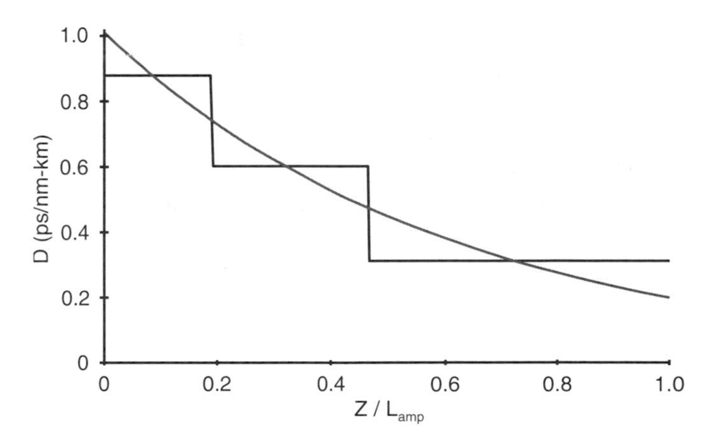

Fig. 12.37 Ideal exponential taper of D and the best three-step approximation to it for a fiber span with $L_{amp} = 33.3$ km and a loss rate of 0.21 dB/km.

5.5 *FOUR-WAVE MIXING IN WDM*

In Section 5.2, we were able to ignore the terms (in $\pm 3\Omega$) that result from four-wave mixing. In general, however, pseudo phase matching from the periodic intensity fluctuations associated with the use of lumped amplifiers can enable the four-wave mixing products to grow to high levels [27]. Very serious timing and amplitude jitter can then result. Thus, it is necessary to consider four-wave mixing carefully.

There are three possible four-wave mixing processes:

$$2\Omega_1 - \Omega_2 \rightarrow \Omega_S \tag{5.11a}$$

$$2\Omega_2 - \Omega_1 \rightarrow \Omega_A \tag{5.11b}$$

$$\Omega_1 + \Omega_2 - \Omega_S \rightarrow \Omega_A. \tag{5.11c}$$

The spectrum associated with these processes is shown in Fig. 12.38. In the first two processes, two photons from one of the strong fields mix with one from the other strong field, to create either a Stokes or an anti-Stokes photon. In the third process, one photon each at Ω_1, Ω_2, and Ω_S (or Ω_A) combine to form a photon at Ω_A (or Ω_S).

Note that the first two of these processes are dominant because three of the fields involved are initially nonzero. For these processes, the phase mismatch is

$$\begin{aligned} \Delta k &= (k_2 + k_S - 2k_1 \text{ or } k_1 + k_A - 2k_2) \\ &= \left.\frac{\partial^2 k}{\partial \omega^2}\right|_{1,2} \Delta\omega^2 = -\frac{\lambda^2 D(\lambda_{1,2})}{2\pi c}\,\Delta\omega^2, \end{aligned} \tag{5.12}$$

where the subscripts 1,2 apply to the processes in Eqs. (5.11a) and (5.11b), respectively. It is important that Eq. (5.12) is completely independent of third-order dispersion and of the choice of zero frequency.

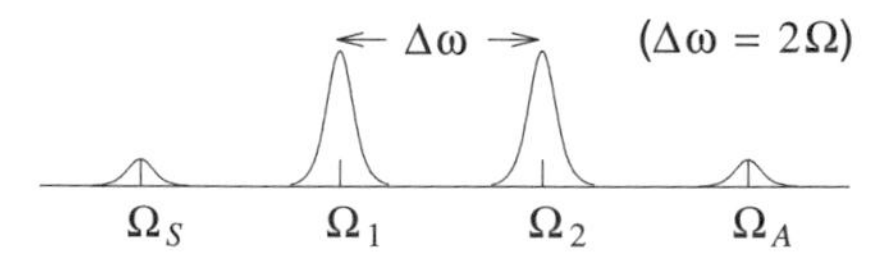

Fig. 12.38 Spectrum of source fields and their four-wave mixing products at Ω_S and Ω_A.

When the fields at Ω_1 and Ω_2 are continuous waves, the four-wave mixing products grow as

$$\frac{dE_i}{dz} \propto E_j^2 E_k^* \exp(i\,\Delta kz), \tag{5.13}$$

where the set of subscripts i,j,k on the field quantities E is either S,1,2 (process in Eq. [5.11a]) or A,2,1 (process in Eq. [5.11b]), and z is the distance along the fiber. Note that the phase of the generated product is periodic in z, with period $L_{res} = 2\pi/\Delta k$, as a result of the phase mismatch. For a lossless fiber with constant dispersion (i.e., where $E_{1,2}(z) = const$, and $\Delta k = const$), Eq. (5.13) is readily integrated to yield

$$E_i(z) \propto \frac{E_j^2 E_k^*}{i\,\Delta k}[\exp(i\,\Delta kz) - 1], \tag{5.14}$$

a field that merely oscillates between zero and a fixed maximum and never grows.

Nevertheless, if the transmission line has periodic perturbations with k_{pert} in resonance with the phase mismatch of the four-wave mixing, i.e., when

$$Nk_{pert} = \Delta k, \qquad (N = 1, 2, 3, \ldots), \tag{5.15}$$

then one has pseudo phase matching, and the four-wave mixing product can grow steadily. The perturbations can correspond to the gain–loss cycle whose period is the amplifier spacing, L_{amp}, and/or to periodic variations of the fiber parameters (dispersion, mode area). For the case of lumped amplifiers, $k_{pert} = 2\pi/L_{amp}$, and the pseudo-phase-matching conditions are met when

$$L_{amp} = NL_{res} \equiv 2\pi N/\Delta k. \tag{5.16}$$

Although four-wave mixing generation during a soliton–soliton collision is more complicated than with continuous waves, the basic features remain the same. Figure 12.39 shows the numerically simulated growth in energy of the four-wave mixing products at $\omega_{A,S}$ during a single collision of two solitons. (The particular parameters represented in Fig. 12.39 are those of recent experiments [28, 29], viz., τ = 20 ps, adjacent channel separation Δf = 75 GHz [$\Delta\lambda$ = 0.6 nm at λ = 1556 nm], and where the path-average dispersion $\overline{D}$ = 0.5 ps/nm-km.) For these parameters, L_{res} = 44.4 km. Note that for the case of lossless fiber of constant dispersion, and for the case of real fiber with exponentially tapered dispersion, the four-wave mixing energy disappears completely following the collision. Also note that because

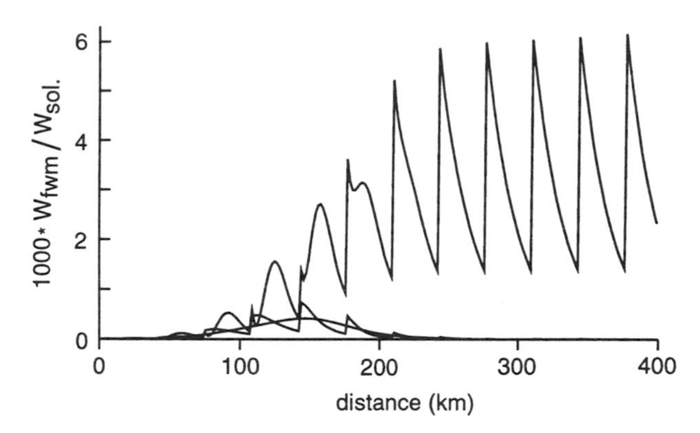

Fig. 12.39 Growth of four-wave mixing energy during a single soliton–soliton collision, for three conditions: lossless fiber with constant dispersion (*small, smooth curve*); real fiber with lumped amplifiers spaced 33.3 km apart and exponentially tapered dispersion (*small, jagged curve*); real fiber with lumped amplifiers spaced 33.3 km apart and constant dispersion (*large, jagged curve*). The four-wave mixing energy is for a single sideband and is normalized to the soliton pulse energy. Note that for the first two cases, the four-wave mixing energy disappears completely following the collision, whereas for the third case, where there is effective pseudo phase matching, the four-wave mixing energy builds to a large residual value.

the solitons have finite temporal and spectral envelopes, and because of the effect of cross-phase modulation (which shifts the pulses' carrier frequencies during the collision), the oscillations of the four-wave mixing energy with the period of L_{res} are almost completely washed out. Finally, for the case of real fiber with constant dispersion, note that the collision produces a residual four-wave mixing energy several times larger than the (temporary) peak obtained with lossless fiber.

Although the residual energy from the pseudo-phase-matched collision of Fig. 12.39 may seem small, the fields from a succession of such collisions can easily build to a dangerously high value. Such uncontrolled growth of the four-wave mixing imposes penalties on the transmission by two mechanisms. First, because the energy represented by the four-wave mixing fields is not reabsorbed by the solitons, the solitons tend to lose energy with each collision. Because the net energy loss of a given soliton depends on the number of collisions it has suffered, and upon the addition of four-wave mixing fields with essentially random phases, it directly creates amplitude jitter. The energy loss leads to timing jitter as well, both through the intimate coupling between amplitude and frequency inherent in filtered systems, and through its tendency to asymmetrize the collision, and hence to induce net velocity shifts. Finally, certain noise fields in the same band with the four-wave mixing products can influence the soliton's frequencies, in a kind of extended Gordon–Haus effect. Thus, even in a two-channel WDM, there can be serious penalties (see Fig. 12.40). Moreover, if the wavelengths of the four-wave mixing products coincide with the wavelengths of other WDM channels (possibly only when three or more channels exist), the runaway four-wave mixing becomes an additional source of noise fields to act on those channels. In that way, the well-known amplitude and timing jitter effects of spontaneous emission are enhanced.

The growth of four-wave mixing can often be controlled adequately with the use of one or another of the N-step approximations to the ideal exponential dispersion taper discussed in Section 5.4. Figure 12.41 plots the residual four-wave mixing intensity resulting from a single collision, as a function of L_{amp}, for various numbers of steps in D per L_{amp}, for the channel separation of 0.6 nm, and for the $\tau = 20$ ps solitons and $\overline{D} = 0.5$ ps/nm-km of Fig. 12.33. Figure 12.42 does the same, but for twice the channel separation (1.2 nm). First, note that the intensity scale in Fig. 12.41 is approximately a factor of $2^5 = 32$ times that of Fig. 12.42, just as implied by Eqs. (5.12) and (5.14) and by the fact that L_{coll} scales inversely as the channel spacing. This scaling is easily generalized; for channels spaced n

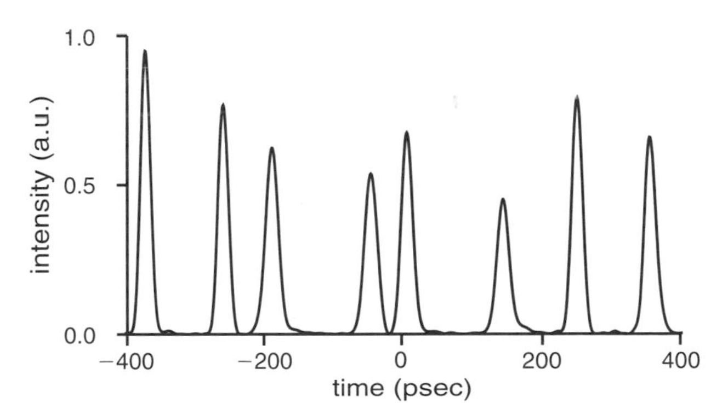

Fig. 12.40 Pulses that have traversed a 10-Mm transmission line with L_{amp} = 33 km and constant D = 0.5 ps/nm-km, and that have undergone collisions with an adjacent channel, 0.6 nm away, containing all ones. In this numerical simulation, there were no guiding filters. A small seed of noise was added, but only in the four-wave mixing sidebands; thus, the large amplitude and timing jitter seen here is from the uncontrolled growth of four-wave mixing alone.

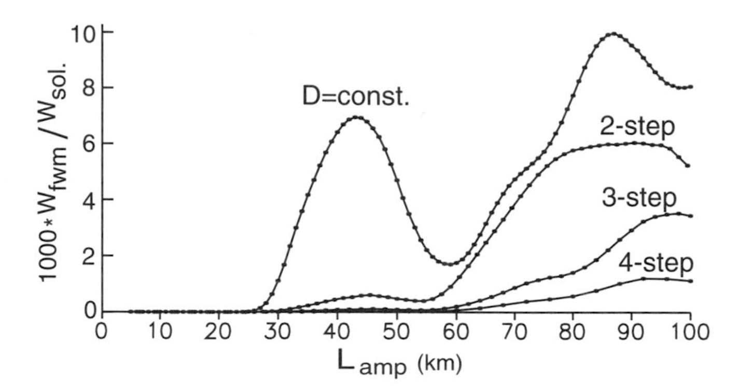

Fig. 12.41 Residual four-wave mixing energy following a single collision of 20-ps solitons in channels spaced 0.6 nm apart in a chain of fiber spans with $\overline{D} = 0.5$ ps/nm-km, as a function of the amplifier spacing, for constant D and for the optimal two-, three-, and four-step approximations to the ideal exponential taper. The four-wave mixing energy is for a single sideband and is normalized to the soliton pulse energy. No noise seed was used in these simulations.

times the adjacent channel spacing, the four-wave mixing intensity should scale as n^{-5}. This apparently rapid falloff in four-wave mixing effect is tempered somewhat by the fact that the number of collisions tends to increase as n, and that it is really the vector addition of residual field quantities from at least several successive collisions that is to be feared in this case. Also note that the number of steps required for total suppression

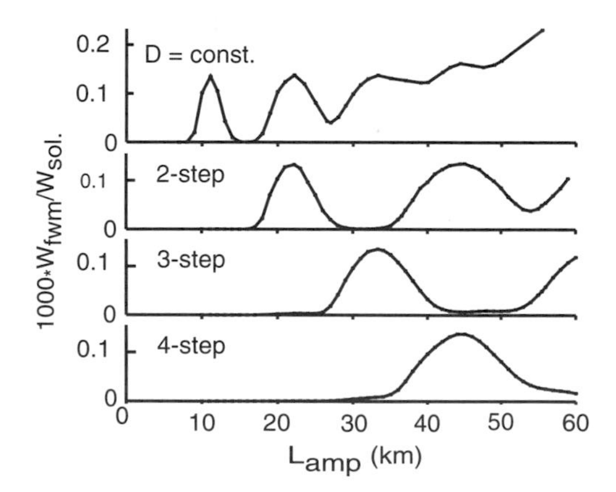

Fig. 12.42 The same setup as Fig. 12.41, except the channel spacing is twice as great; i.e., it is 1.2 nm.

of the four-wave mixing intensity increases with wider channel spacing. For example, in Fig. 12.41, just two steps are required for L_{amp} in the neighborhood of 30 km, whereas four steps are required for the same in Fig. 12.42. Finally, note that because of the finite nature of the pulse widths and collision lengths, the resonances in Figs. 12.41 and 12.42 are fairly broad.

5.6 CONTROL OF COLLISION-INDUCED TIMING DISPLACEMENTS

With the use of dispersion-tapered fiber spans, the only major penalty in WDM comes from the collision-induced timing displacements given by Eq. (5.7) or Eq. (5.8c). Because some pulses undergo dozens of collisions whereas others undergo nearly none in the course of a transoceanic transmission, the resultant timing jitter can be substantial, even when the time displacement from a single collision is no more than a picosecond or two. It so happens, however, that the frequency guiding filters nearly eliminate that jitter as well [30]. The argument can be made as follows.

First, to simplify the notation, let $\delta v_g^{-1} \rightarrow v$, and $\partial v_g^{-1}/\partial z \rightarrow a$. Without filtering, let the colliding solitons accelerate each other by $a_0(z)$, whose first z integral is $v_0(z)$, and whose second z integral is $t_0(z)$. We require only that the completed collision leave no residual velocity shift (see Fig. 12.35, for example). Thus,

$$v_0(\infty) = \int_{-\infty}^{\infty} a_0(z)\, dz = 0. \tag{5.17}$$

For simplicity, we assume that the filters are continuously distributed and that they provide a damping (acceleration) $a_d = -\gamma v = -v/\Delta$, where γ and Δ are the damping constant and characteristic damping length, respectively (see Eq. [4.8]). The equation of motion then becomes

$$\frac{dv}{dz} = a_0(z) - v(z)/\Delta. \tag{5.18}$$

Equation (5.17) can be rewritten as

$$v(z) = \left[a_0(z) - \frac{dv}{dz}\right]\Delta. \tag{5.18a}$$

To get t, we simply integrate Eq. (5.18a):

$$t(z) = \int_{-\infty}^{z} v(x)\, dx = \Delta \times \int_{-\infty}^{z} \left[a_0(x) - \frac{dv}{dx} \right] dx. \tag{5.19}$$

We are primarily interested in $t(\infty)$:

$$t(\infty) = \Delta \times \int_{-\infty}^{\infty} a_0(x)\, dx - \Delta \times \int_{-\infty}^{\infty} dv. \tag{5.19a}$$

The first term on the right is equal to zero by virtue of Eq. (5.17). The second term is just $v(\infty)$, which for a filtered system with no excitation beyond a certain point must equal zero. Figure 12.43 shows the quantities $a_0(z)$, the δf corresponding to $-v(z)$, and $t(z)$, numerically simulated for the case of lumped filters. One can see from this figure how the timing displacement is nulled: the filters reduce the maximum frequency (velocity) shift, so that the acceleration in the second half of the collision causes an overshoot in frequency; the area under the long tail of the frequency curve thus produced just cancels the area under the main peak.

Real filters, such as etalons, do not always perform exactly as in the simplified model just discussed. First, with real filters the damping force is not always strictly proportional to $-v$, or, equivalently, to δf. Second, the time delay through the filters exhibits a certain dispersion as the signal frequency moves off the filter peak. Nevertheless, numerical simulation shows that the etalons used in the theoretical examples and in the experi-

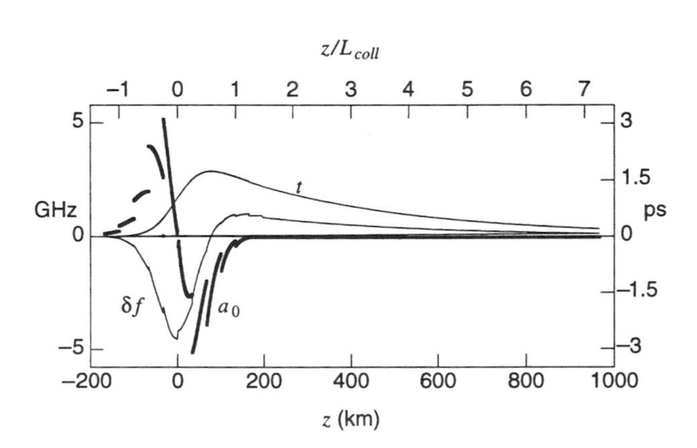

Fig. 12.43 The frequency and time shifts, as a function of distance, resulting from a collision in a transmission line with lumped amplifiers and lumped filters spaced 33.3 km apart. The other parameters are as follows: D = 0.5 ps/nm-km; τ = 20 ps; channel spacing, $\Delta\lambda$ = 0.6 nm; damping length, Δ = 400 km.

ments cited here tend to cancel out at least 80% of the timing displacements. That large improvement has turned out to be sufficient for most purposes.

5.7 EFFECTS OF POLARIZATION

Thus far in the discussion, we have assumed that the colliding pulses are copolarized. It is important to note, however, that all the effects discussed so far (cross-phase modulation and four-wave mixing) are significantly affected by polarization, and all in the same way. Although we defer a full-blown discussion of polarization until the following seciton, the state of affairs can be summarized as follows: First, just as for the path-average solitons (Section 2), the residual birefringence of even the highest quality transmission fibers available at present is large enough that over the (tens or hundreds of kilometers long) path of a single collision, the Stokes vectors representing the polarization states of the individual pulses tend to rotate, more or less at random, many times over and around the Poincaré sphere. Thus, in that way, the polarization tends to be well averaged over a representative sample of all possible states during a collision. On the other hand, the relative polarization — i.e., the angle between the Stokes vectors — of the colliding pulses tends to be only mildly affected. (The relative polarization is affected by two factors: the dispersion in the fiber's linear birefringence and, as is detailed in Section 6, a nonlinear birefringence induced by the collision itself.) Thus, even in the worst case, to first order at least, one can treat the relative polarization during a collision as a constant, so polarization does not significantly affect the symmetries of the collision. Nevertheless, under those conditions (of thorough averaging over absolute polarization states, while relative polarization states are preserved), the frequency shifts induced by cross-phase modulation and the intensities of four-wave mixing products are both just half as great for orthogonally polarized pulses as they are for copolarized pulses. This fact is of obvious practical importance.

5.8 GAIN EQUALIZATION WITH GUIDING FILTERS

The inevitable variation of amplifier gain with wavelength presents a problem for WDM, one that becomes ever more serious with increasing system length. In linear transmission (such as NRZ), where no self-stabilization of the pulse energies is possible, custom-designed, wavelength-dependent loss elements must be inserted periodically along the line to try to compensate for the variable amplifier gain. Even then, however, in practice it has proven difficult to maintain even approximately equal signal levels among the various channels. For soliton transmission using guiding filters, however, the guiding filters themselves provide a powerful built-in feedback mechanism for controlling the relative strengths of the various signal channels in the face of variable amplifier gain. As should be evident from the discussion of guiding filters already presented (Section 4), the control stems from the fact that the guiding filters provide a loss that increases monotonically with the soliton bandwidth, and hence with the soliton energy. Thus, for a channel having excess amplifier gain, a modest increase in soliton energy quickly creates a compensating loss, and the signal growth is halted.

Because the soliton bandwidth scales as $\tau^{-1} \propto W/D$, where W is the soliton pulse energy (see Eq. [2.16]), the soliton's energy loss from the guiding filters can be written as a monotonically increasing function of (W/D). For Gaussian filters, the relation $f(W/D)$ is quadratic; for the shallow etalon filters used in practice, however, $f(W/D)$ is more nearly linear. The energy evolution of N WDM channels in a soliton transmission line with sliding filters can then be described by the following system of coupled nonlinear equations [31]:

$$\frac{1}{W_i}\frac{dW_i}{dz} = \frac{\alpha_i}{1 + mR(W_1 + W_2 + \ldots + W_N)/P_{sat}} - \alpha_{Li} - f(W_i/D_i). \tag{5.19}$$

The subscript $i = 1, \ldots, N$ identifies the particular channel, $W_i(z)$ is its energy, α_i is its small-signal gain coefficient, α_{Li} is its linear loss rate, m is the mark-to-space ratio (usually one-half), R is the per-channel bit rate, P_{sat} is the saturation power of the amplifiers, and $D_i(z)$ is the dispersion at the ith channel wavelength (the dispersion could change with distance z as a result of the combined action of the frequency sliding and third-order dispersion). Note that the Eq. (5.19) fixes the equilibrium values of W_i/D_i according to the various α_i. Thus, when the α_i are all nearly the same, the various channel energies will scale in direct proportion to $D(\lambda)$. In the usual situation where the third-order dispersion is essentially a constant, the channel energies will be in direct proportion to their separation in wavelength from λ_0, the wavelength of zero D. It should also be noted that Eq. (5.19) is essentially the same as Eq. (4.5a), with the frequency offset term omitted. (Recall that in soliton units, one has $W = A$ [not $W = A^2$], and recognize that the quantity α in Eq. [4.5a] is just a compact way of writing the sum of the first two terms on the right-hand side of Eq. [5.19]. Finally, for Gaussian filters, $f[W/D] = const. \times W^2$.) Thus, just as

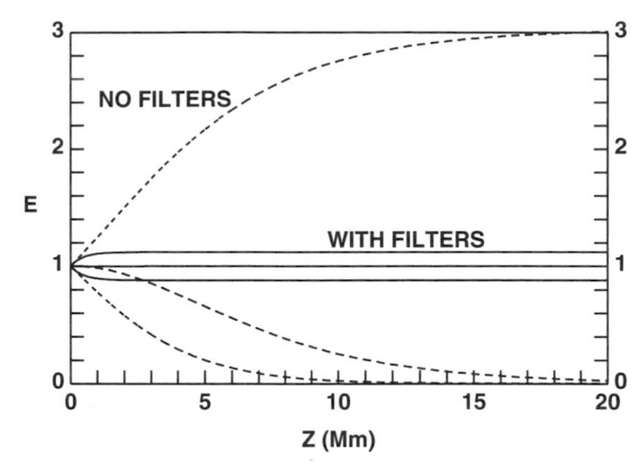

Fig. 12.44 Growth with distance of signal pulse energies in WDM channels having relative gain rates of +1 and 0 dB/Mn, respectively, for a system using sliding-frequency filters (damping length Δ = 400 km), and for no filters at all.

a linearized version of Eq. (4.5a) was shown to be a damping equation, the linearized version of Eq. (5.19) is essentially the same damping equation, with $W\, df/dW$ as the damping constant. Thus, one has

$$W \frac{df}{dW} = \frac{1}{\Delta}, \tag{5.20}$$

where, at least for Gaussian filters, Δ is the same characteristic damping length as discussed in Section 4.3.

Figure 12.44 shows the solution to Eq. (5.19) for the case of three channels with significantly different small-signal gain rates, both with and without filters. Note how the filters quickly lock the signal energies to tightly clustered equilibrium values. By contrast, the large divergence in channel energies that occurs when no filters are used will clearly lead to large penalties and disastrous error rates. Thus, the ability of the guiding filters to regulate the relative channel energies constitutes a major and very important advantage for solitons over all other possible modes of WDM.

5.9 EXPERIMENTAL CONFIRMATION

Recently, the ideas in this section have been put to extensive test in transmission involving massive WDM at per-channel rates of 5 and 10 Gb/s (with the major emphasis on the greater rate) [28, 29]. Whereas many aspects of the transmissions were monitored, the ultimate criterion of success was to achieve measured BER rates of 1×10^{-9} or less on all channels over at least the trans-Pacific distance of 9 Mm.

Figure 12.45 is an overall schematic of the signal source. The soliton pulse shaper, based on a $LiNbO_3$ Mach–Zehnder-type modulator, both carves out the pulses and provides them with a controlled and desired chirp [32]. (A simple phase modulator, driven sinusoidally at the bit rate, was also used successfully, as an alternative to the pulse shaper. The idea in this case is that whenever the upper or lower sidebands created by the phase modulator are selected by a suitable filter, one obtains a train of nearly sech-shaped pulses suitable for soliton transmission [33]. In this example, the filtering action of the transmission line itself provides the necessary sideband selection.) After a second modulator imposes the data (a 2^{15}-bit, random pattern), the 4-km length of standard fiber ($D \sim 17$ ps/nm-km at 1557 nm) compresses the pulses to about 20 ps; it also separates the bits of adjacent channels by 40 ps. This separation prevents the occurrence of half-collisions at the input to the transmission line for all but the most widely spaced channels. Finally, the 3.7-m length of polarization-maintaining fiber, used as a multiple half-wave plate, enables adjacent channels to be launched with orthogonal polarizations.

The orthogonal polarizations provide two significant benefits: First, they reduce the interchannel interaction by a factor of 2 over that obtaining with copolarization states. Second, at least where the number of channels is even, the net optical power in the transmission line is essentially unpolarized, so the amplifiers exhibit no significant polarization-dependent gain from polarization hole burning. Thus, there is no need for polarization scrambling of the individual channels. There is more here than just the

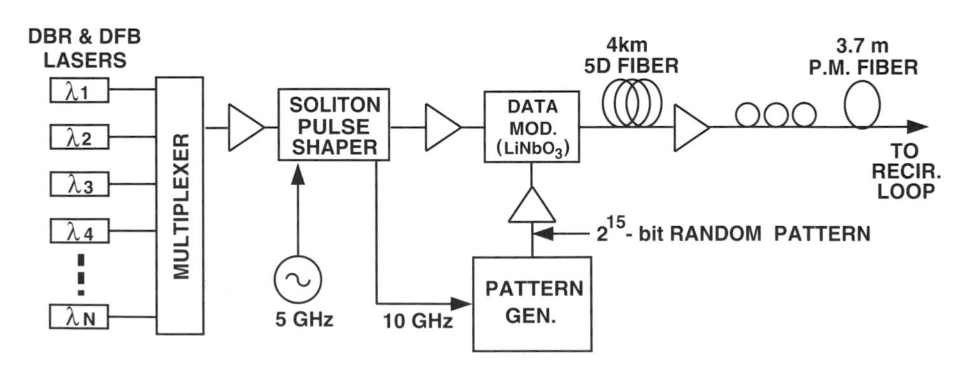

Fig. 12.45 Source for soliton WDM experiments at 10 Gb/s per channel. DBR, distributed Bragg reflector; DFB, distributed feedback; P.M., polarization maintaining.

avoidance of unnecessary and expensive hardware, however, because the fiber's birefringence tends to convert polarization scrambling into timing jitter. Therefore, unless it is needed (as with NRZ) to avoid serious polarization hole burning, polarization scrambling is a detriment and is thus to be avoided.

It should also be noted that the order of the pulse shaper and the data modulator is of no consequence. In a real system, of course, prior to multiplexing, each continuous-wave (CW) laser source would be followed by its own individual data modulator. Thus, for soliton transmission, schematically one just has a WDM NRZ source, followed by a common soliton pulse shaper. In a source for true NRZ transmission, however, one must use expensive $LiNbO_3$, Mach–Zehnder-type modulators, symmetrically driven to avoid significant chirping at the transition points between ones and zeros. For soliton transmission, on the other hand, one could just as well use a set of semiconductor absorption modulators, with their absence of significant bias drift and their potentially much lower cost. This exchange is possible because in removing all but the center of each NRZ bit, the soliton pulse shaper also tends to get rid of all chirping induced by the data modulators. Thus, in this way the soliton transmission has an advantage over NRZ that is significant both economically and in the strictly technical sense.

The recirculating loop contains six spans of 33.3 km each between (Er fiber) amplifiers, each span dispersion-tapered typically in three or four steps, with span path-average value $\overline{D} = 0.5 \pm 0.05$ ps/nm-km at 1557 nm. Two of every three loop amplifiers are immediately followed by piezo-driven, Fabry–Perot etalon filters, each having a 75-GHz (0.6 nm at 1557 nm) free spectral range and mirror reflectivities of 9%; this combination provides the optimum path-average filter strength of $\eta \sim 0.4$ at 1557 nm (Fig. 12.46).

At the receiver, the desired 10-Gb/s channel is first selected by a wavelength filter and is then time-division demultiplexed to 2.5 Gb/s by a polarization-insensitive electrooptic modulator having a 3-dB bandwidth of 14 GHz [34] and driven by a locally recovered clock (Fig. 12.47) The demultiplexer provides a nearly square acceptance window in time, one bit period wide.

Figure 12.48 shows a typical set of BER data. Note the tight clustering of the BER performance for all channels.

Figure 12.49 plots the measured error-free distances versus the number, N, of 10-Gb/s channels. For each of these points, the BER was better than 1×10^{-9} on all N channels. Note that most of the data points correspond

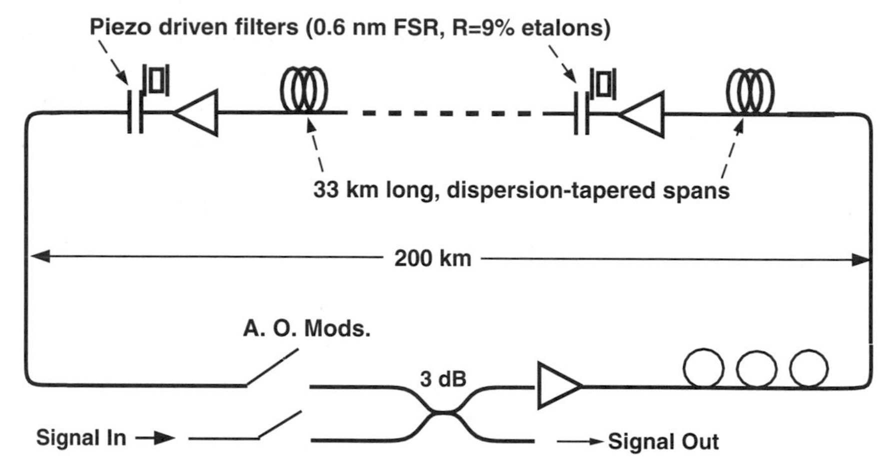

Fig. 12.46 Recirculating loop with sliding-frequency filters and dispersion-tapered fiber spans. The A.O. (acoustooptic) modulators act as optical switches with very large on–off ratios to control the sequencing of each transmission. That is, initially the lower switch is held closed and the upper switch open long enough for the source to fill the recirculating loop with data and to bring the amplifier chain to equilibrium. The conditions of the two switches are then simultaneously reversed, so that the loop is closed on itself, and there is no longer an external source. During each such transmission, a linear ramp voltage is applied to the piezo-driven filters to produce the desired sliding of the filter frequencies. Samples of the signal, corresponding to successive round-trips, emerge more or less continuously from the signal out port.

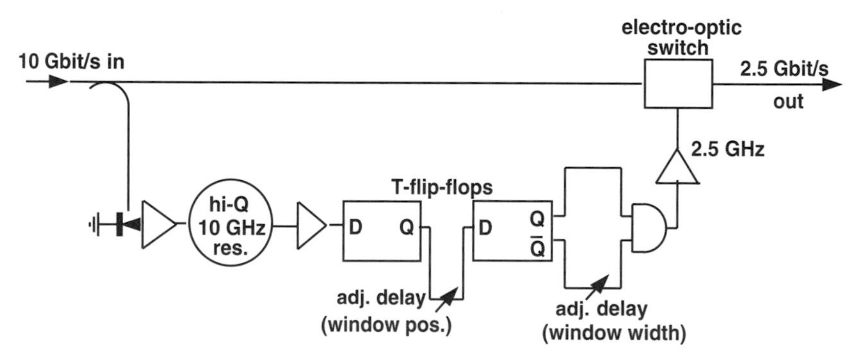

Fig. 12.47 Time-division demultiplexer.

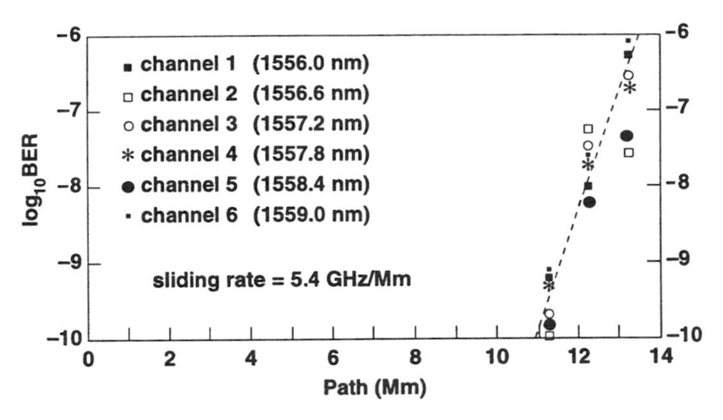

Fig. 12.48 Measured BER versus distance for a 6 × 10 Gb/s WDM transmission.

to loop amplifiers pumped at 1480 nm, with corresponding high noise figure (~6 dB) and narrow gain bandwidth. The last point corresponds to pumping at 980 nm, however, where the noise figure is much closer to 3 dB and the gain bandwidth is improved, at least toward shorter wavelengths. The error-free distances represented in Fig. 12.49 tend to be determined by a low-

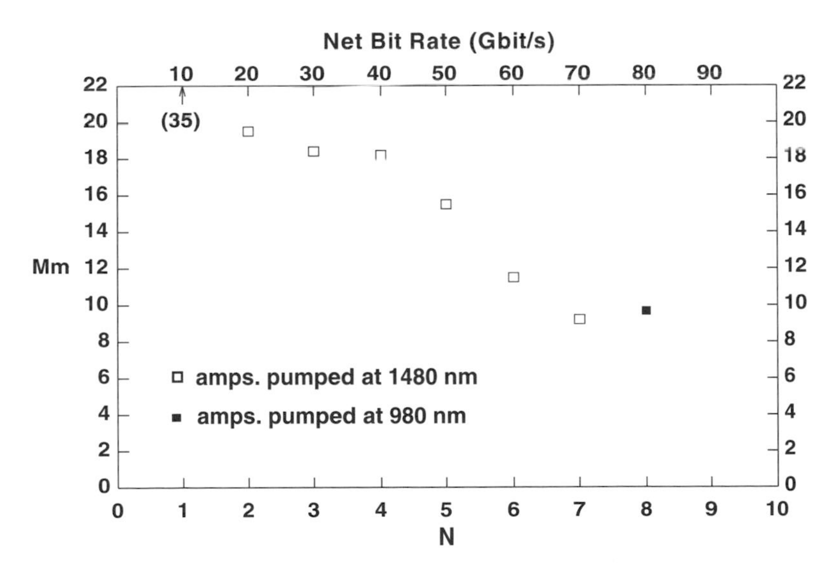

Fig. 12.49 Measured error-free distances versus the number, N, of channels at 10 Gb/s per channel.

level error floor, which is only very weakly dependent on distance. This dependence is especially noticeable for the largest values of N. Thus, even small future improvements should enable both the error-free distances and the maximum number of allowable channels to be increased.

An experiment was also performed at 5 Gb/s per channel (six channels), with the same apparatus, simply by programming the pattern generator to eliminate every second pulse of the otherwise 10-Gb/s data. As the only other substantial change, consistent with the lower bit rate, the time-acceptance window of the demultiplexer (Fig. 12.47) was opened up by a factor of 2. For that experiment, the error-free distance was greater than 40 Mm on all channels. The great increase in error-free distance was due primarily to two factors: First, at half the bit rate, the rate of collisions was decreased by half. Second, the doubled time-acceptance window greatly increased the tolerance to timing jitter.

Finally, Fig. 12.50 shows an example of the spectrum of the WDM transmission. Although the example corresponds to a particular distance (10 Mm), the spectrum looks the same at any other but the very shortest distances. In all the experiments carried out so far, regardless of the number of channels, the spectra all had the same feature, viz., that the spectral peaks could all be joined by a straight line that passed through the zero intensity axis at λ_0, the wavelength of zero dispersion. All this occurred in the face of considerable amplifier gain variation over the total wavelength span. Thus, in these spectra we have direct and complete confirmation of

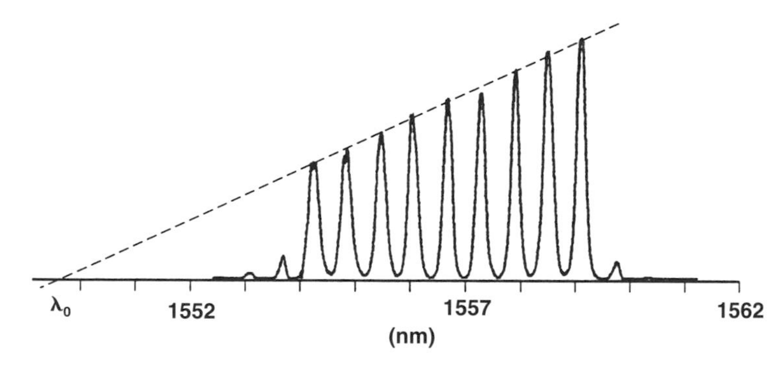

Fig. 12.50 Optical spectrum of 9 × 10 Gb/s WDM transmission, as measured at 10 Mm. Note that the *dashed line* connecting the spectral peaks of the individual channels passes through the wavelength of zero dispersion, λ_0. This behavior results from the strong regulation of the soliton pulse energies provided by the sliding-frequency filters. Following initial adjustment, it becomes independent of distance.

the ideas presented in Section 5.8, where the guiding filters were shown to provide a very tight control over the relative signal strengths of the various channels.

6. Polarization Effects

6.1 APOLOGIA

Thus far in this chapter, we have tended to gloss over polarization and effects of the fiber's birefringence on it. That is, for the most part we have been content to note the thorough polarization averaging, created by the fiber's random residual birefringence, over those distances for which the nonlinear term has significant effects. We have used that averaging to justify further neglect of polarization. Nevertheless, there are certain important polarization phenomena that require examination. One of these is called *polarization mode dispersion* (*PMD*), which makes the transit time for a pulse dependent on its polarization history and produces some dispersive-wave radiation in the process. The other is the fact that colliding solitons in WDM alter each other's polarization states. Therefore, we now examine the linear birefringence of fibers and its statistical properties, the birefringence induced nonlinearly by the pulse itself, and the effects of both on transmission.

6.2 POLARIZATION STATES AND THE STOKES–POINCARÉ PICTURE

If z is the propogation direction and $\hat{x}$ and $\hat{y}$ are unit vectors in the x and y directions, respectively, a unit normalized polarization vector can be written as $\hat{\mathbf{u}} = (r\hat{x} + s\hat{y})$, where r and s are complex numbers with $|r|^2 + |s|^2 = 1$, so that $\hat{\mathbf{u}}^* \cdot \hat{\mathbf{u}} = 1$. A corresponding normalized real field component at frequency ω has the form

$$\mathrm{Re}(\mathbf{u}) = \mathrm{Re}[(r\hat{x} + s\hat{y})e^{i\Psi}] \quad : \quad \Psi = kz - \omega t + \Phi. \tag{6.1}$$

If the phases of r and s are both changed by the same amount $\delta\Phi$, then Φ in Eq. (6.1) simply changes to $\Phi + \delta\Phi$. With no loss of generality we can therefore express r and s in polar form by $r = \cos(\theta)\exp(-i\phi)$ and $s = \sin(\theta)\exp(i\phi)$, where $0 \leq \theta \leq \pi/2$ so that $\cos(\theta) = |r|$ and $\sin(\theta) = |s|$. Thus, the real field can be written as

$$\mathrm{Re}(\mathbf{u}) = \hat{x}\cos(\theta)\cos(\Psi - \phi) + \hat{y}\sin(\theta)\cos(\Psi + \phi). \qquad (6.1a)$$

One way to visualize the polarization state of the field is to plot the motion of the vector $\mathrm{Re}(\mathbf{u})$ in an x–y coordinate system as Ψ varies from 0 to 2π. The resulting figure is generally an ellipse. The state of polarization of the field is determined by the shape of the ellipse and by the direction of motion of the vector around the ellipse. It is independent of the constant Φ in the phase Ψ. In general, a change from ϕ to $-\phi$ reverses the direction of motion around the same ellipse. When $\phi = 0$ or $\phi = \pi/2$, the ellipse degenerates into a straight line making an angle θ or $-\theta$, respectively, with the x axis. This represents linear polarization. When $\phi = \pi/4$, the axes of the ellipse are the x and y axes, and they have lengths of $2\cos(\theta)$ and $2\sin(\theta)$, respectively. When $\theta = \pi/4$, so that $\cos(\theta) = \sin(\theta) = 1/\sqrt{2}$, the axes of the ellipse are rotated by $\pi/4$ from the x and y axes, and they have lengths of $2|\cos(\phi)|$ and $2|\sin(\phi)|$, respectively. When both ϕ and Ψ are equal to $\pi/4$, the ellipse degenerates to a circle and we have circular polarization.

A more useful tool for visualizing the state of polarization as it varies during transmission is the real three-dimensional Stokes vector. A Stokes vector $\hat{\mathbf{S}}$ of unit length is derived from the normalized polarization vector in Eq. (6.1) or Eq. (6.1a). It has the components

$$\begin{aligned} S_1 &= |r|^2 - |s|^2 = \cos(2\theta) \\ S_2 &= 2\mathrm{Re}(r^*s) = \sin(2\theta)\cos(2\phi) \\ S_3 &= 2\mathrm{Im}(r^*s) = \sin(2\theta)\sin(2\phi). \end{aligned} \qquad (6.2)$$

We see that the angles 2θ and 2ϕ are, respectively, the polar and aximuthal angles representing the vector $\hat{\mathbf{S}}$ in a spherical coordinate system with the S_1 axis as the polar axis. It is worth noting that S_1 is proportional to the difference in the powers that would emerge from linear polarizers in the x and y directions, and that S_2 is likewise proportional to the difference in the powers that would emerge from linear polarizers rotated from the previous two by an angle of $\pi/4$, thus bisecting the x and y directions, whereas S_3 is the difference between the powers that would emerge from left and right circular polarizers. Hence, the Stokes vector can be measured directly, and such machines are now fairly common. One can verify that the (S_1, S_2) plane represents plane polarized fields ($2\phi = 0$ or π), whereas the S_3 axis represents circularly polarized fields. Note that the Stokes vectors corresponding to the orthogonal polarization states $\hat{x}$ ($r = 1$, $s = 0$) and

$\hat{y}$ ($s = 1$, $r = 0$) lie antiparallel along the S_1 axis. More generally, two polarization vectors $\hat{\mathbf{u}}_a$ and $\hat{\mathbf{u}}_b$ are orthogonal if $\hat{\mathbf{u}}_a^* \cdot \hat{\mathbf{u}}_b = 0$. Thus $(s^*\hat{x} - r^*\hat{y})$ represents the state orthogonal to $(r\hat{x} + s\hat{y})$, and the Stokes vectors of these two polarization states always lie antiparallel along the same axis. Because the Stokes vectors of concern here have unit length, their ends always lie on a sphere of unit radius, the Poincaré sphere, as shown in Fig. 12.52.

6.3 LINEAR BIREFRINGENCE OF TRANSMISSION FIBERS

6.3.1 Birefringence Element and Its Effects

Consider the effect on $\hat{\mathbf{S}}$ of a short length dz of birefringent fiber (Fig. 12.51). There are many reasons for such birefringence, principal among them being a slight ellipticity of the fiber, or some strain on it. Suppose first that the principal states of the birefringence are the x and y linear polarizations ($\theta = 0$ in Fig. 12.51). If δk is the corresponding difference in wave number, then in the course of traversing the piece of fiber, a phase shift $\delta\phi = \delta k\, dz$ will develop in the quantity rs^*. Consequently, the Stokes vector precesses through the angle $\delta\phi$ around the S_1 axis, marking out a cone. The corresponding generalization is that for any birefringence, the Stokes vector precesses through an angle $\delta\phi = \delta k\, dz$ around the axis in Stokes space that corresponds to the two principal states of the birefringence. If $\boldsymbol{\beta}$ is a vector whose length is δk and that lies along this axis of birefringence, then $\hat{\mathbf{S}}$ precesses around $\boldsymbol{\beta}$ according to the equation

$$\frac{d\hat{\mathbf{S}}}{dz} = -\boldsymbol{\beta} \times \hat{\mathbf{S}}. \tag{6.3}$$

Figure 12.52 illustrates this behavior. When $\delta\phi$ reaches 2π, $\hat{\mathbf{S}}$ has swept out a complete cone. An optical element of this sort is called a *full-wave plate.*

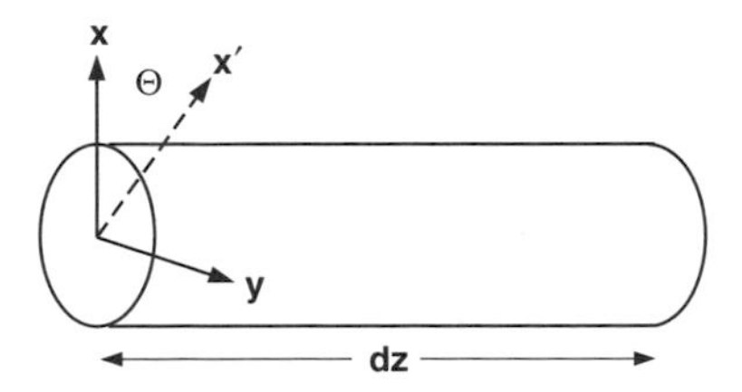

Fig. 12.51 Element of fiber with birefringence axis x'.

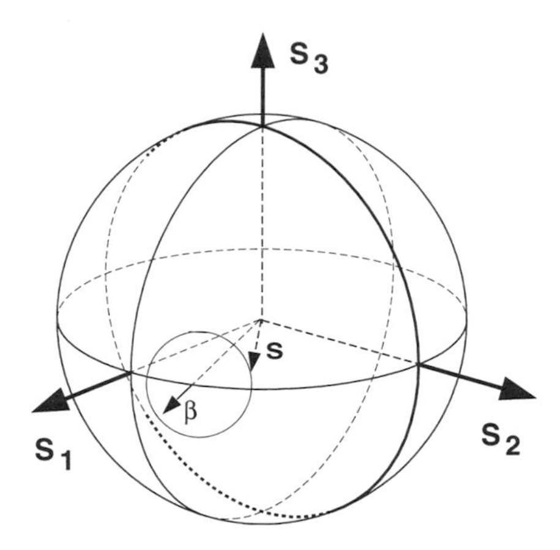

Fig. 12.52 The birefringence element β causes **S** to precess in a cone around it.

In a fiber, the needed length $2\pi/\delta k = \lambda(k/\delta k)$ is called the *beat length*. It is typically very short, a few meters to tens of meters. If the birefringence axis varies in a random way along the fiber, the Stokes vector soon comes to have a random direction on the Poincaré sphere.

Along with the wave-number birefringence just discussed comes a birefringence in the inverse group velocity. Let $\mathbf{b} = d\boldsymbol{\beta}/d\omega$. If $\boldsymbol{\beta}$ does not change direction with frequency (in practice the change tends to be negligibly small), then we have

$$b = \frac{d\beta}{d\omega} = \frac{d\delta k}{d\omega} = \delta v_g^{-1}, \tag{6.4}$$

where b, for example, is the magnitude (length) of the vector **b**, and the last equality is so because $dk/d\omega = v_g^{-1}$ holds for each of the two principal states of polarization. If we refer again to our piece of fiber with x–y birefringence, we can determine that the average time delay for the energy of a field in the polarization state $(r\hat{x} + s\hat{y})$ is proportional to $(|r|^2 - |s|^2)$ and thus to S_1. In general, the time delay is proportional to the projection of $\hat{\mathbf{S}}$ on the birefringence axis, so we get

$$dt_d = \tfrac{1}{2}\,\hat{\mathbf{S}} \cdot \mathbf{b}\; dz. \tag{6.5}$$

6.3.2 Calculus for Long Fibers

In a typical transmission fiber, both the strength and the axis of the birefringence vary along the length of the fiber, which causes the Stokes vector to wander more or less randomly around the Poincaré sphere. The motion of $\hat{\mathbf{S}}$ in going between any two points in the fiber can be expressed formally by

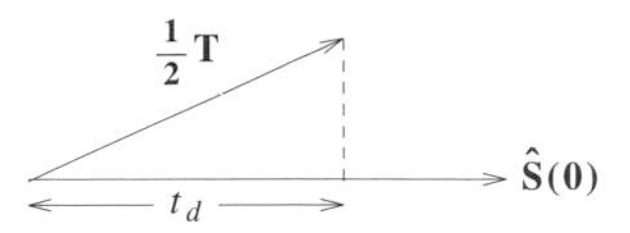

Fig. 12.53 Relative time delay, in a long fiber, of a pulse with polarization vector $\hat{\mathbf{S}}$.

$$\hat{\mathbf{S}}(z_2) = \mathbf{M}(z_2, z_1) \cdot \hat{\mathbf{S}}(z_1), \tag{6.6}$$

where $\mathbf{M}$ is a rotation matrix, a (3×3) form of a Mueller matrix, and $\mathbf{M}(z_1, z_2)$ is the inverse of $\mathbf{M}(z_2, z_1)$. For a long fiber, of length Z, the total delay caused by the birefringence can be written as

$$\begin{aligned} t_d &= \frac{1}{2}\int_0^Z \hat{\mathbf{S}}(z) \cdot \mathbf{b}(z)\, dz = \frac{1}{2}\, \hat{\mathbf{S}}(0) \cdot \int_0^Z [\mathbf{M}(0, z) \cdot \mathbf{b}(z)]\, dz \\ &= \tfrac{1}{2}\, \hat{\mathbf{S}}(0) \cdot \mathbf{T}, \end{aligned} \tag{6.7}$$

where we have used the inverse Mueller matrix to project the vector dot product at location z back to the input of the fiber, and the last equality defines the vector $\mathbf{T}$. We call $\mathbf{T}$ the *polarization time-dispersion vector* [35].* Its length is the difference between the maximum and minimum delay times, and the Stokes vectors pointing in its positive and negative directions represent the principal states of polarization for which the delay times are, respectively, longest and shortest. It behaves very much like the local birefringence, except its magnitude and direction on the Poincaré sphere are more rapidly frequency dependent (Fig. 12.53). In the case of linear propagation, any input pulse can be resolved into a linear combination of components in the two principal states, which have different delays. As a result, one sees the output pulse width vary as a function of the input polarization, its mean delay time being given by Eq. (6.7). In the case of soliton propagation, if the PMD is not too large, the nonlinear effects hold the pulse together, so that the pulse width does not change, but its mean delay time also obeys Eq. (6.7). Soliton propagation is discussed further later in this chapter.

6.3.3 Growth of T with Increasing Fiber Span Length

We can think of a long fiber as a concatenation of two shorter fibers, joined at some point $z = z_1$. Accordingly, the vector $\mathbf{T}$ for the whole fiber can be split into two pieces as

* Prior to the appearance of Ref. 35, however, the polarization dispersion vector and its statistical properties had already been thoroughly explored by Poole and coauthors; see, for example, Refs. 36 and 37. Note that the vector Ω in those papers is the same as $\mathbf{M}(Z, 0) \cdot \mathbf{T}$ here, where Z is the length of the fiber.

$$\begin{aligned}\mathbf{T} &= \int_0^{z_1} \mathbf{M}(0, z) \cdot \mathbf{b}(z)\, dz + \mathbf{M}(0, z_1) \cdot \int_{z_1}^{Z} \mathbf{M}(z_1, z) \cdot \mathbf{b}(z)\, dz \\ &= \mathbf{T}_1 + \mathbf{M}(0, z_1) \cdot \mathbf{T}_2,\end{aligned} \tag{6.8}$$

where $\mathbf{T}_1$ and $\mathbf{T}_2$ are the $\mathbf{T}$ vectors for the two sections taken individually. We are primarily interested in fiber spans that are long with respect to the characteristic distance (typically only a few meters) for the reorientation of $\mathbf{b}$. In that case, the direction of $\mathbf{T}_2$ is at random with respect to that of $\mathbf{T}_1$; furthermore, because there is no correlation between $\mathbf{M}(0, z_1)$ and $\mathbf{T}_2$, the second term in Eq. (6.8) is still oriented at random with respect to the first. This division can be iterated, giving

$$\mathbf{T} = \mathbf{T}_1 + \sum_{i=2}^{N-1} \mathbf{M}(0, z_i) \cdot \mathbf{T}_i, \tag{6.9}$$

until the $\mathbf{T}$ vectors of the individual sections cease being uncorrelated. From this exercise it should be obvious that the growth of T (the magnitude of $\mathbf{T}$) is a random walk process, where T is expected to grow as $z^{1/2}$. Another way of seeing this is to look at the quantity

$$\begin{aligned}T^2 = \mathbf{T} \cdot \mathbf{T} &= \int_0^Z \int_0^Z [\mathbf{M}(0, z) \cdot \mathbf{b}(z)] \cdot [\mathbf{M}(0, z') \cdot \mathbf{b}(z')]\, dz'\, dz \\ &= \int_0^Z \int_0^Z \mathbf{b}(z) \cdot \mathbf{M}(z, z') \cdot \mathbf{b}(z')\, dz\, dz'.\end{aligned} \tag{6.10}$$

Beyond the length over which $\mathbf{b}(z)$ is correlated with $\mathbf{M}(z, z') \cdot \mathbf{b}(z')$ this double integral is expected to grow noisily but roughly linearly with Z. To take a simple example, imagine a fiber, initially with a constant linear birefringence, which is cut into short sections of length l_{sect} and put back together after each section has been rotated through a random angle around its cylindrical axis. Then, referring to Eq. (6.10), we can see that because $\mathbf{b}(z)$ is oriented along the local birefringence axis, $\mathbf{M}(z, z')$ reduces to the identity matrix as long as z and z' are both in the same section. There is no correlation between the directions of $\mathbf{b}(z)$ in the different sections, so in getting the expected value of T^2, the first integral over z' reduces simply to $b^2 l_{sect}$, and the second integral produces the factor Z.

An expected value of T is usually inferred from measurements made over a wide band of optical frequencies. As the fiber is cut back, the data [38] indeed tend to fit a curve of form

$$T(z) = D_p z^{1/2}. \tag{6.11}$$

D_p is known as the PMD parameter. For the highest quality transmission fibers available at present, $0.06 \leq D_p \leq 0.1$ ps/km$^{1/2}$.

6.3.4 Statistical Properties of T

The PMD parameter D_p for a length of fiber is found from averaging the values of T obtained over a rather broad range of frequencies. The distribution of T values so obtained tends to be Maxwellian. Changing the temperature of a fiber leads to similar results. This distribution can be understood as follows: The value of T is sensitive to the wave-number birefringence of the fiber, because this is what causes precession of the Stokes vector, and so determines the values of the Mueller matrices. We have seen that $\mathbf{T}$ can be considered to be the sum of a large number of independent vectors $\mathbf{T}_i$ from fiber sections, each projected to the beginning of the fiber by the appropriate Mueller matrix. As these Mueller matrices are changed by changing frequency, say, one would expect the components of $\mathbf{T}$ to have nearly independent Gaussian distributions.

If each of the three components of $\mathbf{T}$ in Stokes space (T_1, T_2, T_3) has an independent Gaussian distribution with standard deviation σ, then the distribution of $\mathbf{T}$ has spherical symmetry and the magnitude of $\mathbf{T}$ will have a Maxwellian distribution, as illustrated in Fig. 12.54. The probability that T lies between T and $T + dT$ is

$$p(T)\, dT = \sqrt{\frac{2}{\pi}} \frac{1}{\sigma^3} T^2 e^{-1/2(T/\sigma)^2}\, dT. \tag{6.12}$$

The most probable value of T/σ is $\sqrt{2} \approx 1.414$, its mean value is $\sqrt{8/\pi} \approx 1.596$, and its standard deviation is $\sqrt{3} \approx 1.732$. By definition, D_p is the mean value of T divided by the square root of the length of the fiber, so that $\sigma = D_p \sqrt{\pi Z/8}$. As an example, if $D_p \sim 0.1$ ps/km$^{1/2}$ and

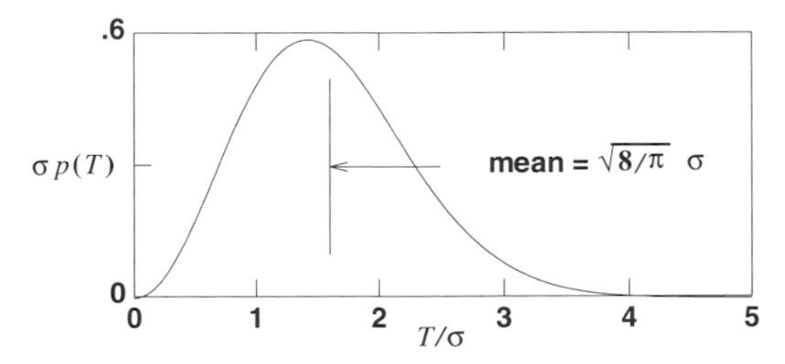

Fig. 12.54 Maxwellian probability distribution of T.

$Z = 10{,}000$ km, then $\sigma \sim 6.3$ ps. As a final bit of housekeeping here, note from Eq. (6.7) that the delay time for a pulse is half the component of $\mathbf{T}$ along the direction of $\hat{\mathbf{S}}(0)$. Thus the expected standard deviation of the pulse delay is $\sigma_{delay} = \sigma/2 = D_p\sqrt{\pi Z/32}$.

6.4 SOLITON PROPAGATION

We now consider soliton pulse propagation in a randomly birefringent fiber. First, it is known that the nonlinear coefficient n_2 is a function of polarization, varying, relative to its polarization average, from $\frac{9}{8}$ for linear polarization to $\frac{3}{4}$ for circular polarization (on average, linear polarization is twice as likely as circular polarization). The changes in n_2 as a soliton's polarization state varies during propagation cause some radiative loss of energy. However, this loss is much smaller than a similar loss due to PMD, and it may be ignored. To avoid unnecessary complication, we shall therefore continue to treat n_2 as though it were polarization independent.

We now come to consider the residual birefringence. On the often used assumption that the local form of the birefringence is not very important overall, we shall invoke a well-worn model similar to that used previously, in which the fiber is composed of short sections of constant linear birefringence, whose axes are assumed randomly directed, and whose magnitude may also have some random distribution. Consider the effect of one such section. We can denote the slow and fast axes of this piece of fiber by the orthogonal unit vectors $\hat{x}$ and $\hat{y}$. The entering soliton will be in some linear combination of these two polarization states and will in general have the form

$$\mathbf{u}(z, t) = (r\hat{x} + s\hat{y})\,\mathrm{sech}(t), \tag{6.13}$$

where the polarization state vector has unit length, as before, satisfying $(|r|^2 + |s|^2 = 1)$. After traversing the section of length l, this soliton will have changed to the form

$$\mathbf{u}(z + l, t) = e^{i\theta}[re^{i\phi}\hat{x}\,\mathrm{sech}(t - \varepsilon) + se^{-i\phi}\hat{y}\,\mathrm{sech}(t + \varepsilon)], \tag{6.14}$$

where the angle $\phi = (\frac{1}{2}\,\delta kl$ arises from the local wave-number birefringence δk, while $\varepsilon = (\frac{1}{2})\,bl$ is half the time delay birefringence of the fiber section. For good fibers, ε is a very small number. Using this, we can expand the sech functions to first order in ε [$d\,\mathrm{sech}(t)/dt = -\mathrm{sech}(t)\tanh(t)$], and with a bit of manipulation arrive at

$$\mathbf{u}(z + l, t) \approx e^{i\theta}[(re^{i\phi}\hat{x} + se^{-i\phi}\hat{y})\text{sech}(t - \varepsilon(|r|^2 - |s|^2)) \\ + \varepsilon\, 2rs(s^* e^{i\phi}\hat{x} - r^* e^{-i\phi}\hat{y})\text{sech}(t)\tanh(t)]. \tag{6.15}$$

We have split the terms proportional to ε into two parts, one having the same polarization state as the soliton, which we put back together with the soliton, and the other having a polarization state orthogonal to the soliton. We see that the soliton's polarization state and PMD time delay have changed in accordance with our previous general discussion. The field scattered into the orthogonal polarization state, proportional to sech(t)tanh(t), can be shown to be a dispersive field and so represents a loss mechanism for the soliton. The fractional energy loss due to this section of fiber is the ratio of the time integrals of $|u|^2$ in the two orthogonal polarization states, which yields $\delta E/E = (\frac{4}{3})|rs|^2\varepsilon^2$, independent of ϕ. If we assume that the birefringent time delay stays small with respect to the pulse width of the soliton, then the scattered fields will be uncorrelated, and the total loss will be the sum of the energies scattered from each section. To find the expected loss, it is appropriate to do an average over polarizations. From Eq. (6.2) one can see that $4|rs|^2$ is the sum of the squares of the S_2 and S_3 components of the unit Stokes vector $\hat{\mathbf{S}}$, and therefore its polarization average is $\frac{2}{3}$. Using this, and evaluating ε, we get

$$\alpha_{pmd}l = \frac{\delta E}{E} = \frac{1}{18}b^2l^2, \tag{6.16}$$

where α_{pmd} is the exponential energy loss coefficient. In comparison, the time delay for this section, $\delta t_d = \varepsilon(|r|^2 - |s|^2)$, is proportional to the S_1 component of $\hat{\mathbf{S}}$, so its polarization average is zero, but its variance is

$$\sigma_d^2 = \tfrac{1}{3}\,\varepsilon^2 = \tfrac{1}{12}\,b^2l^2. \tag{6.17}$$

The loss due to PMD is therefore related to the variance of the time delay by $\alpha_{pmd}l = (\frac{1}{3})\sigma_d^2$. Because the loss and the variance of the time delay both grow linearly with the number of sections, the loss for a length z of fiber will be given by

$$\alpha_{pmd}z = \frac{2}{3}\,\sigma_d^2(z) = \frac{\pi}{48}\,D_p^2 z. \tag{6.18}$$

This equation is written in soliton units. It can be made dimension free by dividing its right side by t_c^2, or $(\tau/1.7627)^2$; removing the common factor of z as well, one gets

$$\alpha_{pmd} = 0.2034 D_p^2 / \tau^2. \tag{6.18a}$$

To take a good fiber as an example, with $D_p = 0.1$ ps/km$^{1/2}$, then a pulse with $\tau = 20$ ps yields $\alpha_{pmd} = 5 \times 10^{-6}$/km, or 0.005/Mm. Although the loss rate in soliton energy due to PMD appears small in this case, note that the total energy lost in 10 Mm is 5% of the soliton's energy. Because this lost energy is scattered into dispersive waves, it can add significantly to the total noise, especially in a broadband transmission line. On the other hand, sliding-frequency filters (see Section 4) control the growth of this noise just as effectively as they control the growth of spontaneous emission noise.

It should also be noted from Eq. (6.18a) that the loss rate from PMD increases rapidly with decreasing pulse width. For example, for the maximum pulse width ($\tau \approx 2$ ps) permitting a single-channel rate of 100 Gb/s, and again for $D_p = 0.1$ ps/km$^{1/2}$, α_{pmd} rises to 0.5/Mm, a value large enough to cause very serious problems. This is yet another reason why, for the attainment of very large net bit rates, massive WDM (as described in Section 5.9, for example) is by far the better choice.

If the value of D_p becomes too large, solitons can become unstable. For distances of the order of z_c, essentially linear propagation occurs, so a soliton has a chance of being split into its two principal state components. Some years ago, a criterion for stability was established by numerical simulation, using a kind of worst case scenario [39]. In soliton units, the result was simply

$$D_p \leq 0.27 t_c / z_c^{1/2}. \tag{6.19}$$

Using Eq. (6.19) with the equality sign to establish the largest allowable D_p, and setting $D_p z^{1/2} \approx t_c$, note that a linear pulse would split into two pieces spaced by t_c in a distance $(0.27^{-2} \approx 14) \times z_c$. Under those same conditions, however, nonlinear effects hold the soliton together over an indefinitely long distance of propagation. From Eq. (2.15a), we recall that $t_c / z_c^{1/2}$ scales with the dispersion constant D, so in standard units Eq. (6.19) becomes

$$D_p \leq 0.3 D^{1/2}. \tag{6.19a}$$

Note that for $D_p = 0.1$ ps/km$^{-1/2}$, Eq. (6.19a) is satisfied for D as small as about 0.11 ps/nm-km. For an example of what can happen when the criterion of Eq. (6.19a) is violated, see Ref. 40.

It is interesting that even at the stability border, the loss calculated from Eq. (6.18) seems very small. If we use Eq. (6.19) with an equals sign, then from Eq. (6.18) we get $\alpha_{pmd} = 0.048/z_c$, which does not seem large enough to signal impending doom for the soliton. The answer to this conundrum is probably in the statistics. The loss was calculated on the basis of polarization averaging, while the split-up of a soliton can occur in any section of the transmission fiber a few z_c long if the **T** vector for that section is large enough. Simulations also support this conclusion. As long as the soliton transmission is stable, the loss (per z_c) is small.

6.5 POLARIZATION SCATTERING BY SOLITON–SOLITON COLLISIONS

As already noted in the introduction to this section, colliding solitons in WDM alter each other's polarization states. Although this polarization scattering was described in 1973 by Manakov [41], its consequences were not generally appreciated until they became manifest in a fairly recent experiment [42]. In the experiment, each of several 10-Gb/s WDM channels was subdivided into two polarization- (and time-) division multiplexed, 5-Gb/s subchannels. As had been known for some time [43], such polarization-division multiplexing works well, at least in the absence of WDM. Indeed, in the experiment, with only one such polarization-multiplexed, 10-Gb/s channel present, the orthogonality of the 5-Gb/s subchannels was well maintained over transoceanic distances, and the transmission was error free. As soon as a second WDM channel was added, however, polarization scattering from the collisions destroyed the orthogonality of the subchannels, and the error rate became high for all but very short distances. To confirm that polarization scattering was to blame, the degree of polarization of each 10-Gb/s channel (this time with all pulses initially copolarized) was measured as a function of distance. The results are summarized in Fig. 12.55. With only one channel present (no WDM), as expected, the degree of polarization (DOP) was only slightly reduced in 10 Mm, from the mild effects of spontaneous emission [43]. With just one other channel present, however, the degree of polarization of either channel was reduced nearly to zero in the same distance.

The origin of the polarization scattering in WDM transmission is not difficult to understand. It comes about because the magnitude of the cross-phase modulation between waves of different frequencies is dependent on their relative polarizations, being twice as large for copolarized waves as for orthogonally polarized waves. One can think of this as a nonlinearly induced effective birefringence, keeping in mind that the birefringence seen

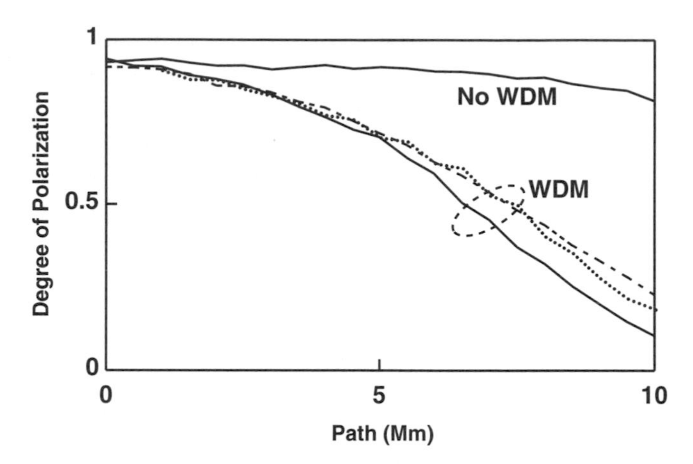

Fig. 12.55 Experimentally measured degree of polarization for a given wavelength channel as a function of distance. *No WDM*: Only one channel is present on the transmission line. *WDM*: Two channels are present on the line. Channel separations: 0.6 nm (*solid line*); 1.2 nm (*dashed line*); 1.8 nm (*dotted line*). In all cases, the channels were initially copolarized.

by each of the waves is different from that seen by the other. We discuss this effect in more detail later. Consider now a collision between solitons in two frequency channels of a WDM system. Label the solitons a and b. The cross-phase modulation phase shift given to the component of soliton a copolarized with soliton b is twice as large as that given to the component of soliton a orthogonal to soliton b. The result is a change in the polarization state of soliton a. In terms of the three-dimensional Stokes vector representation of a soliton's polarization state, the first-order result is a precession of soliton a's Stokes vector around that of soliton b. Soliton b is similarly influenced by soliton a, so their Stokes vectors precess around each other. The cross-phase modulation occurs only while the solitons overlap, and after a completed collision the differential phase shift (which, as we shall show, equals the precision angle of the Stokes vectors) is approximately equal to $L_{coll}/(1.76z_c)$. Note that there is no change in the polarizations of either soliton if they are in the same or orthogonal polarization states.

To analyze the polarization scattering in detail, we describe the optical field envelope in a fiber, in the manner of Eq. (6.13), using

$$\mathbf{u} = u_x\hat{x} + u_y\hat{y}. \tag{6.20}$$

Here u_x and u_y are the x and y components of the field, normalized so that $\mathbf{u}^* \cdot \mathbf{u} = |u_x|^2 + |u_y|^2$ is the optical power in the fiber. Instead of the Stokes vector of unit length used in Section 6.2, we shall here use an instantaneous Stokes vector with the components $(S_1, S_2, S_3) = [|u_x|^2 - |u_y|^2, 2\mathrm{Re}(u_x^* u_y), 2\mathrm{Im})u_x^* u_y)]$, so that the length of the Stokes vector is the optical power. Using a unitary transformation (the Jones matrix), we can mathematically eliminate the rapid motion of the polarization of the optical field caused by the fiber's wave-number birefringence. Then, taking into account that the collision length for the solitons of interest here is short enough ($L_{coll} \leq 200$ km), that the effects of polarization dispersion on the relative polarizations of the two fields can be neglected, and that the nonlinear term is averaged over all polarizations, we are left with a much simplified version of the propagation equation, known as the Manakov equation [41]:

$$-i\frac{\partial \mathbf{u}}{\partial z} = \frac{1}{2}\frac{\partial^2 \mathbf{u}}{\partial t^2} + (\mathbf{u}^* \cdot \mathbf{u})\mathbf{u}. \tag{6.21}$$

Now if $\mathbf{u}$ is composed of two fields with distinctly separable frequency ranges (e.g., pulses in a WDM system), we can isolate the terms of Eq. (6.21) in each frequency range. If the two frequencies are identified by subscripts a and b, then for the frequency of $\mathbf{u}_a$, we find the equation

$$-i\frac{\partial \mathbf{u}_a}{\partial z} = \frac{1}{2}\frac{\partial^2 \mathbf{u}_a}{\partial t^2} + (\mathbf{u}_a^* \cdot \mathbf{u}_a)\mathbf{u}_a + (\mathbf{u}_b^* \cdot \mathbf{u}_b)\mathbf{u}_a + (\mathbf{u}_b^* \cdot \mathbf{u}_a)\mathbf{u}_b. \tag{6.22}$$

The equation for the frequency of $\mathbf{u}_b$ is obtained by interchanging indices. There are three nonlinear terms on the right-hand side of Eq. (6.22). The first is the self-phase modulation term, while the second and third constitute the polarization-dependent cross-phase modulation. For copolarized fields, the second and third terms become identical (compare with Eq. [5.3]), while for orthogonally polarized fields, the third term goes to zero.

If we expand Eq. (6.22) into its field components, it is straightforward to show that the cross-phase modulation terms of Eq. (6.22) modify the Stokes vector of the a field according to the equation

$$\frac{\partial \mathbf{S}_a}{\partial z} = \mathbf{S}_a \times \mathbf{S}_b. \tag{6.23}$$

This equation and that for $\mathbf{S}_b$ (exchange subscripts) show that the nonlinear term causes the Stokes vectors of the two fields to precess around each other, so that $\partial(\mathbf{S}_a + \mathbf{S}_b)/\partial z = 0$. It is important to note from Eq. (6.23) that when the pulses are either exactly copolarized or exactly orthogonally polarized, there is no scattering. Thus, for example, in the experiment of Fig. 12.55, there was no scattering between the initially copolarized channels until the fiber's PMD gradually opened up the angle between their Stokes vectors.

To apply Eq. (6.23) to solitons, we have only to integrate it over the course of a collision. In first order, if we neglect the simultaneous precession of soliton b, then the integration of Eq. (6.23) gives an effective precession angle of $\mathbf{S}_a$ around $\mathbf{S}_b$. Single solitons of Eq. (6.21) have the general form

$$\mathbf{u}(z, t) = \hat{\mathbf{u}} A \operatorname{sech}[A(t + \Omega z)] \exp[iz(A^2 - \Omega^2)/2 - i\Omega t], \quad (6.24)$$

where $\hat{\mathbf{u}}$ is a unit normalized polarization vector ($\hat{\mathbf{u}}^* \cdot \hat{\mathbf{u}} = 1$). To evaluate the precession angle most easily, let $\mathbf{u}_a$ be a stationary soliton ($\Omega_a = 0$), and let $\mathbf{u}_b$ be a soliton substantially displaced in frequency ($|\Omega_b| \gg 1$). Then the integration over the collision involves the integration over z of $A_b^2 \operatorname{sech}^2[A_b(t + \Omega_b z)]$, which gives a precession angle of $2A_b/\Omega_b$. This formula uses soliton units. Noting that the full spectral width of the soliton in Eq. (6.24) is $\Delta\omega = (2/\pi) \times 1.763\, A$, we can reexpress the polarization angle in the dimensionally independent form

$$\Delta\theta_a \simeq 1.78 \frac{\Delta\nu_b}{\Delta\nu_{ab}}, \quad (6.25)$$

where $\Delta\theta_a$ is the precession angle for the stationary solition, $\Delta\nu_b$ is the full (spectral) width at half maximum of the passing soliton, and $\Delta\nu_{ab}$ is the frequency separation of the two solitons. Note the inverse dependence of the change in polarization angle on channel separation ($\Delta\nu_{ab}$). Because the number of collisions in a given distance is in direct proportion to $\Delta\nu_{ab}$, however, the net spread in polarization vectors tends to be independent of channel separation, as observed experimentally (Fig. 12.55). Note also that Eq. (6.25) is valid whether or not the colliding solitons are equal in amplitude. Because the bandwidth of a soliton is proportional to its amplitude, the collision of two unequal solitons will yield unequal precessions for the two. In the WDM experiments described in Section 5, the ratio of the channel spacing to the soliton's spectral FWHM was about 5, which would make the precession angle per collision about 0.35 radians. This is a small enough angle to make the previously described theory applicable, and yet is large enough that just a few collisions are enough to prohibit the use of polarization-division multiplexing.

The argument just presented gives results applicable to a WDM communications system, where the channel spacing is much greater than the soliton bandwidth, so that the precession angle in a collision is small. Manakov [41] was able to show that Eq. (6.21) supports polarized solitons in the strict sense. A collision of two solitons therefore does not give rise to any scattered radiation. The solitons emerge from the collision with their energies and velocities unchanged, but their polarizations change as well as their positions and phases. A translation of Manakov's result gives the following Stokes–Poincaré picture. If colliding solitons a and b have amplitudes A_a and A_b, and normalized Stokes vectors $\hat{\mathbf{S}}_a$ and $\hat{\mathbf{S}}_b$, then the vector $\mathbf{A} \equiv A_a\hat{\mathbf{S}}_a + A_b\hat{\mathbf{S}}_b$ remains the same before and after the collision. As a result of the collision, the two Stokes vectors $\hat{\mathbf{S}}_a$ and $\hat{\mathbf{S}}_b$ precess around the axis defined by $\mathbf{A}$ through an angle ϕ whose tangent is given by $\tan \phi = 2A\Omega/(\Omega^2 - A^2)$, where Ω is the difference between the solitons' frequencies and A is the length of the vector $\mathbf{A}$. With a little practice in geometry, one can show that when $\Omega^2 \gg A^2$, the exact result reduces to the approximate one given previously.

Another consequence of the polarization scattering from collisions, more fundamental than the simple prohibition of polarization-division multiplexing, is a jitter in pulse arrival times, mediated by the fiber birefringence. Reference 35 describes a qualitatively similar birefringence-mediated jitter, as initiated by the (relatively small) noise-induced scatter in polarization states. Because the spread in polarization states from collisions tends to be much larger (note that it eventually tends to spread the Stokes vectors over a large fraction of the Poincaré sphere), the jitter is correspondingly greater, and in a typical case can easily add at least a few tens of picoseconds to the total spread in arrival times over transoceanic distances. This represents a significant reduction in safety margin for individual channel rates of 10 Gb/s or more. Nevertheless, note that in the WDM results reported in Section 5.9, that reduction, although undoubtedly present, was not fatal.

References

[1] Mollenauer, L. F., J. P. Gordon, and M. N. Islam. 1986. Soliton propagation in long fibers with periodically compensated loss. *IEEE J. Quantum Electron.* QE-22:157–173.

[2] Mollenauer, L. F., M. J. Neubelt, S. G. Evangelides, J. P. Gordon, J. R. Simpson, and L. G. Cohen. 1990. Experimental study of soliton transmission over more than 10,000 km in dispersion shifted fiber. *Opt. Lett.* 15:1203–1205.

[3] Mollenauer, L. F., S. G. Evangelides, and H. A. Haus. 1991. Long distance solition propagation using lumped amplifiers and dispersion shifted fiber. *J. Lightwave Tech.* 9:194–197.

[4] Blow, K. J., and N. J. Doran. 1991. Average soliton dynamics and the operation of soliton systems with lumped amplifiers. *Photon. Tech. Lett.* 3:369–371.

[5] Hasegawa, A., and Y. Kodama. 1990. Guiding-center soliton in optical fibers. *Opt. Lett.* 15:1443–1445.

[6] Gordon, J. P., and L. F. Mollenauer. 1991. Effects of fiber nonlinearities and amplifier spacing on ultra long distance transmission. *J. Lightwave Tech.* 9:170–173.

[7] Gordon, J. P., and L. F. Mollenauer. 1990. Phase noise in photonic communications systems using linear amplifiers. *Opt. Lett.* 15:1351–1353.

[8] Gordon, J. P., and H. A. Haus. 1986. Random walk of coherently amplified solitons in optical fiber. *Opt. Lett.* 11:665–667.

[9] Smith, K., and L. F. Mollenauer. 1989. Experimental observation of soliton interaction over long fiber paths: Discovery of a long-range interaction. *Opt. Lett.* 14:1284–1286.

[10] Dianov, E. M., A. V. Luchnikov, A.N. Pilipetskii, and A. N. Starodumov. 1990. Electrostriction mechanism of soliton interaction in optical fibers. *Opt. Lett.* 15:314–316.

[11] Dianov, E. M., A. V. Luchnikov, A. N. Pilipetskii, and A. M. Prokhorov. 1991. Long-range interaction of solitons in ultra-long communication systems. *Soviet Lightwave Commun.* 1:235–246.

[12] Dianov, E. M., A. V. Luchnikov, A. N. Pilipetskii, and A. M. Prokhorov. 1992. Long-range interaction of picosecond solitons through excitation of acoustic waves in optical fibers. *Appl. Phys. B* 54:175–180.

[13] Mollenauer, L. F., E. Lichtman, G. T. Harvey, M. J. Neubelt, and B. M. Nyman. 1992. Demonstration of error-free soliton transmission over more than 15,000 km at 5 Gbit/s, single-channel, and over 11,000 km at 10 Gbit/s in a two-channel WDM. *Electron. Lett.* 28:792–794.

[14] Mecozzi, A., J. D. Moores, H. A. Haus, and Y. Lai. 1991. Soliton transmission control. *Opt. Lett.* 16:1841–1843.

[15] Kodama, Y., and A. Hasegawa. 1992. Generation of asymptotically stable optical solitons and suppression of the Gordon–Haus effect. *Opt. Lett.* 17:31–33.

[16] Mollenauer, L. F., J. P. Gordon, and S. G. Evangelides. 1992. The sliding-frequency guiding filter: An improved form of soliton jitter control. *Opt. Lett.* 17:1575–1577.

[17] Golovchenko, E. A., A. N. Pilipetskii, C. R. Menyuk, J. P. Gordon, and L. F. Mollenauer. 1995. Soliton propagation with up- and down-sliding-frequency guiding filters. *Opt. Lett.* 20:539–541.

[18] Mamyshev, P. V., and L. F. Mollenauer. 1994. Stability of soliton propagation with sliding frequency guiding filters. *Opt. Lett.* 19:2083–2085.

[19] Mollenauer, L. F., P. V. Mamyshev, and M. J. Neubelt. 1994. Measurement of timing jitter in soliton transmission at 10 Gbits/s and achievement of 375 Gbits/s-Mm, error-free, at 12.5 and 15 Gbits/s. *Opt. Lett.* 19:704–706.

[20] Mecozzi, A. 1995. Soliton transmission control by Butterworth filters. *Opt. Lett.* 20:1859–1861.

[21] LeGuen, D., F. Fave, R. Boittin, J. Debeau, F. Devaux, M. Henry, C. Thebault, and T. Georges. 1995. Demonstration of sliding-filter-controlled soliton transmission at 20 Gbit/s over 14 Mm. *Electron. Lett.* 31:301–302.

[22] Nakazawa, M., E. Yamada, H. Kubotra, and K. Suzuki. 1991. 10 Gbit/s soliton transmission over one million kilometers. *Electron. Lett.* 27:1270–1271.

[23] Widdowson, T., and A. D. Ellis. 20 Gbit/s soliton transmission over 125 Mm. *Electron. Lett.* 30:1866–1867.

[24] Forysiak, W., K. J. Blow, and N. J. Doran. 1993. Reduction of Gordon–Haus jitter by post-transmission dispersion compensation. *Electron. Lett.* 29:1225–1226.

[25] Suzuki, M., I. Morita, N. Edagawa, S. Yamamoto, H. Taga, and S. Akiba. 1995. Reduction of Gordon–Haus timing jitter by periodic dispersion compensation in soliton transmission. *Electron. Lett.* 31:2027–2028.

[26] Mollenauer, L. F., S. G. Evangelides, and J. P. Gordon. 1991. Wavelength division multiplexing with solitons in ultra long distance transmission using lumped amplifiers. *J. Lightwave Tech.* 9:362–367.

[27] Mamyshev, P. V., and L. F. Mollenauer. 1996. Pseudo-phase-matched four-wave mixing in soliton WDM transmission. *Opt. Lett.* 21:396–398.

[28] Mollenauer, L. F., P. V. Mamyshev, and M. J. Neubelt. 1996. Demonstration of soliton WDM transmission at 6 and 7 × 10 GBit/s, error-free over transoceanic distances. *Electron. Lett.* 32:471–473.

[29] Mollenauer, L. F., P. V. Mamyshev, and M. J. Neubelt. 1996. Demonstration of soliton WDM transmission at up to 8 × 10 GBit/s, error-free over transoceanic distances. In *OFC '96, San Jose, CA,* Postdeadline paper PD-22, PD22-2–PD22-5.

[30] Mecozzi, A., and H. A. Haus. 1992. Effect of filters on soliton interactions in wavelength-division-multiplexing systems. *Opt. Lett.* 17:988–990.

[31] Mamyshev, P. V., and L. F. Mollenauer. 1996. WDM channel energy self-equalization in a soliton transmission line using guiding filters. *Opt. Lett.* 21:1658–1660.

[32] Veselka, J. J., and S. K. Korotky. 1994. Optical soliton generator based on a single Mach–Zehnder generator. In *Technical Digest 1994 OSA & IEEE integrated photonics research topical meeting, San Francisco, CA, 17–19 February,* 190–192.

[33] Mamyshev, P. V., 1994. Dual-wavelength source of high-repetition-rate, transform-limited optical pulses for soliton transmission. *Opt. Lett.* 19:2074–2076.

[34] Yamada, K., H. Murai, K. Nakamura, H. Satoh, Y. Ozeki, and Y. Ogawa. 1995. 10-Gbit/s EA modulator with a polarization dependence of less than

0.3 dB. In *Technical Digest 1995 OSA & IEEE Optical Fiber Conference, San Diego, CA, 26 February–3 March,* 24–25.

[35] Mollenauer, L. F., and J. P. Gordon. 1994. Birefringence-mediated timing jitter in soliton transmission. *Opt. Lett.* 19:375–377.

[36] Poole, C. D. 1988. Statistical treatment of polarization dispersion in single-mode fiber. *Opt. Lett.* 13:687–689.

[37] Foschini, G. J., and C. D. Poole. 1991. Statistical theory of polarization dispersion in single mode fibers. *J. Lightwave Tech.* 9:1439–1456.

[38] Poole, C. D. 1989. Measurement of polarization-mode dispersion in single-mode fibers with random mode coupling. *Opt. Lett.* 14:523–525.

[39] Mollenauer, L. F., K. Smith, J. P. Gordon, and C. R. Menyuk. 1989. Resistance of solitons to the effects of polarization dispersion in optical fibers. *Opt. Lett.* 14:1219–1221.

[40] Wai, P. K. A., C. R. Menyuk, and H. H. Chen. 1991. Stability of solitons in randomly varying birefringent fibers. *Opt. Lett.* 16:1231–1233.

[41] Manakov, S. V. 1973. On the theory of two-dimensional stationary self-focusing of electromagnetic waves. *Zh. Eksp. Teor. Fiz.* 65:505. Also published in 1974 in *Sov. Phys. JETP* 38:248–263.

[42] Mollenauer, L. F., J. P. Gordon, and F. Heismann. 1995. Polarization scattering by soliton–soliton collisions. *Opt. Lett.* 20:2060–2062.

[43] Evangelides, S. G., L. F. Mollenauer, J. P. Gordon, and N. S. Bergano. 1992. Polarization division multiplexing with solitons. *J. Lightwave Tech.* 10:28–35.

Chapter 13 | A Survey of Fiber Optics in Local Access Architectures

Nicholas J. Frigo

AT&T Laboratories–Research, Holmdel, New Jersey

1. Introduction

In the United States, telecommunications is pervasive in our personal and professional lives: we routinely consume video services on television sets, can engage in a few minutes of transcontinental conversation for the price of a cup of coffee, exchange faxes in business and commerce, and, more recently, exchange electronic mail or browse on the Internet®. As of this writing, telecommunications legislation and the concomitant industry re-alignments are foretelling significant changes, at least in the United States.

Nowhere is this change more clear than in the access* arena, the "last mile" of the communication links to subscribers' residences.† This chapter reviews research from the mid-1980s to the mid-1990s on access architectures using fiber optics, and although it reflects a bias toward the U.S. situation, the work we review is truly global in scope [1–217]. The drive to privatize the public telecommunications networks is clearly evident in Asia and Europe, but the pace is quickening in the United States. The once-stable divisions among entertainment providers, long-distance network operators, local telephone companies, cable television network owners, computer software vendors, wireless service providers, and so forth are blurring as legal barriers to competition are removed. Each industry

* *Access* refers to the equipment and methods used to connect customers to the larger network.

† This is sometimes called the *loop*, variously attributed either to the twisted pair copper loop that has traditionally been connected to telephone sets, or to an acronym for "Local equipment Outside Of Plant" — i.e., the equipment outside the telephone company buildings.

OPTICAL FIBER TELECOMMUNICATIONS, VOLUME IIIA

ISBN: 0-12-395170-4

sees direct access to telecommunications consumers as important for selling services [159], yet although there are a variety of network options available [84, 85] (see Chapter 15 in Volume IIIA), the primary communications channels to most residences in the United States are the twisted pair of the telephone network and the coaxial cable of the CATV* network. These service networks are essentially complementary in both bandwidth and delivery method: telephony is the epitome of a *narrowband* service (64 kb/s) that is *switched* (i.e., user-specific data), whereas CATV is the epitome of a *broadband* service (≈GHz) that is *broadcast* (i.e., all users receive the same data). Unfortunately, there is a perceived need for future networks to be both broadband and switched. But since the two current networks have evolved over time to carry their respective services efficiently, changing from one type to the other requires major modifications and large investments.

Much has changed since the publication of *Optical Fiber Telecommunications II* in 1988. At that time, the delivery of individual video channels to the home was considered to be the most likely candidate for future service expansion [198]. However, that market has not materialized, and since that time, a video compression technology has evolved that has enabled 100-fold reductions in the bandwidth required to provide video, we have seen dramatic improvements in the quantity and quality of video delivery possible over coaxial cable, and we have been exposed to a plethora of optical delivery methods. The intent of this chapter is to provide a snapshot of the current architectural situation, how we arrived here, and where we may be headed.

As general background, the reader is referred to recent texts [61, 88, 117] describing fundamental technologies and optical architectures; several special issues [74–78] on components, architectures, and systems; and more general articles on networks [84, 85] and Passive Optical Networks [52, 102, 157]. Section 2 presents a simple telecommunications access model based on telephony, focusing on the outside plant (OSP) architecture, and this model forms the basis for describing other telecommunications architectures.† Interest in loop architectures arises from the suspicion that existing networks will be inadequate for future services: they will have to

* The first two letters of this acronym are sometimes taken to mean either *community* or *common* and *access* or *antenna*, reflecting the historical development of the service.

† It also describes other distribution systems, such as food store chains, in which there is a hierarchical distribution with warehousing, local distribution, and so forth. The "information superhighway" metaphor makes this similarity clear.

be modified or rebuilt. Current and potential network providers will have different views of how to proceed, however, and in Section 3 we list some of the considerations faced by a prospective network provider in deciding on the type of architecture to install. The list is intended to illuminate the complexity, uncertainty, and economic pressures that network providers must face in making this choice. Network options that are currently being considered (e.g., fiber and coaxial cable alternatives) are briefly surveyed in Section 4, with the exception of passive optical networks (PONs) that use wavelength-division multiplexing (WDM). This last option was proposed shortly after the previous volume of this book was published in 1988, and we explore it in more detail in Section 5. Although WDM PONs were proposed in the late 1980s, we argue that only recently has component, architecture, and system research caught up to the vision of that early work [107, 108, 118, 186–190].

2. Models of Current Access Architectures

2.1 TELEPHONY DISTRIBUTION MODEL

Figure 13.1, adapted from the PON literature [20, 22, 187], illustrates the distribution architecture from a telephony perspective and orients us to terminology that we use for all the distribution architectures. In the telephony model, a central office (CO) serving one or several telephone ex-

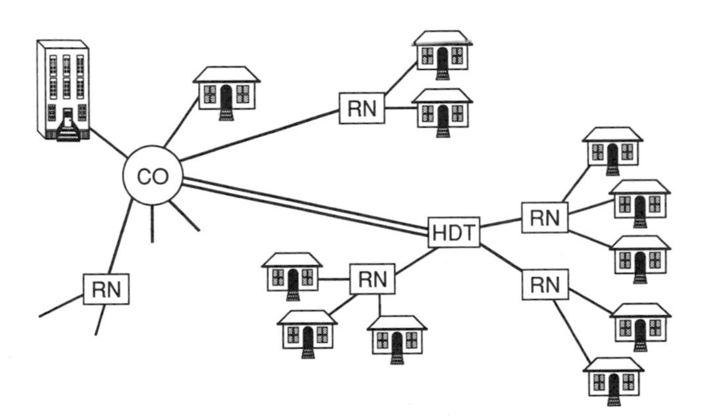

Fig. 13.1 Telephony distribution model. The central office (CO) communicates to subscribers with direct connections, or with multiplexed connections through remote nodes (RNs). A host digital terminal (HDT) may be remotely located.

changes (e.g., sets of telephone numbers such as 432–XXXX) of up to 10,000 subscribers each is connected either directly or through remote nodes (RNs) to the subscribers. The plant (historically copper) between the CO and the RN is usually called the *feeder*, whereas the plant between the RN and the subscribers is usually called the *distribution plant.* Racks of electronics, which comprise switching, multiplexing, and test circuits, terminate direct lines or feeders and process calls for on the order of 2000 subscribers [20, 22]. These electronics are usually referred to as *host digital terminals* (*HDTs*) and are often located in the CO. In some cases it is more economical to locate the HDT at a remote site and carry multiplexed traffic to it over much higher speed lines. The location of the HDTs is determined by balancing transmission costs versus feeder costs. The RNs, depending on function, are often called *remote terminals* (*RTs*) and may support another layer: wires from RTs are often run to distant terminals (DTs) or pedestals from which drops are made to individual subscriber residences. Geographic diversity and business–residence mixes make different layers more or less important.

The feeder plant, by carrying multiplexed traffic out to the RN, permits a reduction in the overall OSP, albeit at increased cost of the RNs. This reduction is often called *pair gain*, a reference to the twisted-pair lines saved by multiplexing and was a primary motivation for the development of digital loop carrier (DLC) approaches [9]. Economic considerations of the geographic extent, anticipated traffic volume, potential expansion, right-of-way access, relative costs of plant and transmission equipment, and so forth determine the relative amounts of feeder–distribution investments, and these costs are straightforward to calculate. High terminal costs in the earliest DLC systems made them economical only on the longest loop lengths and required changes in the operations procedures [66], which created a resistance to early deployment over time. Savings from automated test procedures and decreasing terminal costs, however, have resulted in more widespread deployment. An interesting historical aside is that although the advantages of digital transport now seem obvious, it was initially resisted on the grounds that it squandered bandwidth: the traffic carried in a 1.5-Mb/s digital signal, level 1 (DS-1) stream (24 independent telephone calls of 64 kb/s each) could be carried on only 96 kHz of analog bandwidth [9]. It took time to recognize that in addition to the earlier network metric (bandwidth) longer term advantages had to be considered. Today, similar debate is going on about installation costs versus upgrade and maintenance savings [156, 157].

Finally, as mentioned previously, the feeder plant terminates at an HDT that is in, or connected to, a CO. The CO, in turn, is part of the larger telephony network and is connected through interexchange trunks to other COs operated by a local exchange carrier (LEC), Bell Operating Company (BOC), or PTT (post, telephone, and telegraph), and these are linked to the long-distance networks.

2.2 CATV NETWORK MODEL

The CATV networks [26] (see Chapter 14 in Volume IIIA) that deliver television are also distribution networks, but they can take advantage of the fact that they are primarily unidirectional.* The television signals that they carry are common or broadcast — each subscriber receives the same information — in contrast to telephony. A model layout is shown in Fig. 13.2, illustrating the tree-and-branch layout that has been typical. Long chains of trunk amplifiers transport video signals to local branches for distribution. The high bandwidth of the entire televison spectrum (up to 400 or 750 MHz in current systems) causes high losses in the coaxial cables; these losses are compensated for by linear RF amplifiers. This situation is in contrast to the low losses in the telephony plant [66] and therefore requires electrical power at the amplifier sites. Main trunks require larger diameter low-loss coax cables and more linear trunk amplifiers (open), whereas bridger and distribution amplifiers (solid triangles) distribute the signal to subscriber residences, idealized as a

* Considerations of upstream bandwidth channels are too detailed for this brief sketch.

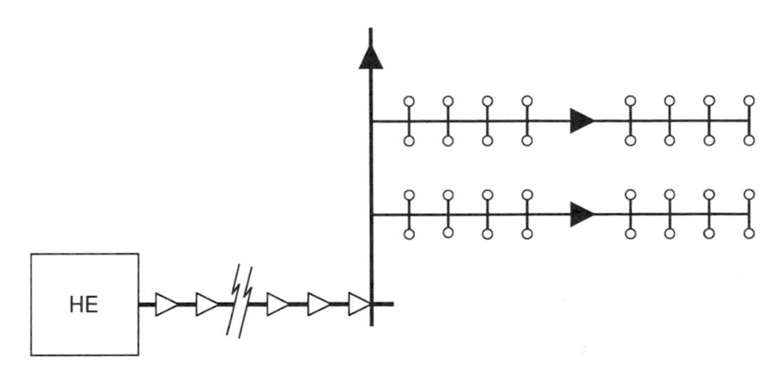

Fig. 13.2 CATV distribution model. A head end (HE) distributes video signals over coaxial cable with a tree-and-branch feeder and drop plant. Trunk coax lines and amplifiers (*open triangles*) are replaced with fiber in newer plant.

lattice in Fig. 13.2. The common television signal that is distributed makes new subscriber hookups relatively easy (compared with telephony), because a line extender amplifier can be used to increase the reach of plant that is already installed.

In terms of Fig. 13.1, the CO is replaced with a head end (HE) that receives television programming from a variety of sources, the feeder plant is low-loss coax cable with RF amplifiers, and the distribution is the cable drops, as shown in Fig. 13.2. There is little similarity to the signal processing done at a telephony RN. Improvements in fiber optic analog video transmission since the mid-1980s have changed the delivery method: in most new and rehabilitated installations, the coax trunks have been eliminated and replaced with fiber optic links to the serving area. At a fiber node the optical signal is received, amplified, and put onto the local coax plant (we return to this topic in Section 4). The high-quality fiber optic links increase the performance and reliability of the network, while reducing maintenance costs and outage by eliminating the traditional cable trunk. Current trends [26] in the cable industry are moving toward reducing the size of the area served by a fiber node. This tactic both increases the quality of service by driving fiber deeper into the loop (reducing the number of amplifiers traversed) and permits more specialized programming to be delivered. Note that this trend does not significantly change the drops to the houses, the majority of the capital expense of the plant. Because conventional television sets receive amplitude modulated–vestigial sideband (AM-VSB) signals, that is the format that is carried over current cable networks, but the network is certainly capable of delivering the more robust digital formats of current satellite services.

As a technological aside, it is difficult to overestimate the impact made by the seemingly incremental improvements in distributed feedback (DFB) laser linearity in the mid-1980s that enabled practical analog television delivery over fiber. This technology led to an explosive growth in CATV fiber installation in the late 1980s and thus created a dramatically improved service platform. In turn, the new high-capacity infrastructure made competition with the telephone companies credible, moving the "invisible hand" of the market to fashion alliances and legislative imperatives that will influence how we will experience commerce, education, and leisure in the 21st century. Such a perspective presages the tremendous potential impact of technical developments such as those that we discuss later, in Section 5.

2.3 LOOKING FORWARD

The distribution architecture models that we just discussed are generic, but they help to sort out access alternatives that are currently under discussion. Notwithstanding the recent introduction [26] of CATV fiber "backbones," the present architectures have evolved gradually in response to technological improvements and incremental changes in the relative economic weights of the network components [137]. Until recently, the evolutionary pace has been sustained by a regulatory environment that promoted relative stability in the form and capacity of the delivered services. This stability is disappearing.

The advent of new services, new service types, and competition has the potential to change both the hardware and the delivery architectures now used in the local loop: for the first time since the late 1800s it is clear that the telephony plant is inadequate for clearly observable services, for example. Traditional network providers and new entrants are examining current and future architectures to position themselves now for the decades to come. The sudden emergence of popular World Wide Web® browsers has introduced the Internet to millions of new users and is beginning to make an impact on the traffic in the telephone network. Although today its commercial usefulness is limited (wags have described the service as "click and wait"), it offers a clear view of how information exchange, education, entertainment, and economic transactions can occur. Regardless of whether this is the form that will dominate telecommunications in the future, it shows that a nontraditional service can emerge rapidly that does not fit naturally into current telecommunications models: surfing the Web is not a natural fit to either the telephone industry or the cable industry. Alternatives to traditional circuit switching such as the asynchronous transfer mode (ATM) have been developed to provide a flexible transition toward eventual packet transport, but the actual physical plant that is used to deliver communications will almost certainly have to evolve as well, not just the switches. For instance, commercial exploitation of the new palette of services will require low-latency, high-bandwidth channels for user efficiency, and guarantees of high quality of service (QoS) in availability and reliability. Because the network planning horizon is on the order of a decade [140], the emergence of such services is a signal that new network capabilities must be developed soon for services that may be unimagined today. This is a formidable task, laden with opportunities, uncertainties, and risks. The architecture choices facing providers are plentiful and diverse, but these choices are only part of the difficulty. The attributes that the network must have (other than low cost!) are unclear. Idealized attributes for the network include the abilities to provide narrow band, wideband, and broadband two-way services* with various rates and formats as demanded; to provide broadband entertainment video; to permit flexible provisioning of new services as they emerge; and to provide simple and cost-effective operations, administration, and maintenance (OAM) procedures on a long-lived robust OSP. In short, the network should be "future proof." Current network providers would like these attributes in the form of a "magic pill" that can be added inexpensively to their current plant, whereas new providers would like to be able to grow such a network economically and gradually. The next section examines the types of trade-offs involved in making such an architectural choice.

3. Considerations for Network Providers

In addition to facing the technical choices, a network provider who is planning to deliver services in the 21st century is confronted with a large set of forces and constraints. If the provider already owns a network, he or she has shareholders, stakeholders, employees, public utility commissions, financiers, and a running operation that will provide inertia. If the provider does not own a network, he or she must contend with current providers who, disliking the new provider's intention to share in their livelihood, may strive to see that the new business does not achieve a viable start. Pugh and Boyer [140] have given a detailed example of some of the considerations facing a network provider in choosing an access network. These researchers show that there is a complicated evaluation between the technical choices available and the corporate business plan. In brief, there are trade-offs to be made in light of the *plan view* (the expected view of developments), the *opportunity view* (the possibilities for additional revenue growth), and the *risk view* (the stakes that can be lost in trying to implement the plan). In this section, we show some of the choices that a provider must make that will affect these different views. Architectural considerations are perhaps the easiest part to deal with because they can be analyzed in some way.

* Generally, narrowband means one to several times the conventional DS-0 rate of 64 kb/s, wideband means rates comparable to the DS-1 rate of 1.5 Mb/s, and broadband means rates near and exceeding 10 Mb/s.

We illuminate the types of considerations with a representative list of 12 trade-offs to show the scope of complication: even without considering wireless and satellite alternatives, for instance, there would be 2^{12}, or 4096, assessment profiles if each choice were binary. However, each trade-off is a continuum, and the provider's view of customers' desires, his or her competitive position, and his or her financial strength will all change. In considering architectural alternatives, the provider may want to consider the following model choices among others.

3.1 DISTRIBUTIVE VERSUS INTERACTIVE SERVICES

Will the successful network of the future (say, in the year 2010) primarily be carrying entertainment video from a sparse set of large servers, or will there be a significant percentage (e.g., 10%) of the residences and small businesses that are effectively servers themselves? How much upstream versus downstream bandwidth (i.e., asymmetry) is needed, and in what form [140, 211]?

3.2 BROADCAST VERSUS SWITCHED NETWORK

In a related vein, what is the mode of the network — broadcast (point-to-multipoint), switched (point-to-point), or a combination? Entertainment services may favor broadcast delivery, whereas data communications and telephony may favor switched delivery. To what extent will capacity in one mode interfere with the operations of the other mode? At some point, 64-kb/s switches will become untenable, and a migration must be made to a more flexible partitioning [147, 199]. Can the relative weights be easily changed at a later date? How will subscriber privacy or network integrity* be compromised by the choice? At some point, traffic to a group of subscribers is merged: Where is the bus? Is it a medium to which subscribers have access (and therefore can corrupt it), or is it a backplane under centralized network control and management?

3.3 CURRENT VERSUS FUTURE CHANNEL CHARACTERISTICS

Current CATV and telephone distribution networks transport analog channels in an increasingly digital world. Presumably, the fragility and expense of analog transport of increasingly complicated traffic will yield to the

* *Network integrity* or *security* refers to the robustness with which a node prevents unauthorized upstream transmissions from compromising other, authorized, upstream traffic.

robustness of digital transmission and its capacity for compression, processing, encryption, switching, and storage [197]. However, the location of the analog–digital boundaries for these functions has architectural consequences and costs for interface designs: a network with ATM switches in each node may look different than one that seeks to emulate point-to-point circuit connectivity [10].

3.4 CENTRALIZED VERSUS DISTRIBUTED INTELLIGENCE

A more complex node (e.g., RN or RT) can offer processing and powering options that allow switching, grooming, and multiplexing to take place, which may also permit simpler subscriber electronics at the residence. Because the node cost is shared and the subscriber electronics are not, there are economic advantages to this approach. However, this comes with a penalty of power and maintenance costs required to support the more intelligent RN, and it makes future service changes more difficult to retrofit because they can make the RN obsolete. For a stable service mix (e.g., telephony) this is not a consideration, but it is of utmost concern for provisioning of future (i.e., currently uncertain) services.

3.5 CURRENT TECHNOLOGY VERSUS FUTURE TECHNOLOGY

Should a provider install a less than ideal network now or delay in the hopes of installing a more ideal network in the future? On one hand, a powerful argument for immediate installation of current technology is the desire to fully participate in today's service opportunities, from both the standpoint of increasing current revenue streams and that of protecting a customer base for future market entries. On the other hand, a less costly but more inflexible architecture installed today may have higher continuing operations costs [66], be more difficult to upgrade, and possess an embedded base more vulnerable to competitive bypass when the putative market opportunities arrive [146]. To what extent should perceived future needs be traded for current technical constraints? There are unmistakable technological trends in place today: lower cost digital storage and processing, lower costs for optical power, and higher functionality in optoelectronic integrated circuits, to name a few. If it had been clear that in the time between the introduction of DLC and 1988 that fiber optic terminal costs would decrease by a factor of 15 while copper costs would rise by a factor of 3, there would have been much less opposition to its installation [66]. To what extent should trends like these [173] be factored into today's plans?

3.6 BACKWARD COMPATIBILITY VERSUS NEW BUILDS

The two networks described in Section 2 have an embedded base of plant and customers. For instance, the current CATV format is AM-VSB, and the telephone network provides operations interfaces (BORSCHT [88]) that are well known and expected by an extensive consumer premises base. On the other side of that interface, such compatibility issues prompted the development of *pair gain numbers* as DLCs were introduced, in order to make the multiplexed traffic look like individual 64-kb/s circuits to the operations systems [9, 66]. Presumably, at some time in the future it will be advantageous to look at traffic in a more general way, but at what point should the transition be made? To what extent should the architecture be able to integrate large investments in current plant and electronics [140]? Is there a way to remove these current interfaces at a later date? To what extent should legacy network owners incorporate rehabilitation of existing plant versus new builds into their long-range business plans? When and how should the differences between optics and electronics be exploited?

3.7 EXTENSIVELY SHARED OSP VERSUS UPGRADE POTENTIAL

One way of reducing installation costs is to heavily share equipment in the field, for instance, although this has the potential to create future bandwidth or service bottlenecks. (This is a trade-off of the risk view versus the opportunity view.) How can such bottlenecks be opened or bypassed at a later time for a reasonable cost (without disturbing the rest of the network) [27, 29], or will the provider be vulnerable to alternative provider competition [30]? This point is closely related to the previous two points.

3.8 WIDESPREAD VERSUS LOCALIZED UPGRADES

Will the network experience ubiquitous upgrades in bandwidth across a service area [99] or upgrade "spikes" in which only a relatively small fraction of subscribers demand bandwidth [25, 193]? How economically elastic are the upgrade paths? How flexibly can the network provision new types of localized (low take rate) but potentially lucrative high-bandwidth services on short time scales? This point balances the plan view against the opportunity view.

3.9 OSP COSTS VERSUS TERMINAL COSTS

The system requires an OSP, or transmission medium, as well as terminals at the HDT, the subscriber, and perhaps in the field. However, the terminals comprise electronics processors and optoelectronics, which are on much different learning curves than that of the OSP [130]. That is, it is more likely that ongoing developments will change the cost, design, and capabilities of the terminal equipment than that parallel developments will change the OSP. Although the initial system costs have both OSP and terminal components, it is more likely that changes in the terminal equipment rather than changes in the OSP will be desired* and cost-effective: i.e., the terminal is more likely to be a short-term investment and the OSP is more likely to be a long-term investment [66]. This factor makes the economic decision more complicated because two investment classes are bundled into one initial system cost. Essentially, current and future terminal costs are mediated by near-term OSP decisions.

3.10 UNIVERSAL SUBSCRIPTION VERSUS LOW TAKE RATES

In some views, the services must precede the infrastructure [99] in order to justify the expense. However, in this case there may be a variety of concurrent networks that arise to provide competing services. For instance, broadcast television can be delivered through free space, over coax cable, by terrestrial microwave (MMDS, LMDS), or by satellite, and telephony may be delivered over copper pairs, coax networks, or cellular radio. Although in these cases the services are clear enough to make a detailed business plan, they are also clear enough to foster vigorous competition. Others view a commitment to the infrastructure as necessary to enable the new services that will be necessary for the industries to grow [120], but such investment carries a risk and an opportunity cost. This decision is closely related to the three previous points, and trades off the risk view versus the plan view of the business case [42, 140].

* It may even be argued that a measure of the success of a system is that services and capacities grow sufficiently large that the terminals *need* to become obsolete!

3.11 FUTURE OPERATIONS VERSUS TRADITIONAL MAINTENANCE

How much different (compared with current networks) will the maintenance be for this type of network [8, 9, 20, 29, 66, 156, 157]? Can operations and administration savings be relied upon to control service and maintenance costs in the future? Are there cost penalities up front? If optical network technology is to be used, can the difference between optics and electronics be exploited to obtain a service advantage [208]? Is there a point or complexity level at which techniques useful in larger networks can be used to organize the network [148] or improve maintenance and diagnostic procedures [205]? Can current operational procedures be adjusted toward the types of procedures that will become inevitable in the future?

3.12 ULTIMATE BANDWIDTH VERSUS INSTALLATION COSTS

One way of expressing the issue of ultimate bandwidth versus installation costs is to ask "How future proof should the network be, and when?" Every network runs into a capacity limitation beyond which it becomes inordinately costly to increase throughput [67, 169, 170]. Paying too much today for tomorrow's capacity is an opportunity cost that must be borne in a competitive market. On the other hand, having insufficient capacity in today's network to carry tomorrow's services will also be penalized in a competitive market.

From a practical standpoint, perhaps the most important question facing a current network provider is "Even if I can determine exactly where I want my network to be in a decade, how do I get there from here?" It is likely that a telephone operator company will have a strong preference to deliver as many services as possible over twisted pairs, and cable operators will prefer architectures with coax drops, even if both believe that another architecture will prove to be the most effective for the long term. It is understood that the usual business view is that one will keep one's core business intact while gradually encroaching on a competitor's business [140], but in a rapidly changing environment [29, 42] this may not be true: current network owners may want to reevaluate the value of their embedded base [156, 157]. Whether one believes that we are near a technological transition like the advent of telegraphy [8], analog telephony [66], or the DLC [9] or believes that we are merely entering a more competitive era for telecommunications services, current and future network providers face challenging technical and economic decisions. In the next section, we survey some of the architectural options under current consideration.

4. Examples of Current Access Architecture Models

4.1 FIBER IN THE LOOP

By the mid-1980s, the successful deployment of fiber optic communication in the long-haul network motivated investigations into the feasibility of introducing fiber into the plant [20], so-called Fiber in the Loop (FITL). In the context of Fig. 13.1, the feeder and distribution plant are fiber based rather than copper based. Systems tests and demonstrations were performed in which the feeder plant was replaced with fiber, followed by trials in which both the feeder and the distribution plant were fiber.

4.2 FIBER TO THE HOME

The early goal was Fiber to the Home (FTTH), in which the links (even the drops to the subscriber residences) would be replaced with fiber. The tremendous strides in digital compression had not occurred yet, and the forward-looking view at the time was that video delivery would require 150-Mb/s rates to the home [198]. This perceived need for bandwidth to the home tended to force fiber as the transmission medium, even for the early trials, because, clearly, the copper medium would be inadequate. On the other hand, because direct point-to-point links between the CO and each subscriber were seen as too expensive, architectures with fiber gain were considered.

The early trials used RNs that were electrically powered: they terminated feeder fiber, electronically demultiplexed the signals, and distributed them to the subscribers over fiber. At the subscriber residence, the fiber was terminated at an optical network unit (ONU), which performed the optical-to-electrical and electrical-to-optical conversions and provided a telephone interface. Figure 13.3 shows* an implementation in which a fiber feeder

* In figures with both fiber and copper, we show fiber as a dashed line and copper as a heavier dark line. In figures showing only fiber, the fiber is represented by a solid line.

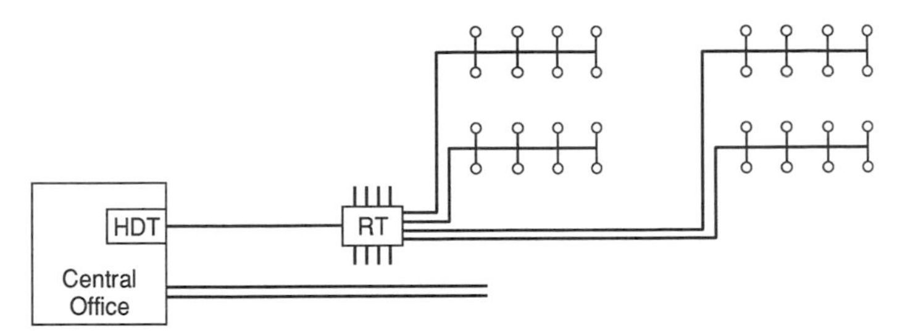

Fig. 13.3 Fiber to the home (active double star). A remote terminal (RT) terminates optical fiber from the HDT, electronically demultiplexes traffic, and distributes it over individual fibers to subscribers (*open circles*).

trunk delivers signals to RNs, which terminate the feeder, demultiplex the subscriber circuits, and send them to the subscriber ONUs [22]. This is an example of what is sometimes referred to as the *double star configuration*, because the fiber terminations of the primary (feeder) star are the centers of the secondary (distribution) star [111]. Because the RT contains actively powered demultiplexing electronics and optoelectronic components, this configuration is usually referred to as an *active double star* (*ADS*). These developments were viewed as outgrowths of the DLC developments that had been initiated a decade earlier [9].

An interesting example of a recent FTTH architecture, EROS (Fig. 13.4), unlike the ADS architecture just discussed, does not have fiber gain in the OSP [138, 139]. It comprises a broadcast star (power splitter, P/S) in the CO, with each subscriber connected by a single-mode fiber, and is based on the observation that it is unlikely that all N subscribers on a given node will require service at the same time. Thus a set of M ($<N$) transmitters at distinct wavelengths are coupled into a passive star, so that their output is sent to all subscribers in a broadcast-and-select network. Through scheduling on a separate control channel, a session between the CO and a subscriber is set up, and the subscriber tunes to the correct source's wavelength (λ_k for the session illustrated in Fig. 13.4). A portion of the downstream light (that remaining after tapping off a signal for the receiver, R) is looped back to the CO after amplification and modulation by a semiconductor amplifier* as the upstream traffic. Advantages of this approach are that it requires fewer lasers than links and uses wavelength-division multiple access (WDMA) to ensure virtual point-to-point links

* The modulation is performed by switching the gain of the amplifier on and off with the data stream.

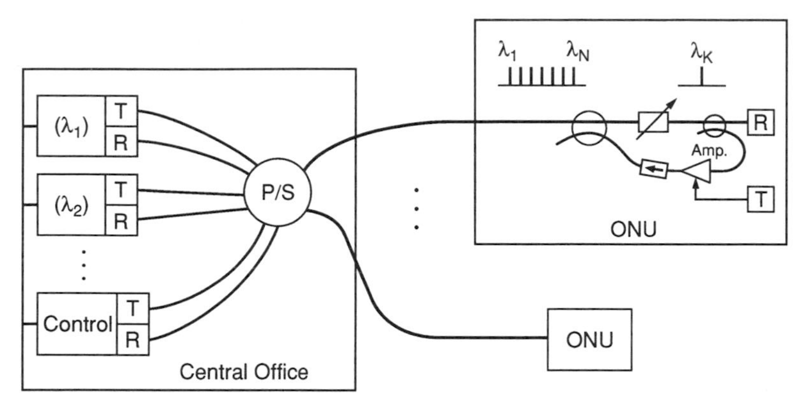

Fig. 13.4 EROS architecture (single star fiber to the home). A passive power-splitting (P/S) star at the CO couples M users to N transceivers ($N < M$) and a control transceiver. An optical filter selects the assigned wavelength at the ONU. Upstream signals are returned by amplifying and remodulating a portion of the downstream signal. ONU, optical network unit; R, receiver; T, transceiver.

with high capacity. Disadvantages include its lack of fiber gain (which favors short loop applications) and the fact that the transceiver costs are likely to require high-speed links to justify the costs.

Another recent architecture, called *LOC-Net* [5], is specifically a single star point-to-point network: each subscriber has a fiber running directly to the CO–HDT (Fig. 13.5). Like EROS, it does not use a source at the subscriber location, but instead impresses upstream information on a downstream carrier by means of a reflective modulator. In this way, its ONU is potentially less expensive than that for the EROS architecture. A feature that we see again later in this chapter is the specific inclusion of a broadcast

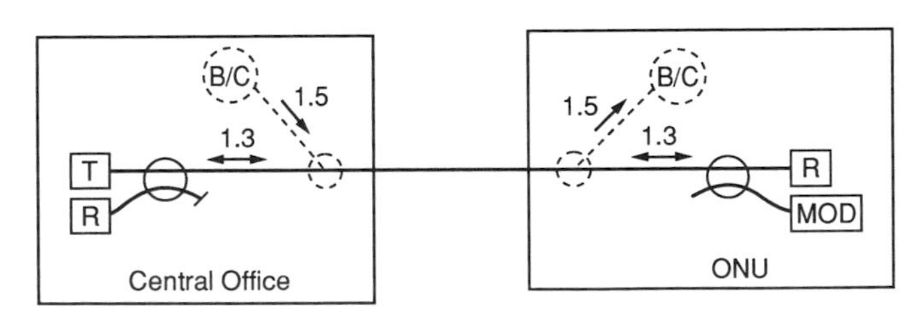

Fig. 13.5 LOC-Net architecture (single star fiber to the home). This is a point-to-point architecture with a low-cost modulator (MOD) providing upstream signals. High-capacity broadcast (B/C) and/or switched services are delivered with a wavelength-division multiplexed overlay (*dashed lines*).

overlay at a different wavelength (dashed lines) than that of the narrow-band services.

4.3 FIBER TO THE CURB

High initial costs of the optical equipment in early FTTH trials led to the idea of sharing the cost of the optics over several (e.g., four to eight) subscribers [185] and because the terminated fiber served several residences, this architecture became known as *fiber to the curb* (*FTTC*). In Fig. 13.6, for example, each ONU (and thus each fiber) is shared by eight subscribers. Although the FTTC ONU and RN capabilities (and cost) are greater than those in the FTTH case, the increased sharing of optical components brings the per-subscriber cost down. The splitting (S) is done by an active RT in the ADS configuration, but when the splitting is done by a passive optical splitter, it is called a *passive optical network,* or *PON* (see Section 4.4).

Although less costly than the FTTH trials, the FTTC trials also indicated that the costs were more than those of an equivalent copper plant [155]. The watchword became *copper parity:* it was believed that when these FITL approaches were as economically attractive as installing copper plant, the LECs would begin new and rehabilitative builds with them, subject to the restrictions of local public utility commissions. Subscriber loop carrier (SLC) systems by AT&T are early examples of this approach [22]. Meanwhile, as mentioned in Section 3.5, steadily reducing costs have made DLCs prove-in economically at shorter and shorter distances [9, 160], and recent

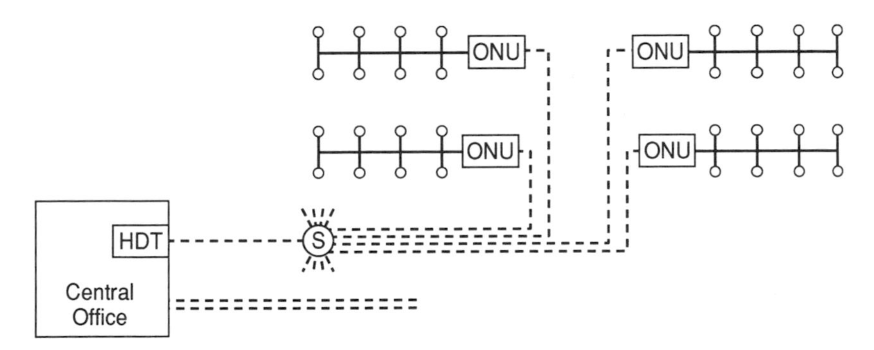

Fig. 13.6 Fiber to the curb. An ONU terminates a fiber carrying traffic for a group (e.g., $N = 8$) of subscribers, then delivers signals over twisted-pair or coax cable to the subscribers. Splitting (S) can be done with an active remote terminal (active double star) or with a passive optical splitter (passive double star).

estimates show that copper parity has been reached for FTTC narrow band systems based on PONs [156, 157], as described in the next section.

In assessing the FTTC characteristics in the framework of Section 3, we view it primarily as a switched network because each customer sees an individual connection to network equipment (the ONU). That network equipment may be on a logical bus (as for the PONs discussed in the next section) or a logical star (as for the DLC). The switched connection permits desirable diagnostic features common to telephony networks, with distributed intelligence (e.g., at an RT and/or the ONU that terminates the fiber) taking advantage of the multiplexing opportunities. The technology, which was forward-looking at the time it was proposed, is currently realizable and forms the basis of most international deployment plans, as discussed subsequently. Generally, compared with FTTH, FTTC is viewed as economically superior for narrowband services, although sharing and powering issues may muddy the comparison somewhat in the United States [157, 178]. Localized high-bandwidth upgrades are possible because of the processing power at the RN (for ADS) and ONU, although at some point the terminals would impose an electronic bottleneck [140]. Power consumption at the ONU and RN is a drawback but permits interface compatibility with existing telephones. This compatibility is not trivial, because in the United States there are on the order of 100 million analog telephones in existence.

4.4 PASSIVE OPTICAL NETWORKS

The FTTH–FTTC division described previously for FITL systems discriminates on whether the terminal point for the fiber is at or near the subscriber residence, respectively. Another division is the number of layers of active electronics between the HDT and the subscriber: in the ADS FTTH example of Fig. 13.3 there are two levels (RT and ONU), whereas in the FTTC version of Fig. 13.6 there is one — namely, the ONU if S is a passive splitter. Thus, this division of electronic terminal layers can be made independently of the division of where the fiber terminates. In an effort to remove the cost, powering, and complexity of the active electronics in the field, the use of totally passive RNs was proposed in a series of papers by workers at British Telecom Research Labs (BTRL) [e.g., 43, 44, 74, 135, 142, 168]. In schematic form, a possible implementation is shown in Fig. 13.7: the fiber terminates at the home, which has the only layer of electronics, even if upgraded [96, 202]. Feeder fibers transport signals from an HDT to the RN, a passive optical power splitter. The RN output fibers either could be

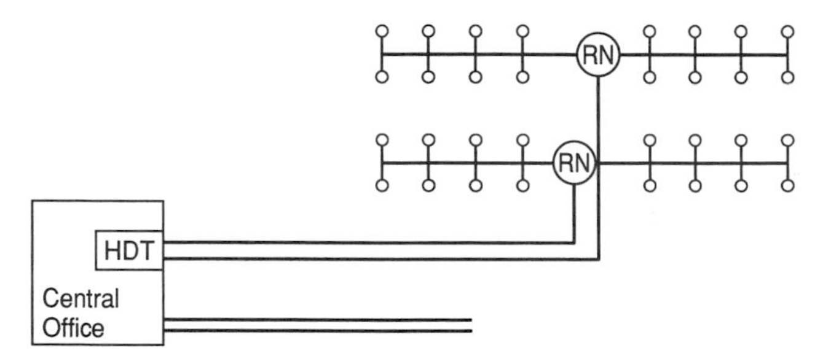

Fig. 13.7 Fiber to the home (passive optical network). A passive optical splitter in the RN distributes light on individual fibers directly to subscribers.

further split by another layer of passive splitters (not shown) or could be connected directly to the subscriber ONUs (as in Fig. 13.7). This approach differs from that in Section 4.2 in that fiber gain is achieved without consuming power in the field. Thus, as mentioned before, these networks were called *PONs*.* They are the "most optical" of the networks discussed in this chapter. Typical power budgets for standard Bellcore specifications [17] are about 24 dB for a 16-way split: 12 dB for the intrinsic split (only $\frac{1}{16}$th of the light reaches each user) and 12 dB for the fiber and excess losses. Although this effectively precludes broadband AM-VSB television delivery without additional optical amplification, current packaged receiver prototypes with auxiliary peak detectors [14] are capable for narrowband services with conventional lasers, as are burst mode receivers [131–133, 174] that have been demonstrated recently. Several groups are investigating further integration, to transceiver modules [79, 87, 115, 119, 177, 204]. Because the subscriber terminal costs are unshared, such levels of integration are needed to reduce individual terminal costs to cost-effective levels.

The architecture is dominated by its broadcast structure, which is used to advantage as far as cost, availability, and maturity of components are concerned. A ranging protocol, necessary for establishing the time-division multiple access (TDMA), has been demonstrated [44], as has a time-compression multiplexing ("Ping-Pong") scheme [87]. There are optimal

* This point has engendered numerous "religious" discussions that center around the doctrinal question "How can a PON consume power in the FTTC implementation and still be considered a *passive* network?" The common usage seems to be that, when viewed as a *distribution* system, the FTTC configuration is a PON in which the ONU happens to serve more than one subscriber, whereas purists argue that PONs can refer only to FTTH systems with passive RNs. This author is an agnostic on the issue.

split ratios (for any given service set) for installed first costs (IFCs) and for operations costs. As the split ratio increases, the IFC per subscriber drops quickly, reaches a broad minimum, and then slowly increases as the drop lengths and power requirements grow [111, 156, 178, 187]. Ultimately N increases to a point for which network management and medium control costs begin to increase rapidly. In general, the community seems to think that IFCs are minimized in the $N = 8$ to $N = 32$ range. The split ratio also strongly influences upgrade considerations, insofar as the ONUs are on a logical bus [101].

Recent discussion by BTRL workers has emphasized the idea of using the PON as a sort of "ether" in which there is optical amplification, a much larger split ratio, and much more intelligence on the periphery [29, 194]. In this architecture, there is a need for optical amplification: although a costly proposition today, current research indicates that we will likely see lower costs for high-performance optical amplifiers in the future [63, 64, 207]. That is, the fiber is used as a flexible medium to set up calls that can be separated by tunable components on the periphery. This wholesale migration of intelligence to the periphery takes advantage of current trends in processing power and costs, but it leaves network management and control issues to be resolved. Such issues are being addressed in high-capacity local area network (LAN) research test beds such as LAMBDA-NET [57], RAINBOW [37, 62], and STARNET [89], although these test beds are in the context of communicating in concurrent sessions between peers with high-quality terminal equipment. The transition of these concepts to massive deployment of ubiquitous, inexpensive, yet intelligent nodes will be a formidable technical challenge [29, 136], and there are issues of failure group size that may make it difficult to penetrate the U.S. market [17]. Coherent communication has also been proposed, using this basic configuration, to take advantage of the sensitivity and selectivity improvements possible [90], or as an overlay that upgrades an initial narrowband system to high capacity [96].

Recent PON research has centered on ways to (1) reduces optical impairments, (2) improve the utility of the architecture, and (3) make the networks more cost-effective. A well-known optical impairment arises from the fact that light from two sources will produce beating tones on a receiver; the intended reception will be impaired if the sources' optical frequencies are closer than the receiver's bandwidth. This can occur with unintended scattering or reflections from a single source [200, 201] or through coincidence with several sources [152], even when subcarrier modulation [32] is used as a multiple access technique. Techniques to reduce the coherence of lasers by overmodulation [45, 46] or the use of low-coherence sources [154] have been shown to reduce this impairment. An approach to increase the power of light-emitting diodes (LEDs) has been demonstrated [109], monolithically integrated [110], and successfully tested in a system experiment [47].

Architecture improvements include broadcast overlays in the wavelength domain [43], the use of filters to segregate services, and the use of two fibers to reuse wavelength bands [16, 136]. The use of subcarrier multiplexing was proposed and demonstrated [154] as a technique to create channels operating at a lower data rate than possible with TDMA and time-compression multiplexing [87], and solutions to impairments arising from compact use of the RF spectrum have been reported [46]. Multiplexing services in the RF domain [154] have been demonstrated in the context of a television broadcast service with the robust quaternary phase shift keying (QPSK) [151] delivery of current satellite services [143]. In this latter case (Fig. 13.8), more than 150 channels of high-quality digital television were delivered with an inexpensive laser, showing that node-by-node programming selectivity is possible. That is, it is economically feasible to have an electrical rather than an optical backplane for broadcast video delivery. An advantage of this approach is that the digital format of the satellite signals makes encryption and tailored "narrowcasting" to individual service areas possible, while a disadvantage is that, like the popular satellite broadcast services that it emulates, it requires a set-top box to convert the QPSK signals to analog form for today's television receivers. However, recent history has shown a tremendous willingness to pay for this form of service, even with a set-top box and a dish antenna: the demonstration shows that

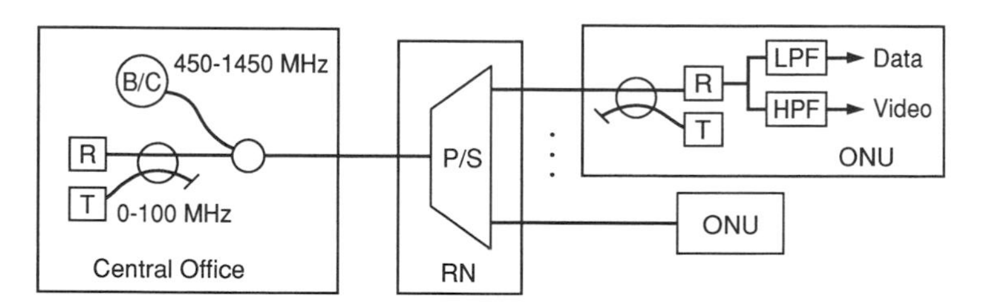

Fig. 13.8 Broadcast passive optical network with broadcast video overlay. Narrow band service is delivered over a power-splitting passive optical network. Broadband broadcast service is optically coupled with (and RF segregated from) the narrow band service. HPF, High-pass filter; LPF, low-pass filter.

a similar type of service can be added to a PON system for not much more than the cost of an RF filter, while eliminating the need for an external parabolic dish antenna.

Demonstrations intended to show cost-effectiveness of the networks have also been performed. Subcarrier multiplexing [32] allows lower speed electronics to be used in transmitting data (at the cost of more RF equipment) [154, 203], permitting services to be segregated, while coarse WDM (CWDM) for directional duplexing [88] with 1.3-/1.5-μm devices [e.g., 104] relaxes the power budget (and thus the cost) for the ONU components and permits the RF spectrum to be reused. Further reductions in component costs are also being pursued with a variety of techniques to permit more efficient coupling between devices and fibers [21, 53], micromechanical approaches to packaging [162] and component placement [105, 106], and more cost-effective splicing [176].

PONs have begun to be deployed around the world, and the overwhelming majority of PON deployments at this point [3, 13, 30, 42, 121, 145, 157, 179, 191, 195] are FTTC, although Nippon Telephone and Telegraph (NTT) has proposed a national FTTH infrastructure [120]. Deployment in the United States has been somewhat slower, perhaps in part because of privacy and security issues related to the bus architecture [187], in part because of the greater emphasis on AM-VSB entertainment video [170], and in part because of the component costs. The first issue is intrinsic, but it can be expected to improve with increasing electronic component reliability and restricted physical access to ONUs. The second issue may be addressed by optical amplifiers (see Chapter 14 in Volume IIIA) or by the overlay of Fig. 13.8, and the third will be resolved with time. A case can now be made that FTTC PONs are less expensive than copper, once life-cycle maintenance issues are included [157].

The strengths and limitations of the architecture are well understood. Because the conventional PON is a broadcast bus, there is an inherent factor of N loss due to the optical splitting. In addition, sharing the CO laser over all downstream channels requires the bit rate to be a factor of N higher than the subscriber data rate. These disadvantages are balanced by the passive plant, the component availability, and the fiber infrastructure. Emerging ATM, packet switching, and medium access control technology makes flexible bandwidth partitioning on a bus easier, and there is vigorous research looking into network management issues [10, 31, 62, 92, 206]. In terms of our checklist in Section 3, the conventional PON is a broadcast architecture that could exploit existing technology (e.g., lasers and couplers) in a highly shared plant. Localized high-bandwidth upgrades might be difficult without upgrading the entire network or the management scheme, and there is an expectation of operational savings due mainly to the fiber plant.

In summary, the broadcast PONs can be seen to attack various components of the cost involved in creating a network. The strictly power-splitting PONs can be inexpensive in the OSP. Their power budgets for baseband operation scale as N^2, whereas with additional RF circuitry, this scaling can be reduced to the N-fold splitting loss. In general, they use technology that is close at hand and currently deployable.

4.5 HYBRID FIBER/COAX

The CATV model of Section 2.2 contains the rudiments of what has become known as the *hybrid fiber/coax (HFC) system* and is treated in more detail in Chapter 14 (Volume IIIA). The difference between the two is that *HFC* usually describes systems that deliver other types of traffic in addition to television signals and has explicit provisions for carrying upstream data [33, 112] (see Chapter 14 in Volume IIIA). The broadcast nature of CATV networks does not lend itself naturally to bidirectional communication because such networks are designed to send a broadband common signal to a large number (500–2000) of subscribers. In general, RF diplexers are used in the network to break the spectrum into two bands: 5- to 40-MHz signals are used for the upstream direction, whereas the spectrum exceeding 50 MHz is used for the downstream video delivery. Newer architectures are planned that use RF bands above the conventional television bands for additional services such as digital television and communications links to personal communication services base stations. The advantage of the CATV network, relatively inexpensive delivery of high-bandwidth video to a large number of users, is something of a disadvantage when the same number of users must share the narrow bandwidth in the upstream direction. In current proposals, a fiber node terminates several fibers that serve approximately 500 subscribers. One of the fibers carries the downstream composite video signal and another carries data and narrowband telephony. At the fiber node, the optical–electronic conversions are made so that narrowband services are received (upstream) in the 5- to 40-MHz band from the coax and sent (downstream) in a band above the television channels. Each quadrant of 120 subscribers shares a 35-MHz upstream band on its own coax trunk: statistical multiplexing and the use of efficient modems can provide on the order of 1-Mb/s upstream capability to each household. Because the four quadrants each occupy the same 35-MHz RF band for upstream communications on their respective coax trunks, the fiber node block-converts each quadrant's band into a composite signal, which modulates a laser connected to the CO–HE by an upstream fiber.

There are several strong points that make the HFC architecture look attractive to network providers, especially the cable television multiple service operators (MSOs) [130]. For instance, the wide experience from the CATV industry reduces some of the technical risks of infrastructure deployment, because most of the components are currently in production, and high-bandwidth cable modems are imminent. Such modems are especially attractive for a current cable provider offering services with low take rates. That is, in the early stages of deployment, only a few of the 120 users on a trunk will take advantage of the 40 MHz or so that will be available. This architecture also provides an immediate high-bandwidth path for the LECs to deliver downstream video services to compete with the other broadcast video providers (CATV, MMDs, LMDs, direct broadcast satellite (DBS), etc.) in the near term. Although the network is intrinsically broadcast, it is hoped that the large-bandwidth "pipes" afforded by coax will suffice for upstream needs on early service rollouts. Some of the drawbacks include fears about the quality of the upstream channel against ingress noise, issues of coax reliability and lifetime, powering, and the fact that many subscribers share the upstream bus. Some of the switching functions are accomplished by time and frequency assignments for the subscriber's modem, so the intelligence at the fiber node is not extreme: the signals are essentially backhauled from the fiber node to the HDT for processing. Although this is an advantage from the standpoint of fiber node cost and power, the bus architecture introduces a vulnerability to node failures caused by improper modem operation, and there will be complications due to network administration [140]. Additionally, the extensive sharing may limit certain upgrades: if one of the subscribers desired a large upstream bandwidth (e.g., 45 Mb/s for a small business a decade from now), there could be congestion and contention problems due to bandwidth limitations and segmentation [170]. The metallic OSP life is a matter of research, and corrosion retardation schemes are being investigated. The extensive sharing makes HFC desirable for low take-rate services, as current CATV deployment indicates, whereas it is more undesirable for localized high-bandwidth upgrades.

A novel scheme for improving the upstream capacity has been proposed by Lu and coworkers [113]. In this architecture, mini-fiber nodes ("m" in Fig. 13.9) are placed at amplifier locations, and they may be connected to the usual fiber node over a fiber optic link. This layout reduces the number of subscribers who effectively share a link, increases the bandwidth of the coax (shorter lengths), and can increase the upstream bandwidth by using some of the unused high-frequency "downstream" coax bandwidth. The mini-fiber nodes can be placed at the time of network installation and activated as needed or added to existing plant as an upgrade. This approach greatly increases the capacity of the HFC and would appear to be a natural enhancement that uses conventional, mature, and low-cost components. In some ways, this approach simulates a CATV network with an FTTC overlay: the "ONU" is the mini-fiber node in this view, and the "drop" is a shared bus with the CATV coax instead of a switched circuit.

In summary, the HFC architecture is probably the nearest term approach to the problem of delivering both entertainment video and narrowband data and telephony. The low cost, mature technology, and early rollout potential make this architecture a strong contender for providers who need an early video deployment. The intrinsic bus architecture is an advantage for low-cost and low take-rate service scenarios where concentration is

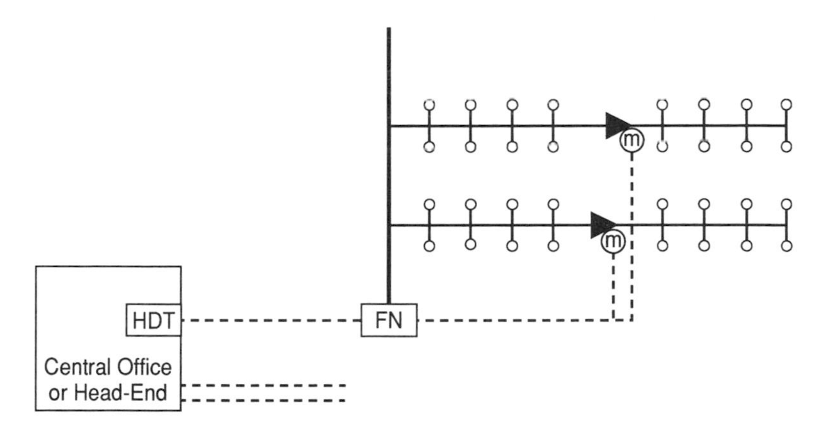

Fig. 13.9 Hybrid fiber-coax architecture (with mini-fiber nodes [m]. This is a modern CATV architecture using feeder fiber to a fiber node (FN), which then delivers narrow band (telephony, data) and video services to subscribers over coaxial cable (*heavy lines*). Downstream RF spectrum can be used for upstream transmission over a fiber link connecting the mini-fiber nodes (at the RF amplifier) and the FN.

possible, such as bursty data. Potential problems are upstream network capacity, security, and availability on the bus, as well as the metallic medium's maintenance and lifetime limitations.

4.6 SWITCHED DIGITAL VIDEO

Technically, *switched digital video (SDV)* describes a service rather than an OSP architecture. It is usually meant to convey the idea that the architecture is capable of delivering wideband or broadband digital signals (such as video on demand) *without* putting subscribers on a broadcast medium. Although the name seems to be antithetical to *broadcast, analog* video, it is often assumed that SDV also provides a CATV overlay. That is, a CATV overlay provides broadcast video service while switched telephony and digital services are delivered from a high-capacity ONU. In effect, it is two interpenetrating networks: an FTTC architecture with a high-capacity ONU delivering digital services to individual subscribers, and a CATV overlay that delivers entertainment video and powers the FTTC nodes [82, 140]. A current proposal suggests delivering switched services over several twisted pairs with high-speed modems and delivering telephony over another twisted pair, while the CATV overlay is delivered on coax [82]. One way of viewing SDV is that it is a high-capacity FTTC PON that solves the ONU powering problem by drawing power from an overlay CATV network. A diagram would look like Fig. 13.9, only the solid lines would carry *both* broadcast and switched signals. A comparison between HFC, HFC with mini-fiber nodes, and SDV may be useful at this point. HFC uses fiber to transport signals to and from a large fiber node that supports a serving area's coaxial subscriber buses. HFC with mini-fiber nodes also delivers all services on a coaxial bus, but the mini-fiber nodes push fiber deeper into the network so that each subscriber's spectrum is shared over a smaller "neighborhood." This gives the mini-fiber node architecture an advantage in flexibility and upstream capacity. SDV, on the other hand, has approximately the same fiber coverage, but it is brought to a more complicated and more expensive node than the mini-fiber node: the node supports a $\approx$Gb/s switch that feeds services as high as 50 Mb/s to individual lines rather than a block of upconverted signals as the mini-fiber node does. Intuitively, one expects that HFC is the least expensive, HFC with mini-fiber nodes is the most efficient, and SDV is the most capable of the three. SDV is very rich in service possibilities because it has AM-VSB television and high-capacity digital services: 50 Mb/s downstream and 3 Mb/s upstream are commonly quoted values [82]. On the other hand, there are explicit bottlenecks (because of the electronics in the field) which might inhibit later upgrades, precisely the position that the BOCs intended to avoid. The cost of construction is expected to be somewhat more expensive than, but comparable to, that for HFC and mini-fiber node architectures [82, 130, 140], and the compelling attraction of these architectures is that they are near term: they use technology that is at hand with coax plant that is well understood, and they can deliver entertainment video immediately. Furthermore, the upstream capacity is presumably enough for near-term demands until a clearer view emerges of the future needs. For MSOs, the HFC or HFC with mini-fiber nodes seems a most natural fit, whereas for LECs the SDV is more natural.

4.7 CONCLUSION

A complete analysis of current access alternatives might, at this point, evaluate the various architectural examples in Section 4 against a set of criteria (as in Section 3, for example) to weigh the desirability of various options and highlight the expected difficulties. Such an analysis is beyond the scope of this chapter. Each of the architectures has arisen from perceived needs (bandwidth for broadcast, bandwidth for downstream digital delivery, low costs, low maintenance, etc.) and perceived limitations (technology, performance, time, etc.). Each has a distinct profile against checklists like that of Section 3. It is commonly believed that HFC and SDV are the current favorites because they match up with existing operations [140]: HFC is similar to the MSOs' CATV plants, and the FTTC part of SDV looks similar in some respects to the newer DLC plant. Both networks support broadcast video from the beginning and thus can be counted on to keep cash flowing in while the network is built. However, both networks also have limitations when viewed from a long-term perspective. The metallic plant of both networks is not as long-lived as fiber is, both will have increasingly power-hungry remote electronics in the field, both will require refurbishing electronic plant in the field to meet varying provisioning estimates (refurbishing costs for localized upgrades in both networks may not be directly billable to the customers and may need utility commission review) and ultimately both run into electronic bottlenecks in one or both directions.

In the next section, we examine WDM PONs, architectures that are complementary in many senses to HFC and SDV. They are unable to instantly provide broadcast AM-VSB television service, and they do not instantly match up with either LEC or MSO current capital plant. Nonetheless, they have extremely compelling long-term properties advantageous to access networks, and they provide a target or yardstick for future evolutionary plans, if for no other reason than the common opinion that fiber optics will undoubtedly go to the home: the question is "When?"

5. Use of WDM in Passive Optical Networks for Access

5.1 INTRODUCTION

In this section, we review the general characteristics of, the technology for, and representative architectures of WDM PONs. The component technology for WDM PONs, although not yet mature, is the subject of vigorous research and development, with test beds [69] and commercial products just beginning to be fielded, and is reviewed in Sections 5.2 and 5.3. Consequently, this architecture, which is perhaps more capable and flexible than those previously reviewed, is not appropriate for immediate deployment. However, we argue in Section 5.4 that the *infrastructure* can be deployed before it is used as an actual WDM PON and can be upgraded as demanded.

The WDM PON idea as an access architecture was pioneered at Bellcore by Wagner, Lemberg, Menendez, and others in a set of articles [e.g., 107, 108, 118, 186–190] that appeared in the late 1980s and early 1990s, shortly after the BTRL PON proposal. WDM in this context allows the CO or HDT to send multiplexed signals, each at a different wavelength, to the RN. The RN, in turn, contains a wavelength-division (de)multiplexer that passively splits the light by wavelength, directing each color to an assigned subscriber. The multiplexer–demultiplexer used to establish the channels is commonly called a *dense* WDM, a term signifying that the wavelength channel spacings are on the order of 1 nm, as opposed to *coarse* (1.3-/1.5-μm) WDMs with more than 100-nm separations. The WDM acts as a switch: electronic control over the color of a light packet, for example, in effect throws a remote switch at the RN (i.e., it changes the packet's destination). However, the change in the network character effected by this is more important than the switching:* WDM can transform a point-to-multipoint network to a "virtual" point-to-point network. This is *not*

* The switching function can be performed electronically. For instance, a packet switch in an ONU on a broadcast PON will switch desired packets to the ONU's output.

the same thing as the "logical" point-to-point network one would have in a broadcast PON, in which an ONU electronically reads only the channels or packets intended for it. The virtual (as opposed to logical) point-to-point character permits a host of network advantages: *privacy* (downstream information for user X is directed to only user X), *security* (the source of unauthorized transmissions can be identified unambiguously by wavelength and prevented from corrupting the node served by the WDM) *efficiency* (more of the light directed toward user X is received by user X, with only incidental, not splitting losses, which permit higher data rates), *separability* (each subscriber communicates with the CO independently of other subscribers on the node), and *diagnostics* (wavelength-specific tests can check the health of individual lines) are all attributes at the physical level that distinguish WDM PONs from broadcast PONs. These attributes help to make WDM PONs future proof.

The initial passive photonic loop (PPL) is depicted in Fig. 13.10, with a multiwavelength source providing the downstream communication. For communication in the upstream (ONU to CO) direction, there were several proposals. The one shown in Fig. 13.10 uses a power-splitting PON in the upstream direction. Several multiplexing schemes are available, such as WDM (for which a WDM receiver is needed at the CO) and time-division multiplexing (TDM) (for which a broadcast-like protocol is needed) [107]. In the most aggressive version [186] the upstream splitter at the RN was

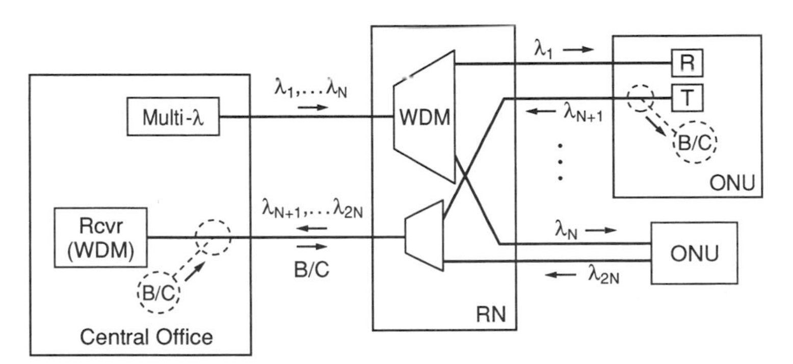

Fig. 13.10 Passive photonic loop (PPL). A multiwavelength laser source sends downstream information to a wavelength-division multiplexer (WDM), which directs light to ONUs as a function of wavelength. Upstream information may be assigned wavelengths (shown) and multiplexed with either a WDM (not shown) or a power splitter, or upstream wavelengths may be unassigned. A broadcast overlay (*dashed lines*) for hybrid PPL (HPPL) is over a power splitter.

also a WDM, with frequency-selected DFB lasers at each subscriber location aligned to hit the correct wavelength channel. These sources were estimated to cost in the range of $200–$400 in volumes of millions [187], inexpensive by today's low-volume prices, but a significant fraction of the allowable expense for the per-user plant costs. This version has all the WDM PON features.

To justify the additional terminal and plant expenditures such an architecture suggested, a broadcast overlay called *hybrid PPL, or HPPL*, was proposed [86, 118] that provided the capability of delivering additional services that would presumably generate additional income streams to cover the installation costs. This is shown as the dashed overlay in Fig. 13.10. The passive splitter would pass both 1.3- and 1.5-μm light, allowing CWDMs to be used to multiplex the broadcast service onto and off the same fiber used for the upstream service.

The original PPL has all the standard WDM advantages (privacy, security, efficiency, separability, and diagnostics), whereas HPPL has reduced security, efficiency, and diagnostics in the upstream direction as a result of the broadcast star. Nonetheless, the component costs and network management made the costs prohibitive.

To address the issue of high WDM source costs, another approach was proposed, called *spectral slicing* [141], which is shown in Fig. 13.11. A low-cost broadband optical source, such as an LED, is used to establish a point-to-point link by communicating over only the wavelength established by the ports on the device. That is, the broad spectrum on the left side of the WDM in Fig. 13.11 is "sliced" into the spectral segments shown on the right side, each corresponding to the passband appropriate for that port.

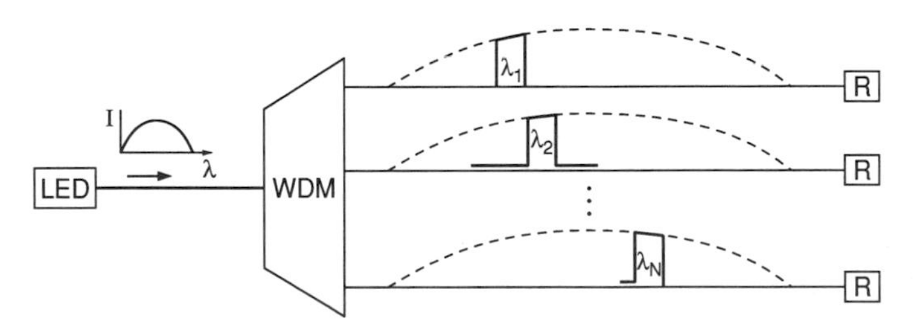

Fig. 13.11 Spectral slicing with a WDM. If the spectrum of a light-emitting diode (LED) is wider than the span ($\lambda_N - \lambda_1$) of a WDM, the light can be "sliced" into each output port. Each output port carries a replica of the signal that modulates the LED.

Similarly, a set of identical LEDs on the *right* side of the WDM could "reconstruct" the spectrum on the left. Two WDMs, connected by a single fiber and operating back to back with synchronized channel wavelengths, could be used to establish point-to-point links [141, 190, 217]. The spectral extent of the LED must be wide enough to cover all channels under all operating conditions, but not too much wider (because the spectral excess represents additional loss). Spectral slicing was also implemented in a network demonstration [190] in which DFBs were used for the downstream channels and LEDs were used for the upstream channel. Recently, plans for combining spectral slicing with integrated fiber gratings and modulators have been proposed [19].

The advantages of WDM PONs (privacy, security, point-to-point communication, etc.), although well known since the late 1980s, have not generated any serious deployment plans to date because the components are on later learning curves than those for conventional components, and the *need* for an upgradable and/or high-bandwidth system has not been adequately demonstrated. Both these points are in the process of changing.

Studies show, and CATV experience indicates, that in addition to rapidly dropping prices, significant maintenance savings are possible when fiber optic systems are deployed [157]. Second, as we show in the following sections, the components needed for such networks are in a vigorous state of research, and several are just now coming onto the market. These points address the cost issue. More fundamentally, the explosive growth in Internet traffic, facilitated in large part by World Wide Web browsers with user-friendly graphical interfaces, has fired the first shot of the telecommunications revolution warning us of future network needs. We are poised at a point where high-capacity, flexible, upgradable architectures such as WDM PONs may become more of a necessity for greatly expanded communications capacity in commerce, education, and entertainment.

In the remainder of this section we review some of the recent progress in WDM components that will enable WDM PONs, and some of the recent architectures that can take advantage of that progress. This review is meant to be illustrative rather than exhaustive.

5.2 *WAVELENGTH-DIVISION MULTIPLEXERS*

The heart of the WDM PON is the multiplexer–demultiplexer that distributes the light to the subscribers. The simplest device is the conventional star coupler that simply splits incoming light into all ports equally

(Fig. 13.12a): it is the trivial case of WDM, because no selection is made on wavelength, and is the idea behind the "ether" approaches in LANs and broadcast PONS. Such devices are currently sold as either fused couplers or in silicon optical bench form. Economies of production for the larger split ratios seem to favor the silicon optical bench form.

Two wavelength selective components that have recently been developed and can be adapted to fiber devices such as the star coupler are the tunable fiber Fabry–Perot filter (Fig. 13-12b) [171, 172) and the in-line fiber Bragg grating (Fig. 13.12c) [70, 71, 116]. Fiber Fabry–Perot filters with excellent fitnesses and tunability have been fabricated, which makes them attractive as tunable filters in WDM broadcast-and-select LAN applications. They

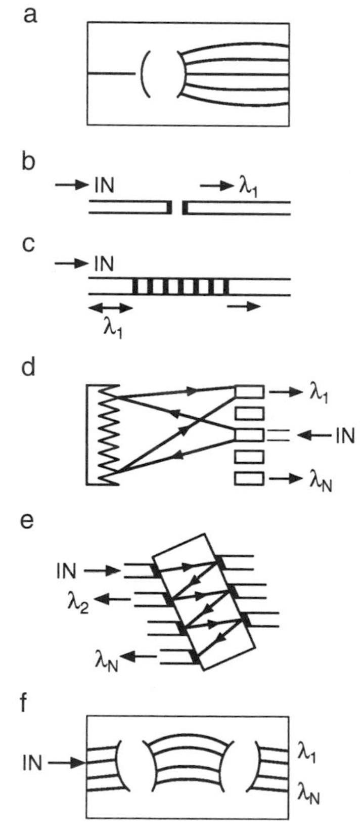

Fig. 13.12 Multiplexers–demultiplexers: (a) power-splitting star, (b) fiber Fabry–Perot filter, (c) in-line fiber Bragg grating, (d) diffraction grating, (e) cascaded filters, and (f) waveguide grating router.

are relatively immune from polarization dependence, and they can be made insensitive to temperature [172]. Fiber Bragg gratings have been used in several sensor and filter applications [18] and have been proposed as a demultiplexer [123] with a "broadcast-then-select" characteristic (Fig. 13.12d). That is, incoming light is broadcast to all output ports, and the undesired wavelengths are rejected by the Bragg filters. In a manner similar to that used in DFB lasers, applying a quarter-wavelength shift to the grating permits transmission in the middle of the stop band [2] so that it is possible to consider applications in which the demultiplexer works with a single grating on each fiber. In another application, the broadband Fabry–Perot character of the FFP can be emulated by using gratings that are chirped over the band of interest [182]. Recent work has demonstrated that these Bragg gratings can also be introduced directly onto the waveguides of the silicon optical bench devices [1, 68].

A hybrid bulk–fiber optic demultiplexer [150] has been demonstrated (Fig. 13.12e) in which dielectric mirrors deposited onto a transparent substrate permit a sequential coupling of wavelength pass channels through a cascaded set of mirrors and filters into the corresponding output fibers. This device has the promise of robust construction, although large split ratios have not yet been demonstrated. One hybrid device [122] has been shown to have a temperature coefficient an order of magnitude lower than that of glass.

Another robust device, based on a reflective bulk grating design, has been demonstrated in the multistripe array grating integrated cavity (MAGIC) devices. Like many of the silicon optical bench components, it uses a planar waveguide structure to implement optical functions: in this case, illumination and beam forming of a Rowland circle diffraction grating [163–165]. An interesting feature of this technology is that it lends itself to a device platform technology [62, 164].

Perhaps the most promising, in terms of diverse functionality is the waveguide grading router (WGR) (Fig. 13.12f) [38–40]. It, too, is based on diffraction, but the diffraction takes place as a result of patterned waveguides [158] rather than ruled gratings. Its other names, *arrayed waveguide grating* (*AWG*) and *phased array* (*PHASAR*), also indicate its basic operating principle, which is described more carefully in Chapter 8 in Volume IIIB, and it has been fabricated successfully on the silicon optical bench platform [39, 127, 128, 183] as well as on InP [7, 215, 216]. Like the other devices mentioned, the WGR has the WDM property: it splits incoming light into spectral constituents, launching them onto a set of output fibers, as shown in Fig. 13.13. However, two other attractive properties increase its usefulness in networks: the routing property and the periodicity property. These properties have interesting system [56, 129] and network [15] consequences and greatly increase the architectural flexibility of WDM PONs, as we now show.

The routing property is a generalization of the WDM property for more than one input port: not only does each input port possess the WDM property, but each optical frequency gives routing instructions that are *independent* of the input port. Thus, f_k's routing instruction is to exit the output port that is $k - 1$ ports below the corresponding input port. In Fig. 13.13 this is shown for $k = 1$ for f_1 going from input port 1 to exit port 1, and from input port 5 to exit port 5. Similarly, f_3 incident on input port 1 is directed to output port 3, whereas if f_3 were incident on port 5, it would be directed to output port 7.

As even more unusual property is the periodicity property. By choice of design parameters, higher diffraction orders can be made to overlap lower orders to some extent. Thus, it is possible to design the grating in such a way that if optical frequency f_9 (i.e., one free spectral range higher in frequency than f_1) enters port 1, for example, it will "wrap around" and exit port 1 instead of being lost, as with f_{17}, and other frequencies separated

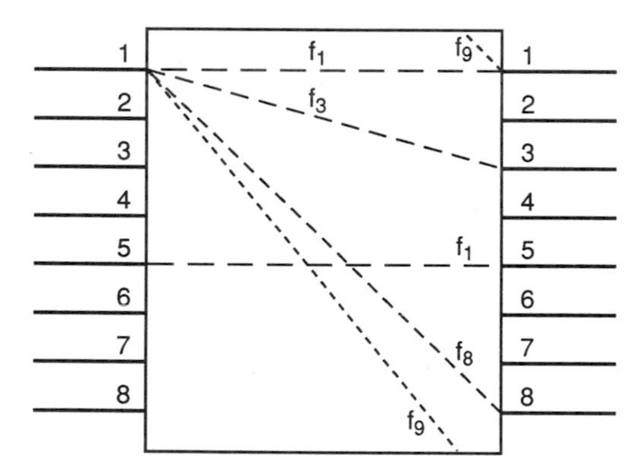

Fig. 13.13 Waveguide grating router frequency connection plan. The WDM property is the connection of input port 1 to output port k for optical frequency f_k. The routing property is the WDM property for other input ports (e.g., f_1 connects input 1 to output port 1 and input port 5 to output port 5). The periodicity property permits frequencies that would normally be dropped (e.g., f_9 on input port 1) to cyclically "wrap around" to other output ports.

by an integral number of free spectral ranges.* This periodicity property and the routing property enable more flexible and efficient architectures than conventional WDMs, although other devices emulate some of the WGR characteristics [95, 171].

As an example of increased efficiency, consider spectral slicing of an LED's spectrum. A conventional WDM necessarily slices inefficiently: the LED spectrum must be much wider than the N channels to cover them reliably, yet the light that falls outside the channels is lost (Fig. 13.11). On the other hand, the WGR's periodicity samples the spectrum at many points for each port (Fig. 13.14), equalizing each port's power and utilizing the entire optical spectrum.

Another advantage of the periodicity property is service segregation by using additional wavelengths for a given port: services could be added later through the use of "intermediate" WDM in which channels are separated by free spectral ranges. Additionally, the WGR can eliminate upstream–downstream wavelength collisions because adjacent input ports are coupled to adjacent output ports for the same wavelength. This feature permits a single wavelength to carry both upstream and downstream traffic over separate fibers [49] without optical beat interference [152] or coherent Rayleigh backscattering [201].

* Limitations, due to dispersion and diffraction effects, are beyond the scope of this discussion.

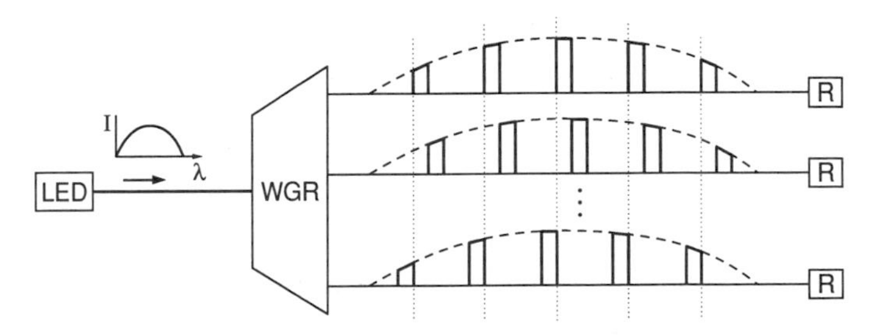

Fig. 13.14 Spectral slicing with the wavelength grating router (WGR). The periodicity property of the WGR permits efficient harvesting of the broadband optical spectrum. If the spectrum is wider than the free spectral range of the WGR (dotted vertical lines), each output port receives approximately the same power (about 1/4 of the input power in this example).

In summary, there are several devices capable of performing the WDM function. The WGR is certainly an attractive candidate for flexible architectures in the future.

5.3 SOURCES AND RECEIVERS

WDM sources and receivers are also the subject of a great deal of research, and, first, in this section we concentrate on WDM sources (Fig. 13.15). One category (Fig. 13.15a) is tunable lasers [6, 91, 100, 153]. These lasers essentially have one or two frequency-dependent mirrors. Examples include the tunable distributed Bragg reflector laser with two or three sections [91]. If current is injected across a junction containing a Bragg grating, the effective pitch can be changed to create a wavelength-dependent reflectivity that is controlled electronically. This, in turn, permits the laser to operate on one of the Fabry–Perot modes defined by the laser cavity [100]. In more recent years, there have been several variations on this theme. The Bragg gratings can be spatially modulated in amplitude as in the sampled grating [81], in spatial frequency as in the superstructure grating [80, 180], or both, sampled superstructure grating. When more than one grating is used (as in Fig. 13.15a), the spatial frequencies of the gratings need not coincide, which permits a vernier effect that increases the tuning range. In addition to these examples of contradirectional coupling, codirectional coupling between two dissimilar waveguides has been used to "leverage" index changes, and impressive tuning ranges of greater than 100 nm have been obtained [4, 144] in the vertical coupler filter (VCF) approach. Current research is addressing system issues [55, 93, 166, 175] and exploring tuning [34] and integrated modulation [98] options. Single-current tuning is preferred for access applications. Research is active in this area and the first devices appear near commercial realization.

A different approach is the class of asymmetric Y-branch lasers (Fig. 13.15b). In this case, optical feedback in the laser cavity has two components with different free spectral ranges. With changing currents, a vernier effect selects a lasing frequency over a wide tuning range [94, 149], and single-current tuning has been demonstrated [97]. As in the case of the VCF laser, this leverages the refractive index change caused by the injection current. Recent research has demonstrated the feasibility of three-section lasers with improved tuning stability.

The aforementioned sources are tunable: they lase at a single wavelength at a time, but that wavelength can be changed. Progress has also been

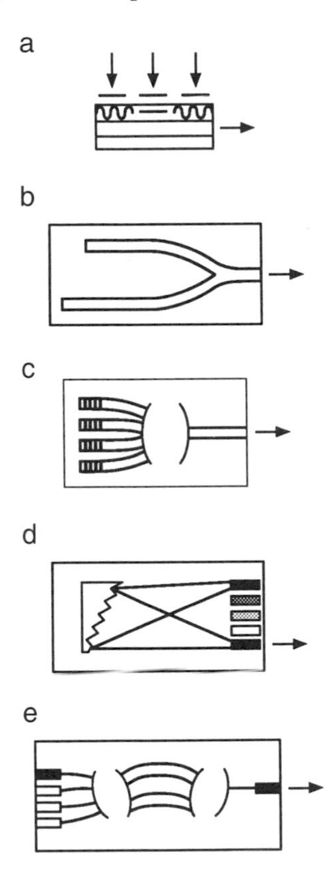

Fig. 13.15 Laser sources. Whereas (a) tunable (distributed Bragg reflector, sampled grating, superstructure grating, or vertical coupler filter) and (b) asymmetric *Y*-branch lasers operate on one wavelength at a time, others such as (c) an array of distributed feedback or distributed Bragg reflector lasers, can operate at several wavelengths simultaneously. The optical output of integrated arrays of independent lasers is combined by a passive splitter. (d) Multistripe array grating integrated cavity (MAGIC) and (e) waveguide array (multifrequency, arrayed waveguide grating, or phased array) lasers share an intracavity turning element.

made on sources that can emit on several wavelengths simultaneously. Such sources would have the functionality envisioned in the PPL [118, 186] architectures but would not require *N* independently packaged lasers and a WDM. The expectation is that if several sources are integrated, the packaging effort and the total physical size can be minimized, thus the costs of the device can be reduced. This savings is expected to more than offset the increased complexity involved with bringing N electronic signals to a single device.

In Fig. 13.15c, an array of DFB [103, 210, 212] or distributed Bragg reflector [209] lasers is shown. The optical output of each laser is coupled immediately into a waveguide that transports it to a passive splitter to be combined with the output of the other lasers. This sort of approach was precisely the original motivation for photonic integrated circuits. In this case, the simplicity of the coupling brings a $(1/N)$ power penalty with it, but even so, there are potential advantages to integration because numerous external interfaces are avoided, and there is even the potential for a final stage of amplification after the output coupler [210]. Feed-forward techniques have been proposed to ameliorate the inevitable cross-coupling of independent sources [36]. High channel accuracy has been reported on recent devices, heterodyne applications have been demonstrated [65], and arrays have been bonded to Silicon Optical Bench motherboards [114].

The last two devices shown in Fig. 13.15 operate on a different principle: they share a bulk laser cavity between several independent lasers. Figure 13.15d shows the MAGIC laser [163], which shares a common output amplifier section (lower stripe) with N distinct amplifier sections. A diffraction grating (left) efficiently couples light of a unique wavelength between the output amplifier and one of the stripes. The right-most facet is the reflective surface for both ends of the laser cavity, but without sending current through a given junction stripe, there is insufficient optical gain for lasing. Thus, independently exciting individual stripes while continuously driving the output amplifier allows multiple wavelengths to be emitted simultaneously. Monolithic construction enables a compact, rugged platform for a variety of devices. Recent work has utilized the same platform in fabricating a tunable WDM receiver [165, 181], showing that the potential exists for large-scale manufacturing savings that could accrue from similar production facilities [62].

The multifrequency laser (Fig. 13.15e) also shares a common optical cavity between several lasers. In this case, the shared cavity is an integrated version of the WGR [214, 216]. As in the MAGIC laser, a common output amplifier and a discrete amplifier stripe form the cavity for each wavelength. Again, the macroscopic form of the cavity determines the lasing wavelengths, and there are manufacturing savings possible from sharing a common platform. The same basic multifrequency laser structure has been demonstrated as a digitally tunable laser with an output modulator section [83].

Research in this area is active and the design trade-offs among device size, insertion loss, fabrication accuracy, and diffraction grating mechanism [12, 28] are currently being addressed.

Another class of sources that figure prominently in WDM networks are those capable of emitting light into a wide optical spectrum. We already mentioned the LED (Fig. 13.16a) when discussing spectral slicing. For this device, the quality of the optical cavity is degraded in some way to frustrate the optical feedback that would result in lasing on a Fabry–Perot mode. As a consequence, these sources characteristically emit into many longitudinal and transverse optical modes, which results in inefficient coupling to single-mode fibers. Recent work has shown that by coupling the light from an LED into an optical amplifier, the desirable spectral qualities can be maintained while optical powers comparable to those of single-mode lasers are delivered [109, 110].

As discussed for LOC-Net, optical modulators (Fig. 13.16b) can also be sources, and they have been proposed as a method for communicating with an RT [24, 41, 126, 161, 196, 200]. The appeal of such an approach is that if the modulator can be made inexpensively, there are several advantages that accrue from not having an optical source that needs to be powered continuously. Recent modulator work has included both waveguide devices, such as a split Mach–Zehnder interferometer [5] and a polarization-independent $LiNbO_3$ Y-branch switch [124], as well as surface-normal devices based on electroabsorption [60, 134] and a movable membrane [58,

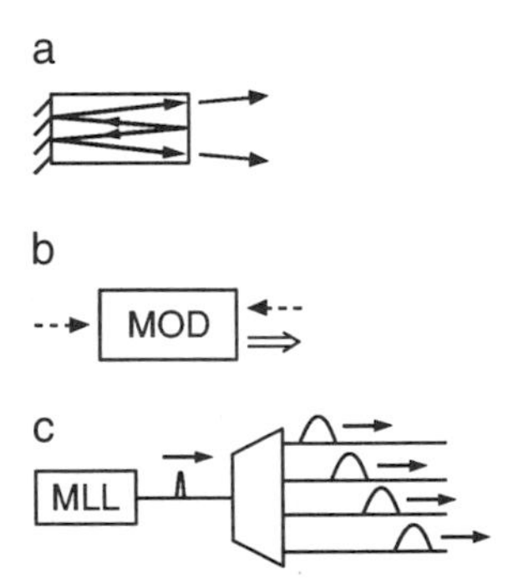

Fig. 13.16 Broadband sources. (a) The LED has a wide spectral output. (b) Modulators can be operated in the transmissive (*left dashed arrows*) or reflective (*right dashed arrows*) mode, and they can either be passive or be amplifiers. *Open arrow* represents output. (c) The optical spectrum of pulses from a mode-locked laser (MLL) is sliced by WDM.

59]. The latter holds the promise of inexpensive component costs for low (≈Mb/s) and moderate (≈10 Mb/s) rates [192] because its surface-normal design relaxes considerably the usual single-mode fabrication tolerances. Amplified modulators can also be broadband sources, for which the optical gain is switched on and off to impress data, while the optical carrier at the input determines the optical frequency of the signal [48]. Semiconductor amplifiers can have gain over the entire communication windows, a feature that makes these truly broadband optical sources.

Another approach to spectral slicing (Fig. 13.16c) has been proposed that utilizes the wide spectrum implicit in ultrashort optical pulses [35, 125]. The optical spectrum of ultrashort laser pulses is spectrally filtered and each spectral component is modulated. In this way, N independent sources operating at a bit rate of B bits/s can be modulated by an integrated set of N modulators, each operating at rate B, rather than needing a single modulator operating at the rate NB.

The richness and diversity of multiplexing approaches and source options available for WDM networks has grown tremendously in the years since the PPL proposals. Although the only component that can be viewed as mature is the LED, recent progress in all these components has been remarkable. Routers are now available in small quantities from several vendors, as are some of the sources (modulators, laser arrays).

In a full-blown WDM PON, a WDM receiver is placed at the CO to separate the upstream signals by wavelength [187]. As a consequence, the CO effectively has a dedicated point-to-point link with each subscriber, yet takes advantage of optical multiplexing on the feeder fiber. Although a WDM receiver may be thought of as the opposite of a WDM source (i.e., wavelength demultiplexing of the input signals, O/E conversion with a photodiode array, and electronic processing with a receiver front end), the additional technical challenges of integration, small signal levels, and the need to suppress cross talk have made this component lag slightly behind the WDM sources. A number of hybrid approaches have been demonstrated recently, aimed at the transmission market that is most likely to use them first. Capabilities in different approaches (high-speed receiver arrays [23, 167], demultiplexers using either a Silicon [181] or InP [54] Rowland grating, an integrated lens and Bragg grating [184], InP Waveguide Grating Router [167, 215], and on-chip optical preamplifiers [184, 215]) have all been demonstrated. The need for WDM receivers in local access architectures will likely come much later than the need for WDM sources, since it is generally assumed that for the foreseeable future the downstream rates will exceed the upstream rates.

In summary, there are strong international efforts in developing WDM components. LEDs are certainly the most mature, WDM splitters and routers are just now coming to market, WDM sources are available in small volumes, and WDM receivers are being developed. The development activity for these components and the availability of research devices has prompted recent work in optical architectures, to which we now turn.

5.4 RECENT ARCHITECTURES

The new component work has allowed a revisiting of the WDM PON architectures described by the Bellcore groups, in the sense of making the necessary components closer to commercial reality and in expanding the network possibilities with new component functions and capabilities. In this subsection, we review several recent architectures enabled by research on the WGR, WDM lasers, and LEDs.

An architecture with modest capacity is shown in Fig. 13.17a: it utilizes spectral slicing in both directions, taking full advantage of the WGR's periodicity [72]. It has the interesting property that it essentially mimics the operation of the broadcast PONs, but it does so over a WDM infrastruc-

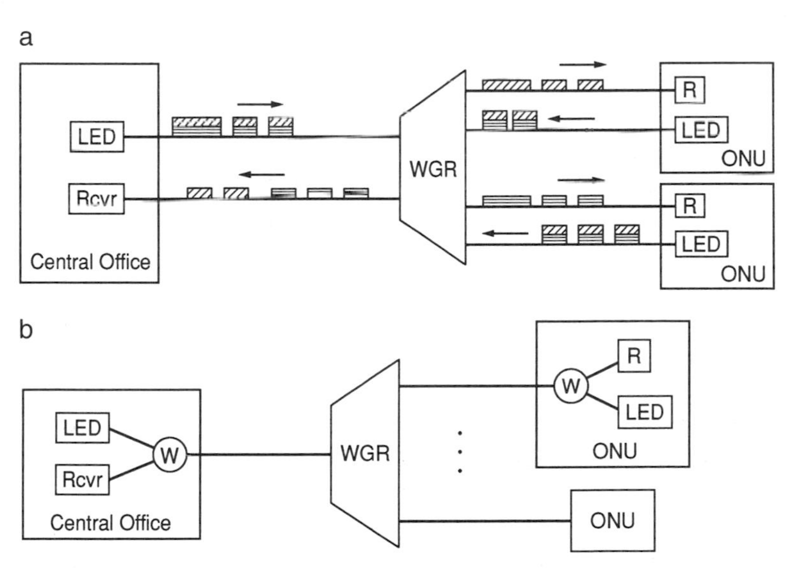

Fig. 13.17 Bidirectional spectral slicing. (a) LEDs and spectral slicing are used to implement a broadcast passive optical network with a WDM infrastructure. (b) Single-fiber implementation of the same idea. A coarse WDM (W) is used for directional multiplexing.

ture. The motivation for this is that in early deployment stages there may be little need for individual high-speed *point-to-point* links because broadband services are likely to be limited to video, and thus a modest broadcast bus capacity (e.g., 50 Mb/s shared by 16 users) may be adequate. The architecture uses spectral slicing to broadcast a downstream signal to all subscribers. The shading in Fig. 13.17a on the downstream signal is meant to illlustrate the wide optical bandwidth that is sliced into components for the two ONUs shown: each receives the same electrical signal, although it is borne on different optical wavelength sets. LEDs are also used in the upstream direction, one for each ONU. The upstream access is TDMA: each ONU sends its information in a preassigned slot, timed so that the packets arrive at the CO without interference. This is shown in Fig. 13.17a as the top ONU sending its message (two packets) to arrive at an earlier time than the three packets sent by the lower ONU. Each ONU's signals occupy a wide optical spectrum, again represented by shading, and different optical components from different subscribers (diagonal for upper ONU and horizontal for lower ONU) ultimately arrive at the CO. If a conventional receiver is used at the CO, the spectral separation is ignored and packets are separated as in a TDMA PON: the electronics for such a PON are indistinguishable from the NTT PON electronics. If the receiver at the CO were a WDM receiver, the point-to-point spectral slicing proposal [141] would result, with more of the WDM PON attributes. Presumably, for low data rates it is less expensive to multiplex *N* upstream circuits with TDMA onto a single receiver than to break the signal into *N* individual receiver circuits.

This approach may seem counterproductive, because its performance will be worse than that of a conventional PON (LEDs have less power than lasers, and WGRs cost more than splitters), and although it has some of the WDM comonents, it loses the WDM PON attributes because of the broadcast nature. However, the central point of this architecture is the infrastructure, not the initial service [72]. The network can be designed such that the instant it is installed one of the subscribers could upgrade (at a sizable cost!) to a bidirectional high-capacity link, because the infrastructure for a PPL-like is there, even if 15 of 16 subscribers cannot use and do not want high-capacity links. This layered entry has an attractive "pay-as-you-go" feature, in which the *subscribers,* not the providers, do the paying. Let us contrast the conventional PON trying to emulate this feature: once the broadcast PON is chosen, the splitting penalty is *always* present, and network controls must be set up to filter and interdict the nonupgraded users because the connections are *never* point-to-point in the optical domain.

Figure 13.17b shows a single-fiber implementation of the same idea. In this case, a coarse (1.3-/1.5-μm) WDM is used to separate the upstream information from the downstream with a minimal power penalty. The single-fiber approach has the advantage of using less optical fiber in the OSP (a cost savings), saves 2 to 3 dB in the slicing power budget, and serves twice as many subscribers for a given WGR size as the two-fiber version. On the other hand, it also requires a WDM in the ONU and is less upgradable because the directional duplexing with CWDM restricts the unplanned overlays that can be made.

A more capable network, LAR-Net (local access router network), is shown in Fig. 13.18. In this case, the upstream information is spectrally sliced, as described previously, but the downstream information is transmitted using a multiwavelength source: a multifrequency laser was proposed [213]. This architecture possesses the WDM attributes in the downstream direction but not in the upstream direction, unless the point-to-point slicing approach is used [141, 186]. An economic advantage of this architecture is that the ONU source is inexpensive and mature, while the more expensive CO source is shared by the N subscribers. The flexibility afforded by the WGR is evident in its behavior as a WDM for downstream signals and a passive splitter for upstream signals. Coarse WDM allows a single fiber to be used for directional duplexing.

A network with bidirectional point-to-point (instead of shared) links that does not require a wavelength-controlled source in the ONU is shown in Fig. 13.19, RITE-Net (remote integration of terminal equipment) [49]. Because it uses WDM circuits exclusively, it has all the WDM PON advantages. A multiwavelength source sends downstream signals to individual

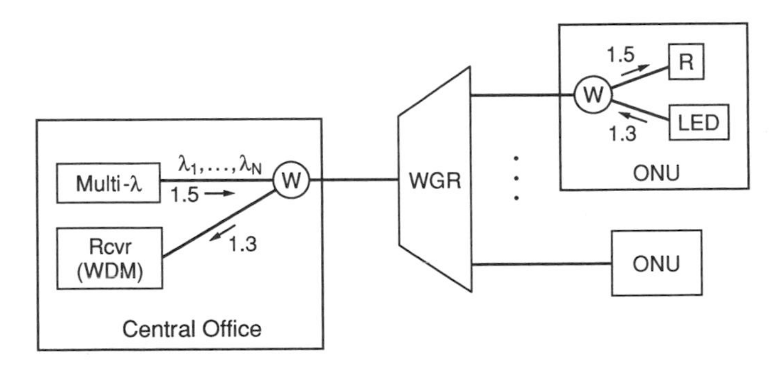

Fig. 13.18 LAR-Net. A WDM downstream and spectral slicing upstream, with a coarse WDM (W) for directional multiplexing.

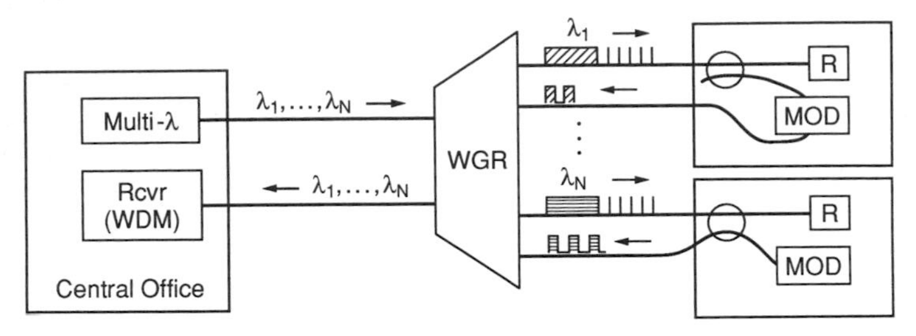

Fig. 13.19 RITE-Net. A WDM downstream with (*top*) transmissive and (*bottom*) reflective modulators imposing upstream data by overmodulation of a portion of the downstream signal. Space division (two fibers) for separating upstream and downstream light avoids interference and facilitates network diagnostics.

users in two components: a downstream message and a blank packet that can be thought of as an "optical chalkboard" that the ONU can write upon to provide an upstream signal. At the ONU, a coupler splits part of the light to a receiver to decode the downstream information, and the remainder of the light to an optical modulator. This modulator can be either transmissive (upper ONU) or reflective (lower ONU), and it can operate passively [51] or with amplification [48, 138]. As a replica of the *downstream* packet traverses the modulator, the modulator blocks it from returning to the CO. As a replica of the blank chalkboard passes by, however, the modulator impresses data on the chalkboard corresponding to the bits for the upstream signal. As in other proposals [5, 138, 139], the downstream data format can differ from the upstream format [50]. After modulation, the upstream output is looped back toward the CO. Although a single-fiber version is possible, as in the EROS architecture [138, 139], two fibers reduce Rayleigh scattering impairments [201] and thus permit longer loops, inexpensive passive modulators, and more service-rich upgrades. At the RN, the WGR routing property (discussed previously) makes multiplexing automatic: if input ports 1 and 2 are used to connect the CO and the WGR, for instance, then f_1 connects ONU 1 through output ports 1 and 2, f_3 connects ONU 2 through output ports 3 and 4, and so forth in *both* the downstream and the upstream directions. As before, a WDM receiver permits true point-to-point communication.

The loopback characteristic has a number of desirable features. First, one achieves a true point-to-point WDM connection without a wavelength-registered source at the ONU (the registration is carried with the downstream light): the WGR routing property ensures that light traversing from the CO to the ONU must make it back from the ONU to the CO. Second, there are potential cost savings associated with the use of inexpensive modulators [59] that can be turned off for idle channels. Third, optical loop back permits the ONU to be *optically* interrogated with the modulator in the "pass" state (but electrically OFF), which enhances maintenance and diagnostic features. On the system level, network management, control, and security are simplified because the ONU can transmit only light supplied by the CO. This feature makes it an architecture particularly adaptable to secure network operation when the ONU becomes customer premises equipment: a low-quality terminal degrades that subscriber's service only, not the services of other subscribers. These advantages are accompanied by several disadvantages: (1) inexpensive modulators are only now becoming available [11], (2) use of a second fiber adds expense and installation complexity, (3) attenuation of unamplified light through round-trips of the network reduces the maximum upstream bit rate, and (4) there is a possible loss of security and diagnostic advantages if amplifiers are used [48].

Finally, the unique properties of the WGR mentioned in Section 5.2 permit not only a natural extension to broadcast overlays (Figs. 13.5 and 13.10), using coarse WDM [5, 118] and spectral slicing, but also architecture evolution. Two examples are shown in Fig. 13.20, in which 81 digital video channels (on 16 QPSK subcarriers) from a commercial satellite service were broadcast over WDM PONs [73] that were operating with either bidirectional spectral slicing [72] or WDM point-to-point [51] communications. Experimental results showed that the satellite signals could be RF down-converted, applied to an inexpensive LED, and transported with no noticeable impairments from counter- or copropagating signals [73]. The robust QPSK modulation of the DBS system and excellent coarse WDM isolation [104] made this possible. The video signals were decoded with a commercial satellite set-top motion pictures expert group (MPEG) decoder. Because the components and terminal equipment (LEDs, WDMs, WGRs, receivers, mixers, the broadcast video system with set-top boxes, and the broadcast PON terminal equipment) are all currently available, the system in Fig. 13.20a, permitting both narrow band telecommunications and broadband entertainment, could be deployed more quickly than the virtual point-to-point architectures [51, 213] that require multiwavelength sources. Once

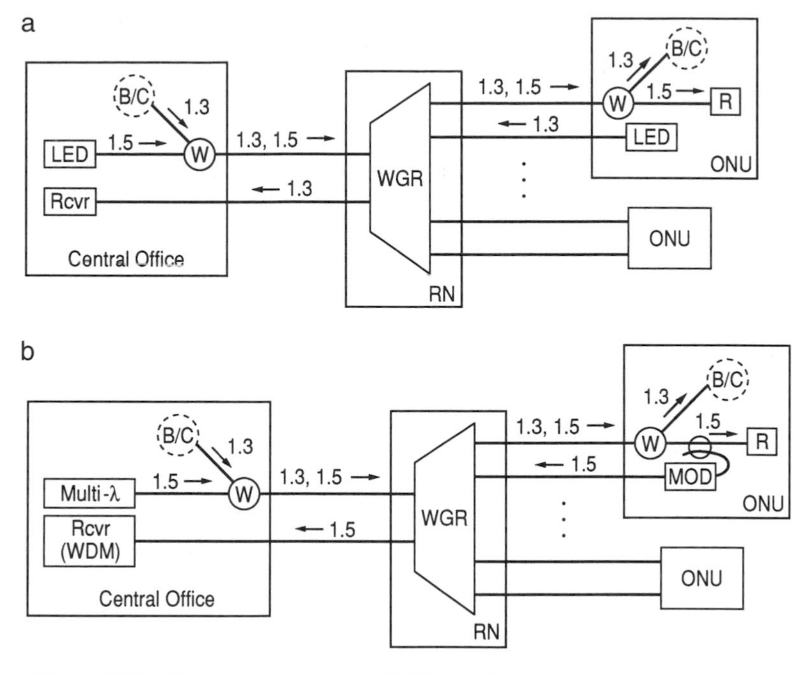

Fig. 13.20 WDM broadcast overlays. WGR periodicity is used to provide spectral slicing for a broadcast service that is multiplexed with coarse WDM. Broadcast overlays (*dashed lines*) at 1.3 μm are used on two different architectures that have the same outside plant. (a) Bidirectional spectral slicing (broadcast) passive optical network. (b) RITE-Net (switched) passive optical network.

installed, the underlying fiber infrastructure will support a natural *demand-driven* evolution to WDM PONs [51, 118, 186, 213] as the need arises, as shown in Fig. 13.20b.

5.5 SUMMARY

In this section, we argued that as a result of vigorous and diverse component research, WDM PONs can be configured to operate as two types of networks: (1) a switched, private, secure network with a clear upgrade path and (2) a flexible broadcast network for the distribution of entertainment video. We also showed that overlays may enable both to be present in the same architecture. Finally, we demonstrated that the WGR is flexible enough to evolve the *type* of delivery (e.g., going from a broadcast PON to a switched PON) as demanded in the future.

6. Summary

The economic advantages resulting from fiber feeder in both the telephony and CATV plant make it clear that fiber optics will continue to play an expanding role in the local loop. The exact nature of that role is less clear.

In this chapter, we examined the current loop architectures and the evolutionary forces that shaped their characteristics. We noted that we are witnessing large-scale changes in the form, content, and environment of communications, changes that current plants may be inadequate to accommodate. Section 3 showed the types of complexity involved in choosing an architecture that would be capable of performing well in the near to long term. A review of current architectural proposals in Section 4 showed the diversity of potential solutions, each with its distinct characteristics, each with its assumptions and approaches toward balancing near-term availability with long-term profitability. We took a more detailed view, in Section 5, of WDM PONs, presumably the architecture with the greatest long-term potential. The network's strengths that we described were in management (privacy, security, diagnostics), ease of provisioning and upgrading (separability), and capacity (optical efficiency), strengths enabled by WDM's virtual point-to-point character. We showed that the components are emerging but technologically robust: some are now coming to market, and research on new components is vigorous. These components now permit Bellcore work of the late 1980s to be realized, and they have enabled new architectures. Finally, we suggested that the robust digital broadcast format used for satellite television delivery enables early installment of simple networks to satisfy current business plans, while using a high-quality future friendly infrastructure that can exploit future component development.

Ultimately, the evolution of the access environment will be driven by economic opportunities and constraints, real and perceived. The excitement in this area derives from the tension between the tremendous opportunities afforded by the immense and growing importance of telecommunications and the formidable technical challenges of providing greatly enhanced future capabilities at current costs.

Acknowledgments

It is a pleasure to thank T. H. Daugherty, E. E. Harstead, and P. P. Bohn for introducing me to local access problems, to thank P. D. Magill and T. E. Darcie for supporting my early forays in passive optical networks, and to thank K. C. Reichmann and P. P. Iannone for fruitful discussions on systems concepts.

References

1. Adar, R., C. H. Henry, R. C. Kistler, and R. F. Kazarinov. 1992. Polarization independent narrowband Bragg reflection gratings made with silica-on-silicon waveguides. *Appl. Phys. Lett.* 60:640–641.
2. Agrawal, G. P., and S. Radic. 1994. Phase-shifted fiber Bragg gratings and their application for wavelength demultiplexing. *IEEE Photon. Tech. Lett.* 6(8):995–997.
3. de Albuquerque, A., A. J. N. Houghton, and S. Malmros. 1994. Field trials for fiber access in the EC. *IEEE Commun. Mag.* 32(2):40–48.
4. Alferness, R. C., U. Koren, L. L. Buhl, B. I. Miller, M. G. Young, T. L. Koch, G. Raybon, and C. A. Burrus. 1992. Broadly tunable InGaAsP/InP laser based on a vertical coupler filter with 57 nm tuning range. *Appl. Phys. Lett.* 60:3209–3211.
5. Altwegg, L., A. Azizi, P. Vogel, Y. Wang, and F. Wyler. 1994. LOCNET: A fiber in the loop system with no light source at the subscriber end. *IEEE J. Lightwave Tech.* 12(3):535–540.
6. Amann, M-C. 1994. Tunable semiconductor lasers. In *Proceedings of the ECOC'94, Florence, Italy,* 1011–1018.
7. Amersfoort, M. R., J. B. D. Soole, H. P. LeBlanc, N. C. Andreadakis, A. Rajhel, C. Caneau, M. A. Koza, R. Bhat, C. Youtsey, and I. Adesida. 1996. Polarization-independent InP-arrayed waveguide filter using square cross-section waveguides. In *Proceedings of the OFC'96, San Jose, CA,* 101–102.
8. Andrews, F. T. 1989. The heritage of telegraphy. *IEEE Commun. Mag.* 27(8):12–18.
9. Andrews, F. T. 1991. The evolution of digital loop carrier. *IEEE Commun. Mag.* 29(3):31–35.
10. Angelopoulos, J. D., I. S. Venieris, and G. I. Stassinopoulos. 1993. A TDMA based access control scheme for APONs. *IEEE J. Lightwave Tech.* 11(5/6): 1095–1103.
11. Battig. R. ASCOM Tech. Ltd., Morgenstrasse 129, Bern, Switzerland. *"REMO" 2Mb/s reflective modulator.* Application note 8.
12. Ashgari, M., *et al.* 1995. High-speed integrated multiwavelength laser source for WDM optical-fiber communication systems. In *Proceedings of the OFC'95, San Diego, CA,* 307–308.
13. Aurelli, G., V. C. Di Biase, F. Caviglia, and C. Guerricchio. 1994. Passive optical networks: Field trials in Italy. In *Proceedings of the ECOC'94, Florence, Italy,* 95–98.

14. Back, N. R., B. R. White, and S. C. Thorp. 1991. Highly integrated 20 Mb/s GaAs optical receiver for use in TPON network termination equipment. *Electron. Lett.* 27:566–568.
15. Barry, R. A., and P. A. Humblet. 1991. Latin routers, design and implementation. *IEEE J. Lightwave Tech.* 11(5/6):891–899.
16. Baskerville, L. J. 1989. Two fibers or one? (A comparison of two-fiber and one-fiber star architectures for FTTH applications). *IEEE J. Lightwave Tech.* 7:1733–1740.
17. Bellcore Technical Advisory. 1993. *Generic requirements and objectives for fiber in the loop systems.* TR-NWT-000909, Issue 2, December.
18. Bilodeau, F., D. C. Johnson, S. Theriault, B. Malo, J. Albert, and K. O. Hill. 1995. An all-fiber dense-wavelength-division multiplexer/demultiplexer using photoimprinted Bragg gratings. *IEEE Photon. Tech. Lett.* 7(4):388–390.
19. Blair, L. T., and S. A. Cassidy. 1993. Impact of new optical technology on spectrally-sliced access and data networks. *BT Tech. J.* 11(2):46–55.
20. Bohn, P. P., M. J. Kania, J. M. Nemchik, and R. C. Purkey. 1992. Fiber in the loop. *AT&T Tech. J.* 71:31–45.
21. Brenner, T., M. Bachmann, and H. Melchior. 1994. Waveguide tapers for efficient coupling of InGaAsP/InP OEIC components to flat-end single mode fibers. In *Proceedings of the ECOC'94, Florence, Italy,* 1031–1034.
22. Carroll, R. L. 1989. Optical architecture and interface lightguide unit for fiber to the home feature of the AT&T SLC series 5 carrier system. *IEEE J. Lightwave Tech.* 7:1727–1732.
23. Chandrasekhar, S., L. M. Lunardi, R. A. Hamm, and G. J. Qua. 1994. Eight-channel p-i-n/HBT monolithic receiver array at 2.5 Gb/s per channel for WDM applications. *IEEE Photon. Tech. Lett.* 6(10):1216–1218.
24. Chapman, D. A., and D. W. Faulkner. 1990. Use of reflected light for low data rate upstream signaling in a single mode passive optical network. *IEEE Proc. J.* 137:108–114.
25. Chen, W. Y., and D. L. Waring. 1994. Applicability of ADSL to support video dial tone in the copper loop. *IEEE Commun. Mag.* 32(5):102–109.
26. Chiddix, J. A., J. A. Vaughan, and R. W. Wolfe. 1993. The use of fiber optics in cable communications networks. *J. Lightwave Tech.* 11(1):154–166.
27. Chidgey, P. J. 1994. Multi-wavelength transport networks. *IEEE Commun. Mag.* 32(12):28–35.
28. Clemens, P. C., G. Heise, R. Marz, H. Michel, A. Reichelt, and H. W. Schneider. 1994. 8-Channel optical demultiplexer realized as SiO_2/Si flat-field spectrograph. *IEEE Photon. Tech. Lett.* 6(9):1109–1111.
29. Cochrane, P. 1994. Future networks. *BT Tech. J.* 12(2):9–13.
30. Cook, A., and J. Stern. 1994. Optical fiber access — Perspectives toward the 21st century. *IEEE Commun. Mag.* 32(2):78–86.
31. Dail, J. E., M. A. Dajer, C-C. Li, P. D. Magill, C. A. Siller, Jr., K. Sriram, and N. A. Whitaker. 1996. Adaptive digital access protocol: A MAC protocol

for multiservice broadband access networks. *IEEE Commun. Mag.* 34(3): 104–112.
32. Darcie, T. E. 1987. Subcarrier multiplexing for multiple-access lightwave networks. *IEEE J. Lightwave Tech.* LT-5:1103–1110.
33. Darcie, T. E. 1993. *Lightwave video transmission.* Short course 123, OFC'93, San Jose, CA, February 23.
34. Delorme, F., H. Nakajima, C. Alletru, S. Slempkes, and B. Pierre. 1994. A new distributed Bragg reflector laser for improved tuning. *IEEE Photon. Tech. Lett.* 6(9):1085–1087.
35. DeSouza, E. A., M. C. Nuss, M. Zirngibl, and C. H. Joyner. 1996. Spectrally sliced WDM using a single femtosecond source. In *Proceedings of the OFC'95, San Diego, CA,* Postdeadline paper PD-16, 16-1–16-5.
36. Doerr, C. R., C. H. Joyner, M. Zirngibl, L. W. Stulz, and H. M. Presby. 1995. Elimination of signal distortion and crosstalk from carrier density changes in the shared semiconductor amplifier of multifrequency signal sources. *IEEE Photon. Tech. Lett.* 7(10):1131–1133.
37. Dono, N. R., P. E. Green, Jr., K. Liu, R. Ramaswami, and F. Tong. 1990. A wavelength division multiple access network for computer communication. *IEEE J. Select. Areas Commun.* 8:983–994.
38. Dragone, C. 1991. An $N \times N$ optical multiplexer using a planar arrangement of two star couplers. *IEEE Photon. Tech. Lett.* 3:812–815.
39. Dragone, C., C. A. Edwards, and R. C. Kistler. 1991. Integrated optics $N \times N$ multiplexer on silicon. *IEEE Photon. Tech. Lett.* 3:896–899.
40. Dragone, C. 1990. Optimum design of a planar array of tapered waveguides. *J. Opt. Soc. Am. A* 7:2081–2093.
41. Duthie, P. J., M. J. Wale, I. Bennion, and J. Hankey. 1986. Bidirectional fibre-optic link using reflective modulation. *Electron. Lett.* 22:517–518.
42. Dyke, P. J., and D. B. Waters. 1994. A review of the technical options for evolving FITL to support small business and residential services. *IEEE J. Lightwave Tech.* 12(2):376–381.
43. Faulkner, D. W., and D. I. Fordham. 1989. Broadband systems on passive optical networks. *Br. Telecom. Tech. J.* 7:115–122.
44. Faulkner, D. W., D. B. Payne, J. R. Stern, and J. W. Ballance. 1989. Optical networks for local loop applications. *IEEE J. Lightwave Tech.* 7:1741–1751.
45. Feldman, R. D., T. H. Wood, G. Raybon, and R. F. Austin. 1995. Effect of optical beat interference on the dynamic range of a subcarrier multiple access passive optical network using Fabry–Perot lasers. *IEEE J. Lightwave Tech.* 14(5):711–715.
46. Feldman, R. D., T. H. Wood, and R. F. Austin. 1995. Operation of a frequency shift keyed subcarrier multiple-access system for a passive optical network in the presence of strong adjacent channel interference. *IEEE Photon. Tech. Lett.* 7(4):427–429.

47. Feldman, R. D., K-Y. Liou, G. Raybon, and R. F. Austin. 1996. Reduction of optical beat interference in a subcarrier multiple-access passive optical network through the use of an amplified light-emitting diode. *IEEE Photon. Tech. Lett.* 8(1):116–118.
48. Feuer, M. D., J. M. Wiesenfeld, J. S. Perino, C. A. Burrus, G. Raybon, S. C. Schunk, and N. K. Dutta. 1996. Single-port laser-amplifier modulators for local access. *IEEE Photon. Tech. Lett.* 8:1175–1177.
49. Frigo, N. J., P. P. Iannone, P. D. Magill, T. E. Darcie, M. M. Downs, B. N. Desai, U. Koren, T. L. Koch, C. Dragone, H. M. Presby, and G. E. Bodeep. 1994. A wavelength-division multiplexed passive optical network with cost-shared components. *IEEE Photon. Tech. Lett.* 6(11):1365–1367.
50. Frigo, N. J., P. P. Iannone, M. M. Downs, and B. N. Desai. 1995. Mixed-format delivery and full-duplex operation in a WDM PON with a single shared source. In *Proceedings of the OFC'95, San Diego, CA,* 55–57.
51. Frigo, N. J., P. P. Iannone, K. C. Reichmann, J. A. Walker, K. W. Goossen, S. C. Arney, E. J. Murphy, Y. Ota, and R. G. Swartz. 1995. Demonstration of performance-tiered modulators in a WDM PON with a single shared source. In *Proceedings of the ECOC'95, Brussels, Belgium,* 441–444.
52. Frigo, N. J. 1996. Passive optical networks in the local loop. *Opt. Photon. News* 7:24–29.
53. Fukano, H., K. Yokoyama, Y. Kadota, Y. Kondo, M. Ueki, and J. Yoshida. 1995. Low cost, high coupling-efficient and good temperature characteristics 1.3 μm laser diodes without spot-size transformer. In *Proceedings of the ECOC'95, Brussels, Belgium,* 1027–1030.
54. Gini, E., M. Blaser, W. Hunziker, and H. Melchior. 1995. Packaged 2.5 Gb/s 4-channel WDM receiver module with InP grating demultiplexer and pin-JFET receiver array. In *Proceedings of the ECOC'95, Brussels, Belgium,* 207–210.
55. Glance, B., J. M. Wisenfeld, U. Koren, and R. W. Wilson. 1993. New advances on optical components needed for FDM optical networks. *IEEE J. Lightwave Tech.* 11(5/6):882–890.
56. Glance, B., I. P. Kaminow, and R. W. Wilson. 1994. Applications of the integrated waveguide grating router. *IEEE J. Lightwave Tech.* 12:957–962.
57. Goodman, M. S., H. Kobrinski, M. P. Vecchi, R. M. Bulley, and J. L. Gimlett. 1990. The LAMBDANET multiwavelength network: Architecture, applications, and demonstrations. *IEEE J. Select. Areas Commun.* 8(6):995–1003.
58. Goossen, K. W., J. A. Walker, and S. C. Arney. 1994. Silicon modulator based on mechnically-active anti-reflection layer for fiber-in-the-loop applications. In *Proceedings of the OFC'94. San Jose, CA,* Postdeadline paper PD10, 10-1–10-5.
59. Goossen, K. W., J. A. Walker, and S. C. Arney. 1994. Silicon modulator based on mechanically-active anti-reflection layer with 1 Mb/s capability for fiber-in-the-loop applications. *IEEE Photon. Tech. Lett.* 6(9):1119–1121.

60. Goossen, K. W., J. E. Cunningham, and W. Y. Jan. 1994. Stacked-diode electroabsorption modulator. *IEEE Photon. Tech. Lett.* 6(8):936–939.
61. Green, P. E., Jr. 1992. *Fiber optic networks.* New York: Prentice-Hall.
62. Green, P. E., Jr. 1994. Toward customer-usable all-optical networks. *IEEE Commun. Mag.* 32(12):44–49.
63. Grubb, S. G. 1995. High power diode-pumped fiber lasers and amplifiers. In *Proceedings of the OFC'95, San Diego, CA,* 41–42.
64. Grubb, S. G., D. J. DiGiovanni, J. R. Simpson, W. Y. Cheung, S. Sanders, D. F. Welch, and B. Rockney. 1996. Ultrahigh power diode-pumped 1.5 micron fiber amplifiers. In *Proceedings of the OFC'96, San Jose, CA,* 30–31.
65. Hamacher, M., D. Trommer, K. Li, H. Schroeter-Janssen, W. Rehbein, and H. Heidrich. 1995. High bandwidth heterodyne receiver OEIC, fabricated with a three stage MOVPE. In *Proceedings of the ECOC'95, Brussels, Belgium,* 215–218.
66. Hawley, G. T. 1991. Historical perspectives on the U.S. telephone loop. *IEEE Commun. Mag.* 29(3):24–28.
67. Henry, P. S. 1989. High capacity lightwave local area networks. *IEEE Commun. Mag.* 27(10):20–26.
68. Hibino, Y., T. Kitagawa, K. O. Hill, F. Bilodeau, B. Malo, J. Albert, and D. C. Johnson. 1996. Wavelength division multiplexer with photoinduced Bragg gratings fabricated in a planar-lightwave-circuit-type asymmetric Mach–Zehnder interferometer on Si. *IEEE Photon. Tech. Lett.* 8(1):84–86.
69. Hill, G. R., P. J. Chidgey, F. Kaufhold, T. Lynch, O. Sahlen, M. Gustavsson, M. Janson, B. Lagerstorm, G. Grasso, F. Meli, S. Johansson, J. Ingers, L. Fernandez, S. Rotolo, A. Antonielli, S. Tebaldini, E. Vezzoni, R. Caddedu, N. Caponio, F. Testa, A. Scavannec, M. J. O'Mahony, J. Zhou, A. Yu, W. Sohler, U. Rust, and H. Hermann. 1993. A transport network layer based on optical network elements. *IEEE J. Lightwave Tech.* 11(5/6):667–679.
70. Hill, K. O., Y. Fujii, D. C. Johnson, and B. S. Kawasaki. 1978. Photosensitivity in optical fiber waveguides: Application to reflection filter fabrication. *Appl. Phys. Lett.* 32:647–649.
71. Hill, K. O., B. Malo, K. A. Vineberg, F. Bilodeau, D. C. Johnson, and I. Skinner. 1990. Efficient mode conversion in telecommunication fibre using externally written gratings. *Electron. Lett.* 26:1270–1272.
72. Iannone, P. P., N. J. Frigo, and T. E. Darcie. 1995. A WDM PON architecture with bi-directional optical spectral slicing. In *Proceedings of the OFC'95, San Diego, CA,* 51–53.
73. Iannone, P. P., K. C. Reichmann, and N. J. Frigo. 1996. Broadcast digital video upgrade for WDM passive optical networks. In *Proceedings of the OFC'96. San Jose, CA,* Postdeadline paper PD-32.
74. 1989. (Special issue on subscriber loop technology.) *IEEE J. Lightwave Tech.* 7(11).

75. 1991. (Special issue on the 21st century subscriber loop.) *IEEE Commun. Mag.* 29(3).
76. 1993. (Special issue on broadband optical networks.) *IEEE J. Lightwave Tech.* 11(5/6).
77. 1994. (Special issue on fiber optic subscriber loops.) *IEEE Commun. Mag.* 32(2).
78. 1994. (Special issue on optically multiplexed networks.) *IEEE Commun. Mag.* 32(12).
79. Ikushima, I., S. Himi, T. Hamaguchi, M. Suzuki, N. Maeda, H. Kodera, and K. Yamashita. 1995. High-performance compact optical WDM transceiver module for passive double star subscriber systems. *IEEE J. Lightwave Tech.* 13(3):517–524.
80. Ishii, H., Y. Tohmori, T. Tamamura, and Y. Yoshikuni. 1993. Superstructure grating (SSG) for broadly tunable DBR lasers. *IEEE Photon. Tech. Lett.* 4:393–395.
81. Jayarman, V., A. Mathur, L. A. Coldren, and P. D. Dapkus. 1993. Extended tuning range in sampled grating DBR lasers. *IEEE Photon. Tech. Lett.* 5:489–491.
82. Jones, J. R. 1993. Video services delivery in FITL systems using MPEG encoding and ATM transport. In *LEOS'93, San Jose, CA,* 122–123. Paper FL1.3.
83. Joyner, C. H. 1995. An 8-channel digitally tunable transmitter with electro-absorption modulated ouptut by selective-area epitaxy. *IEEE Photon. Tech. Lett.* 7(9):1013–1015.
84. Kaminow, I. P. 1989. Photonic multiple-access networks: Topologies. *AT&T Tech. J.* 78(March/April):61–71.
85. Kaminow, I. P. 1989. Photonic multiple-access networks: Routing and multiplexing. *AT&T Tech. J.* 78(March/April):72–86.
86. Kashima, N. 1991. Upgrade of passive optical subscriber network. *IEEE J. Lightwave Tech.* 9(1):113–120.
87. Kashima, N. 1992. Time compression multiplex transmission system using a 1.3 μm semiconductor laser as a transmitter and a receiver. *IEEE Trans. Commun.* 40:584–590.
88. Kashima, N. 1993. *Optical transmission for the fiber loop.* Boston: Artech House.
89. Kazovsky, L. G., and P. T. Poggiolini. 1993. STARNET: A multi-gigabit-per-second optical LAN using a passive WDM star. *IEEE J. Lightwave Tech.* 11:1009–1027.
90. Khoe, G., G. Heydt, I. Borges, P. Demeester, A. Ebberg, A. Labrujere, and J. Rawsthorne. 1993. Coherent multicarrier technology for implementation in the customer access. *IEEE J. Lightwave Tech.* 11(5/6):695–713.
91. Koch, T. L., and U. Koren. 1991. Semiconductor photonic integrated circuits. *IEEE J. Quantum Electron.* 27(3):641–653.

92. Kodama, T., and T. Fukuda. 1994. Customer premises networks of the future. *IEEE Commun. Mag.* 32(2):96–98.
93. Kuwano, S., O. Ishida, N. Shibata, H. Ishii, and S. Suzuki. 1994. Rapidly tunable and fully stable 16 frequency channel packet router employing an SSG-DBR laser and arrayed waveguide grating demultiplexer. In *Proceedings of the ECOC'94, Florence, Italy,* 71–74.
94. Kuznetsov, M., P. Verlangieri, and A. G. Dentai. 1994. Frequency tuning characteristics and WDM channel access of the semiconductor three-branch Y3-lasers. *IEEE Photon. Tech. Lett.* 6(2):157–160.
95. Kuznetsov, M. 1994. Cascaded coupler Mach–Zehnder channel dropping filters for wavelength-division-multiplexed optical systems. *IEEE J. Lightwave Tech.* 12(2):226–230.
96. Labrujere, A. C., M. O. van Deventer, O. J. Koning, J. P. Bekooj, A. H. H. Tan, G. Roelofsen, M. K. de Lange, J-P. Boly, J. A. H. W. Berendschot-Aarts, C. P. Spruijt, M. F. L. van Nielen, R. F. M. van den Brink, K. M. de Blok, and A. K. van Bochove. 1993. COSNET — A coherent optical subscriber network. *IEEE J. Lightwave Tech.* 11(5/6):865–874.
97. Lach, E., D. Baums, K. Daub, K. Dutting, T. Feeser, W. Idler, G. Laube, G. Luz, M. Schilling, K. Wunstel, and O. Hildebrand. 1994. 20 nm Single current tuning of asymmetrical Y-laser. In *Proceedings of the ECOC'94, Florence, Italy,* 805–808.
98. Lammert, R. M., G. M. Smith, J. S. Hughes, M. L. Osowski, A. M. Jones, and J. J. Coleman. 1996. MQW wavelength-tunable DBR lasers with monolithically integrated external cavity electroabsorption modulators with low-driving-voltages fabricated by selective-area MOCVD. *IEEE Photon. Tech. Lett.* 8(6):797–799.
99. Lattner, P. D., R. L. Fike, and G. A. Nelson. 1991. Business and residential services for the evolving subscriber loop. *IEEE Commun. Mag.* 29(3):109–114.
100. Lee, T. P., and C-E. Zah. 1989. Wavelength tunable and single frequency semiconductor lasers for photonic communications networks. *IEEE Commun. Mag.* 27(10):42–52.
101. Lemberg, H. L. 1992. *Passive optical networks.* Short course 125, OFC'92, San Jose, CA, February 3.
102. Lemberg, H. L. 1994. *Hybrid fiber access networks.* Short course 117, OFC'94, San Jose, CA, February 21.
103. Li, G. P., T. Makino, A. Sarangan, and W. Huang. 1996. 16-Wavelength gain-coupled DFB laser array with fine tunability. *IEEE Photon. Tech. Lett.* 8(1):22–24.
104. Li, Y. P., C. H. Henry, E. J. Laskowski, H. H. Yaffe, and R. L. Sweat. 1995. A monolithic optical waveguide 1.31/1.55 μm WDM with -50 dB crosstalk over 100 nm bandwidth. *Electron. Lett.* 31:2100–2101.
105. Lin, L. Y., S. S. Lee, K. S. J. Pister, and M. C. Wu. 1994. Micro-machined three-dimensional micro-optics for integrated free-space optical system. *IEEE Photon. Tech. Lett.* 6(12):1445–1447.

106. Lin, L. Y., J. L. Shen, S. S. Lee, M. C. Wu, and A. M. Sergent. 1996. Tunable three-dimensional solid Fabry–Perot etalons fabricated by surface-micromachining. *IEEE Photon. Tech. Lett.* 8(1):101–103.
107. Lin, Y-K. M., and D. R. Spears. 1989. Passive optical subscriber loops with multiaccess. *IEEE J. Lightwave Tech.* 7:1769–1777.
108. Lin, Y-K. M., D. R. Spears, and M. Yin. 1989. Fiber-based local access network architectures. *IEEE Commun. Mag.* 27(10):64–73.
109. Liou, K-Y., and G. Raybon. 1995. Operation of an LED with a single-mode semiconductor amplifier as a broad-band 1.3 μm transmitter source. *IEEE Photon. Tech. Lett.* 7(9):1025–1027.
110. Liou, K-Y., B. Glance, U. Koren, E. C. Burrows, G. Raybon, C. A. Burrus, and K. Dreyer. 1996. Monolithically integrated semiconductor LED-amplifier for applications as transceivers in fiber access systems. *IEEE Photon. Tech. Lett.* 8(6):800–802.
111. Lu, K. W., M. I. Eiger, and H. L. Lemberg. 1990. System and cost analyses of broadband fiber loop architectures. *IEEE J. Select. Areas Commun.* SAC-8:1058–1067.
112. Lu, X., G. E. Bodeep, and T. E. Darcie. 1995. Broad-band AM-VSB/64 QAM cable TV system over hybrid fiber/coax network. *IEEE Photon. Tech. Lett.* 7(4):330–332.
113. Lu, X., T. E. Darcie, G. E. Bodeep, S. L. Woodward, and A. H. Gnauck. 1996. Mini-fiber node hybrid fiber/coax networks for two-way broadband access. In *Proceedings of the OFC'96, San Jose, CA,* 143–144.
114. Mallecot, F., C. Artigue, F. Pommereau, F. Poingt, A. Bodere, D. Carpentier, T. Fillion, J-L. Gentner, F. Gerard, M. Goix, E. Grard, J-L. Lafragette, L. Le Gouezigou, R. Ngo, A. Pinquier, G. Vendrome, G. Grand, P. Mottier, P. Gidon, A. Fournier, I. Wamsler, and G. Laube. 1995. Hybrid silica multiwavelength optical source realized by passive alignment. In *Proceedings of the OFC'95, San Diego, CA,* 227–228.
115. Matz, R., J. G. Bauer, P. Clemens, G. Heise, H. F. Mahlein, H. Michel, and G. Schulte-Roth. 1994. Development of a photonic integrated transceiver chip for WDM transmission. *IEEE Photon. Tech. Lett.* 6(11):1327–1329.
116. Meltz, G., W. W. Morey, and W. H. Glenn. 1989. Formation of Bragg gratings in optical fibers by a transverse holographic method. *Opt. Lett.* 14:823–825.
117. Mestdagh, D. J. G. 1995. *Fundamentals of multiaccess optical fiber networks.* Boston: Artech House.
118. Menendez, R.C., S. S. Wagner, and H. L. Lemberg. 1990. Passive fiber loop architecture providing both switched and broadcast transport. *Electron. Lett.* 26:273–274.
119. Metzger, W., J. G. Bauer, P. Clemens, G. Heise, M. Klein, H. F. Mahlein, R. Matz, H. Michel, and J. Rieger. 1994. Photonic integrated transceiver for the access network. In *Proceedings of the ECOC'94, Florence, Italy,* 87–90.
120. Miki, T. 1994. Toward the service-rich era. *IEEE Commun. Mag.* 32(2):34–39.

121. Miki, T., and R. Komiya. 1991. Japanese subscriber loop network and fiber optic loop development. *IEEE Commun. Mag.* 29(3):60–67.
122. Miyachi, M., M. Ogusu, and S. Ohshima. 1995. Cross-talk penalty-free dense WDM system with a Littrow-mounted optical wavelength mux. In *Proceedings of the ECOC'95, Brussels, Belgium,* 67–70.
123. Mizrahi, V., T. Erdogan, D. J. DiGiovanni, P. J. Lemaire, W. M. MacDonald, S. G. Kosinski, S. Cabot, and J. E. Sipe. 1994. Four-channel fibre grating demultiplexer. *Electron. Lett.* 30:780–781.
124. Murphy, E. J., T. O. Murphy, and R. W. Irvin. 1995. Low voltage, polarization-independent $LiNbO_3$ modulators. In *Proceedings of the OFC'95. San Diego, CA,* Paper TuK2.
125. Morioka,T., K. Mori, S. Kawanishi, and M. Saruatari. 1994. Multi-WDM-channel, Gbit/s pulse generation from a single laser source utilizing LD-pumped supercontinuum in optical fibers. *IEEE Photon. Tech. Lett.* 6(3):365–368.
126. Nicholson, G. 1990. Use of a fiber loop reflector as downstream receiver and upstream modulator in passive optical network. *Electron. Lett.* 26:827–828.
127. Okamoto, K., and Y. Inoue. 1995. Silica-based planar lightwave circuits for WDM systems. In *Proceedings of the OFC'95, San Diego, CA,* 224–225.
128. Okamoto, K. 1995. Application of planar lightwave circuits to optical communication systems. In *Proceedings of the ECOC'95, Brussels, Belgium,* 75–82.
129. Okuno, M., K. Katoh, S. Suzuki, Y. Ohmori, and A. Himeno. 1994. Strictly non-blocking 16×16 matrix switch using silica-based planar lightwave circuits. In *Proceedings of the ECOC'94, Florence, Italy,* 83–86.
130. Oliver, R. 1995. The role of different architecture alternatives in the access network. In *Proceedings of the ECOC'95, Brussels, Belgium,* 493–499.
131. Ota, Y., and R. G. Swartz. 1990. Burst mode compatible optical receiver with large dynamic range. *IEEE J. Lightwave Tech.* 8(12):1897–1902.
132. Ota, Y., and R. G. Swartz. 1992. DC — 1 Gb/s burst-mode-compatible receiver for optical bus application. *IEEE J. Lightwave Tech.* 10(2):244–249.
133. Ota, Y., R. G. Swartz, V. D. Archer III, S. K. Korotky, M. Banu, and A. E. Dunlop. 1994. High-speed, burst-mode, packet-capable optical receiver and instantaneous clock recovery for optical bus operation. *IEEE J. Lightwave Tech.* 12(2):325–331.
134. Pathak, R. N., K. W. Goossen, J. E. Cunningham, and W. Y. Jan. 1994. InGaAs-InP P-I(MQW)-N surface-normal electroabsorption modulators exhibiting better than 8:1 contrast ratio for 1.55 μm applications grown by gas-source MBE. *IEEE Photon. Tech. Lett.* 6(12):1439–1441.
135. Payne, D. B., and J. R. Stern. 1986. Transparent single-mode optical networks. *IEEE J. Lightwave Tech.* 4:864–869.
136. Payne, D. B. 1993. Opportunities for advanced optical technology in access networks. *BT Tech. J.* 11(2):11–18.

137. Pellegrini, G., P. Passeri, and A. Moncalvo. 1994. Market and regulatory issues impacting the evolution of access network. In *Proceedings of the ECOC'94, Florence, Italy,* 169–176.
138. Perrier, P. A., and R. Boirat. 1993. Wavelength addressing for an efficient photonic subscriber loop architecture. In *Fourth IEEE Conference on Telecommunications, London, England, 18–21 April.*
139. Perrier, P. A., O. Gautheron, C. Coerjolly, S. Gauchard, V. Havard, and E. Leclerc. 1992. A photonic subscriber loop architecture using a single wavelength per bi-directional communication. In *EFOC/LAN'92,* 141–143.
140. Pugh, W., and G. Boyer. 1995. Broadband access: Comparing alternatives. *IEEE Commun. Mag.* 33(8):34–46.
141. Reeve, M. E., A. R. Hunwicks, W. Zhao, S. G. Methley, L. Bickers, and S. Hornung. 1988. LED spectral slicing for single-mode local loop applications. *Electron. Lett.* 24:389–390.
142. Reeve, M. E., S. Hornung, L. Bickers, P. Jenkins, and S. Mallinson. 1989. Design of passive optical networks. *Br. Telecom. Tech. J.* 7(2):89–99.
143. Reichmann, K. C., T. E. Darcie, and G. E. Bodeep. 1996. Broadcast digital video as a low-cost overlay to baseband digital-switched services on a PON. In *Proceedings of the OFC'96, San Jose, CA,* 144–146.
144. Rigole, P-J., S. Nilsson, L. Backbom, T. Klinga, J. Wallin, B. Stalnacke, E. Berglind, and B. Stolz. 1995. Improved tuning regularity over 100 nm in a vertical grating assisted codirectional coupler laser with a superstructure grating distributed Bragg reflector. In *Proceedings of the ECOC'95, Brussels, Belgium,* 219–222.
145. Rocks, M., N. Gieschen, and A. Gladisch. 1995. Germany's fibre in the loop installations as basis for further access network evolution. In *Proceedings of the Australia Conference on Fiber Technology, ACOFT'95, Coolum Beach, QLD, Australia,* 19–22.
146. Rowbotham, T. R. 1991. Local loop developments in the UK. *IEEE Commun. Mag.* 29(3):50–59.
147. Sano, K., and I. Kobayashi. 1994. Access network evolution in Japan. In *Proceedings of the ECOC'94, Florence, Italy,* 143–149.
148. Sato, K., S. Okamoto, and H. Hadama. 1994. Network performance and integrity enhancement with optical path layer technologies. *IEEE J. Select. Areas Commun.* 12:159–170.
149. Schilling, M., W. Idler, D. Baums, G. Laube, K. Wunstel, and O. Hildebrand. 1991. Multifunctional photonic switching operation of 1500 nm Y-coupled cavity laser with 28 nm tuning capability. *IEEE Photon. Tech. Lett.* 3(12):1054–1057.
150. Scobey, M. A., and D. E. Spock. 1995. Passive DWDM components using MicroPlasma® optical interference filters. In *Proceedings of the OFC'95, San Diego, CA,* 242–243.

151. Schwartz, M. 1990. *Information, transmission, modulation, and noise.* New York: McGraw-Hill.
152. Shankaranarayanan, N. K., S. D. Elby, and K. Y. Lau. 1991. WDMA/subcarrier-FDMA lightwave networks: Limitations due to optical beat interference. *IEEE J. Lightwave Tech.* 9(7):931–943.
153. Shankaranarayanan, N. K., U. Koren, B. Glance, and G. Wright. 1994. Two-section DBR laser transmitters with accurate channel spacing and fast arbitrary-sequence tuning for optical FDMA networks. In *Proceedings of the OFC'94, San Jose, CA,* 36–37.
154. Shiozawa, T., M. Shibutani, and J. Namiki. 1993. Upstream-FDMA/downstream-TDM optical fiber multiaccess network. *IEEE J. Lightwave Tech.* 11(5/6):1034–1039.
155. Shumate, P. W., and R. K. Snelling. 1991. Evolution of fiber in the residential loop plant. *IEEE Commun. Mag.* 29(3):68–74.
156. Shumate, P. W. 1994. Access network evolution in the United States: Economics and operations drivers. In *Proceedings of the ECOC'94, Florence, Italy,* 151–158.
157. Shumate, P. W. 1996. What's happening with fiber to the home? *Opt. Photon. News* 7(February):17–21, 75.
158. Smit, M. K. 1988. New focusing and dispersive planar component based on an optical phased array. *Electron. Lett.* 24(7):385–386.
159. Smith, J. M. 1995. The future of competition in long-distance telecommunications. *IEEE Commun. Mag.* 33(12):62–64.
160. Snelling, R. K., J. Chernak, and K. W. Kaplan. 1990. Future fiber access needs and systems. *IEEE Commun. Mag.* 28(4):63–65.
161. Solgaard, O., F. S. A. Sandejas, and D. M. Bloom. 1992. Deformable grating optical modulator. *Opt. Lett.* 17:688–690.
162. Solgaard, O., M. Daneman, N. C. Tien, A. Friedberger, R. S. Muller, and K. Y. Lau. 1995. Optoelectronic packaging using silicon surface-micromachined alignment mirrors. *IEEE Photon. Tech. Lett.* 7(1):41–43.
163. Soole, J. B. D., K. R. Paguntke, A. Scherer, H. P. LeBlanc, C. Chang-Hosnain, J. R. Hayes, C. Caneau, R. Bhat, and M. A. Koza. 1992. Wavelength selectable laser emission from a multi-stripe array grating integrated cavity laser. *Appl. Phys. Lett.* 61:2750–2752.
164. Soole, J. B. D., H. P. LeBlanc, N. C. Andreadakis, R. Bhat, C. Caneau, and M. A. Koza. 1994. Monolithic InP reflection grating multiplexer/demultiplexers for WDM components operating in the long wavelength fiber band. *Int. J. High Speed Electron. Syst.* 5(1):111–133.
165. Soole, J. B., A. Scherer, Y. Silberberg, H. P. LeBlanc, N. C. Andreadakis, C. Caneau, and K. R. Poguntke. 1993. Integrated grating demultiplexer and pin array for high-density wavelength division multiplexed detection at 1.5 μm. *Electron. Lett.* 29:558–560.

166. Staring, A. A. M., J. J. M. Binsma, P. I. Kuindersma, E. J. Jansen, P. J. A. Thijs, T. van Dongen, and G. F. G. Depovere. 1994. Wavelength independent output power from an injection-tunable DBR laser. *IEEE Photon. Tech. Lett.* 6(2):147–149.
167. Steenbergen, C. A. M., L. C. N. de Vreede, C. van Dam, T. L. M. Scholtes, M. K. Smit, J. L. Tauritz, J. W. Pedersen, I. Moerman, B. H. Verbeek, and R. G. F. Baets. 1995. Integrated 1 GHz 4-channel InP phasar based WDM-receiver with Si bipolar frontend array. In *Proceedings of the ECOC'95, Brussels, Belgium,* 211–214.
168. Stern, J. R., J. W. Ballance, D. W. Faulkner, S. Hornung, and D. B. Payne. 1987. Passive optical local networks for telephony applications and beyond. *Electron. Lett.* 23:1255–1257.
169. Stern, J. R., and R. Wood. 1989. The longer term future of the local network. *Br. Telecom. Tech. J.* 7:161–170.
170. Stern, J. R. 1994. Status and future of passive optical networks. In *Proceedings of the ECOC'94, Florence, Italy,* 179–183.
171. Stone, J., and L. W. Stulz. 1987. Pigtailed high-finesse tunable fibre Fabry–Perot interferometers with large, medium, and small free spectral ranges. *Electron. Lett.* 23:781–783.
172. Stone, J., and L. W. Stulz. 1993. Passively temperature-compensated nontunable fibre Fabry–Perot etalons. *Electron. Lett.* 29:1608–1609.
173. Stordahl, K., and E. Murphy. 1995. Forecasting long-term demand for services in the residential market. *IEEE Commun. Mag.* 33(2):44–49.
174. Su, C., L. K. Chen, and K. W. Cheung. 1994. Inherent transmission capacity penalty of burst-mode receiver for optical multiaccess networks. *IEEE Photon. Tech. Lett.* 6(5):664–667.
175. Tada, Y., O. Ishida, N. Shibata, and K. Nosu. 1993. Design considerations on a DBR-laser transmitter for fast frequency-switching in an optical FDM cross-connect system. *IEEE J. Lightwave Tech.* 11(5/6):813–818.
176. Takahashi, K., Y. Aihara, Y. Suzuki, H. Taya, and M. Yoshinuma. 1994. Ribbon fiber splicing with mass axis alignment device. In *Proceedings of the ECOC'94, Florence, Italy,* 103–106.
177. Takeuchi, T., T. Sasaki, M. Yamamoto, K. Hamamoto, K. Makita, K. Taguchi, and K. Komatsu. 1995. A transceiver PIC for bidirectional optical communication fabricated by bandgap energy controlled selective MOVPE. In *Proceedings of the ECOC'95, Brussels, Belgium,* 87–90.
178. Taylor, T. M. 1991. Power and energy in the local loop. *IEEE Commun. Mag.* 29(3):76–82.
179. Tenzer, G. 1991. The introduction of optical fiber in the subscriber loop in the telecommunications networks of DBP TELEKOM. *IEEE Commun. Mag.* 29(3):36–49.

180. Tohmori, Y., Y. Yoshikuni, H. Ishii, F. Kano, T. Tamamura, and Y. Kondo. 1993. Over 100 nm wavelength tuning in superstructure grating (SSG) DBR lasers. *Electron. Lett.* 29(4):352–354.
181. Tong, F., K. Liu, C. S. Li, and A. E. Stevens. 1995. A 32-channel hybridly-integrated tunable receiver. In *Proceedings of the ECOC'95, Brussels, Belgium,* 203–206.
182. Town, G. E., K. Sugden, J. A. R. Williams, I. Bennion, and S. B. Poole. 1995. Wide-band Fabry–Perot-like filters in optical fiber. *IEEE Photon. Tech. Lett.* 7(1):78–80.
183. Verbeek, B. H., and M. K. Smit. 1995. Phased array based WDM devices. In *Proceedings of the ECOC'95, Brussels, Belgium,* 195–202.
184. Verdiell, J-M., T. L. Koch, B. I. Miller, M. G. Young, U. Koren, F. Storz, and K. F. Brown-Goebeler. 1994. A WDM receiver photonic integrated circuit with net on-chip gain. *IEEE Photon. Tech. Lett.* 6(8):960–962.
185. Vogel, M. O., and R. C. Menendez. 1991. Fiber to the curb systems: Architectural alternatives. *Fiber Integr. Opt.* 9:281–297.
186. Wagner, S. S., H. Kobrinski, T. J. Robe, H. L. Lemberg, and L. S. Smoot. Experimental demonstration of a passive optical subscriber loop architecture. *Electron. Lett.* 24:344–346.
187. Wagner, S. S., and H. L. Lemberg. 1989. Technology and system issues for a WDM-based fiber loop architecture. *IEEE J. Lightwave Tech.* 7:1759–1768.
188. Wagner, S. S., and H. Kobrinski. 1989. WDM applications in broadband telecommunications networks. *IEEE Commun. Mag.* 27(3):22–29.
189. Wagner, S. S., and R. C. Menendez. 1989. Evolutionary architectures and techniques for video distribution on fiber. *IEEE Commun. Mag.* 27(12):17–25.
190. Wagner, S. S., and T. E. Chapuran. 1990. Broadband high-density WDM transmission using superluminescent diodes. *Electron. Lett.* 26:696–697.
191. Wakui, Y. 1994. The fiber-optic subscriber network in Japan. *IEEE Commun. Mag.* 32(2):56–63.
192. Walker, J. A., K. W. Goossen, S. C. Arney, N. J. Frigo, and P. P. Iannone. 1995. A silicon optical modulator with 5 MHz operation for fiber-in-the-loop applications. In *Proceedings of Transducers '95/Eurosensors IX, Stockholm, Sweden, June 25–29,* 285–288.
193. Waring, D. L., J. W. Lechleider, and T. R. Hsing. 1991. Digital subscriber line technology facilitates a graceful transition from copper to fiber. *IEEE Commun. Mag.* 29(3):96–104.
194. Warzanskyj, W., and U. Ferrero. 1994. Access evolution in Europe: A view from EURESCOM. In *Proceedings of the ECOC'94, Florence, Italy,* 135–142.
195. Weippert, W. 1994. The evolution of the access network in Germany. *IEEE Commun. Mag.* 32(2):50–55.
196. Wheeler, J. K., J. Ocenasek, and P. P. Boh. 1986. Two-way transmission using electro-optic modulator. *Electron. Lett.* 22:479.

197. Wheeler, T. E. 1995. It's the information, not the highway. *IEEE Commun. Mag.* 33(12):58–61.
198. White, P. E., and L. S. Smoot. 1988. Optical fibers in loop distribution systems. In *Optical fiber telecommunications II,* ed. S. E. Miller and I. P. Kaminow, 911–932. Boston: Academic Press.
199. White, P. E. 1991. The role of the broadband integrated services digital network. *IEEE Commun. Mag.* 29(3):116–119.
200. Wood, T. H., E. C. Carr, B. L. Kasper, R. A. Linke, and C. A. Burrus. 1986. Bidirectional fiber-optical transmission using a multiple-quantum-well modulator/detector. *Electron. Lett.* 22:528–529.
201. Wood, T. H., R. A. Linke, B. L. Kasper, and E. C. Carr. 1988. Observation of coherent Rayleigh noise in single-source bidirectional optical fiber systems. *IEEE J. Lightwave Tech.* LT-6:346–352.
202. Wood, T. H., G. E. Bodeep, T. E. Darcie, and G. Raybon. 1991. Demonstration of broadband-ISDN upgrade of fibre-in-loop system. *Electron. Lett.* 27:2275–2276.
203. Wood, T. H., R. D. Feldman, and R. F. Austin. 1994. Demonstration of a cost-effective, broadband passive optical network system. *IEEE Photon. Tech. Lett.* 6:575–578.
204. Woodward, S. L., and G. E. Bodeep. 1995. A full-duplex optical data link using lasers as transceivers. *IEEE Photon. Tech. Lett.* 7(9):1060–1062.
205. Wu, T-H. 1995. Emerging technologies for fiber network survivability. *IEEE Commun. Mag.* 33(2):58–74.
206. Wu, T-H., N. Yoshikai, and H. Fujii. 1995. ATM signaling transport network architectures and analysis. *IEEE Commun. Mag.* 33(12):90–99.
207. Yamada, M., M. Shimizu, T. Kanamori, Y. Ohishi, Y. Terunuma, K. Oikawa, H. Yoshinaga, K. Kikushima, Y. Miyamoto, and S. Sudo. 1995. Low-noise and high power Pr^{3+}-doped fluoride fiber amplifier. *IEEE Photon. Tech. Lett.* 7(8):869–871.
208. Yamamoto, F., I. Sankawa, S. Furukawa, Y. Koyamada, and N. Takato. 1993. In-service remote access and measurement methods for passive double star networks. In *Fifth Conference on Optical/Hybrid Access Networks, September 7–9.* Paper 5.02, 5.02.01–5.02.06.
209. Young, M. G., U. Koren, B. I. Miller, M. A. Newkirk, M. Chien, M. Zirngibl, C. Dragone, B. Tell, H. M. Presby, and G. Raybon. 1993. A 16 $\times$ 1 wavelength division multiplexer with integrated distributed Bragg reflector lasers and electroabsorption modulators. *IEEE Photon. Tech. Lett.* 5(8):908–910.
210. Young, M. G., T. L. Koch, U. Koren, G. Raybon, A. H. Gnauck, B. I. Miller, M. Chien, K. Dreyer, R. E. Behringer, D. M. Tennant, and K. Feder. 1995. Six-channel, WDM transmitter module with ultra-low chirp and stable wavelength selection. In *Proceedings of the ECOC'95, Brussels, Belgium,* 1019–1022.
211. Zaganiaris, A., M. Tahkokorpi, M. Kalervo, B. T. Olsen, K. Stordahl, U. Ferrero, S. Balzaretti, and M. Drieskens. 1994. Methodology for risk assess-

ment and techno-economic evaluation of optical access networks. In *Proceedings of the ECOC'94, Florence, Italy,* 83–90.

212. Zah, C. E., F. J. Favire, B. Pathak, R. Bhat, C. Caneau, P. S. D. Lin, A. S. Gozdz, N. C. Andreadakis, M. A. Koza, and T. P. Lee. 1992. Monolithic integration of multiwavelength compressive-strained MQW DFB laser array with star coupler and optical amplifiers. *Electron. Lett.* 28(25):2361–2362.
213. Zirngibl, M., C. H. Joyner, L. W. Stulz, C. Dragone, H. M. Presby, and I. P. Kaminow. 1995. LAR-Net, a local access router network. *IEEE Photon. Tech. Lett.* 7:215–217.
214. Zirngibl, M., B. Glance, L. W. Stulz, C. H. Joyner, G. Raybon, and I. P. Kaminow. 1994. Characterization of a multiwavelength waveguide grating router laser. *IEEE Photon. Tech. Lett.* 6:1082–1084.
215. Zirngibl, M., C. H. Joyner, and L. W. Stulz. 1995. WDM receiver by monolithic integration of an optical preamplifier, waveguide grating router, and photodiode array. *Electron. Lett.* 31(7):581–582.
216. Zirngibl, M., C. H. Joyner, L. W. Stulz, U. Koren, M-D. Chien, M. G. Young, and B. I. Miller. 1994. Digitally tunable laser based on the integration of a waveguide grating multiplexer and an optical amplifier. *IEEE Photon. Tech. Lett.* 6(4):516–518.
217. Zirngibl, M., C. R. Doerr, and L. W. Stulz. 1996. Study of spectral slicing for local access applications. *IEEE Photon. Tech. Lett.* 8(5):721–723.

Chapter 14 | Lightwave Analog Video Transmission

Mary R. Phillips
ATx Telecom Systems, Naperville, Illinois
Thomas E. Darcie
AT&T Laboratories–Research, Holmdel, New Jersey

1. Introduction

The ability to transmit multiple analog video channels over tens of kilometers of optical fiber has had a profound impact on the telecommunications industry. Cable systems operators using hybrid fiber coax (HFC) systems have been able to increase the number and quality of the video signals delivered and dramatically reduce the frequency of system failures. In addition, HFC systems offer the capability for transport of telephony and broadband data services, which allows cable operators the opportunity to provide a full spectrum of information services at low cost and with high reliability. This emergence of viable competition for local access, enabled primarily by the analog lightwave technology represented in this chapter, has been a major driving force in shaping procompetitive public policy for the establishment of a broadband information infrastructure.

Given the success of HFC, it has also been embraced by some local exchange carriers (LECs) as a means to supply both traditional telephony services and analog video. Such systems currently provide cost-effective broadband services more easily than the traditional copper twisted-pair network, without the cost impediments of more fiber-intensive alternatives like fiber to the curb (FTTC) and passive optical networks (PONs).

For those who maintain that fiber will ultimately replace twisted-pair and coax as the information conductor to subscribers, analog lightwave

OPTICAL FIBER TELECOMMUNICATIONS,
VOLUME IIIA

ISBN: 0-12-395170-4

systems also play a critical role. Until digital video technology manages to displace a broad base of cable-ready analog consumer electronics, any fiber-based subscriber loop system must deliver the same offering of analog video channels as delivered with ease by HFC or it will not meet with widespread consumer acceptance. Given the strict fidelity requirements for analog video, present technology limitations, and cost constraints for subscriber loop access systems, delivering analog video by means of fiber to individual subscribers is a challenge.

The history of transmission of analog signals over fiber began in the late 1970s. Limitations in laser output power and relative-intensity noise (RIN) restricted early efforts to just a few channels over short distances. As better laser structures resulted in increased laser power, and single-frequency distributed feedback (DFB) lasers provided low RIN, the opportunity for application within cable television distribution systems became apparent. In the late 1980s this motivated considerable activity in understanding and eliminating the nonlinear mechanisms in semiconductor lasers, and an explosive penetration of lightwave systems into analog video systems. Within just a few years, numerous vendors had emerged to supply a diverse assortment of high-performance analog links to an eagerly waiting cable industry.

By the mid-1990s, analog links capable of delivering more than 100 channels of analog video plus a broad spectrum of digital–RF channels over distances in excess of 20 km were available. These systems use a variety of approaches, including direct modulation of DFB lasers at 1.3 μm, external modulation of high-power YAG lasers, external modulation and optical amplification using erbium-doped fiber amplifiers (EDFAs) at 1.55 μm, and performance enhancement by sophisticated predistortion and noise reduction techniques. Performance in many cases is close to the theoretical optimum.

In this chapter we discuss the motivation, technology, and systems issues associated with these analog fiber systems. Section 2 describes the requirements of the different formats of the video signals to be transmitted. Also described are examples of systems that require analog transport capability, like HFC and PONs. Section 3 delves into systems specifics, including receivers (Section 3.1), transmitters (Section 3.2), and optical amplifiers (Section 3.3). Because analog video requires very high signal fidelity, numerous effects within the fiber result in impairment. These include multipath interference, dispersion (chromatic and polarization), and fiber nonlinearities, as discussed in Section 3.4. In Section 3.5, we summarize the categories of systems and trends.

2. Analog Lightwave Systems

The analog lightwave system must take multiple frequency-division multiplexed video channels at the input, convert them to an optical signal, and transport them over several tens of kilometers of single-mode fiber and/or through passive fiber splitters to an optical receiver where they are converted back to RF. The end-to-end link must satisfy strict noise and distortion requirements. Traditional analog lightwave links are intensity modulated/direct detection (IM/DD) systems. In this technique, the analog video channel spectrum simply modulates the optical power such that the intensity spectrum of the light is the same as the original RF signal (plus a DC component). At the receiver a photodiode converts the modulated intensity back into an RF signal. The simplicity of the IM/DD system makes it attractive, although it comes at the cost of very stringent performance requirements of the lightwave components. The focus of this chapter is on components for a simple IM/DD lightwave system that transports a standard band of amplitude modulated–vestigial sideband (AM-VSB) analog video channels. The lightwave system consists of the transmitter, optional optical amplifiers, a fiber transport network, and an optical receiver.

We discuss the technical requirements of each system component for two exemplary systems. The first is an 80-channel system that is typical of a high-performance cable trunk system. The system performance is defined by several parameters that are discussed later. We desire a carrier-to-noise ratio (CNR) of 52 dB or more, composite second-order (CSO) distortion of −60 dBc or less, and composite triple beat (CTB) of −65 dBc or less, and we assume that this is achieved with an optical modulation depth (OMD) of 3.5% per channel. Trunk systems in use today support between 40 and 110 channels with numbers similar to those in our example. These specifications allow additional degradation from a coaxial distribution system with typically three coaxial-cable amplifiers in cascade, while still achieving the following specifications at the television set: a CNR of 46 dB or more, CSO distortion of −53 dBc or less, CTB of −53 dBc or less, as required by the National Association of Broadcasters [1]. The requirements for the European PAL systems are somewhat less stringent.

Our second example is for an FTTC or a PON system where degradation from a coax distribution system need not be budgeted. Also, given that we envision the analog video to be delivered with a broad variety of other digital (including video) services, we assume that 50 channels of analog video are sufficient. In this case, the performance at the output of the optical receiver needs to have only a CNR of 47 dB or more, CSO distortion less than or equal to −56 dBc, and CTB less than or equal to −56 dBc. Given the reduced linearity requirements and reduced channel load, we assume that this is achievable with an OMD of 5% per channel.

2.1 VIDEO FORMATS

The challenge in delivering analog video over fiber systems is in meeting strict noise and linearity requirements. These arrive from the complexity and fragility of the AM-VSB video format. Brilliantly designed many decades ago [2] for high spectral efficiency, the single-sideband (vestigial sideband [VSB]) amplitude-modulated (AM) format requires 6-MHz channel spacing (8 MHz in Europe) between video carriers. Baseband video, including intensity and color information, is used to AM a video carrier. This is VSB filtered and frequency multiplexed with a frequency-modulated (FM) audio signal, which results in a spectrum as shown in Fig. 14.1. The dominant feature in the spectrum is the remaining video carrier. Video information appears between the video carrier and the color subcarrier, at power levels many tens of decibels below the video carrier. Noise and

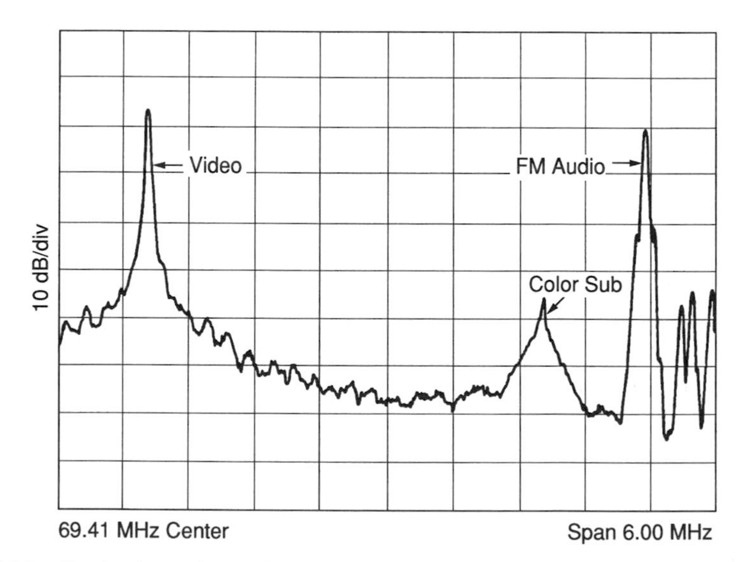

Fig. 14.1 Single-channel amplitude-modulated vestigial-sideband (AM-VSB) video spectrum showing a video carrier, color and audio subcarriers, and time-averaged modulation sidebands at low levels relative to the video carrier.

distortion products must therefore be small in order not to interfere with picture quality.

Multiple video channels are frequency multiplexed according to a particular plan. The most popular in the United States is the standard National Cable Television Association (NCTA) plan, as shown in Fig. 14.2. Video carriers are nominally spaced by 6 MHz, but with irregularities to fit around the FM radio band. As is seen later, the distribution of distortion products that result from the nonlinear mixing between these multiple carriers provides information about the type of nonlinear impairment involved.

The time-varying nature of the live video spectrum makes system diagnostics difficult. Depending on the luminance of the instantaneous point along the image sweep, and the time relative to synchronization and sweep pulses, the magnitude of the instantaneous video carrier varies by up to 6 dB. This makes accurate carrier measurement with a spectrum analyzer difficult. In addition, modulation sidebands obscure distortion products that form the CSO and CTB distortion. In order to perform stable and accurate measurements, the industry uses a set of continuous unmodulated video

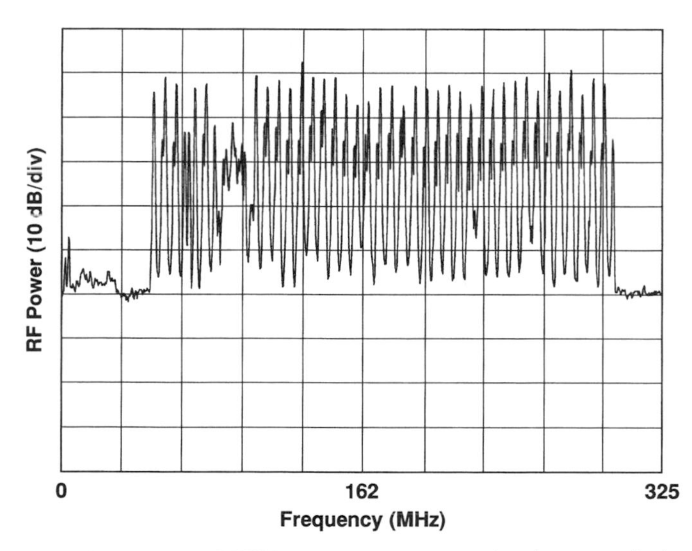

Fig. 14.2 Multichannel AM-VSB spectrum as transmitted on a typical cable television system. The variation in the video carrier level results from instantaneous differences in amplitude modulation (AM) depth. The features resolved in the spectrum are the video carriers and the frequency-modulated (FM) audio subcarriers.

carriers as a test signal. In what follows, we consider the performance parameters in the context of these unmodulated test signals.

Alternative video formats provide much greater immunity to impairment than AM-VSB provides. FM video has been used for decades for satellite and trunk transmission [3], where a high CNR cannot be achieved. The required 15-dB signal-to-noise ratio (SNR) can be achieved easily over a variety of lightwave systems [4, 5]. A typical FM channel requires between 30 and 40 MHz of bandwidth, which makes it unpopular (in the United States) for terrestrial broadcast or cable delivery. Furthermore, the cost of converting between FM and AM-VSB is a disadvantage for systems that deliver FM video to the home. The techniques described in this chapter are applicable to FM video transmission, but many of the impairments that are discussed will not be problematic.

Emerging compressed digital video (CDV) technology will eventually displace AM-VSB and FM. CDV eliminates inter- and intraframe redundancy to compress National Television System Committee (NTSC)-like video into less than 5 Mb/s [6]. When CDV is combined with advanced modem technology, the result is high-quality video with a higher spectral efficiency than that of AM-VSB, and with a much lower required CNR. But, as with FM video, the conversion cost and the embedded base of analog equipment will prevent this new technology from displacing AM-VSB rapidly.

When used with advanced modem technology, CDV allows a trade-off between spectral efficiency and required bandwidth [7, 8]. The required CNR increases from less than 20 dB for simple modems like quadrature phase-shift keying (QPSK) (2 bits/s/Hz), to close to 30 dB for 64-QAM (quadrature amplitude modulation) (6 bits/s/Hz). As even higher spectral efficiencies are employed, the digital video channel becomes more like an analog video channel, in terms of transmission requirements. Much of this chapter is therefore applicable to these digital–RF channels. Transmission of both analog and digital–RF channels simultaneously from the same laser has received considerable attention, primarily because of the onset of clipping-induced impulse noise [9, 10]. This topic is not discussed in this chapter.

2.2 HFC SYSTEMS

As mentioned in the introduction, the availability of high-performance analog lightwave technology has had a great impact on the cable industry. The key was to be able to replace cascades of dozens of coaxial amplifiers, as shown in Fig. 14.3, with fiber, as shown in Fig. 14.4. Rather than suffering the accumulated noise and distortion of the amplifier chain, high-fidelity analog video could be interjected at distributed points throughout the serving area [11]. These fiber nodes (FNs) contain the optical receivers and electronic amplifiers needed to drive relatively short coaxial distribution systems. In addition to improved picture quality, many factors came together to result in the rapid acceptance of this system approach within the cable industry. Because the FNs serve typically between 200 and 2000 subscribers, the lightwave cost per subscriber is small. By eliminating the long amplifier cascades, transmission failure due to amplifier failure is less frequent and affects far fewer subscribers. Finally, because the maximum length of coax serving any subscriber is relatively short, the total bandwidth that can be supported on the coax is increased significantly. Practical amplifier and equalization (to compensate for the frequency-dependent loss of coaxial cable) technology would now enable system bandwidths close to 1 GHz, whereas long amplifier cascades were limited to less than typically 450 MHz.

Various specific HFC systems have been implemented, but the most popular system with both LECs and cable operators has approximately 500

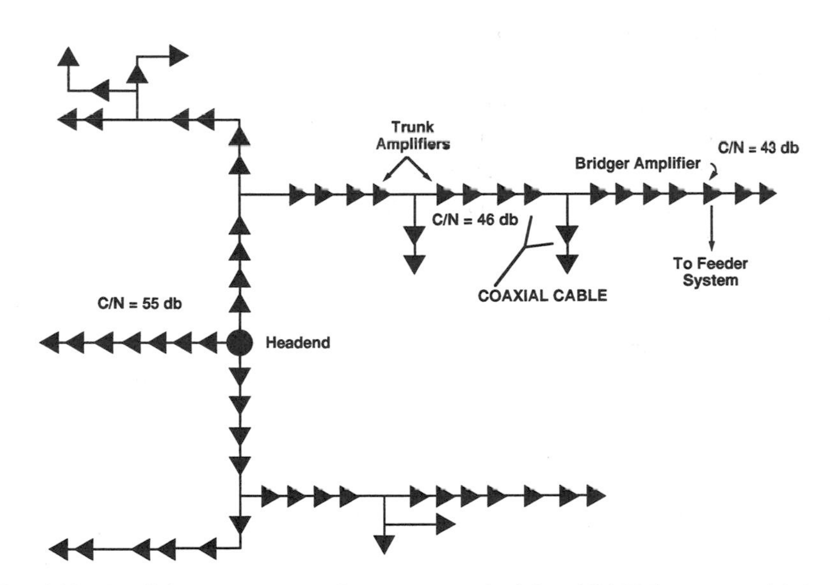

Fig. 14.3 Prelightwave community-antenna television (CATV) system with long cascades of electronic amplifiers and coax, showing a typical carrier-to-noise (C/N) ratio degradation along spans.

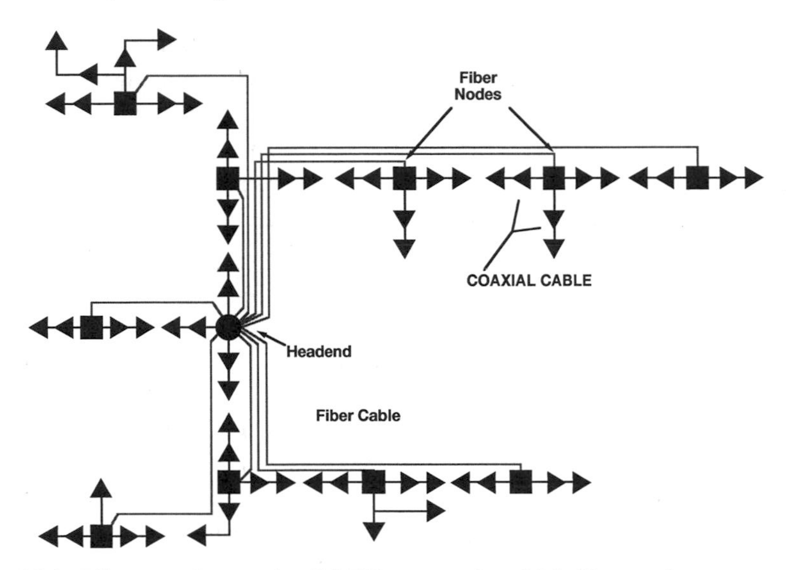

Fig. 14.4 Fiber overlay to the CATV system, in which fiber nodes are used to subdivide serving areas. High-quality signals are delivered far into the serving area using linear lightwave.

subscribers served from each FN. This is achieved with typically three or fewer coaxial amplifiers in cascade. This is a reasonable trade-off between present-day cost and performance. An example is shown in Fig. 14.5. More aggressive system designs seek to eliminate all amplifiers outside the FNs by serving fewer than 200 subscribers per FN. These passive coax systems cost more but have better reliability and lower noise in the upstream band (typically from 5 to 40 MHz).

Analog lightwave is used in three primary manners in HFC systems. First, the cost of converting from digital or FM video formats that can be delivered over long distances and satellite networks to the AM-VSB formats for cable transmission is high. It is therefore desirable to minimize the number of head ends or central offices in which this is done. This head-end consolidation requires that the multichannel AM-VSB spectrum be distributed over long distances between head ends. This requires extremely high-fidelity performance over distances in the range of 50 km, which can be achieved using linearized high-power lasers and/or optical amplifiers.

The second class of analog application in HFC is for the trunk systems between head ends and FNs, as shown in Fig. 14.5. This requires transmission over usually less than 30 km, with performance as discussed in our

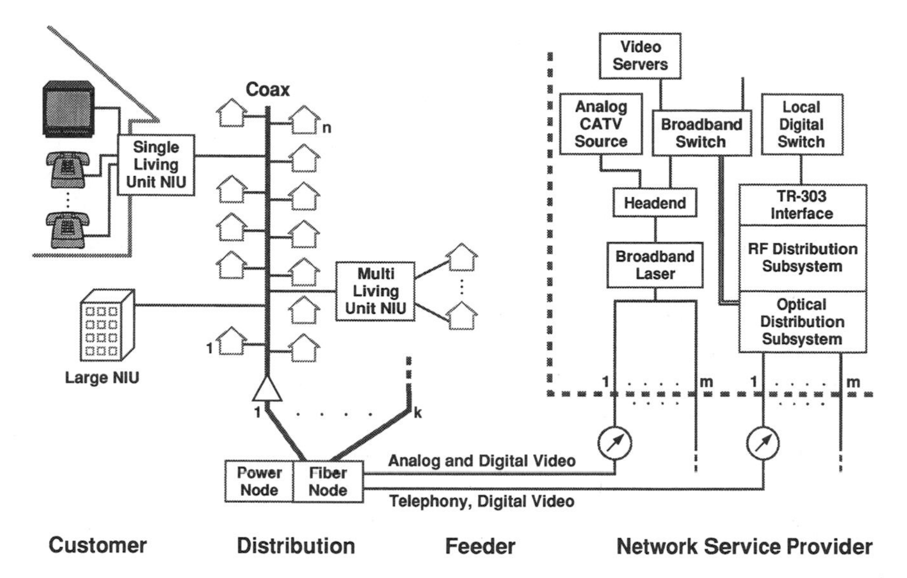

Fig. 14.5 Hybrid fiber coax (HFC) broadband access system. Broadband analog video, broadband digital video, and switched voice and data services are delivered by analog lightwave to fiber nodes. Fiber nodes convert analog optical signals to electrical and drive coaxial distribution systems that serve between 200 and 2000 subscribers. Various network interface units (NIUs) provide the required digital-to-RF conversion.

80-channel example system. Directly modulated DFB lasers, usually at 1.3 μm, are generally used for these links. One of these lasers can support multiple (typically four) FNs using passive optical splitters.

The third application is the delivery of narrow band digital information between the head end and the FNs. This includes telephony or switched digital services. These are generally converted to RF channels using modems at the head end or subscriber and transported as RF through the fiber and coax. Separate lasers are often used for these services, as shown in Fig. 14.5, so that spare lasers can be provisioned in the event of failures. Robust modem techniques like QPSK are used to ensure immunity to noise, particularly in the upstream band, so that a high CNR (compared with AM-VSB) generally is not required. These narrowcast lasers can be relatively inexpensive, because the requirements can be met by low-performance DFB lasers or in some cases even Fabry–Perot lasers. Low cost is critical because many more of these lasers are required than are required for AM-VSB delivery.

2.3 LOOP FIBER SYSTEMS

HFC provides a low-cost medium for combined analog video and narrow band digital services, but problems with ingress noise and cable deterioration lead many to prefer more fiber-intensive alternatives. Various systems that deploy FTTC or fiber to the home (FTTH) are attractive, including PONs and several FTTC systems in which a curbside switch serves multiple subscribers. In all cases, one challenge is to deliver the broadcast spectrum of multiple analog video channels to the optical network unit (ONU) at the curb or home.

Figure 14.6 shows an FTTC system that uses a point-to-point fiber feeder, and a point-to-multipoint PON, in which a fiber feeder is shared by multiple ONUs. Either system can be FTTC or FTTH, if feasible economically. The difficulty for analog lightwave in these applications is that each optical receiver serves a small number of subscribers (1–24), so that the cost of analog lightwave components is not shared by as many subscribers as with HFC (typically 500). This constraint is offset somewhat by the reduced performance requirements, as described in our 50-channel example system.

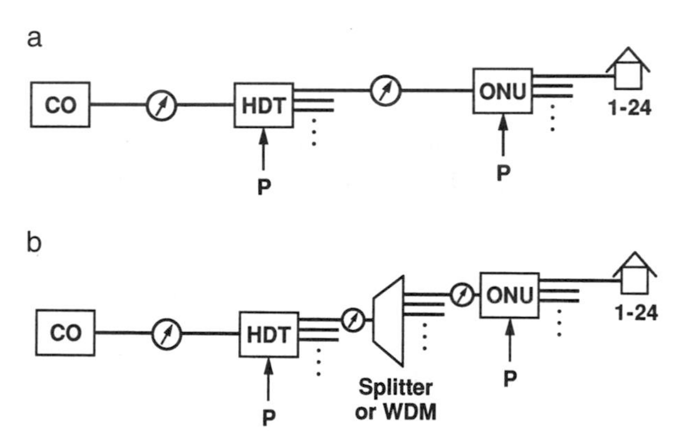

Fig. 14.6 Typical (a) point-to-point and (b) point-to-multipoint subscriber loop access systems. *P* represents points in the network at which power must be supplied (in addition to the central office [CO]). Each optical network unit (ONU) connected to the host digital terminal (HDT) can serve multiple subscribers in a fiber-to-the-curb configuration, or just one subscriber, for fiber to the home. The point-to-multipoint system uses a fiber splitter or wavelength-division multiplexer (WDM).

3. Analog Lightwave Technology

In this section we describe the basic analog link, which consists of a transmitter and a receiver. Receivers are discussed in Section 3.1, from the perspective of the required CNR performance. In Section 3.2 the various options for transmitters are discussed, considering linearity, noise, and spectral characteristics. Optical amplifiers are discussed in Section 3.3. Impairments introduced by transmission through fiber are discussed in Section 3.4, and overall system design is considered in Section 3.5.

3.1 RECEIVERS

The standard optical receiver is a photodiode that converts optical power into electrical current. Although carefully designed PIN photodiodes are inherently linear at the required optical power levels, the coupling of the detector to a preamplifier is a balance between noise and distortion. Noise from the preamplifier combines with shot noise and RIN (from the transmitter and fiber) to limit the link performance. The CNR at the receiver is given by

$$\mathrm{CNR} = \frac{\frac{1}{2}m^2 I_o^2}{B_e(2eI_o + n_{th}^2 + \mathrm{RIN}\ I_o^2)}. \tag{14.1}$$

The electrical power per channel is $\frac{1}{2}m^2 I_o^2$, where m is the OMD per channel, and I_o is the average received photocurrent. The thermal noise of the receiver is $B_e n_{th}^2$, where B_e is the noise bandwidth per channel (4 MHz for NTSC channels) and n_{th} is the thermal noise introduced by the receiver referred to the photocurrent. The noise current of a typical analog receiver falls in the range of 5–12 pA/$\sqrt{\mathrm{Hz}}$. The shot noise is given by $2B_e eI_o$, where e is the electron charge. The last noise term in the denominator is intensity noise, $B_e\mathrm{RIN}\ I_o^2$, where RIN is the relative intensity noise (dB/Hz). The CNR due to this term is independent of the received photocurrent. The noise and quantum efficiency of the receiver determine the necessary received optical power to satisfy the CNR specification. For example, to achieve a CNR of 49 dB for an NTSC channel with 5.0% OMD in the presence of receiver shot noise and thermal noise (8 pA/$\sqrt{\mathrm{Hz}}$), the average photocurrent must be 0.17 mA (0.68 mA for 3.5% OMD and a CNR $\geq$ 54 dB). The received optical power is related to the photocurrent by

$$P_o = \frac{h\nu}{\eta e} I_o, \tag{14.2}$$

where $h\nu$ is the photon energy, and η is the quantum efficiency of the photodiode. The required optical power for 0.17-mA photocurrent is 0.22 mW (1.3 μm) or 0.19 mW (1.55 μm). (A quantum efficiency of 75% is assumed.) The large required received power limits the optical power budget of an AM-VSB lightwave system.

Reducing the receiver noise can improve the link performance, especially in systems where the received photocurrent is low (implying a lower CNR). For higher received powers the shot noise and RIN generally dominate. Receiver noise can be reduced by improving the impedance matching between the photodiode, which is an infinite-impedance current source, and the low-noise amplifier. Building the appropriate impedance-matching network can improve the magnitude of the signal appearing at the amplifier input, relative to the noise, but these high-impedance receivers have two limitations. First, the bandwidth is limited by the input RC time constant, where C is the total input capacitance of the detector and amplifier and R is the input impedance. Second, increasing the input impedance increases the magnitude of the signals that must be amplified with high linearity by the preamplifier. Hence, input impedances greater than about 300 Ω may encounter linearity limitations.

3.2 *TRANSMITTERS*

The transmitter of analog lightwave systems is the most challenging component to realize. It must provide high average power with low noise and have an extremely linear light–electrical input characteristic. As shown in the previous section, a CNR of 49 dB requires about −6.5 dBm (0.22 mW) of optical power at the photodiode, just considering shot noise and thermal noise (8 pA/$\sqrt{\text{Hz}}$) of the receiver, and the CNR due to all other noise sources is assumed to be 51 dB. If we allow for a 10-dB link loss due to coupling loss and fiber propagation, the transmitter–amplifier fiber-coupled power per receiver must be at least 3.5 dBm (2.2 mW). The challenge has been to achieve high transmitter power and high linearity for low cost.

Transmitters fall into two categories: direct and external modulation. Either can be used in either the 1.3- or the 1.55-μm window of low-loss fiber transmission. Motivation for selecting various options is discussed in detail throughout this chapter. Direct modulation takes advantage of the intrinsically linear current-intensity transfer characteristic of diode lasers to provide simple, compact, and low-cost transmitters. Externally modulated transmitters offer extremely good performance with high power and low chirp, as may be required for use with optical amplifiers or specialty (i.e., head-end consolidation) applications. We discuss both types of transmitters within the context of linearity, noise, and spectral properties. The spectral properties determine which types are suitable for certain applications.

3.2.1 Linearity

The transmitter light versus electrical input characteristic must be extremely linear because of the fragile nature of the AM-VSB video signal. This fragility is due partially to the large number of second- and third-order distortion products that are generated within the multichannel multioctave NTSC band. For example, a 50-channel standard NTSC frequency plan (55.25, 61.25, 67.25, 77.25, 83.25, 109.25, 115.25, . . . , 373.25 MHz) has a maximum CSO distortion product count of 39 occurring at 54.0 MHz. The maximum third-order count is 786 and occurs at 229.25 MHz. For a simple (frequency-independent or memoryless) nonlinear characteristic, the optical power can be expressed as a Taylor-series expansion of the electrical signal,

$$P \propto X_o(1 + x + ax^2 + bx^3 + \cdots), \tag{14.3}$$

where X_o is the DC bias (voltage or current) and x is the modulation signal:

$$x(t) = \sum_{i}^{N_{ch}} m_i(t) \cos(2\pi f_i t + \phi_i), \tag{14.4}$$

where $m_i(t)$ is the normalized modulation signal for channel i, f_i is the subcarrier frequency, and N_{ch} is the number of channels. For a system evaluation using a test signal consisting of multiple unmodulated video carriers, the AM channels are simulated by tones of certain modulation depth—i.e., $m_i(t) = m$. The CSO distortion and CTB distortion for a characteristic described by Eq. (14.3) are given by

$$\mathrm{CSO} = 10 \log[N_{\mathrm{CSO}}(am)^2] \tag{14.5}$$

$$\mathrm{CTB} = 10 \log[N_{\mathrm{CTB}}(\tfrac{3}{2}bm^2)^2], \tag{14.6}$$

where N_{CSO} is the second-order product count, and N_{CTB} is the third-order product count. The product counts for several NTSC channel loads are given in Tables 14.1 and 14.2. For a 50-channel NTSC system, in order to achieve a CSO distortion of -59 dB or less, the coefficient $a \leq 3.6 \times 10^{-3}$, and for a CTB of -59 dB or less, the coefficient $b \leq 1.07 \times 10^{-2}$. (For an

Table 14.1 **Two-Tone Product Count, N_{CSO}**

		Channel Load			
		42	**50**	**60**	**80**
		N_{CSO}			
Ch 2 (max.)	55.25 MHz	31	39	49	69
Ch 11	139.25	7	15	25	45
Ch 40	313.25	12	12	12	25
Ch 50	373.25		16	16	16
Ch 60	433.25			21	21
Ch 80	553.25				31

80-channel community-antenna television [CATV] system, $a \leq 2.43 \times 10^{-3}$ and $b \leq 4.65 \times 10^{-3}$.)

With the listed product counts in Tables 14.1 and 14.2, and Eqs. (14.5) and (14.6), we can determine the trade-off between channel load and the modulation depth per channel. For example, if an 80-channel transmitter, which is limited by the CTB described by Eq. (14.6), is operated as a 50-channel transmitter, the modulation depth per channel can be increased by 30%, or 2.2 dBe. The increase in the OMD improves the CNR by 2.2 dB.

Various nonlinear processes limit the usable modulation depth. Linearity of directly modulated lasers has been the subject of many investigations. Distortion is generated by the nonlinear coupling of gain (carrier density)

Table 14.2 **Three-Tone Product Count, N_{CTB}**

		Channel Load			
		42	**50**	**60**	**80**
		N_{CTB}			
Ch 3	61.25 MHz	289	438	669	1282
Ch 11	139.25	530	771	1117	1960
Ch 40	313.25	377	681	1127	2170
Ch 50	373.25		527	1001	2134
Ch 60	433.25			783	1998
Ch 80	553.25				1446
Maximum count		531	786	1172	2170

and optical power (photon density) within the active layer of semiconductor lasers. This resonance distortion is small at low frequencies but large as modulation frequencies approach half the laser relaxation resonance frequency [12, 13]. This distortion cannot be described by the simple memoryless polynomial model described in Eq. (14.3). For high-power lasers where the resonance frequency is near 10 GHz, this distortion is not a problem. In the analog video band, other distortion mechanisms are generally more important. These include longitudinal spatial hole burning [14, 15], nonlinear gain [16], and nonlinear current leakage [17].

For external modulation, the voltage-transmission (V-T) characteristic of the modulator must be linear. Yet Mach–Zehnder modulators have sinusoidal transfer functions [18]. By properly biasing the modulator to a point of inflection in the V-T curve, one can eliminate second-order distortion. A variety of techniques, including feed-forward and predistortion linearization [19, 20], are then used to minimize third-order distortion.

Even for transmitters with "perfect" light versus electrical input characteristics, there is a clipping limit [21, 22]. If the electrical signal at any instant drives the laser below the threshold current, the output optical power "clips," and creates broadband distortion. The statistics of adding numerous uncorrelated signals allows the transmitter to be overdriven (i.e., $N_{ch}m > 1$). The distortion from clipping is largely a function of the root-mean square (RMS) modulation depth of the total signal, $\mu = \sqrt{\frac{1}{2}N_{ch}m^2}$. As seen in Fig. 14.7, the distortion (CSO, CTB, and higher order) rises rapidly around $\mu = 0.25$. This is the origin of the 5.0% modulation depth for 50 channels that is used throughout this chapter. If a less than perfect transmitter is used, the modulation depth must be reduced in order to meet the distortion requirements.

3.2.2 Noise

Any optical transmitter will transmit intensity and phase noise in addition to the intended signal. First we consider the intensity noise that degrades the SNR directly. For an IM/DD system, only the intensity noise falling within the frequency band of the channels, in this case 50–400 MHz, is important. The CNR arising from the RIN of a transmitter is given by

$$\mathrm{CNR_{RIN}} = \frac{m^2}{2B_e\mathrm{RIN}}. \tag{14.7}$$

For a 5.0% OMD and a $\mathrm{CNR_{RIN}}$ of 50 dB or more, the RIN must be less than −145 dB/Hz over the entire modulation bandwidth (for a 3.5% OMD

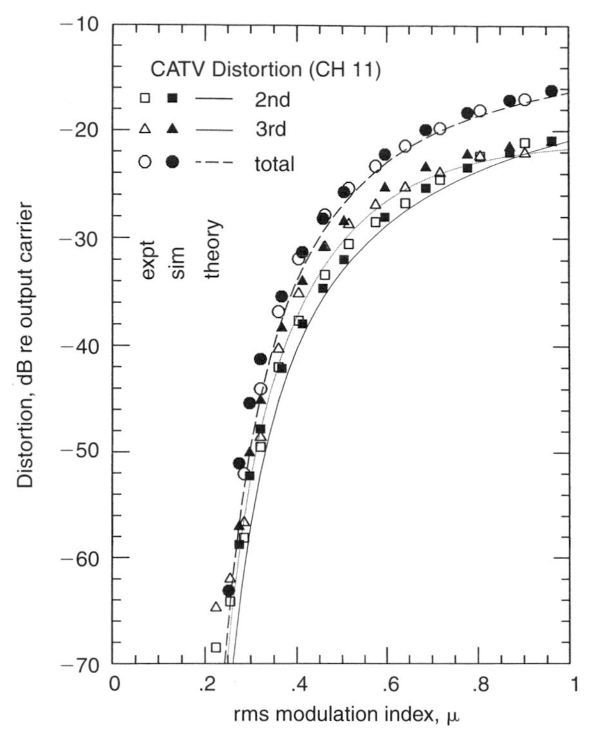

Fig. 14.7 Distortion from clipping. Mu (μ) is the root-mean square (rms) modulation depth, $m\sqrt{N_{ch}/2}$.

and a CNR of 55 dB or more, RIN is less than or equal to −153 dB/Hz). In a real system, the RIN requirement may be more stringent because other noise sources are a larger fraction of the noise budget. Also, we discuss later how fiber reflections or double Rayleigh backscatter adds intensity noise, which can be more problematic than laser RIN.

3.2.3 Spectral Characteristics

Next, we consider the spectral characteristics of the transmitter. These are important because linear and nonlinear effects of the transmission modify the optical field spectrum and can result in intensity distortion or noise at the output.

The optical field spectrum of the transmitter is determined by the FM efficiency (lasers); chirp, or α parameter (lasers and modulators); and phase noise. The term *chirp* is also used to refer to FM efficiency in lasers. The two extremes of transmitter spectra are a chirp-dominated spectrum of a directly modulated semiconductor laser, and a narrow spectrum of an externally modulated transmitter, which is dominated by the continuous-wave (CW) source. The spectral regime can be determined approximately by considering the spectrum of the optical field,

$$E(t) = A\sqrt{1 + x(t)}\,\exp\left[i\nu_o t + i\phi_n(t) + i2\pi\beta_{\mathrm{FM}} I_b \int^t x(\tau)d\tau + i\frac{\alpha}{2}x(t)\right], \tag{14.8}$$

where ν_o is the optical carrier frequency, ϕ_n is the phase noise, β_{FM} is the FM efficiency (MHz/mA), I_b is the bias current level above threshold, and α is the linewidth enhancement factor. For transmitters with significant FM efficiency (e.g., directly modulated semiconductor lasers), the optical spectrum is dominated by the FM efficiency. The time-average spectrum when the transmitter is driven by multiple subcarriers can be approximated by a Gaussian with standard deviation

$$\sigma_\nu = m\beta_{\mathrm{FM}} I_b \sqrt{\frac{N_{ch}}{2}}. \tag{14.9}$$

The standard deviation of a Gaussian is related to the half width at half maximum (HWHM) linewidth by

$$\mathrm{HWHM} = \sqrt{2\ln 2}\,\sigma_\nu \approx 1.18\,\sigma_\nu. \tag{14.10}$$

For bulk DFB lasers, the FM efficiency is 200–600 MHz/mA [23], and the HWHM is 2–9 GHz. For quantum well (QW) lasers, the FM efficiency is 50–200 MHz/mA [23, 24], which results in a spectral halfwidth of 0.5–3.0 GHz. For externally modulated transmitters, the spectral width is a combination of the CW source linewidth and the modulated intensity and chirp spectrum. The frequency deviation due to the α factor is

$$\sigma_\nu = \left[\sum_i^{N_{ch}} \frac{1}{2}\left(\frac{m\alpha f_i}{2}\right)^2\right]^{1/2}, \tag{14.11}$$

where f_i is the subcarrier frequency. The α factor is close to 0 for a Mach–Zehnder interferometer, and 1–2 for an electroabsorption modulator [25]. For an α factor of 2 and 50 NTSC channels of 5% OMD each, the standard deviation, σ_ν, is 60 MHz. Even for this relatively large α factor, the chirp-generated bandwidth is narrow. For interferometric–modulators in which the α factor is close to 0, most of the optical power is in the optical carrier. For example, an intensity-modulated signal with an RMS modulation depth of 25% has 98.4% of its optical power in the CW carrier, so the bandwidth is well approximated by that of the CW source. In Section 3.4, we consider the effects of transport over fiber and their dependence on the spectrum of the signal.

3.3 OPTICAL AMPLIFIERS

One way to increase the loss budget of the lightwave system is to deploy optical amplifiers. With enough additional power, the analog signal can be split among multiple receivers, as required in the PON discussed previously. Also, for head-end consolidation or supertrunk applications, additional power may be required to reach distant nodes. In this section, we consider the noise and distortion associated with several types of optical amplifiers.

Distortion can arise from several mechanisms in an optical amplifier. One is gain saturation, which can lead to intermodulation distortion through signal-induced gain modulation. The gain modulation is determined by [26]

$$G/G_o = 1 - \frac{P_{out}}{P_{sat}} \sum_{i}^{N_{ch}} m_i \frac{\cos(\omega_i t + \phi_i)}{\sqrt{1 + (2\pi f_i \tau_c)^2}}, \tag{14.12}$$

where τ_c is the lifetime of the upper state, P_{out} is the average output power, and P_{sat} is the amplifier output saturation power. The modulated gain of Eq. (14.12) mixes with the input signal and generates second-order distortion. The gain modulation and subsequent intermodulation distortion decrease as the upper state lifetime increases. An upper bound on the CSO distortion is determined by setting f_i in Eq. (14.12) to that of the lowest subcarrier, $ch1$.

$$\text{CSO} \leq 20 \log \sqrt{N_{\text{CSO}}} \frac{mP_{out}}{P_{sat}} \frac{1}{\sqrt{1 + (2\pi f_{ch1} \tau_c)^2}}. \tag{14.13}$$

For a semiconductor optical amplifier, τ_c is on the order of 300 ps, and $2\pi f_i \tau_c$ ranges from 0.09 to 0.75 over the 50-channel NTSC band. If $P_{out} = 0.1P_{sat}$, the induced CSO distortion for 50 channels of 5% OMD is −30 dBc. This is 30 dB worse than the requirement for AM-VSB video signals. EDFAs, on the other hand, have an upper state lifetime of about 10 ms [27]. At the lowest frequency channel, $2\pi f \tau_c = 3.1 \times 10^6$. When the EDFA operates at $P_{out} = P_{sat}$, the CSO distortion for the same channel load is less than −140 dB. This property of the EDFA makes it a good candidate for amplification of analog signals. Other optical amplifier options include Raman amplification and Brillouin amplification. The bandwidth of Brillouin gain is 20 MHz, which is too narrow for the analog signal, but the bandwidth of Raman gain is several terahertz. The properties of EDFAs, rare-earth-doped fiber amplifiers (at 1.3 μm), and Raman fiber amplifiers are discussed further next.

3.3.1 Erbium-Doped Fiber Amplifiers

EDFAs and co-doped erbium-ytterbium doped fiber amplifiers are available commercially with gains of 30–40 dB and 25 dBm of output saturation power. Although optical gain can overcome the limitations of receiver noise, the noise contributed by the amplifier must be considered [28]. In the limit of high gain, the optical noise power per optical mode (standard single-mode fiber supports two polarization modes) from the amplified spontaneous emission (ASE) of an EDFA is given by

$$P_{ASE} = h\nu G n_{sp} B_o, \tag{14.14}$$

where G is the gain of the amplifier and B_o is the optical bandwidth (about 25 nm with no optical filtering). The spontaneous emission factor is n_{sp}, which can be calculated from the noise figure, F, of the amplifier, $n_{sp} = F/2$. The ASE noise is manifest as intensity noise at the receiver in the form of signal–spontaneous (sig–sp) beat noise and spontaneous–spontaneous (sp–sp) beat noise. Consider a system with a cascade of amplifiers and losses as shown in Fig. 14.8. The interamplifier loss includes transmission and splitting loss. The beat noise power at the receiver derived from Ref. 27 is

$$i^2_{\text{sig-sp}} = 4 I_o I_{ASE} \frac{B_e}{B_o} \tag{14.15}$$

and

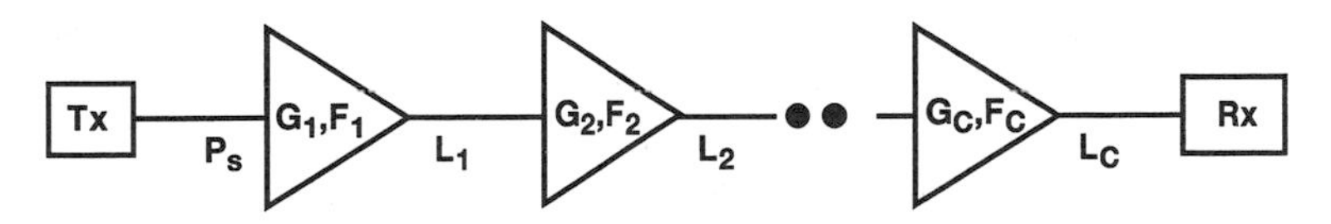

Fig. 14.8 Amplifier cascade. F_i is the noise figure of the ith amplifier, G_i is the optical gain, L_i is the postamplifier loss (splitting and transmission), and P_s is the average optical power at the input of the first amplifier.

$$i^2_{\text{sp-sp}} = M I^2_{ASE} \frac{B_e}{B_o^2}(2B_o - 2f - B_e), \tag{14.16}$$

where I_o is the average signal photocurrent, I_{ASE} is the average ASE photocurrent, B_e is the electrical bandwidth (4 MHz for NTSC video), M is the number of optical modes, and f is the subcarrier frequency. The signal and ASE photocurrents referred to the receiver are

$$I_{ASE} = \frac{\eta e B_o}{2} \sum_{i=1}^{C} \left(F_i \prod_{j=i}^{C} G_j L_j \right), \tag{14.17}$$

and

$$I_o = P_s \frac{\eta e}{h\nu} \prod_{i=1}^{C} G_i L_i, \tag{14.18}$$

where η is the receiver efficiency; F_i, G_i, and L_i are the noise figure, gain, and postamplifier loss of the ith amplifier; C is the number of cascaded amplifiers; and P_s is the average optical power at the input of the first amplifier. The amplifier beat noise is included in the CNR calculation:

$$\text{CNR} = \frac{m^2 I_o^2}{2B_e} \left[\text{RIN } I_o^2 + 2eI_o + 4\frac{I_{ASE}}{B_o} I_o + \frac{2(I_{ASE})^2}{B_o}\left(2 - 2\frac{f}{B_o} - \frac{B_e}{B_o}\right) + n_{th}^2 \right]^{-1}. \tag{14.19}$$

For standard NTSC channels, f, $B_e \ll B_o$, and those terms can be neglected. For amplifier input powers necessary to achieve the required CNR of AM-VSB systems, the sp–sp beat noise is negligible compared with the sig–sp beat noise. For example, in a single high-gain amplifier with a noise figure $F = 5$ dB, the sig–sp beat noise—which is proportional to the input power—is equal to the sp–sp beat noise when the amplifier input power is −32 dBm. Compare this with the expected input power of at least −5 dBm. This also justifies the omission in Eq. (14.19) of shot noise due to the ASE-induced photocurrent. For a large enough receiver photocurrent, RIN and sig–sp beat noise are the dominant sources of noise. For an analog signal with a 5% OMD, the sig–sp noise CNR is 50 dB for a 5-dB noise figure amplifier when the input power is 0.26 mW (−5.8 dBm). (For a 3.5% OMD and a CNR of 55 dB or more, the input power must be 1.7 mW [2.3 dBm].) As expected, this input power is large compared with that necessary for digital transmission systems, but it is achievable with present-day transmitters [29, 30].

3.3.2 Gain Tilt

Gain tilt in the optical amplifier or any wavelength-dependent loss in the system can cause distortion of the analog signal. The worst effect is second-order distortion, which arises from the conversion of incidental frequency modulation (caused by chirp in the transmitter) to intensity modulation (IM). The CSO distortion from the gain tilt and laser chirp is derived by Kuo and Bergmann [31]:

$$\mathrm{CSO} = 20 \log \sqrt{N_{\mathrm{CSO}}} \frac{1}{G} \frac{\partial G}{\partial \nu} m I_b \beta_{\mathrm{FM}}, \tag{14.20}$$

where $G^{-1}\partial G/\partial \nu$ is the power-transmission tilt per optical hertz, β_{FM} is the FM efficiency of the transmitter, and I_b is the bias current above threshold. For 50 channels, a 5% OMD, the product of the gain tilt, FM efficiency, and bias current must be less than 0.45 GHz/nm for a CSO of −59 dB or less. For $\beta_{\mathrm{FM}} I_b = 12.5$ GHz (a typical bulk semiconductor laser), the maximum allowed gain–loss tilt of the system is 0.16 dB/nm. (For an 80-channel CATV system, it is 0.10 dB/nm.) For a directly modulated low-chirp laser (e.g., some QW lasers), $\beta_{\mathrm{FM}} I_b = 2.5$ GHz, and the gain tilt is limited to 0.73 dB/nm for the 50-channel FTTH system (0.50 dB/nm for an 80-channel CATV system).

3.3.3 Rare-Earth-Doped Fluoride Fiber Amplifiers

Rare-earth-doped fluoride fiber amplifiers such as praseodymium-doped fluoride fiber amplifiers (PDFAs) are being developed for amplification of 1.3-μm analog signals [32–34]. The upper state lifetime of this fiber is on the order of 110 μs [32]. To limit the CSO distortion due to gain saturation effects to −59 dBc for the 50-channel FTTH system, the average output power is limited to about 0.70 of the saturated output power. Other issues for these amplifiers are the noise figure, which seems to be higher than that of an EDFA, and gain flatness. Also, the loss and reflection at splices between the fluoride fiber and the silica fiber can degrade the overall performance of the amplifier.

3.3.4 Raman Fiber Amplifiers

A Raman fiber amplifier is another candidate for amplification of the analog signal. Recent advances in laser sources at the appropriate wavelengths make this more promising than in the past. Raman amplification occurs from resonant coupling of pump photons and signal photons through optical phonons [35]. The Raman gain coefficient is shown in Fig. 14.9 as a function of the frequency separation between the pump and the signal. The gain is

$$G = \exp\left(\frac{\gamma_R(\Delta\nu)P_p L_{eff}}{A_{eff}}\right), \tag{14.21}$$

where $\gamma_R(\Delta\nu)$ is the Raman gain coefficient as illustrated in Fig. 14.9, P_p is the pump power, L_{eff} is the effective length of the fiber at the pump wavelength, and A_{eff} is the mode effective area. The bandwidth of the Raman gain at, for example, a 20-dB peak gain is 3.6 THz, which is more than adequate for an analog signal. For a gain of about 10 dB for a 10-km effective length of standard fiber, the pump power must be about 250 mW. In order to avoid saturation of the propagated pump by stimulated Brillouin scattering (SBS), the effective bandwidth of the pump must be much broader than 20 MHz (see Section 3.4.2.1).

The inherent noise from the Raman amplification process is equivalent to a 3-dB noise figure [36]. Because the scattering is a fast process, however, intensity noise of the pump is impressed on the signal. This effect is mitigated by the fact that the Raman fiber amplifier is distributed along the fiber, and group-velocity mismatch between the pump and signal fields results in an averaging effect that attenuates high-frequency noise. In the

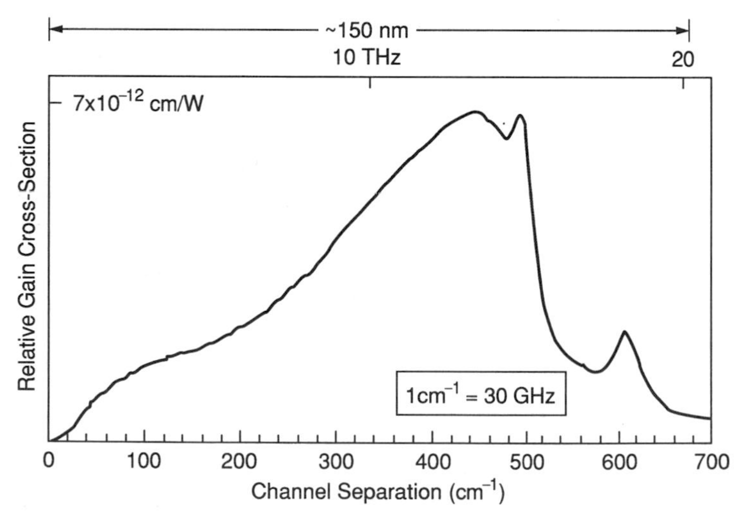

Fig. 14.9 Raman gain in fused silica fibers [23].

case of reverse-direction pumping, the effective averaging is over a longer time interval and the intensity noise is reduced more.

Gain saturation in a Raman amplifier occurs when the pump power is depleted by amplifying the signal. Because there is essentially no upper state lifetime to mitigate the effect of gain saturation, as per Eq. (14.12), the second-order distortion can be significant. Like the case of the pump intensity noise, however, the CSO distortion is less than the prediction of Eq. (14.13) because of pump–signal walk off, which causes a pump averaging effect as seen by the signal [37]. Forghieri, Tkach, and Chraplyvy [37] analyzed a 30-km Raman amplifier with two wavelength-division multiplexed input signals for pump and signal cross talk. At a 100-MHz modulation frequency in standard fiber, the cross-talk-induced OMD was 18 dB less than that predicted using the simple formula of Eq. (14.12). If we assume that the induced CSO distortion is adjusted by the same factor, the Raman amplifier output power for a 50-channel NTSC system with CSO distortion less than −59 dBc must be below 25% of the saturated output level (for an 80-channel CATV system the output power is limited to 15% of saturation). The saturated output power is approximately equal to the pump power. If the Raman amplifier is pumped in the reverse direction, however, the cross talk is greatly diminished; at 100 MHz it is 37 dB less than the forward-pump case. The CSO distortion due to gain saturation in the reverse-pump case is less than −100 dB and can be neglected. A logistical problem with reverse pumping is that the pump source must be located remotely from the signal source.

If the wavelength separation between the signal and the pump is not centered on the peak of the Raman gain (13.2 THz), distortion from gain-tilt effects can occur. For example, if the pump and the signal are separated by 10 THz and the amplifier gain is 20 dB, the gain tilt is 0.34 dB/nm. For pump–signal separation of 15.3 THz, which is greater than the separation at the peak Raman gain, the gain tilt for 20 dB of total gain is 2.7 dB/nm. These gain-tilt values will cause unacceptable CSO distortion if a directly modulated laser is used as the transmitter. If the Raman amplifier has less total gain, the gain tilt is reduced. For example, at 10 dB of gain, the gain tilt for 10-THz separation is 0.17 dB/nm and for 15.3-THz separation is 1.55 dB/nm. Clearly, if directly modulated lasers are used in conjunction with Raman fiber amplifiers with appreciable gain, the frequency separation between the pump and the signal must be limited to within a range of 1.2 THz (9.6 nm at 1.55 μm). For externally modulated sources, the distortion caused by gain tilt is negligible.

3.4 FIBER TRANSPORT

Even though the transmission distance is typically less than 20 km, linear and nonlinear effects of fiber transport can violate the stringent noise and distortion requirements of the AM-VSB video signals. The received intensity spectrum can be modified by changes in the phase or amplitude of the optical field spectrum. In the analyses that follow, the transport fiber is assumed to be standard single-mode fiber unless otherwise noted: The effective mode area is 80 μm^2, and the dispersion is zero at 1.3 μm and 17 ps/nm-km at 1.55 μm. The assumed loss coefficient is 0.35 dB/km at 1.3 μm and 0.22 dB/km at 1.55 μm.

3.4.1 Linear Effects

3.4.1.1 Multipath Interference

The first linear effect that we consider is multipath interference (MPI). Reflections in the optical path can lead to interference between the signal and a doubly reflected version of itself arriving at the receiver. In addition to discrete reflections at splices and so forth, inherent microreflections occur due to Rayleigh backscatter in the fiber [38]. An expression for the equivalent reflectivity due to double Rayleigh scatter is

$$R_{eq}(z) = \frac{S\alpha_s}{2\alpha}\sqrt{2\alpha z - 1 + e^{-2\alpha z}}, \tag{14.22}$$

where S is the fraction of scattered light captured by the fiber ($\approx 1.5 \times 10^{-3}$), α is the total loss, α_s is the loss due to Rayleigh scattering, and z is the fiber length. For 20 km of standard fiber, R_{eq} is approximately -29.5 dB for 0.35 dB/km (1.3 μm), and -30.9 dB for 0.22 dB/km (1.55 μm). This is the inherent R_{eq} for a length of fiber. The reflectivity from nonideal splices and other optical components must also be added to determine the equivalent reflectivity of the system. If there is optical gain between two reflections, the equivalent reflectivity increases linearly with the optical power gain.

For reflections spaced by a distance greater than the coherence length of the transmitter, there is incoherent beating of the field with itself. For a directly modulated laser in which the optical spectrum is approximately Gaussian because of the dominance of the FM laser chirp (see Section 3.2.3), the MPI-induced relative intensity noise is [39]

$$\mathrm{RIN} = \frac{2pR_{eq}^2}{\sqrt{\pi}\sigma_\nu}\exp\left(-\frac{f^2}{4\sigma_\nu^2}\right), \tag{14.23}$$

where p is the square of the polarization overlap of the interfering fields (one-half is the average for a long length of non-polarization-maintaining fiber). If the signal spectrum is Lorentzian, which is a good approximation for most transmitters in which a typical CW single-wavelength source is externally modulated, the RIN is as follows [38]:

$$\mathrm{RIN} = \frac{8\Delta\nu}{2\pi(f^2 + \Delta\nu^2)}pR_{eq}^2, \tag{14.24}$$

where $\Delta\nu$ is the full width at half maximum (FWHM) linewidth of the source. Example RIN spectra for various linewidths are shown in Fig. 14.10. For example, in order to keep the RIN in the channel band below -145 dB/Hz (CNR = 50 dB for a 5.0% OMD) for 20-km transmission at 1.3 μm, the FWHM of a Lorentzian source must be less than 10 MHz or greater than 250 MHz. For a Gaussian spectrum, the standard deviation σ_ν must be greater than 240 MHz. These spectral targets are easily achievable with standard sources. (For an 80-channel CATV system,

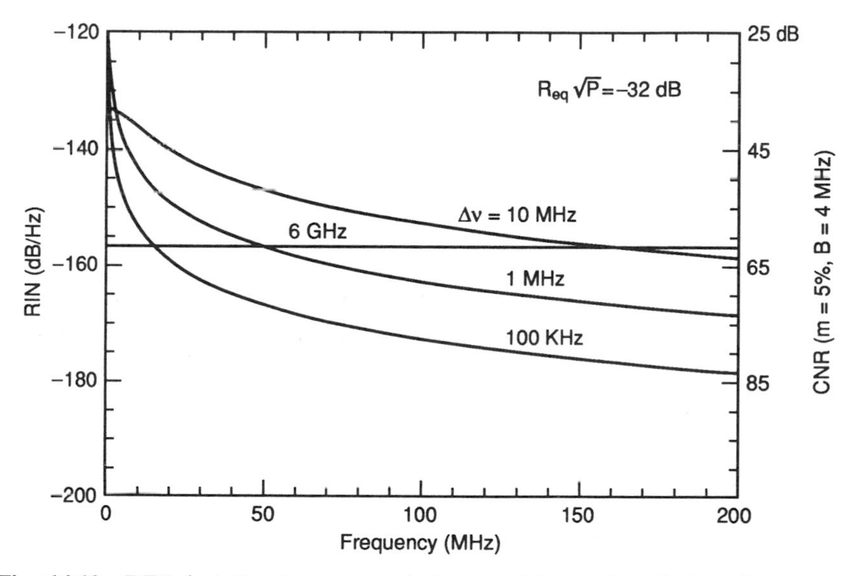

Fig. 14.10 RIN (relative-intensity noise) caused by multipath interference for various spectral linewidths. The equivalent reflectivity, -32 dB, is approximately the equivalent of double Rayleigh backscatter in 20 km of standard fiber.

the FWHM of a Lorentzian source must be less than 1.6 MHz, or the standard deviation of a Gaussian source must be greater than 1.4 GHz.) If the optical bandwidth of the modulated source falls within the range of unacceptable values, measures such as artificial spectral broadening can reduce the in-band MPI noise [40].

3.4.1.2 Dispersion

A second linear effect in fiber transmission is dispersion, both chromatic and polarization mode. The most debilitating effect occurs when dispersion is coupled with a highly chirped transmitter, and the incidental frequency modulation is converted to IM. The mixing of the induced IM with the original IM leads to intermodulation distortion.

A theoretical and experimental study of the effect of chromatic dispersion on intermodulation distortion has been done [41]. An analytic expression for the CSO is given by

$$\mathrm{CSO} = 10\log\left\{(m\ddot{\beta}z\Omega_d)^2\left[\gamma^2 N_{\mathrm{CSO}} + \left(\frac{\ddot{\beta}z\Omega_d}{8}\right)^2 \sum_{i,j:\Omega_d=\Omega_i\pm\Omega_j}^{N_{ch}} (\Omega_i\Omega_j)^2\right]\right\}, \tag{14.25}$$

where

$$\ddot{\beta} = \frac{-\lambda^2}{2\pi c}D. \tag{14.26}$$

In these equations, D is the dispersion coefficient (ps/nm-km), Ω_d is the frequency (rad/s) of the intermodulation distortion, z is the fiber length, and γ is the chirp parameter $\gamma = 2\pi\beta_{\mathrm{FM}}(I_b - I_{th})$, where β_{FM} is the FM efficiency (MHz/mA) as described in Section 3.2.3. The second term inside the brackets of Eq. (14.25) is the CSO distortion attributed to dispersion of an intensity-modulated signal with no chirp. It is not a simple function of the product count and the distortion frequency, but because of the strong dependence of the CSO distortion on the distortion frequency, the worst CSO distortion will occur at the highest channel and an upper limit can be determined:

$$\mathrm{CSO(at\ highest\ channel)} \le 10\log\left\{N_{\mathrm{CSO}}(m\ddot{\beta}z\Omega_d)^2\left[\gamma^2 + \left(\frac{\ddot{\beta}z\Omega_d^3}{32}\right)^2\right]\right\}. \tag{14.27}$$

To keep the CSO distortion at less than -59 dBc for a 50-channel system, the product $\gamma Dz \leq 2.99 \times 10^2$ GHz-ps/nm. For example, a directly modulated bulk laser with $\beta_{\mathrm{FM}} I_b = 12.5$ GHz, Dz must be less than 23.9 ps/nm, whereas for a low-chirp laser (multiple QW, or MQW) with $\beta_{\mathrm{FM}} = 2.5$ GHz, Dz must be less than 120 ps/nm. These parameters correspond to 1.4 and 7.0 km of standard fiber at 1.55 μm. The limit for an externally modulated source with negligible chirp is $Dz < 6.05 \times 10^4$ ps/nm, which is over 3500 km at $D = 17$ ps/nm-km. (The limit for an 80-channel CATV system is $Dz \leq 10.4$ ps/nm for a directly modulated bulk laser, and $Dz \leq 52.2$ ps/nm for an MQW laser.) To transmit over an appreciable distance using a directly modulated bulk DFB laser, therefore, the wavelength must be near the dispersion zero or the dispersion must be compensated for. Electronic dispersion compensation has been demonstrated in which second-order distortion of the appropriate phase is added to the signal to cancel that generated by the dispersion. Optical dispersion compensation has also been demonstrated and dispersion-compensating fibers have recently become commercially available.

An expression for the third-order distortion, CTB, is

$$\mathrm{CTB} \approx 20 \log\left[(m\ddot{\beta} z \Omega_d)^2 \left(\frac{3}{4}\gamma^2 \sqrt{N_{\mathrm{CTB}}} + \frac{1}{16}\sqrt{\sum_{i \neq j \neq k}^{\Omega_d = \Omega_i + \Omega_j \pm \Omega_k} (\Omega_i \Omega_j \pm \Omega_i \Omega_k \pm \Omega_j \Omega_k)^2}\right)\right]. \tag{14.28}$$

The expression includes only the nondegenerate mixing terms in the second-term sum. As in the case of CSO distortion, the largest distortion occurs at the highest frequency channel because of the strong dependence on the distortion frequency. At the highest channel, an upper limit for the CTB can be determined by

$$\mathrm{CTB(at\ highest\ channel)} \leq 20 \log[(m\ddot{\beta} z \Omega_d)^2 \sqrt{N_{\mathrm{CTB}}}(\tfrac{3}{4}\gamma^2 + \tfrac{1}{48}\Omega_d^2)]. \tag{14.29}$$

For multioctave frequency bands like the NTSC CATV band, the CSO distortion is much larger than the CTB. For $\gamma Dz = 2.99 \times 10^2$ GHz-ps/nm, which gives a CSO distortion of -59 dBc for the 50-channel system, the CTB is -117 dBc. The limitation on system performance is clearly determined by the CSO distortion.

If the fiber has significant *polarization mode dispersion* (*PMD*), distortion and increased intensity noise can occur [42, 43]. The distortion and noise fluctuate in time because of time-dependent polarization mode coupling. The average CSO distortion due to PMD interacting with transmitter chirp is [42]

$$\langle \text{CSO} \rangle = 10 \log\left[N_{\text{CSO}} \gamma^2 m^2 \left(\frac{\pi^2 \Omega_d^2 \langle \Delta\tau \rangle^4}{256} + \frac{\pi \Delta T^2 \langle \Delta\tau \rangle^2}{48} \right) \right], \quad (14.30)$$

where γ is the chirp, as before; $\langle \Delta\tau \rangle$ is the expected value of the PMD (ps); and ΔT is the polarization-dependent loss (PDL). A margin of at least 10 dB should be allowed between the system CSO distortion specification and the average CSO distortion given by Eq. (14.30). The first term in the parentheses describes CSO distortion due to a purely dispersive effect similar to CSO distortion caused by chromatic dispersion. The second term depends on the PDL. The presence of PDL converts polarization modulation into IM. To keep the average CSO distortion at less than -69 dBc for 50 channels in a system with no PDL, the product $\gamma\langle \Delta\tau \rangle^2 \leq 614$ GHz-ps^2 or $\langle \Delta\tau \rangle \leq 7.0$ ps for $\gamma = 12.5$ GHz (DM bulk) or $\langle \Delta\tau \rangle \leq 15.7$ ps for $\gamma = 2.5$ GHz (DM MQW). If there is a PDL of 2.0% in the system, the product $\gamma\langle \Delta\tau \rangle \leq 3.04$ GHz-ps, which means $\langle \Delta\tau \rangle \leq$ 0.24 ps for $\gamma = 12.5$, and $\langle \Delta\tau \rangle \leq 1.22$ ps for $\gamma = 2.5$ GHz. Clearly, even a small amount of PDL is detrimental to system performance. (The requirements for an 80-channel CATV system are $\gamma\langle \Delta\tau \rangle^2 \leq 268$ GHz-ps^2 for no PDL, and $\gamma\langle \Delta\tau \rangle \leq 1.3$ GHz-ps for a PDL of 2.0%.) The PMD of most currently available production fiber is less than 0.04 ps/$\sqrt{\text{km}}$ for single-mode fiber, and 0.10 ps/$\sqrt{\text{km}}$ for dispersion-shifted fiber. Therefore, for a 20-km span of standard fiber, the maximum PMD is 0.18 ps, which is acceptable for all the 50-channel examples mentioned previously. This is also acceptable for the 80-channel examples, except the one with a 2% PDL and a highly chirped source. Most analog lightwave systems with current production fiber, therefore, will not be significantly degraded by PMD.

3.4.2 Nonlinearities

In contrast to distortion caused by linear effects of fiber transport, nonlinear effects depend on the optical power and can be ignored at low optical power levels. As the power available from analog transmitters and optical amplifiers increases, the effect of these nonlinearities on analog systems should be evaluated. The system design must conform to any power–distance limits imposed by the fiber nonlinearities. Three nonlinear effects that have been evaluated for digital lightwave systems are presented [44]: SBS, stimulated Raman scattering (SRS), and self-phase modulation (SPM).

3.4.2.1 Stimulated Brillouin Scattering

The most prevalent nonlinear effect in analog lightwave systems is SBS, which arises from the interaction between the optical field and acoustic phonons in the fiber [35]. For narrow band sources, SBS can be observed for powers on the order of a few milliwatts. SBS scatters light into the backward direction and increases intensity noise in the forward direction [45]. At the Brillouin threshold power, the intensity noise in the forward direction rises dramatically, degrading the received CNR. See Fig. 14.11. The threshold power is determined by [44]

$$P_{SBS} = \frac{21 A_{eff} \alpha}{g_B p(1 - e^{-\alpha z})}, \tag{14.31}$$

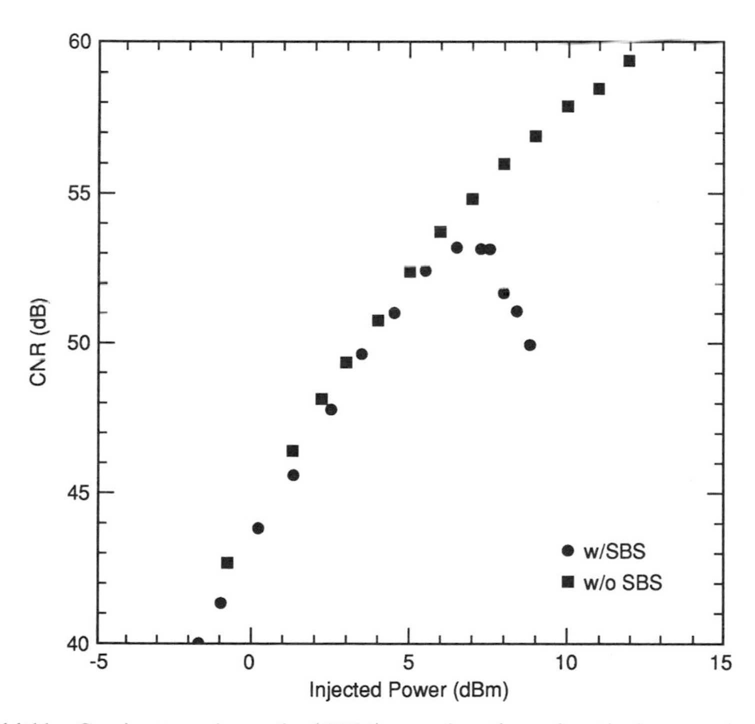

Fig. 14.11 Carrier-to-noise ratio (CNR) as a function of optical power input to 13 km of dispersion-shifted (DS) fiber. The source is an externally modulated narrow-linewidth continuous-wave (CW) laser [45]. SBS, stimulated Brillouin scattering.

where A_{eff} is the effective mode area of the fiber, g_B is the effective Brillouin gain coefficient, p is the polarization overlap (one-half for long lengths of non-polarization-maintaining fiber), and z is the fiber length. The peak Brillouin gain coefficient is 5×10^{-11} m/W. This is the effective gain coefficient if the optical spectrum of the signal is narrower than the Brillouin bandwidth of approximately 20 MHz [35]. For 20-km transmission over standard fiber ($A_{eff} = 80\ \mu m^2$, $p = \frac{1}{2}$), the Brillouin threshold for a narrow-linewidth source is 6.8 mW at 1.3 μm and 5.3 mW at 1.55 μm. If the optical spectrum is much broader than the Brillouin linewidth, the effective gain is reduced by $\Delta \nu_B / \Delta \nu_s$. The Brillouin threshold for directly modulated lasers, therefore, is much greater than for narrow band sources. For example, a directly modulated bulk laser with a spectral width of 6 GHz has a Brillouin threshold of 2 W, which explains why SBS is generally not observed for directly modulated sources. A narrow-spectrum source can be artificially broadened using phase or frequency modulation, and the Brillouin threshold power increased [46].

3.4.2.2 Stimulated Raman Scattering

SRS is a phenomenon similar to SBS except that optical phonons rather than acoustic phonons are involved. The peak gain coefficient, 7×10^{-14} m/W at 1.55 μm, is much smaller than that for SBS, but the bandwidth, 10 THz, is much broader [35]. The gain coefficient scales inversely with wavelength. The power threshold for a signal to lose half its power to spontaneous Raman scattering is

$$P_{SRS} = \frac{16 A_{eff} \alpha}{p g_R (1 - e^{-\alpha z})}, \tag{14.32}$$

where g_R is the peak Raman gain coefficient. For 20-km transmission over standard fiber, the Raman threshold is 3.1 W at 1.3 μm and 2.9 W at 1.55 μm. This threshold power well exceeds the power available from today's optical amplifiers and analog sources. Unlike SBS, however, the effects of SRS on analog system performance have not been measured, and they might occur at powers less than the threshold power.

A type of SRS degradation seen in wavelength-division multiplexing systems is amplification of longer wavelength channels at the expense of shorter wavelength channels [47]. This is not expected to be a factor for an analog system, because it is typically a single channel with a spectral width of less than 1 GHz. For separation of less than 1 GHz, there is no appreciable SRS coupling between parts of the signal spectrum. It appears, therefore, that the effects of SRS can be ignored for analog short-haul systems.

3.4.2.3 Self-Phase Modulation

SPM arises from the nonlinear index of refraction of the fiber. The index of refraction of the fiber is given by

$$n = n_o + n_2 I, \tag{14.33}$$

where n_2 is the nonlinear refractive index, $n_2 = 3.2 \times 10^{-20}$ m^2/W, and I is the optical intensity (W/m^2) [35]. Through the nonlinear index, intensity modulation of the signal induces a phase variation. When the signal is transmitted through a dispersive medium, the induced phase modulation leads to additional IM and therefore distortion at the output. The expression for the induced CSO distortion is [41]*

$$\text{CSO} = 20\log\left(\sqrt{N_{\text{CSO}}}\frac{1}{2}m\ddot{\beta}\Omega_d^2\frac{2\pi n_2 P}{\lambda A_{eff}}\frac{1}{\alpha}\left\{L - \frac{1}{\alpha}[1 - \exp(-\alpha L)]\right\}\right), \tag{14.34}$$

where P is the launched optical power into the fiber, A_{eff} is the effective mode area, and α is the loss coefficient. Unlike the CSO distortion, the CTB is not a simple function of the fiber loss or the distortion frequency. The intermodulation distortion depends not only on the intermodulation distortion frequency, but also on the frequencies that mix to generate it. The product count and distortion frequency give an upper bound of the distortion, however, and this is used in the equations that follow. The CTB is given for two limits—fiber length much greater than the effective length, and fiber length equal to the effective length (no loss):

$$\text{CTB} \leq 20\log\left[\sqrt{N_{\text{CTB}}}\frac{1}{12}\left(m\ddot{\beta}\Omega_d^2\frac{2\pi n_2 P}{\lambda A_{eff}}L\right)^2\alpha^{-2}\right],\ L \gg \alpha^{-1} \tag{14.35a}$$

$$\text{CTB} \leq 20\log\left[\sqrt{N_{\text{CTB}}}\frac{1}{12}\left(m\ddot{\beta}\Omega_d^2\frac{2\pi n_2 P}{\lambda A_{eff}}L\right)^2\frac{L^2}{3}\right],\ \alpha \to 0 \tag{14.35b}$$

* The expressions for the distortion due to self-phase modulation in the fiber discussed in Ref. 41 are for the second and third harmonics in the case of no loss in the fiber. The expressions here include loss and in the case of CTB are given by the two limits: (1) the effective length is approximately equal to $1/\alpha$, and (2) the effective length is equal to the physical length.

For a standard NTSC frequency plan, it is the CSO distortion rather than the CTB that limits the optical power. For standard fiber ($A_{eff} = 80\ \mu m^2$), 50 channels, and a 5.0% modulation depth, the CSO distortion is -59 dBc for $PD\alpha^{-1}\{L - \alpha^{-1}[1 - \exp(-\alpha L)]\} = 1000$ W-km-ps/nm. For example, in 20 km of standard fiber at 1.55 μm, the launch power must be less than 400 mW. (For the same link parameters, the power for an 80-channel CATV system is limited to 120 mW.) The power to induce SPM degradation is higher than the SBS threshold power for a narrow source and lower than the SRS threshold. For a broad-bandwidth source, SPM effects impose the power limitation.

3.5 SYSTEMS DESIGN SUMMARY

Having discussed a multitude of linear and nonlinear fiber effects and component limitations, we may find it useful to summarize these in the context of systems design. It is convenient to treat directly modulated (high-chirp) transmitters separately from externally modulated (low-chirp) transmitters. For direct modulation, which offers simplicity and low cost, one must be concerned with impairments from both chromatic and polarization mode dispersion. Chromatic dispersion effects force operation at close to the zero fiber dispersion, unless optical or electrical dispersion compensation is employed. Because these compensation techniques are either expensive or limited in performance, this generally results in the use of 1.3-μm directly modulated transmitters with standard fiber. PMD can be a problem unless low-PMD fiber is used, and components with PDL should not be used after any fiber with appreciable PMD. Also, for direct modulation, multipath beat noise results in intensity noise that increases as laser chirp decreases. In this case, in contrast to the dispersions discussed previously, more chirp is desirable. The broad linewidth of the directly modulated laser eliminates any problem with SBS, and other fiber nonlinearities (nonlinear refractive index and SRS) are not generally problems for typical power levels.

For external modulation, the transmitter is generally more complex, but the chirp is close to zero and noise from the transmitter can be close to the shot-noise limit. The low chirp eliminates problems with chromatic and polarization mode dispersion. Multipath beat noise is not a problem for source linewidths less than a few megahertz. However, because of the low source linewidth, special techniques must be used to overcome SBS.

Both external and direct modulation can be used with optical amplifiers at either a 1.3- or a 1.55-μm wavelength. For direct modulation, chirp can interact with gain tilt to create second-order distortion. Doped-fiber or Raman amplification can provide the high linearity and the high power needed for analog video transmission. Distortion from saturation-induced gain modulation limits the usefulness of semiconductor amplifiers.

Continuing advances in analog transmission technology aim to reduce transmitter cost and increase transmitter power:

- Analog transmitters using directly modulated DFBs are a commercial product today. The cost of the optics in the transmitter is largely driven by the low yield of lasers that meet the noise and linearity requirements, the necessary packaging costs of cooling and isolation, and testing. Some ways to reduce the cost are to increase the yield significantly through production processes or facet phase control, and to design DFB lasers that do not require isolation and temperature control.
- Directly modulated Fabry–Perot lasers can be very low cost. They are limited in analog performance, however, by a sublinear L-I curve, high RIN (required RIN is −150 dB/Hz), and mode-partition noise, which when transmitted through dispersive media creates intensity noise [48]. Even when the total system dispersion is only 20 ps/nm, the induced RIN for a multimode Fabry–Perot laser can be more than −130 dB/Hz. A Fabry–Perot laser is not likely to be able to carry 50 AM-VSB channels with high performance.
- Distributed Bragg reflector (DBR) lasers have RIN values on the order of those for DFB lasers and ostensibly can have more reproducible performance than that of DFBs because they are insensitive to facet phase. The linearity of the L-I curve must be improved, but recent work shows promising distortion performance [49]. As in the case of the DFB, the packaging costs will be reduced if no isolation or temperature control is needed.
- Integrated electroabsorption modulator–lasers are being developed for analog and digital links. Given that they are intrinsically lower chirp than a directly modulated laser diode, they might allow the requirement on gain tilt and dispersion to be relaxed. The major challenge is that the light–voltage characteristic of an electroabsorptive modulator is very nonlinear. Methods for linearizing this characteristic have been explored [50]. This may become competitive with existing technology if the cost of linearization is decreased and output power is increased.

4. Summary

An overview of lightwave technology for analog video transmission has been presented. Currently, the majority of systems deployed primarily into fiber coax video distribution systems use direct modulation of DFB lasers at a 1.3-μm wavelength. Transmission at 1.55 μm is feasible despite dispersion-induced complications, and it is most applicable in systems where long distances or splitting requires the additional power of EDFAs. Although this is uncommon for fiber coax architectures, new applications in passive optical networks or other fiber-intensive subscriber-loop distribution systems will require much more optical splitting. These applications will require challenging reductions in the cost of high-power and optically amplified analog lightwave systems. Continuing advances in integrated laser–modulators, fiber amplifiers, Raman amplification, dispersion-compensation techniques, predistortion techniques, and the cost and performance of many photonic components will continue to enhance the capability of video transmission technology.

As digital video compression gains in popularity, and as analog video systems take on additional channels dedicated to narrowcast digital services, the need for RF modems and the transmission of multiple digital–RF channels (like QPSK or M-QAM) will increase. The technology described in this chapter is applicable to the transmission of such signals, and the comparatively robust digital–RF channels eliminate many of the difficulties encountered for analog video transmission. This opens opportunities for high volumes of transmitters with lower performance than that required for multichannel analog systems.

References

[1] National Association of Broadcasters. 1992. *NAB engineering handbook.* 8th ed. Washington, DC: National Association of Broadcasters.

[2] Westman, H. P., ed. 1956. *Reference data for radio engineers.* 4th ed. New York: International Telephone and Telegraph Corp., American Book—Stratford Press.

[3] Pratt, T., and C. W. Bostian. 1986. *Satellite communications.* New York: Wiley.

[4] Way, W., C. Zah, C. Caneau, S. Menocal, F. Favire, F. Shokoochi, N. Cheung, and T. P. Lee. 1988. Multichannel FM video transmission using traveling wave amplifiers for subscriber distribution. *Electron. Lett.* 24:1370.

[5] Olshansky, R., V. Lanzisera, and P. Hill. 1988. Design and performance of wideband subcarrier multiplexed lightwave systems. In *Proceedings of the ECOC '88, Brighton, U.K.* 143–146.

[6] Netravali, A. N., and B. G. Haskel. 1988. *Digital pictures.* New York: Plenum Press.

[7] Feher, K. 1981. *Digital communication-microwave applications.* Englewood Cliffs, NJ: Prentice-Hall.

[8] Feher, K., ed. 1987. *Advanced digital communications.* (Englewood Cliffs, NJ: Prentice-Hall.

[9] Maeda, K., N. Ishiyama, H. Nakata, and K. Fujito. 1993. Analysis of BER of 16QAM signal in AM/16QAM hybrid optical transmission system. In *OFC'93, San Jose, CA.* Paper THL2.

[10] Lu, X., G. E. Bodeep, and T. E. Darcie. 1994. Impulse-induced bit-error impairment in AM-VSB/64QAM hybrid optical transmission systems. In *OFC'94, San Jose, CA.* Paper WH3.

[11] Chiddix, J. A., H. Laor, D. M. Pangrac, L. D. Williamson, and R. W. Wolfe. 1990. AM video on fiber in CATV systems, need and implementation. *IEEE J. Select. Areas Commun.* 8:1229.

[12] Lau, K. Y., and A. Yariv. 1984. Intermodulation distortion in a directly modulated semiconductor injection laser. *Appl. Phys. Lett.* 45:1034–1036.

[13] Darcie, T. E., R. S. Tucker, and G. J. Sullivan. 1984. Intermodulation and harmonic distortion in IaGaAsP lasers. *Electron. Lett.* 21:665–666. Erratum, 1986, *Ibid.* 22:619.

[14] Orfanos, J., T. Sphicopoulos, A. Tsigopoulos, and C. Caroubalos. 1991. A tractable above-threshold model for the design of DFB and phase-shifted DFB lasers. *IEEE J. Quantum Electron.* 27(4): 946–956.

[15] Takemoto, A., H. Watanabe, Y. Nakajima, Y. Sakakibara, S. Kakimoto, U. Yamashita, T. Hatta, and Y. Miyake. 1990. Distributed feedback laser diode and module for CATV systems. *IEEE J. Select. Areas Commun.* 8:1359.

[16] Tucker, R. S., and D. J. Pope. 1983. Circuit modeling of the effect of diffusion on damping in a narrow-stripe semiconductor. *IEEE J. Quantum Electron.* QE-19:1179.

[17] Agrawal, G. P., and N. K. Dutta. 1986. *Long-wavelength semiconductor lasers.* New York: Van Nostrand Reinhold.

[18] Gnauck, A. H., T. E. Darcie, and G. E. Bodeep. 1992. Comparison of direct and external modulation for CATV lightwave transmission at 1.55 μm wavelength. *Electron. Lett.* 28(20):1875–1876.

[19] Bertelsmeier, M., and W. Z. Schunke. 1984. Linearization of broadband optical transmission systems by adaptive predistortion. *Frequenz* 38(9):206.

[20] Nazarathy, M., J. Berger, A. J. Ley, I. M. Levi, and Y. Kagan. 1993. Progress in externally modulated AM CATV transmission systems. *IEEE J. Lightwave Tech.* 11(1):82–105.

[21] Frigo, N. J., M. R. Phillips, and G. E. Bodeep. 1993. Clipping distortion in lightwave CATV systems: Models, simulations, and measurements. *IEEE J. Lightwave Tech.* 11(1):138–146.

[22] Saleh, A. A. M. 1989. Fundamental limit of number of channels in subcarrier-multiplexed lightwave CATV system. *Electron. Lett.* 25(12):776.

[23] Agrawal, G. P., and N. K. Dutta. 1993. *Semiconductor lasers.* New York: Van Nostrand Reinhold, 468.

[24] Cebulla, U., J. Bouayad, H. Haisch, M. Klenk, G. Laube, H. P. Mayer, R. Weinmann, and E. Zielinski. 1993. 1.55-μm Strained layer quantum well DFB-lasers with low chirp and low distortions for optical analog CATV distribution systems. In *Conference on Lasers and Electro-Optics '93.* Paper CWA5.

[25] Koyama, F., and K. Iga. 1988. Frequency chirping in external modulators. *J. Lightwave Tech.* 6(1):87–92.

[26] Saleh, A. A. M., T. E. Darcie, and R. M. Jopson. 1989. Nonlinear distortion due to optical amplifiers in subcarrier-multiplexed lightwave communications systems. *Electron. Lett.* 25:79.

[27] Desurvire, E. 1994. *Erbium-doped fiber amplifiers: Principles and applications.* New York: Wiley.

[28] Habbab, I. M. I., and L. J. Cimini. 1991. Optimized performance of erbium-doped fiber amplifiers in subcarrier multiplexed lightwave AM-VSB CATV systems. *J. Lightwave Tech.* 9(10):1321–1329.

[29] Phillips, M. R., A. H. Gnauck, T. E. Darcie, N. J. Frigo, G. E. Bodeep, and E. A. Pitman. 1992. 112 Channel split-band WDM lightwave CATV system. *IEEE Photon. Tech. Lett.* 4(7):790–792.

[30] Kuo, C. Y., D. Piehler, C. Gall, J. Kleefeld, A. Nilsson, and L. Middleton. 1996. High-performance optically amplified 1550-nm lightwave AM-VSB CATV transport system. In *OFC'96, San Jose, CA.* Paper WN2.

[31] Kuo, C. Y., and E. E. Bergmann. 1991. Erbium-doped fiber amplifier second-order distortion in analog links and electronic compensation. *IEEE Photon. Tech. Lett.* 3:829.

[32] Kikushima, K., M. Yamada, M. Shimizu, and J. Temmyo. 1994. Distortion and noise properties of a praseodymium-doped fluoride fiber amplifier in 1.3 μm AM-SCM video transmission systems. *IEEE Photon. Tech. Lett.* 6(3):440–442.

[33] Yoshinaga, H., M. Yamada, and M. Shimizu. 1995. Fiber transmission of 40-channel VSB-AM video signals by using a praseodymium-doped-fluoride fiber amplifier with direct and external modulation. In *Optical Fiber Communication '95, February.* Paper TuN5.

[34] Shimizu, M., M. Yamada, T. Kanamori, Y. Ohishi, Y. Terumuma, S. Sudo, H. Yoshinaga, and Y. Miyamoto. 1995. High-power, low-noise praseodymium-doped fluoride-fiber amplifiers. In *Optical Fiber Communication '95.* Paper WD5.

[35] Agrawal, G. P. 1989. *Nonlinear fiber optics.* San Diego: Academic Press.
[36] Olsson, N. A., and J. Hegarty. 1986. Noise properties of a Raman amplifier. *J. Lightwave Tech.* 4(4):396–398.
[37] Forghieri, F., R. W. Tkach, and A. R. Chraplyvy. 1994. Bandwidth of crosstalk in Raman amplifiers. In *Optical Fiber Communication '94.* Paper FC6.
[38] Judy, A. F. 1989. The generation of intensity noise from fiber Rayleigh backscatter and discrete reflections. In *European Conference on Optical Communications.* Paper TuP-11.
[39] Darcie, T. E., G. E. Bodeep, and A. A. M. Saleh. 1991. Fiber-reflection induced impairments in lightwave AM-VSB CATV systems. *IEEE J. Lightwave Tech.* 9(8):991–995.
[40] Woodward, S. L., and T. E. Darcie. 1994. A method for reducing multipath interference noise. *IEEE Photon. Tech. Lett.* 6(3):450–452.
[41] Phillips, M. R., T. E. Darcie, D. Marcuse, G. E. Bodeep, and N. J. Frigo. 1991. Nonlinear distortion generated by dispersive transmission of chirped intensity-modulated signals. *IEEE Photon. Tech. Lett.* 3(5):481–483.
[42] Poole, C. D., and T. E. Darcie. 1993. Distortion related to polarization-mode dispersion in analog lightwave systems. *J. Lightwave Tech.* 11(11):1749–1759.
[43] Phillips, M. R., G. E. Bodeep, X. Lu, and T. E. Darcie. 1994. 64-QAM BER measurements in an analog lightwave link with large polarization-mode dispersion. In *Conference on Optical Fiber Communication,* Technical Digest Series, vol. 4, 112. Washington, DC: Optical Society of America.
[44] Chraplyvy, A. R. 1990. Limitation on lightwave communication imposed by optical-fiber nonlinearities. *J. Lightwave Tech.* 8(10):1548–1557.
[45] Mao, X. P., G. E. Bodeep, R. W. Tkach, A. R. Chraplyvy, T. E. Darcie, and R. M. Derosier. 1992. Brillouin scattering in externally modulated lightwave AM-VSB CATV transmission systems. *IEEE Photon. Tech. Lett.* 4(3):287.
[46] Mao, X. P., G. E. Bodeep, R. W. Tkach, A. R. Chraplyvy, T. E. Darcie, and R. M. Derosier. 1993. Suppression of Brillouin scattering in lightwave AM-VSB CATV transmission systems. In *Conference on Optical Fiber Communication,* Technical Digest Series, 141–142.
[47] Chraplyvy, A. R. 1984. Optical power limits in multi-channel WDM systems due to stimulated Raman scattering. *Electron. Lett.* 20:58.
[48] Wentworth, R. H., G. E. Bodeep, and T. E. Darcie. 1992. Laser mode partition noise in lightwave systems using dispersive optical fiber. *J. Lightwave Tech.* 10(1):84–89.
[49] Muroya, Y., T. Okuda, H. Yamada, T. Torikai, and T. Uji. 1995. Multiple-wavelength partially corrugated LD array for high-capacity CATV applications. In *Optical Fiber Communication '95.* Paper FC2.
[50] Wilson, G. C., T. H. Wood, J. L. Zyskind, J. W. Sulhoff, S. B. Krasulick, J. E. Johnson, T. Tanbun-Ek, and P. A. Morton. 1996. Suppression of SBS and MPI in analog systems with integrated electroabsorption modulator/DFB laser transmitters. In *Optical Fiber Communication '95.* Paper ThR1.

Chapter 15 | Advanced Multiaccess Lightwave Networks

Ivan P. Kaminow

AT&T Bell Laboratories (retired), Holmdel, New Jersey

A. Introduction

Novel component technology allows enhanced network architectures, and, conversely, novel architectures require enhanced component functionality. This duality has driven much of the innovation in lightwave networks at the physical layer. The initial opportunity for lightwave was in point-to-point, long-haul, broadband transport systems. The new opportunities lie in multiple access networks, bringing wideband digital multimedia services to end users. The access application, although offering a wider market, is much more difficult to realize than point-to-point transport. In the latter, the challenge is to go longer distances at higher bit rates. The costs of sophisticated hardware can be shared among many users, and the customer is a knowledgeable network provider. The access application is considerably more nebulous. Much of the equipment is dedicated to an individual user and the costs cannot be shared. The investment in a local distribution plant often must be made without knowing what fraction of accessible users will actually take the service (called the *take rate*) within a time suitable to defray the capital costs of installation. And, most daunting, no one can predict the applications that will lead to widespread use. Because the applications determine the required network performance and functionality requirements, a great financial leap of faith is needed to choose and install novel access facilities. The safest approach is to choose a flexible architecture that provides wanted services now and can be gracefully upgraded as appetities for interactive bandwidth develop.

Wavelength-division multiplexing (WDM) techniques can provide the flexibility and scalability needed in a variety of network architectures. Ear-

OPTICAL FIBER TELECOMMUNICATIONS,
VOLUME IIIA

ISBN: 0-12-395170-4

lier chapters discussed commercial long-haul applications of WDM (Chapters 8, 9, and 10 in Volume IIIA) and near-commercial WDM passive optical networks (PONs) (Chapter 13 in Volume IIIA). The novel WDM components required, such as passive demultiplexers, waveguide grating routers (WGRs), and optical add–drop multiplexers (OADMs) (Chapters 7 and 8 in Volume IIIB), and active tunable distributed feedback (DFB) lasers and multifrequency lasers (MFLs) and receivers (MFRs) (Chapter 4 in Volume IIIB), and erbium-doped fiber amplifiers (EDFA) (Chapter 2 in Volume IIIB), were also discussed.

In this chapter, we consider some speculative wavelength-routed networks and discuss the features of optical transparency that they provide.

B. Optical Transparency

The terms *all optical* and *optically transparent* have been used loosely to define networks that transport and switch optical signals without reducing the data to the electronic state by optical-to-electrical (O/E) conversion at intermediate points. Neither term stands up to close scrutiny, but the concept is a useful one. *All optical* is too restrictive, because electrons and electronics are always lurking nearby, but we can define *optical transparency* by examples of its utility.

The AT&T long-haul WDM system described in Chapter 9 in Volume IIIA has transparency features on the link between electronic nodes, as illustrated in Fig. 15.1. The optical amplifiers (OAs) are transparent to optical carrier frequency within a 1.5-THz-wide band near 1555 nm, and they are transparent to the modulation format and bit rate impressed on each optical carrier. These transparency characteristics must be limited by nonlinear and noise effects as noted in Chapter 8 in Volume IIIA, but in

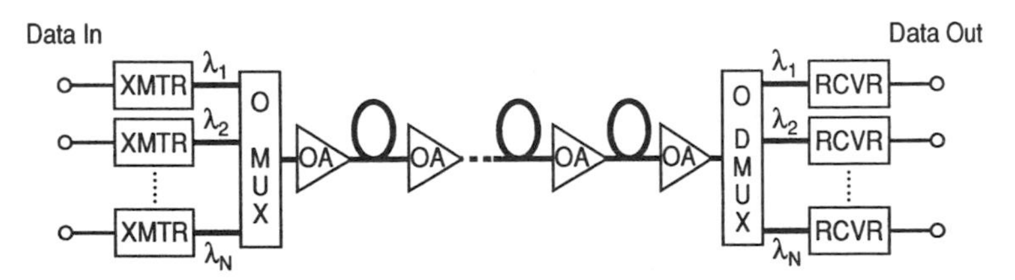

Fig. 15.1 AT&T long-haul wavelength-division multiplexing (WDM) system (OLS-2000). OA, optical amplifier.

some sense we can say that the link is bit-rate and wavelength transparent within wide bit-rate and wavelength bands. It is not transparent in the strict sense of having the same photons pass from end to end because of the intermediate amplifiers.

The optical multiplexer and demultiplexer provide a *wavelength-routing* function in the sense that data impressed on the top transmitter in Fig. 15.1 at wavelength λ_1 are routed by the demultiplexer to the top O/E receiver. (The multiplexer in Fig. 15.1 could be a wavelength-sensitive WGR in order to conserve power. However, an $N \times 1$ star combiner is usually employed for simplicity.) On the next link to the right (not shown), the baseband data are electronically multiplexed and then impressed on the following laser (electrical-to-optical [E/O]) transmitter, and so on. The electronic multiplexer adds and drops local traffic and regenerates, reshapes, and retimes the data bits at a fixed clock rate. Thus, the path through the O/E, electronic multiplexer, and E/O is *not* transparent in any optical sense. The baseband data must be formatted in the manner prescribed by the electronic multiplexer, which, for example, might satisfy a synchronous optical network (SONET)–synchronous digital hierarchy (SDH) standard with a line rate of 2.5 Gb/s (optical carrier [OC]-48) and a channel rate of 155 Mb/s (OC-3). All end-to-end data must be in blocks of OC-3, although in principle a signal could be synchronously concatenated to OC-48.

Telecommunications-oriented customers are satisfied, for the most part, with network transmission standards like SONET–SDH because they are required for universal connectivity. Computer-oriented customers are not, or think they are not. They have legacy networks operating under a variety of protocols and bit rates: Ethernet, analog video, and so forth. They would like to plug these networks into "dumb pipes" for transport and extract the information untainted at the other end. It is this notion that provided the original driving force for transparent optical networks, the idea being that the computer network people at the ends could do the smart, high-profit-margin tasks and the telecommunications people could do the rest. In reality, the laws of physics, networking, and economics conspire to make this approach impractical.

1. CONVENTIONAL NETWORKS

A conventional telecommunications network is illustrated in Fig. 15.2a, where fibers interconnect *opaque* electronic nodes that provide multiplexing, cross-connect, and regeneration functions. Because the electronic bit

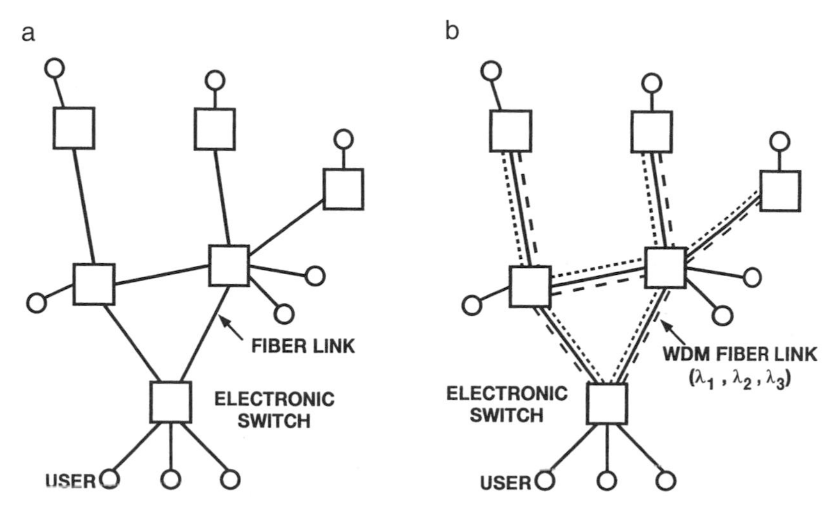

Fig. 15.2 (a) Conventional telecommunications network with electronic nodes providing opaque multiplexing, cross-connect, and regeneration functions. (b) WDM network with electronic nodes.

stream must satisfy the format, both optically and electronically, for which the node is designed, there is no transparency, and the word *opacity* has been used (Bala, Cordell, and Goldstein 1995). Although one fiber per connection is shown for simplicity, it is understood that two-way (duplex) transmission is provided by a second parallel fiber.

WDM can be added on the fiber links as shown in Fig. 15.2b to increase the capacity carried by each fiber. In this case, λ_1, λ_2, and λ_3 are three independent channels carried on each fiber. The electronic nodes now have one more degree of freedom to deal with — i.e., wavelength. The cross-connect function, which provided routing of multiplexed channels on input fibers to multiplexed channels on output fibers in Fig. 15.2a, must now also cross-connect arbitrary input wavelengths on an input fiber to arbitrary output wavelengths on the output fibers. The cross-connect function generally takes place at time-division multiplexing (TDM) rates below the line rate — e.g., for a line rate of 2.4 Gb/s (OC-48), the intermediate TDM channels may be cross-connected at 155 Mb/s (OC-3) or 622 Mb/s (OC-12). Because all 2.4-Gb/s channels are fully demultiplexed to this intermediate rate at every node, end-to-end transmission at higher rates, say 1 Gb/s, is not possible on a conventional telecommunications network: hence, the perceived need for transparent networks by supercomputer users.

2. *WAVELENGTH-ROUTED NETWORKS*

A transparent, wavelength-routed network is shown schematically in Fig. 15.3. A set of local wavelengths, $\lambda_{\ell i}$, are reused in the local area networks (LANs), shown shaded on the periphery, for interconnecting optical terminals within a common locality. A set of global wavelengths, λ_{gi}, interconnect optical terminals in different local areas across a regional or metropolitan area network (MAN). The intermediate nodes consist of passive wavelength routers, such as the $N \times N$ WGR (Dragone, Edwards, and Kistler, 1991) discussed in Chapter 8 in Volume IIIB.

For example, the same local wavelength $\lambda_{\ell 1}$ can be reused to simultaneously connect terminals **a** and **f**, and **h** and **k**. Global wavelength λ_{g1} simultaneously connects **a** and **b**, and **c** and **d**. These single-wavelength paths between terminals are called *wavelength paths*. The maximum number of simultaneous global connections in a passive wavelength-routed network is given by (Gallager 1992)

$$N \leq F^2, \tag{15.1}$$

where F is the number of available transmitter global wavelengths.

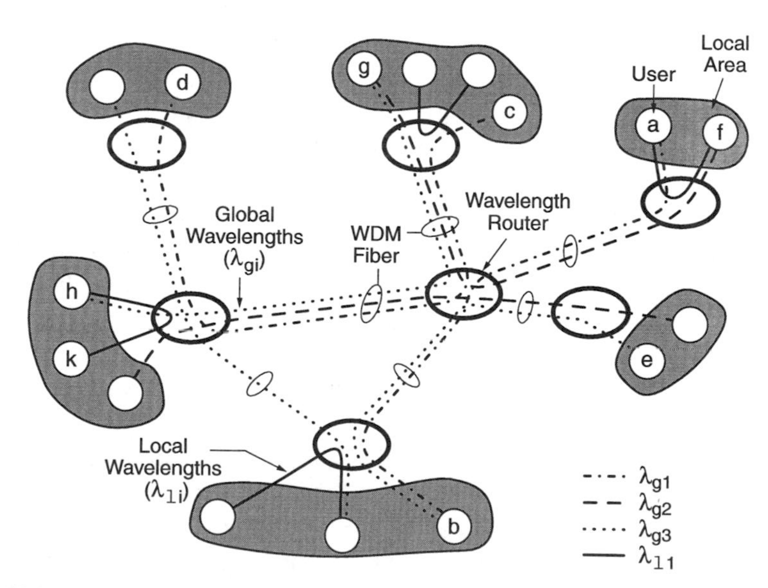

Fig. 15.3 Transparent wavelength-routed network. $\lambda_{\ell i}$, local wavelengths; λ_{gi}, global wavelengths.

If, instead of continuous circuit connections, time is separated into frames and the frames into T time slots, the maximum number of connections per frame is increased to (Gallager 1992)

$$N \leq TF^2. \tag{15.2}$$

However, all terminals must be synchronized and a scheduling protocol must be defined to efficiently interleave time slots at transmitters, at receivers, and along common fiber paths. We give examples of network test beds employing simple architectures that have been built to demonstrate wavelength routing on a circuit-switched and time-slotted basis. The bounds on N given in Eqs. (15.1) and (15.2) apply under special simplifying conditions for very low blocking probability; more sophisticated treatments are available elsewhere (Gallager 1992; Barry 1993; Barry and Humblet 1996; Jeong and Ayanoglu 1996; Karasan and Ayanoglu 1996). However, the order of magnitude of the bounds in Eqs. (15.1) and (15.2) provides a guide to network limits.

Because of the limitation in the number of connections, the passive wavelength-routed network is not scalable to wide-area networks (WANs). Furthermore, as N and the geographic extent increase, the number of routers on a wavelength path increases and the problems of scheduling and synchronization become unmanageable, especially for the time-slotted case. Hence, passive wavelength-routed networks will likely find application in a short-range campus or regional environment where the advantages of format transparency and distributed control are desirable. The network can be extended, at the expense of greater complexity, by incorporating rearrangeable routers, optical wavelength changers, and/or optical space division switches in order to reduce blocking probability along a *light path*, which is defined as a concatenation of wavelength paths along which the wavelength may change. These limitations will become clearer as we present examples of wavelength-routed networks.

Before leaving the topic, we should note that format transparency does not mean that any modulation format can be carried over a wavelength path of given length. For example, analog systems require a higher signal-to-noise ratio at the receiver than that of digital systems, and high bit-rate digital signals are more susceptible to dispersion and nonlinear optical effects than lower bit-rate signals are (see Chapter 8 in Volume IIIA).

C. WDM Point-to-Point Links

A WDM point-to-point link is illustrated in Fig. 15.1. The optical erbium-doped fiber amplifiers (EDFAs) have a relatively flat gain over a range of about 1.5 THz (12 nm) in the neighborhood of 1555 nm. The amplifiers can thus serve about $N = 8$ channels spaced by 200 GHz (1.6 nm) (or $N =$ 16 channels spaced by 100 GHz [0.8 nm]). The span between electronic regenerators is determined by the accumulation of amplified spontaneous emission (ASE) noise from the amplifiers, and fiber dispersion and nonliner optical effects. These effects in turn are determined by the spacing between amplifiers and the fiber dispersion. (See Chapter 8 in Volume IIIA). Typically, the regenerator span is about 400 to 600 km with amplifier (repeater) spacing of 120 to 80 km, respectively, for dispersion-tailored fiber with 2.5 Gb/s (OC-48) per channel. The aggregate throughput is 20 Gb/s (or 40 Gb/s).

In the usual case that one or more channels need to be dropped along the way in a telecommunications network, OADMs can provide flexibility and cost savings when placed strategically along the optical span as shown in Fig. 15.4. The WDM spectrum is groomed at the upstream node so that one or more wavelengths can be dropped and added at an intermediate point. The alternative is to provide an electronic add–drop multiplexer, at the aggregate bit rate, for each intermediate point or to back haul with a separate fiber, carrying one or more channels, from the downstream regenerator.

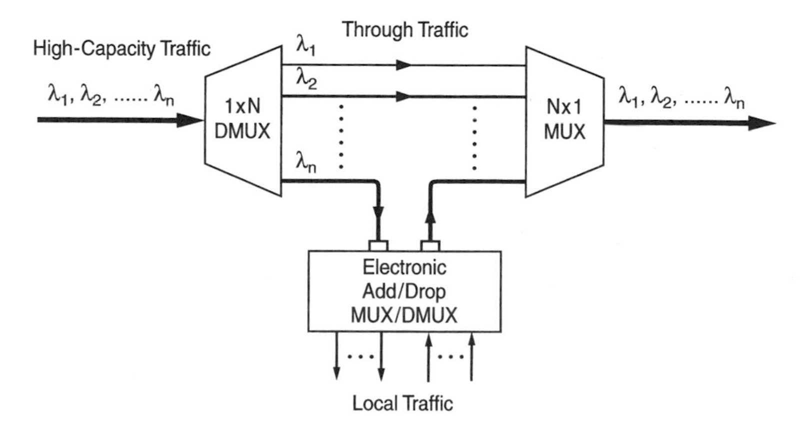

Fig. 15.4 Passive WDM add–drop multiplexer.

The performance specifications of the OADM are critical. In particular, the degradation of the span transmission in a series of filters caused by insertion loss, cross-talk rejection, and flatness of the passbands of individual OADMs increases in geometric progression with their number. Several technologies for realizing the 1 × N optical multiplexers in Fig. 15.4 are under investigation. (See Chapters 7 and 8 in Volume IIIB) The performance targets are an insertion loss of less than 3 dB, cross talk less than −40 dB, and a flat passband width greater than 0.7 × channel spacing.

The network planning is also an issue with passive (or fixed) optical add–drops: (1) Because traffic patterns change with time, the number and wavelengths of the channels to be dropped will change; (2) less than one full channel add–drop may be required; and (3) some wavelength paths may pass through greater or fewer OA sections than required before electronic regeneration, with consequent signal–noise impairment. The first problem at least can be addressed, at the expense of greater complexity, by designing rearrangeable add–drop multiplexers. A number of research examples of rearrangeable multiplexers that require optical switches have been reported (e.g., Kobayashi and Kaminow 1996). The second problem of add–drop granularity can be addressed by taking advantage of bit-rate transparency and reducing the line rate on selected wavelength channels. The third problem has not yet been addressed.

D. WDM Rings

1. *SONET RINGS*

Among the most important features of a telecommunications network are its reliability and survivability. In order to survive a component failure or a cable cut, redundancy must be provided, and for fast recovery the switching in of the protection equipment or protection path must be automatic. A dual-ring architecture, with one ring providing service and the other protection, affords a simple architectural redundancy. And the SONET standards provide the automatic fault detection and protection switching at the electronic add–drop nodes on the dual ring (Wu 1992). A number of self-healing optical SONET ring architectures have been proposed (Hersey and Soulliere 1991). The simplest is the two-fiber, unidirectional, path-switched ring illustrated in Fig. 15.5. The four nodes shown are electronic SONET add–drop multiplexers. The outer ring carries the service

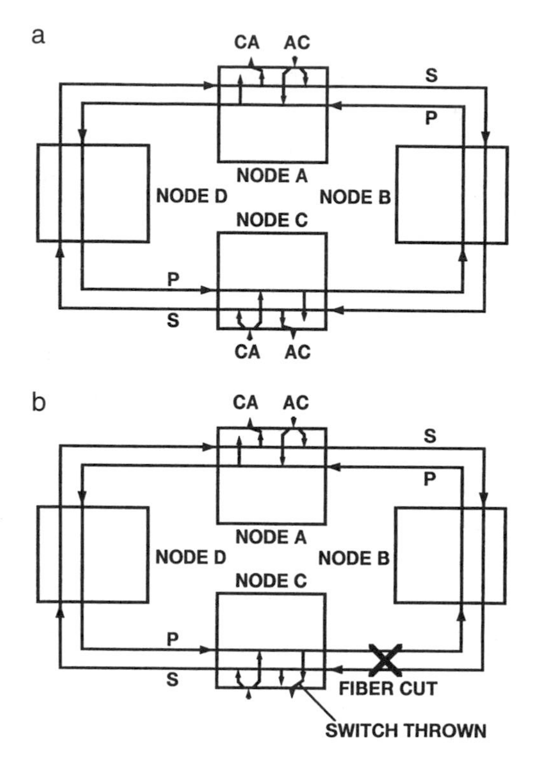

Fig. 15.5 Two-fiber, unidirectional, path-switched synchronous optical network (SONET) ring. (a) Normal operation. (b) Protection after a fiber cut. P, protection line; S, service line; AC, switch set for A to C connection; CA, switch set for C to A connection.

traffic in a clockwise sense, and simultaneously the inner ring carries the duplicate protection traffic in a counterclockwise sense. In the case of a cable cut, as indicated in Fig. 15.5b, only the switch in node *C* needs to be thrown to restore service. This architecture is well-suited to access applications where most traffic homes on a single node, such as an interexchange carrier point of presence (POP) or the hub in a client–server campus network. For interoffice connectivity or peer-to-peer applications, other SONET ring configurations may be preferable (Chapter 9 in Volume IIIA; Hersey and Soulliere 1991).

With WDM on the rings in Fig. 15.5, the transport capacity of each fiber can be enhanced as in the point-to-point links in Fig. 15.1.

2. *WDM RINGS WITH OPTICAL ADD–DROP*

A number of broadband applications call for a client–server restorable ring architecture. The hub server might be a long-distance POP or an interactive multimedia server, and the clients might be cable television, competitive access, or wireless head ends. A simple undirectional ring with optical add–drop is shown in Fig. 15.6 (Wu 1992; Elrefaie 1991, 1993; Tomlinson and Wagner 1993). A counterpropagating ring with redundant add–drops and hub can be provided for path restoration, as in Fig. 15.5. Inputs from each node travel simultaneously in opposite senses on both the service and protection rings. When a loss of signal is detected at a drop, the receiver switches automatically to the protection ring. The inset in Fig. 15.6 shows that the physical ring, with its restoration capability, behaves like a logical star. With an asynchronous transfer mode (ATM) switch at the hub, services can be requested by node A on λ_A and switched to the server on λ_D, and the required ATM cells transmitted on λ_A back to node A, all in the counterclockwise sense on the fiber ring. Peer-to-peer communication on the unidirectional ring is also possible. A connection from node C to node

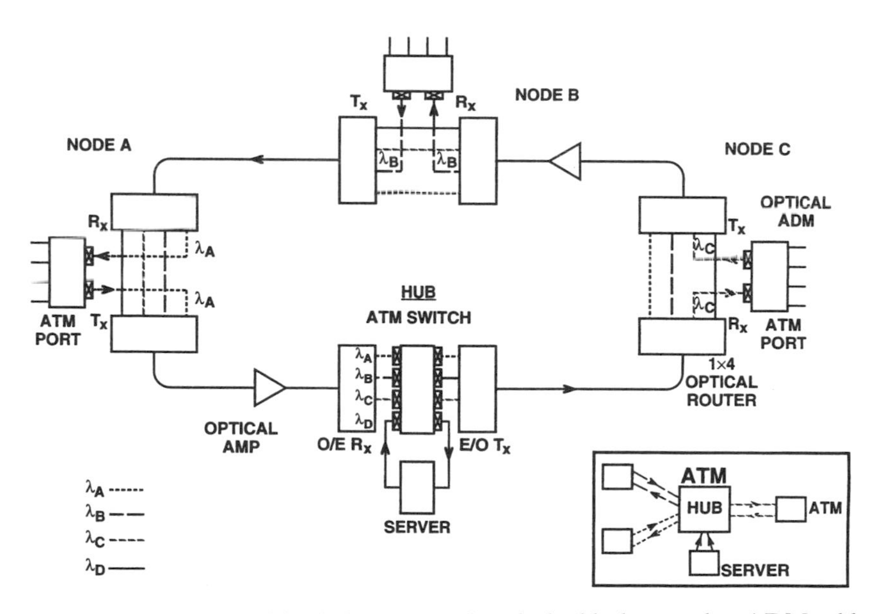

Fig. 15.6 WDM ring with a hub server and optical add–drop nodes. ADM, add–drop multiplexer, ATM, asynchronous transfer mode; E/O, electrical-to-optical; O/E, optical-to-electrical.

A would travel counterclockwise around the ring on λ_C to the ATM switch, be switched to λ_A, and be retransmitted counterclockwise to node *A*. Thus, the same signal passes node *B* twice; in campus-size networks, this transmission inefficiency may be compensated for by the network simplicity. A simple example of a restorable ring was demonstrated recently by Glance *et al.* (1996).

A restorable ring could be deployed by an independent hosting provider in a metropolitan region to offer interactive and one-way services from a variety of independent content providers to a variety of independent access providers. As an example, consider Fig. 15.7. A regional ring connects a hub server to a number of cable television head ends, each providing interactive multimedia services through hybrid fiber coax (HFC) technology to 10,000 homes passed (HP). Logical point-to-point duplex connectivity to the hub is provided by specific wavelengths to each head end. Each head end contains an OADM and a modem to provide both analog broadcast and interactive digital services to residences and businesses. If only 2.5% of the HP are actively using digital services at 10 Mb/s, each head end will require 2.5 Gb/s for interactive digital services. As in the linear WDM system of Fig. 15.1, the WDM ring might support 10 OC-48 channels over

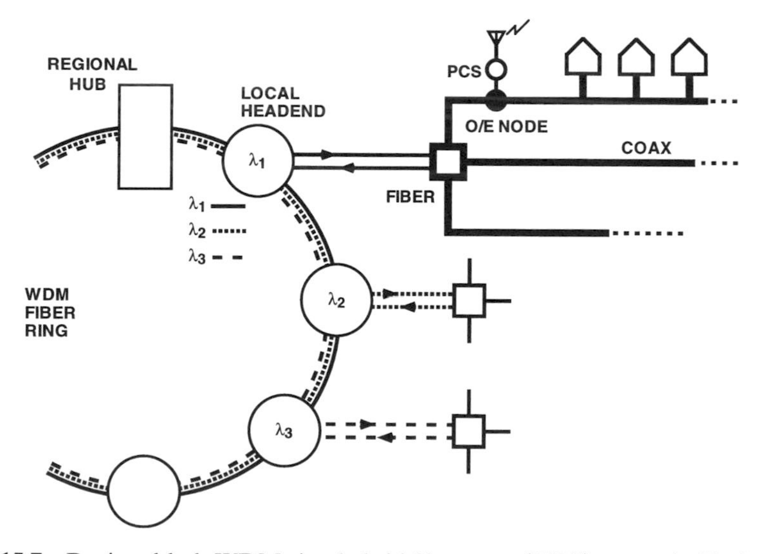

Fig. 15.7 Regional hub WDM ring hybrid fiber coax (HFC) network. Each local head end contains an optical add–drop multiplexer at λ_i. PCS, personal communication services.

a perimeter of 400 km, serving a total of 100,000 HP with an aggregate bandwidth of 25 Gb/s.

3. *AfricaONE*

The most ambitious example of an optical add–drop WDM ring is the AfricaONE (Africa optical network) proposal (Marra and Schesser 1996) of AT&T and Alcatel. The submarine cable ring will circumscribe the African continent, with a perimeter of 30,000 km (Fig. 15.8). Optical add–drops from the WDM submarine ring provide independent access at 2.5 Gb/s (European synchronous transport multiplex, level 16 [STM-16]) to more than 40 coastal countries. The fact that sovereign traffic need not pass through neighboring countries is especially attractive to the African nations. A 30,000-km WDM ring with 40-wavelength channels is not feasible for a variety of physical reasons. Thus, the design includes six or more onshore electronic nodes, or central offices (COs), that provide electronic multiplexing, switching, regeneration, and network management functions. Some of

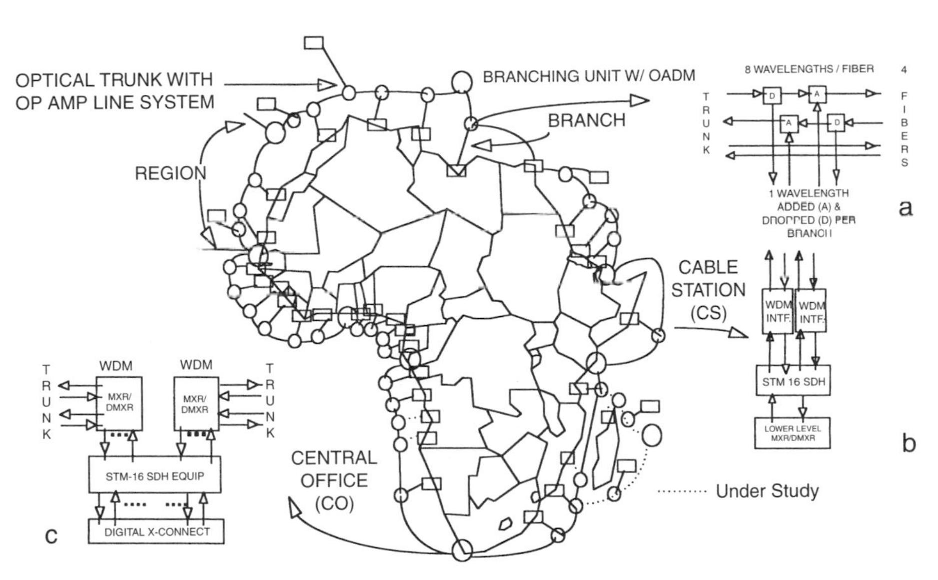

Fig. 15.8 AfricaONE schematic. Insets: (a) Undersea optical add–drop branching unit (BU), (b) onshore cable station (CS), and (c) onshore central office (CO). Note that the number and selection of CSs and COs are subject to change, as are the cable landing sites and cable route. OADM, optical add–drop multiplexer; SDH, synchronous digital hierarchy. (Reproduced with permission of AT&T Submarine Systems, Inc.)

these nodes provide gateways to other global networks. The undersea span between CO nodes comprises four fibers of up to 5000 km in length. Each fiber pair provides eight duplex WDM channels at 2.5 Gb/s for a total trunk capacity of $2 \times 8 \times 2.5$ Gb/s = 40 Gb/s. In one fiber pair, four channels support express traffic between adjacent COs, and four provide access by WDM add–drop multiplexers in the undersea branching units (BUs) to four onshore cable stations (CSs) located between the COs. The second pair of fibers provides protection for the four express channels and access to four additional CSs. Thus, up to eight CSs can be supported between COs.

The sector between the centers of adjacent COs defines a region; traffic between CSs within a region is intraregional, or local; traffic between CSs in different regions is interregional, or express. In normal operation, the traffic between COs comprises four 2.5-Gb/s duplex operational express channels on pair 1, four protection express channels (which may be empty or may carry four preemptible channels) on pair 2, four local channels on pair 1 connecting four CSs to both adjacent COs, and four local channels on pair 2 connecting the other four CSs to both adjacent COs. All traffic between CSs is connected through a CO.

Figure 15.9 shows the BUs with OADMs. Four branch fibers connect the BU and the CS. A CS is normally connected by one pair of branch fibers and the BU to its nearer or home CO. In case of a trunk cable cut between the CS and the home CO, the CS switches to the other pair of branch fibers to provide connectivity to the other CO, from which connection may be restored either intraregionally or interregionally by traveling the long way around Africa to an adjacent CO if need be. Express traffic

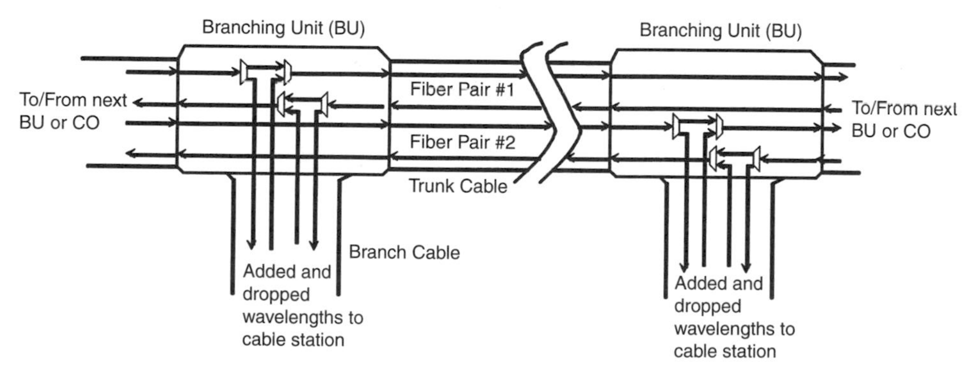

Fig. 15.9 BUs connecting trunk cable and branch cable. (Reproduced with permission of AT&T Submarine Systems, Inc.)

interrupted by a trunk cable cut is restored by switching in the CO to the protection fibers and taking the long way around the continent.

E. WDM Local Access Networks

The newest opportunity for lightwave technology is to provide access to broadband interactive digital services to homes and businesses by bringing the fiber closer to the desktop — i.e., providing fiber to the curb (FTTC) or fiber to the home (FTTH). As a massive consumer application, introduction of this architecture depends on an economical infrastructure, a large take rate, and cheap customer premises components in order to compete with the existing broadband architectures employing fiber and copper, described in Chapter 14 in Volume IIIA. The WDM PON architecture has many attractive features, as discussed in Chapter 13 in Volume IIIA. We mention its salient features here in order to complete our catalog of wavelength-routed networks and their dependence on novel WDM components.

Three alternative fiber distribution networks are shown in Fig. 15.10. The first, in Fig. 15.10a, requires a fiber pair from the CO to each optical network unit (ONU), serving one or more residences or offices, in order to provide interactive digital services. Although simple and convenient, this approach has many defects: (1) fiber costs are not shared, (2) trunk maintenance requires splicing many fibers, (3) many expensive line cards are required at the CO, and (4) it is costly to keep track of numerous circuits and fibers. The power splitting PON of Fig. 15.10b overcomes these defects but has its own problems: (1) the passive splitter reduces the optical power in either direction by $1/N$; (2) because TDM must be used to provide individual circuits, the power from the CO assigned to each ONU is reduced by a further $1/N$ factor; (3) the ONU transceivers must operate at N times the channel bit rate; and (4) it is necessary to synchronize upstream time slots at the splitter to take account of variable delays between the splitter and the ONUs.

The WDM PON shown in Fig. 15.10c overcomes these problems and provides virtual point-to-point connectivity. The fiber trunk between the CO and the passive $1 \times N$ router carries N wavelengths, and the WDM router sends each wavelength to its ONU without intrinsic loss. The independent wavelength paths offer transparent point-to-point connections. If the CO is connected to a WDM ring, each ONU can have direct access by a dedicated wavelength path to the hub server on the ring. Alternatively,

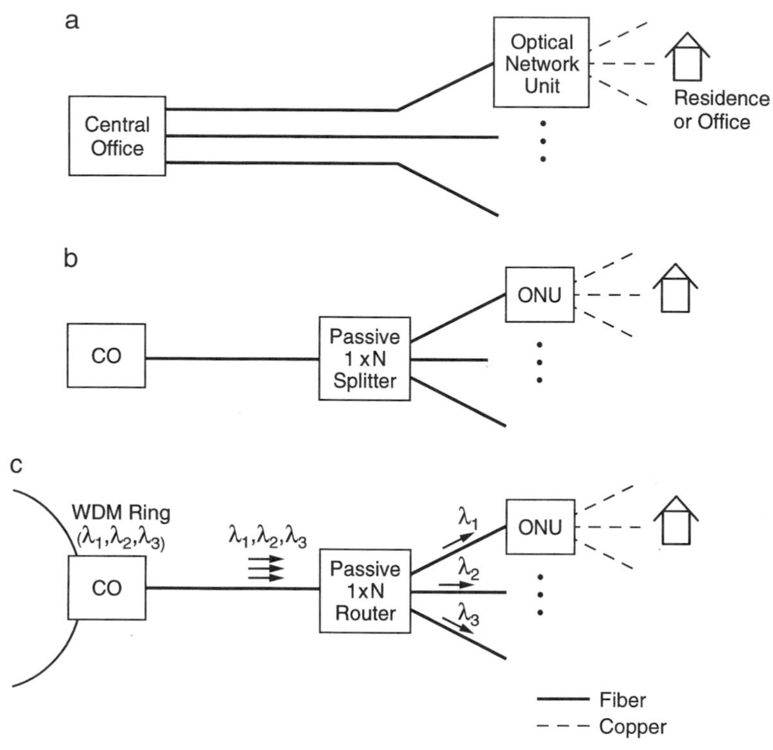

Fig. 15.10 Fiber distribution networks: (a) Multifiber point-to-point network. (b) Power splitting passive optical network (PON). (c) WDM PON, with optional connection to a metropolitan WDM ring. ONU, optical network unit.

the WDM ring can interact with the CO, which can provide electronic regeneration and cross-connecting of WDM channels.

Several proposals have been offered for realizing WDM PONs using inexpensive components in the ONU, where they are not widely shared as with components in the CO and the wavelength router (see Chapter 13 in Volume IIIA). For example, wavelength-controlled components are not generally cost-effective. Two architectures are shown in Fig. 15.11. Both employ MFLs (Glance, Kaminow, and Wilson 1994; Zirngibl, Joyner, and Stulz 1994) and MFRs (Glance, Kaminow, and Wilson 1994; Zirngibl, Joyner, and Stulz 1995), discussed in Chapter 4 in Volume IIIB, in the CO, and the WGR (Dragone, Edwards, and Kistler 1989), discussed in Chapter 8 in Volume IIIB. In the first, a $2 \times 2N$ WGR routes each wavelength of the downstream traffic to a separate ONU. The ONU receives part of the

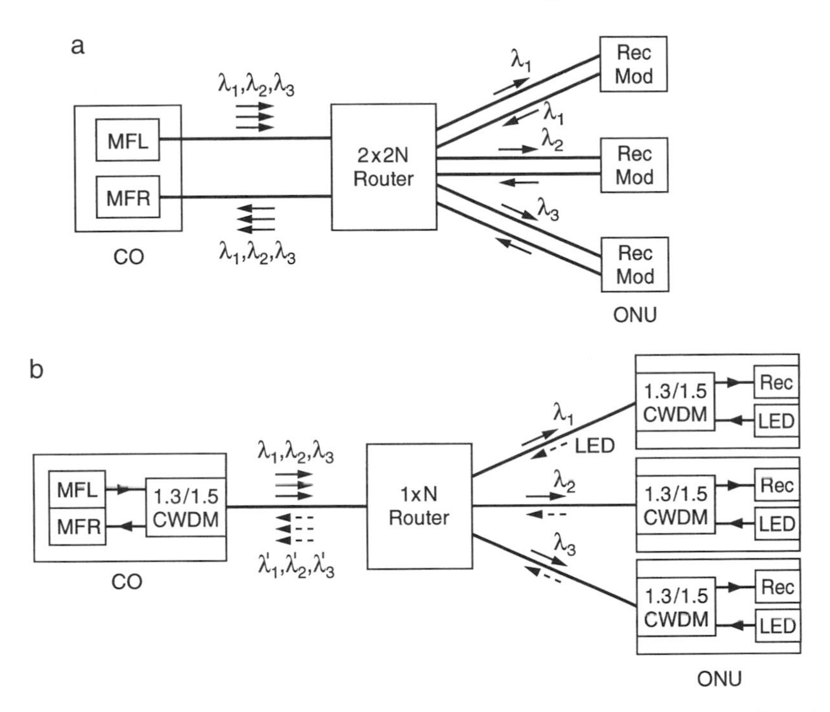

Fig. 15.11 WDM PON architectures employing a multifrequency laser (MFL), a multifrequency receiver (MFR), and a waveguide grating router (WGR). (a) Loop back with an optical modulator at the ONU. (b) Single fiber with light-emitting diodes (LEDs) and a 1300-/1500-nm coarse WDM (CWDM) at the ONU.

optical signal and returns the remainder through a modulator that impresses the upstream traffic on the same optical carrier. The return fiber reenters the WGR displaced by one port. The routing property of the WGR then displaces the upstream output port by one from the downstream input port, as shown in Fig. 15.11a. For initial access applications, bit rates for each channel can be in the 10- to 100-Mb/s range with a downstream–upstream asymmetry likely. It depends on the availability of an inexpensive modulator. This architecture is based on a RITEnet (remote integration of terminal equipment network) proposal (Frigo *et al.* 1994).

The architecture shown in Fig. 15.11b is called *LARNet* (local access router network) (Zirngibl *et al.* 1995). The downstream traffic is carried by a WDM spectrum from the MFL in the 1500-nm band and routed to individual ONUs by a $1 \times N$ WGR. The upstream traffic from each ONU is provided by a light-emitting diode (LED) in the 1300-nm band. It is spectrally sliced by the same WGR, designed to handle the two bands. The routing properties of the WGR select different upstream slices for the MFR in the CO. The 1500- and 1300-nm signals are separated in the ONU by a coarse WDM (CWDM). This architecture depends on the availability of an inexpensive directly modulated LED that can provide sufficient power into a single-mode fiber. A smaller dimension WGR router and fewer fibers are required compared with those needed in Fig. 15.11a.

F. Wavelength-Routing Network Test Beds

A number of wavelength-routing network test beds have been demonstrated recently. The U.S. DARPA (Defense Advanced Research Projects Agency) of the Department of Defense has given partial support to these projects. In all cases, the aim has been to package the components and provide a convenient user interface and network management that go beyond the earlier optical-bench-type laboratory demonstrations. Three of these projects have been completed, and a fourth, with greater expectations for commercialization, is under way. They are described briefly in the next sections. Detailed reports are given in the references.

1. RAINBOW

A group at IBM has worked on the simple broadcast-and-select star architecture (Kaminow 1988) known as *Rainbow* (Janniello, Ramaswami, and Steinberg 1993) and shown schematically in Fig. 15.12. Each optical terminal

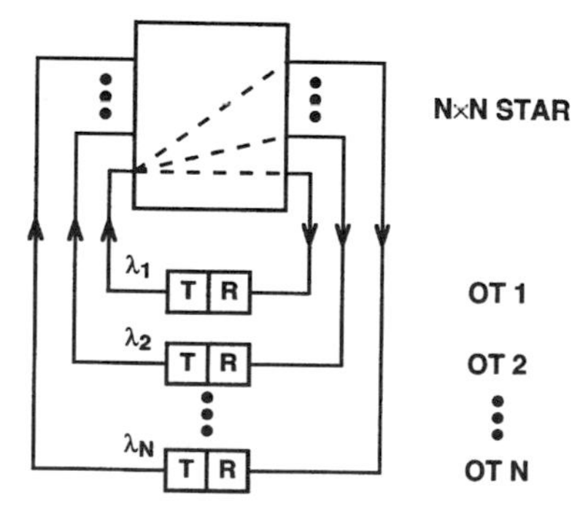

Fig. 15.12 IBM Rainbow broadcast-and-select star network. OT, optical terminal.

has an assigned wavelength for transmission by a DFB laser transmitter. The star coupler broadcasts this signal equally to the receivers of all terminals with an intrinsic $1/N$ splitting loss. In order to establish a connection to a specific receiver, which is based on a piezoelectrically tuned fiber Fabry–Perot (FFP) filter, the receiver must be tuned to the proper wavelength. Then a transparent optical path exists between the transmitter and the receiver.

For the call setup, a simple media access control (MAC) protocol called *circular search* is employed. Circuit cards were developed in order to implement the protocol from an IBM PS/2 workstation. With a keystroke identifying the destination, calling terminal 1 sends out a coded signal on its wavelength λ_1. All idle terminals tune their receivers continuously through the network spectrum. When a terminal, say terminal 2, senses an incoming call, it locks the FFP onto the calling wavelength to complete the circuit. Meanwhile, the originating terminal scans its receiver to set up the reverse connection on λ_2. Because of the slow tuning speed of the FFP, about 50 ms is required to set up a connection. Although a passive 32×32 star, composed of 2×2 fused couplers, is employed, fewer than 10 wavelengths were active in the Rainbow demonstrations. This system, called *Rainbow-1,* is now deployed between IBM locations in Westchester, New York, at rates of 300 Mb/s.

A follow-on project, called *Rainbow-2* (Hall *et al.* 1996) is designed to extend the operation to 1-Gb/s channels, employing stand-alone optical network adapter boxes rather than circuit cards, for interconnecting supercomputers at Los Alamos using three active wavelengths through the 32×32 star coupler. Although the optical transport layer is rather simple and provides only circuit switching, the emphasis in this experiment is on the higher layer communication protocols. Eventually, the aim is to realize optical packet switching with fast-tuning receivers.

The lessons learned from Rainbow are as follows:

(1) Even simple optical components are not yet robust enough for commercial use.
(2) Research-type optical components are prohibitively expensive.
(3) Optical packet switching will require fast-tuned receivers or transmitters.
(4) Passive splitting loss and/or OA gain bandwidth will limit the number of optical channels in a star network.

2. *OPTICAL NETWORKS TECHNOLOGY CONSORTIUM*

The Optical Networks Technology Consortium (ONTC) was formed with partial DARPA support in 1992 to investigate the technologies, architecture, and network control of scalable multiwavelength optical networks. A number of organizations contributed to the design and construction of the test bed: Bellcore, Case Western Reserve University, Columbia University, Hughes Research Laboratory, Lawrence Livermore Laboratory, Northern Telecom, Rockwell, and United Technologies.

The test-bed architecture provides wavelength routing through two WDM rings connected by a WDM cross-connect switch, as illustrated in Fig. 15.13 (Chang *et al.* 1996). Four access nodes are connected to the rings by their own rearrangeable OADMs. In the test bed, four wavelengths in the EDFA band are active with 2-nm channel spacing.

The WDM cross-connect, denoted by the number **3** in Fig. 15.13, is fabricated from bulk components: 2 × 2 optomechanical switches and multilayer dielectric filters, as illustrated in Fig. 15.14. Each filter passes one of the four system wavelengths into a 2 × 2 switch that is switched to route the channel to the upper or lower output, corresponding to one ring or the other.

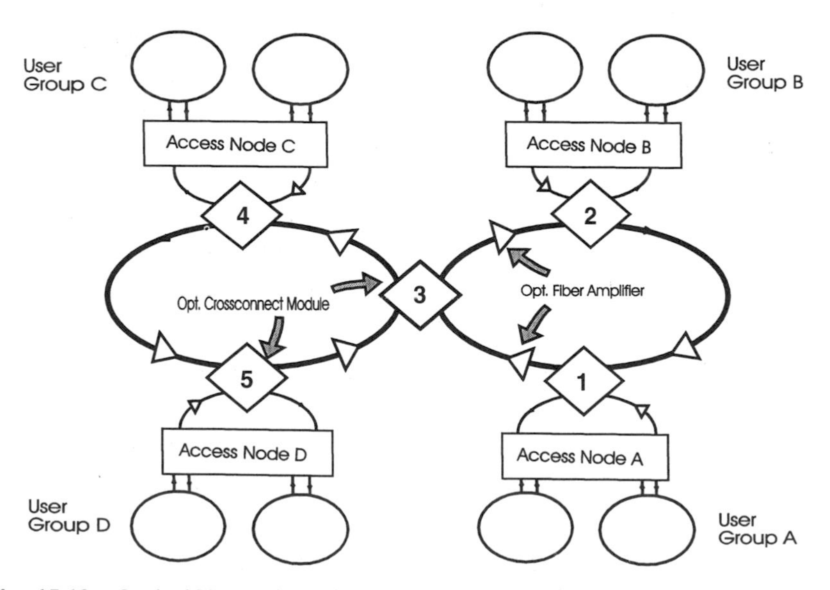

Fig. 15.13 Optical Network Technology Consortium (ONTC) all-optical network test bed. (Reproduced with permission of Bellcore.)

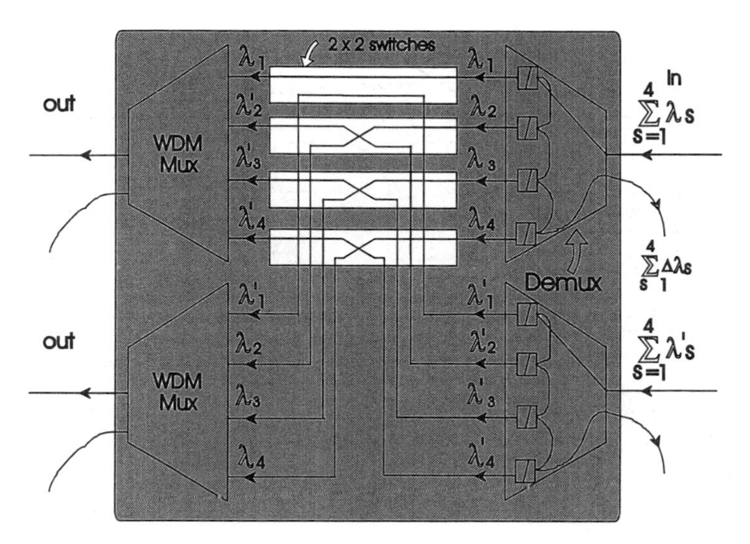

Fig. 15.14 Wavelength selective 2 × 2 WDM cross-connect switch. For the add–drop switch, the lower multiplexer and demultiplexer are replaced by the receiver and transmitter arrays, respectively. (Reproduced with permission of Bellcore.)

The same basic device provides the rearrangeable add–drop function (numbers 1, 2, 4, and 5 in Fig. 15.13). In this case, the lower optical multiplexer and demultiplexer in Fig. 15.14 are eliminated, and the four free fibers connect to the transmitters and receivers in the access nodes as shown in Fig. 15.15. One access fiber and wavelength λ_3 are reserved for transporting, as a group, analog channels (four video and one digital) on microwave subcarriers, according to a satellite standard. The remaining three wavelengths transport baseband digital signals using 155-Mb/s ATM on SONET (OC-3) channels. The remaining three fibers are dropped to a group of receivers that connect to the inputs of a 155-Mb/s ATM switch. The outputs of the ATM switch modulate an array of DFB lasers providing the wavelengths λ_1, λ_2, and λ_4. The ATM switch can serve to translate the optical carrier wavelength of a given data stream in the network by switching to the appropriate laser in the DFB array. EDFAs make up for losses on the rings.

As an example, ATM–SONET data originating at access node *D* in Fig. 15.13 is assigned a wavelength λ_i and is added to its ring by add–drop switch 5, where the network scheduler has ensured that the assigned wavelength is unoccupied. If node *C* is the destination, cross-connect 3 is set to bridge λ_i, and WDM add–drop switch 4 drops the signal to receiver *i*. If

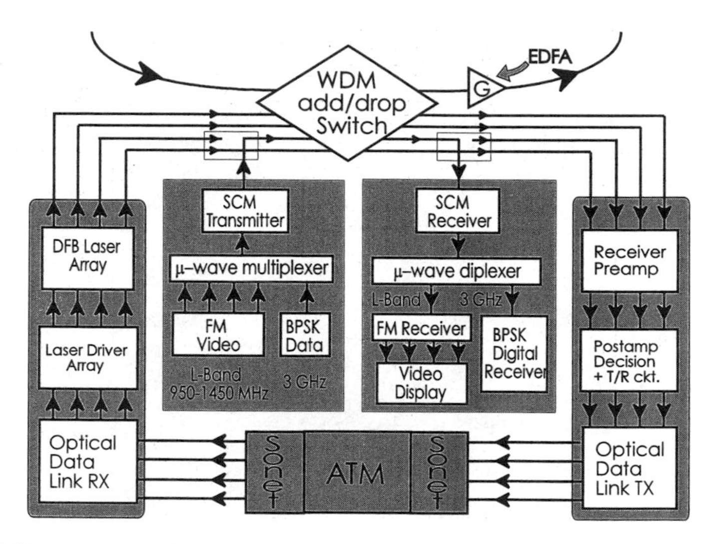

Fig. 15.15 Access node with a subcarrier transmitter–receiver and a multiwavelength ATM–SONET transceiver. BPSK, binary phase shift keying; DFB, distributed feedback; EDFA, erbium-doped fiber amplifier; FM, frequency-modulated; SCM, sub-carrier multiplexed. (Reproduced with permission of Bellcore.)

node *A* is the destination, cross-connect 3 sends the signals to the adjacent ring, switch 2 bridges λ_i, and switch 1 drops λ_i to receiver *i*. If λ_i is the only available access wavelength but is occupied on segments of either ring, the carrier wavelength may have to be translated at intermediate nodes in order to complete the circuit on a multihop basis. The network controller has the job of scheduling the wavelengths throughout the network to avoid contention, and setting the switches and the cross-connect. Some conflicts can be avoided and the number of connections increased by the multihop process, at the cost of added delay. The multihop capability could also provide for network survivability and scalability. A convenient graphical user interface is provided for setting up the connections. The ONTC network test bed was publicly demonstrated at Bellcore and at the Optical Fiber conference in 1995 (OFC'95).

The lessons learned from the ONTC are as follows:

(1) Component cost and performance limit network size and functionality.
(2) OADM cross talk limits scalability.
(3) A flat passband is essential to WDM cross-connects and OADMs.

(4) EDFA cascades add cost and introduce gain equalization problems.
(5) Multihop can provide greater throughput and add survivability and scalability, but it kills transparency.
(6) Global transparency is unlikely.

3. WIDEBAND ALL-OPTICAL NETWORK CONSORTIUM

The wideband All-Optical Network (AON) Consortium was formed with partial DARPA support in 1992 to investigate the technologies, architecture, network control, and applications of scalable multiwavelength optical networks (Alexander *et al.* 1993; Kaminow *et al.* 1996). The AON Consortium comprised AT&T Bell Laboratories, Massachusetts Institute of Technology (MIT) (Campus and Lincoln Laboratories), and Digital Equipment Corporation (DEC). Although the goals were the same, the approach was completely different from the ONTC, which started at about the same time.

The test-bed architecture (Kaminow *et al.* 1996) is hierarchical, as shown in Fig. 15.16. Users, which may be broadband individual workstations or

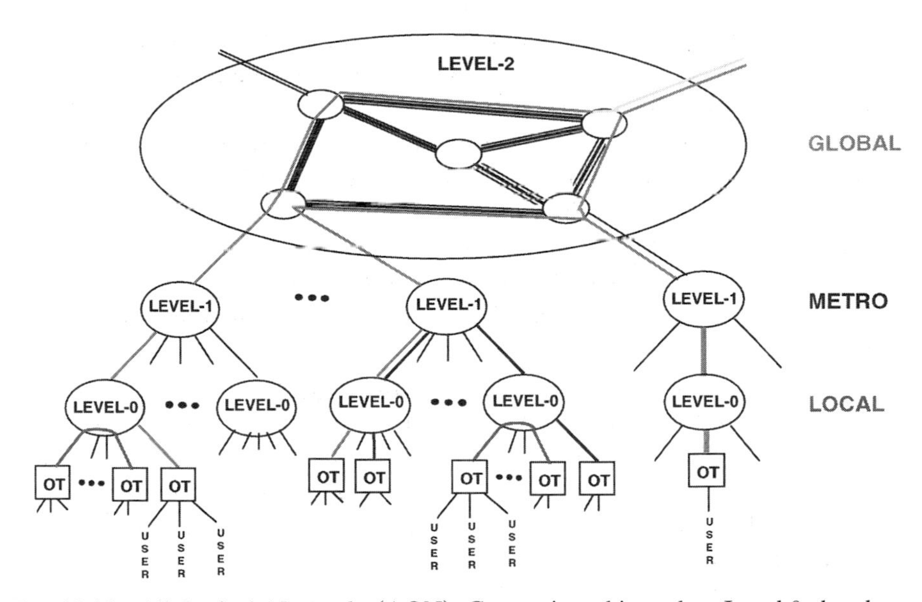

Fig. 15.16 All-Optical Network (AON) Consortium hierarchy. Level-0, local broadcast-and-select level; Level-1, wavelength-routed regional level; Level-2, switched wide-area level. (Reproduced with permission of MIT Lincoln Laboratory.)

servers or aggregations of users on LANs, are connected to optical terminals that are grouped into local subnets denoted *Level-0s*. All the Level-0s reuse the same set of local wavelengths $\lambda_{\ell i}$ in broadcast-and-select subnets. The Level-0s are connected to metropolitan subnets, which are denoted *Level-1s*, by a complementary set of global wavelengths λ_{gi}. The Level-1 subnets are based on passive WGRs that provide connections to other Level-0s under the Level-1 by wavelength routing. A broadcast capability, by means of a bypass star, is also available to allow distribution from one to all optical terminals under a given Level-1 on a subset of λ_{gi}. Another subset of λ_{gi} may also be reserved for providing wavelength-routed connection to Level-2.

Level-2 is a generally non-wavelength-routed WAN that connects Level-1s. AON architectural studies indicate that Level-2 is necessary in order to provide the scalability that is otherwise limited by the number of available wavelength channels, as discussed earlier in connection with Eqs. (15.1) and (15.2). In addition, the test-bed implementation shows that timing and synchronization considerations will limit the geographic reach of time-slotted connections. Level-2 was not implemented in the initial AON phase, although it is clear that it will require configurable combinations of wavelength changers, rearrangeable routers, and space division switches (Kaminow *et al.* 1996; Kobayashi and Kaminow 1996), as well as active regenerators. In Levels-0 and -1, because only one physical path exists between two optical terminals, albeit on several possible wavelengths, there is no restoration capability in the test-bed AON. (Nor do Rainbow and ONTC test beds provide restoration capabilities.) The Level-2 network could have a mesh or ring architecture that could provide for a wide-area restoration capability. Another ARPA project, denoted *MONET* (*multiwavelength optical network*), to be described in Section F.4, is addressing these practical problems.

The AON test bed is operating in the Boston metropolitan area at three sites (MIT, MIT Lincoln Laboratory, and DEC) connected by fiber and separated by about 70 miles (112 km). A fourth site at AT&T Bell Laboratories in Holmdel, New Jersey, operates independently. In each case, a variety of broadband applications communicate through the AON (Kaminow *et al.* 1996).

The Level-0 subnet is shown schematically in Fig. 15.17. The user interface includes a graphical user interface for requesting an AON connection in a user-friendly way. Within Level-0, the Level-0 network controller automatically finds a free wavelength for the transmitter and the receiver, and sets the

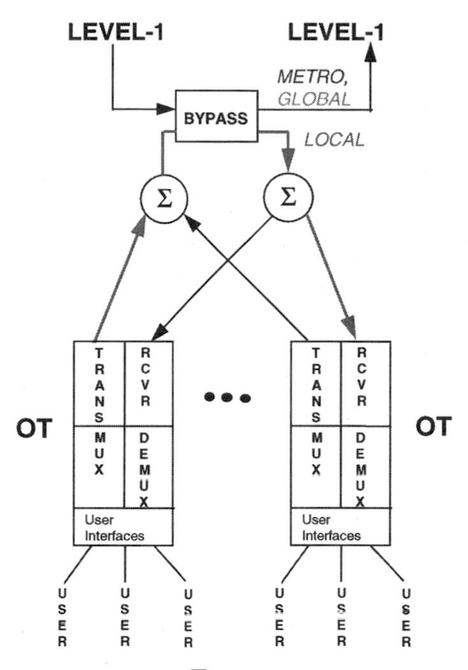

Fig. 15.17 Level-0 AON schematic. Σ, 1 × *N* splitter. (Reproduced with permission of MIT Lincoln Laboratory.)

component voltages to achieve the connection. The tunable laser and receiver in each optical terminal can be tuned to 1 of 20 wavelengths spaced by 50 GHz in the 1555-nm band covered by an EDFA. The laser is a discretely tunable DBR (distributed Bragg reflector) laser (Glance *et al.* 1992) and the tunable receiver a piezoelectrically controlled FFP filter (Stone and Stulz 1991), or, for fast tuning, a coherent receiver using the tunable DBR as a local oscillator (Kaminow *et al.* 1996). The WDM signals to and from all optical terminals are combined and split in passive $N \times 1$ and $1 \times N$ star couplers, respectively. A bypass interferometer routes all local wavelengths $\lambda_{\ell i}$ within the Level-0 and connects all global wavelengths λ_{gi} to Level-1. For convenience, a Mach–Zehnder (MZ) interferometer is used for bypass, so that alternate wavelengths are assigned as global and local, respectively. The use of $1 \times N$ splitters at the transmitter and the receiver introduces a $1/N^2$ circuit loss, as compared with an $N \times N$ star with a $1/N$ splitting loss, as in Fig. 15.12, but allows the use of the simple MZ bypass. Furthermore, this configuration allows the same λ_{gi} to be transmitted and received simultaneously within a given Level-0, doubling the number of possible connections.

The Level-1 subnet is shown schematically in Fig. 15.18. Although the actual Level-1 test-bed implementation is a simplified version, Fig. 15.18 shows how the global wavelengths might be separated to provide Level-2 access, Level-1 point-to-point connection, and Level-1 broadcast, where $1 \times N$ routers provide WDM demultiplexing and multiplexing of these three services. An $N \times N$ star provides the broadcast capability. In the test bed, the $1 \times N$ routers were not used, but broadcasting was demonstrated on certain λ_{gi} by connecting selected outputs of the WGR to a star connected into each Level-0 by 1×2 directional couplers (Kaminow *et al.* 1996).

The signaling required to set up a connection is carried on the same fibers as the data but in a separate band at 1310 nm. Commercial Ethernet on an rf subcarrier provides the required packet communication. The 1310-nm wavelength also carries timing information needed for bit and frame synchronization, as noted next.

The 1310-nm packet signaling network is denoted *C-service*. Once the connection is established, a bidirectional transparent wavelength path exists between optical terminals. In principle, any digital or analog modulation format can be communicated on the assigned wavelengths. In practice, of course, the usual signal–noise and nonlinear system constraints, based on path length and bit rate, must be satisfied for each format. This switched

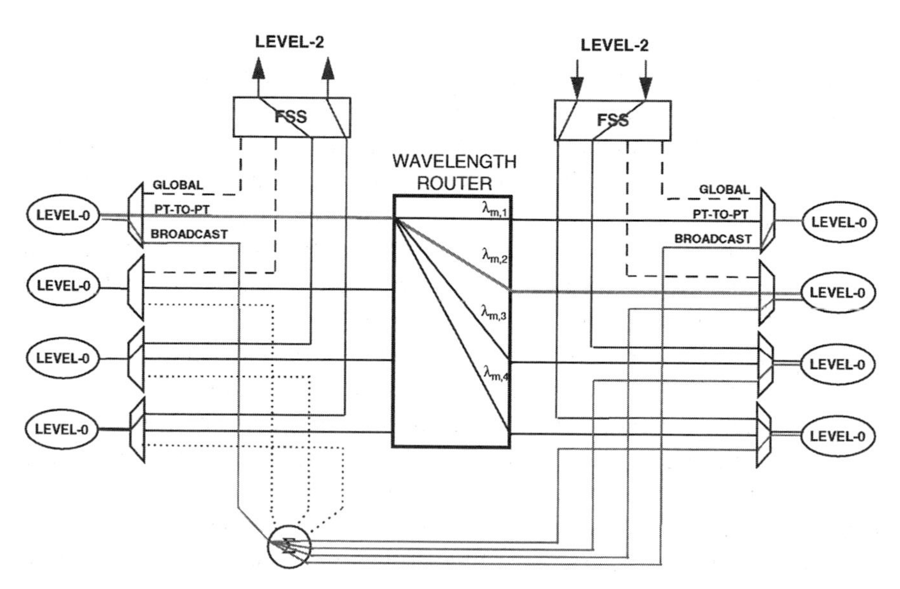

Fig. 15.18 Level-1 AON schematic. Σ, $N \times N$ star; FSS, frequency selective switch. (Reproduced with permission of MIT Lincoln Laboratory.)

circuit connection is denoted *A-service.* The optical terminals are designed to provide any bit rate up to 10 Gb/s, including 0.155-, 1.2-, 2.4-, and 10-Gb/s (OC-3, -24, -48, -192) digital services.

A time-slotted service, denoted *B-service*, allows several users to share a laser transmitter and receiver. The on–off-keyed 1.2-Gb/s rate was demonstrated in the test bed. The concept is illustrated in Fig. 15.19. The B-service shares the 20 WDM channels spaced by 50 GHz with the A-service, as coordinated by a computer-controlled scheduler. B-service wavelengths are timed with 250-μs frames divided into 128 time slots. Users can request from 1 to 128 slots per frame and introduce any modulation format. In the test bed, one slot per frame corresponds to about 10 Mb/s at 1.2 Gb/s. Also, the unused slots on a wavelength are relinquished so that the network scheduler may allocate them to other users, not only connected through optical terminals within the same Level-0, but also among distant Level-0s. In this way, bandwidth may be allocated dynamically on the basis of the offered load, so that high efficiency is achieved (Marquis *et al.* 1995).

Figure 15.19 shows an example of scheduling and coordination of A- and B-services for transmitters or receivers. In the test bed, each transmitter

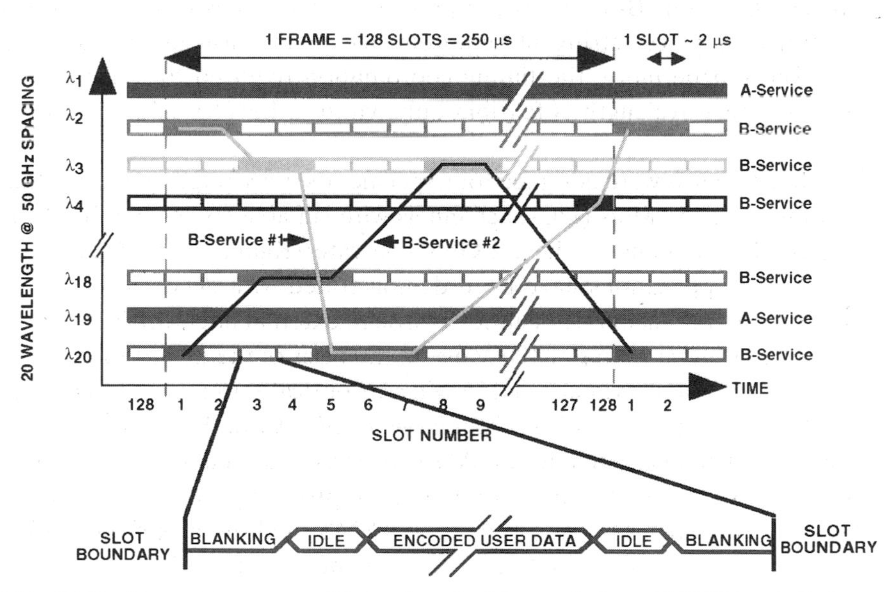

Fig. 15.19 B-service concept. (Reproduced with permission of MIT Lincoln Laboratory.)

or receiver can be tuned to only one wavelength at a time. In future implementations, MFLs and MFRs could be used (Kaminow *et al.* 1996). Figure 15.19 shows the schedule for one transmitter in an optical terminal and one receiver in a second optical terminal tuned to λ_1, and, for example, the corresponding receiver in the first optical terminal and transmitter in the second optical terminal tuned to λ_{19} to provide a duplex A-service connection. Another transmitter can access four separate receivers in sequence on λ_2, λ_3, λ_{20}, and λ_4 by tuning throughout the frame as shown. Receivers in the network would synchronously tune to receive these signals within the time slots. Reverse connections would similarly be provided. The time-slot pattern shown at the bottom of Fig. 15.19 allows for a blanking guard time between slots, an idle period for clock recovery and 8B/10B code-word synchronization, and the encoded data.

An autonomous timing synchronization algorithm was developed (Hemenway 1995) that automatically determines and sets the optimum timing advance and delay offsets within 8 ns to synchronize B-service line cards within an L-1 and L-0 network. The algorithm requires less than 60 s to execute and is performed just once during each optical terminal power-up. There, nonetheless, remains a small overshoot penalty of a fraction of one slot (2 μs) each time a transmitter switches contexts. Scheduling and synchronizing B-service is complex. Although, it was successfully demonstrated, it places stringent requirements on the fast-tuning laser and receiver. Furthermore, the timing coordination required for efficient use of time and wavelength is feasible only within a Level-1 subnet, and not Level-2.

Circuit boards were designed for a specific B-service implementation that combines six 155-Mb/s (OC-3) channels with suitable coding and blanking overhead in a frame at the 1.2-Gb/s line rate (Kaminow *et al.* 1996). A number of applications have been demonstrated with A- or B-service by connecting multimedia workstations through external fiber distributed data interface (FDDI) LANs at 100 Mb/s on the basis of the DEC Gigaswitch® or ATM–SONET switches at 155 Mb/s to optical terminals and communicating through the AON (Kaminow *et al.* 1996). For example, see Fig. 15.20. The idea is to allocate AON resources automatically in response to traffic overload or failure conditions by providing interfaces between the AON network management system (NMS) and the proprietary ATM LAN managers. The AON NMS is implemented over the C-service and is responsible for fault detection, configuration management, performance estimation, and resource allocation (e.g., scheduling wavelengths and time

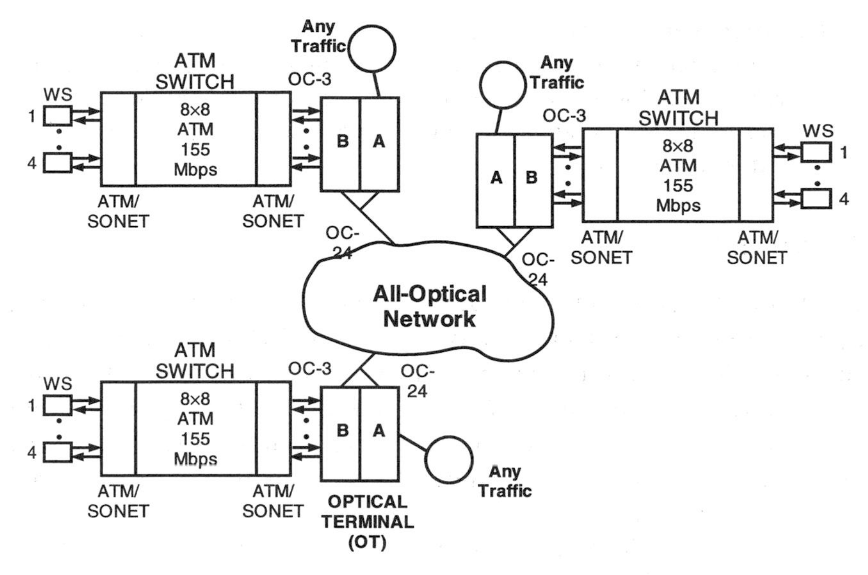

Fig. 15.20 AON ATM applications. WS, work station.

slots). Users request channels from a graphical user interface on a workstation at the optical terminal, using the standard SNMP (simple network management protocol). The appropriate scheduler in Level-0 or -1 responds with wavelength and slot assignments, if available, and directs the appropriate hardware in the optical terminals to tune to the assigned channels and commence communication. The NMS keeps track of ongoing sessions and maintains a routing table of available wavelength paths.

The lessons learned from the AON Consortium are as follows:

(1) The AON can support very large aggregate throughput: e.g., with 10 wavelengths in Level-1 at 10 Gb/s each, the throughput is 1 Tb/s.

(2) The passive wavelength-routing architecture is limited in extent because of the finite number of wavelength channels.

(3) WANs require active WDM cross-connects in Level-2 to achieve scalability; space division switches and/or wavelength converters will be required.

(4) A-service provides line-rate connectivity; traffic must be aggregated at the optical terminal to provide access granularity.

(5) B-service provides effective multiplexing means for multiple users at the optical terminal but is more complex than A-service.

(6) Component performance and cost need to be improved; the requirements are (1) stable and reproducible tunable lasers and receivers; (2) low-cross-talk, flat-passband, low-loss WDM routers and multiplexers; and (3) MFLs and MFRs. Coherent cross talk is a major concern.

4. MONET

The test-bed experiments described previously represent a first effort to evaluate the opportunities and limits of advanced WDM wavelength-routed networks. Clearly, novel components play a critical role in novel network architectures. The next stage of component exploration requires careful attention to robust packaging, reliability, high performance, and low cost. The next-generation network architecture must address the practical features of a commercial system: scalability, reliability, survivability, interoperability, access granularity, network management, and low cost. Some transparency aspects of WDM routing will be needed, as noted in earlier sections of this chapter on OADMs and rings, and some of the novel approaches in various DARPA test beds described previously may not be needed.

The year 1995 saw the inception of an ambitious consortium, with partial DARPA support and the participation of the National Security Agency (NSA) and the Naval Research Laboratory (NRL), devoted to taking the next steps toward defining a commercial WDM network incorporating the lessons learned from commercial WDM systems as well as the experimental DARPA test beds. The consortium includes AT&T, Lucent Technologies, Bellcore, Bell Atlantic, and BellSouth and Pacific Telesis. The 5-year project is denoted *MONET*, which, as mentioned before, stands for multiwavelength optical network (Saleh 1996). The overall architectural approach is illustrated schematically in Fig. 15.21. Various applications are indicated in the top layer and are connected within the local-exchange or regional network at the electronic layer supporting ATM–SONET as well as other LAN and switch standards. The local electronic layer is connected to the local optical WDM (MONET) transport layer. Although local access networks are not addressed directly, such networks would be connected to the long-distance transport network in the MONET layer. The WDM regional network includes router-based networks (R), WDM rings, and WDM cross-connects (X-C) that provide space and wavelength switching. The long-distance network is based on these X-Cs, which are based on lithium niobate electrooptic switch arrays described in Chapter 10 in Volume IIIB. As

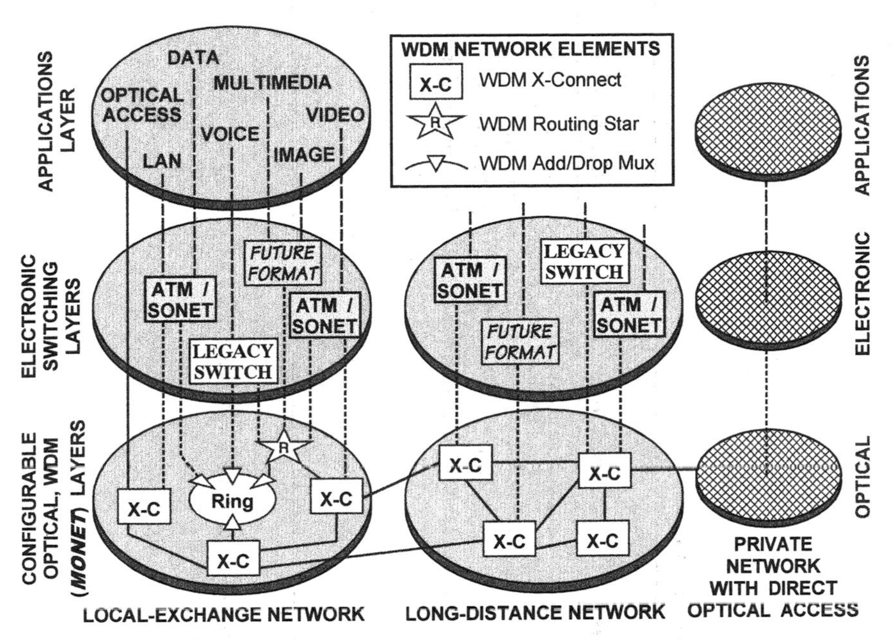

Fig. 15.21 Multiwavelength optical network (MONET) architecture. LAN, local area network. (Reproduced with permission from Saleh, A. A. M. Overview of the MONET Multiwavelength Optical NETworking program. In *Optical Fiber Communication Conference, OFC'96, Symposium on Optical Networks Session,* San Jose, California, February 29, 1996.)

shown in Fig. 15.21, direct fiber optical access to the MONET layer is also available to provide a transparent path. This generalized architecture may satisfy future needs for public and private, transparent, high-capacity networks.

G. Conclusions

We have seen that WDM networks, with their format transparency and wavelength-routing features, have an immediate and a long-range role in telecommunications. As the demand grows for flexible bandwidth for a wide range of interactive services, the deployment of these networks will become more attractive. However, the deployment cannot go forward without practical improvements in the performance, reliability, and cost of critical WDM components, such as (1) configurable WGRs and OADMs

Table 15.1 WDM Test Beds

Test-Bed Name	*Time (Years)*	*N (Active)*[a]	$\Delta\lambda(\Delta F)$[b]	*Bit Rate*	*Laser Diode*[c]	*(D)MUX*[d]
ONTC −1	92–94	4	4 nm	155 Mb/s	DFB LA	Dielectric filter
−2	94–95	8	2 nm	2.5 Gb/s	DFB LA star	DF/integer star
AON −1	92–94	20	0.4 nm	0.155–2.5 Gb/s	DBR-LD	WGR, FFP, heterodyne
−2	94–95	20^+	(50 GHz)	0.155–10 Gb/s	DBR-EML, MFL	WGR, FFP, heterodyne, MFR
Rainbow −1	90	32 (4)	1 nm	300 Mb/s	DFB	FFP
−2	94	32 (4)	1 nm	1 Gb/s		
MONET	95–99	8	1.6 nm (200 GHz)	2.5 Gb/s	DFB LA	WGR/star

[a] Number of channels (number active).
[b] Channel spacing.
[c] DFB, distributed feedback; EML, electroabsorption modulator laser; LA, laser array; LD, laser diode; MFL, multifrequency laser.
[d] DF, dielectric filter; FFP, fiber Fabry–Perot; MFR, multifrequency receiver; WGR, waveguide grating router.

with low loss, low cross talk, and a wide passband; (2) MFLs and MFRs reliably operating at high bit rates with many channels; (3) (optical frequency) wavelength changers; (4) full WDM cross-connects; and so on.

The first steps have been taken with the DARPA test beds described in Section F. We summarize their characteristics in Table 15.1. These characteristics include time duration, number of channels, channel spacing, bit rate, laser diode type, and demultiplexer type.

References

Alexander, S. B., *et al.* 1993. A precompetitive consortium on wideband all-optical networks. *J. Lightwave Tech.* 11:714–735.

Bala, K., R. R. Cordell, and E. L. Goldstein. 1995. The case for opaque multiwavelength optical networks. In *IEEE/LEOS summer topical meeting on technologies for a global information infrastructure, Keystone, CO*, 58–59. Paper WA4. Piscataway, NJ: IEEE.

Barry, R. A. 1993. Wavelength routing for all-optical networks. Ph.D. diss., Massachusetts Institute of Technology, Electrical Engineering and Computer Science Department.

Barry, R. A., and P. A. Humblet. 1996. Models of blocking probability in all-optical networks with and without wavelength changers. *IEEE J. Select. Areas Commun.* 14:858–867.

Chang, G.-K., G. Ellinas, J. K. Gamelin, M. Z. Iqbal, C. A. Brackett. 1996. Multiwavelength reconfigurable WDM/ATM/SONET network testbed. *J. Lightwave Tech.* 14(6):1320–1340.

Dragone, C., C. A. Edwards, and R. C. Kistler. 1991. Integrated optics $N \times N$ multiplexer on silicon. *IEEE Photon. Tech. Lett.* 3:896–899.

Elrefaie, A. 1991. Self-healing ring network architecture using WDM for growth. In *ECOC'91*, 285–288. Paper TuP1.16.

Elrefaie, A. F. 1993. Multiwavelength survivable ring network architectures. In *ICC'93, Geneva.* Paper 48.7.

Frigo, N. J., P. P. Iannone, P. D. Magill, T. E. Darcie, M. M. Downs, B. N. Desai, U. Koren, T. L. Koch, C. Dragone, H. M. Presby, and G. E. Bodeep. 1994. A wavelength-division multiplexed passive optical network with cost-shared components. *IEEE Photon. Tech. Lett.* 6:1365–1367.

Gallagher, R. G. 1992. Spatial scalability of B-service. *MIT lab for information and decision systems.* Unpublished manuscript.

Glance, B., C. R. Doerr, I. P. Kaminow, and R. Montagne. 1996. Optically restorable WDM ring network using simple add/drop circuitry. *J. Lightwave Tech.* 14:2453–2456.

Glance, B., I. P. Kaminow, and R. W. Wilson. 1994. Application of the integrated waveguide grating router. *J. Lightwave Tech.* 12:957–962.

Glance, B., U. Koren, R. W. Wilson, D. Chen, and A. Jourdan. 1992. Fast optical packet switching based on WDM. *IEEE Photon. Tech. Lett.* 4:1186–1188.

Hall, E., J. Kravitz, R. Ramaswami, M. Halvorson, S. Tenbrink, and R. Thomsen. 1996. The Rainbow-II gigabit optical network. *IEEE J. Select. Areas Commun.* 14:814–823.

Hemenway, B. R., D. M. Castagnozzi, M. L. Stevens, S. A. Parikh, D. Marquis, S. G. Finn, R. A. Barry, E. A. Swanson, I. P. Kaminow, U. Koren, R. E. Thomas, C. Ozveren, and E. Grella. 1995. Autonomous timing determination in a time-slotted WDM all-optical network. *LEOS Summer Topical Metting, Keystone, Colorado.*

Hersey, S. H., and M. J. Soulliere. 1991. Architectures and applications of SONET in a self-healing network. In *ICC'91*, Paper 44.3.1, *IEEE*, 1418–1424.

Janniello, F. J., R. Ramaswami, and D. G. Steinberg. 1993. A prototype circuit-switched multi-wavelength optical metropolitan-area network. *J. Lightwave Tech.* 11:777–782.

Jeong, G., and E. Ayanoglu. 1996. Comparison of wavelength-interchanging and wavelength-selective cross-connects in multi-wavelength all-optical networks. *IEEE Infocom '96.*

Kaminow, I. P. 1988. Photonic local networks. In *Optical Fiber Telecommunications II*, ed. S. E. Miller and I. P. Kaminow, 933–972. Boston: Academic Press.

Kaminow, I. P., C. R. Doerr, C. Dragone, T. L. Koch, U. Koren, A. A. M. Saleh, A. Kirby, C. Ozveren, B. Schofield, R. E. Thomas, R. A. Barry, D. M. Catagnozzi, V. W. S. Chan, B. R. Hemenway, D. Marquis, S. S. Parikh, M. L. Stevens, E. A. Swanson, S. G. Finn, and R. G. Gallager. 1996. A wideband all-optical WDM network. *IEEE J. Select. Areas Commun.* 14:780–799.

Karasan, E., and E. Ayanoglu. 1996. Effects of wavelength routing and selection algorithms on wavelength conversion gain in WDM optical networks. *IEEE Globecom '96.*

Kobayashi, H., and I. P. Kaminow. 1996. Duality relationships among space, time and wavelength in all-optical networks. *J. Lightwave Tech.* 14:344–351.

Marquis, D., D. M. Castognozzi, B. R. Hemenway, M. L. Stevens, E. A. Swanson, R. E. Thomas, C. Ozveren, and I. P. Kaminow. 1995. Description of all-optical network testbed and applications. *SPIE Photonics East.* Philadelphia: SPIE.

Marra, W. T., and J. Schesser. 1996. AfricaONE architecture. *IEEE Commun. Mag.* 34:50–57.

Saleh, A. A. M. 1996. Overview of MONET, multiwavelength optical NETworking program. In *Optical Fiber Communication Conference Digest, San Jose, CA.* Symposium on Optical Networks Session, February 29, 1996.

Stone, J., and L. W. Stulz. 1991. High-performance fiber Fabry–Perot filters. *Electron. Lett.* 27:239–240

Tomlinson, W. J., and R. E. Wagner. 1993. Wavelength division multiplexing — Should you let it in your network? In *National Fiber Optics Conference, San Antonio, Session 19,* Book 4, 285–298.

Wu, T-H. 1992. *Fiber network service survivability.* Boston: Artech House.

Zirngibl, M., C. H. Joyner, and L. W. Stulz. 1994. Demonstration of 9 × 200 Mbit/s wavelength division multiplexed transmitter. *Electron. Lett.* 30:1484–1485.

Zirngibl, M., C. H. Joyner, and L. W. Stulz. 1995. WDM receiver by monolithic integration of an optical preamplifier, waveguide grating router and photodiode array. *Electron. Lett.* 31:581–582.

Zirngibl, M., C. H. Joyner, L. W. Stulz, C. Dragone, H. M. Presby, and I. P. Kaminow. 1995. LARNet, a local access router Network. *IEEE Photon. Tech. Lett.* 7:215–217.

Index